THE SCIENCE ALMANAC

THE SCIENCE ALMANAC

1985–1986 Edition

EDITED BY
Bryan Bunch

A Hudson Group Book

Anchor Books
Anchor Press/Doubleday
Garden City, New York
1984

Written by:

ARCHAEOLOGY AND ANTHROPOLOGY
Courtlandt Canby

ASTRONOMY AND SPACE
Bruce Wetterau

BIOLOGY
Karin L. Rhines

CHEMISTRY
Murray Halwer

EARTH SCIENCE
Richard Worth

ENVIRONMENT
Bruce Wetterau

MATHEMATICS
Bryan Bunch and Larry Siegel

MEDICINE
Larry Lorimer, John Harrington, and Frances Barth

PHYSICS
Alexander Hellemans

Library of Congress Cataloging in Publication Data
Main entry under title:
The Science almanac.
"A Hudson Group book."
Includes index.
1. Science — Yearbooks. I. Bunch, Bryan H.
Q9.S23 1984   505   84-20405
ISBN 0-385-19321-1
ISBN 0-385-19320-3 (pbk.)

Printed in the United States of America
FIRST EDITION

Preface

Keeping up with the latest information in science has been a difficult task. So much is happening so rapidly, and it takes a long time for new information to reach reference books.

The computer came to the rescue. With the computer, it was possible to maintain a database of all the information, organize it so that the most important topics could be identified, and produce a book that was as up-to-date as modern computerized publishing techniques allow.

The result is *The Science Almanac*.

There are four interwoven strands that form *The Science Almanac*:

Science Updates form the main body of the book. These are fact-filled reports on the latest events, trends, and research in all the fields of science, written for the nonspecialist.

Feature Articles provide an in-depth look at areas where major new research is taking place.

Tables, Charts, and Lists focus primarily on data where new information has produced a significant change, but also include basic reference material of all types. The lists are subjective guides to books, places, and ideas.

Science Trivia is detailed in short notes that report on new scientific research or events that, while not exactly earth-shaking, nevertheless contribute to the mosaic of science.

Every effort has been made to keep the book accurate as well as up-to-date. Special thanks in this regard go to my editors, James Raimes and Alice Fasano; to Mary and Sally Bunch for editorial assistance; to my copyeditor, Felice Levy, and indexer, Cynthia Crippen; and to the scientists and professionals who reviewed the material: Leonard C. Labowitz (Chemistry Consultant), Gordon Lasher (IBM's Thomas J. Watson Research Center), William V. Mayer (President Emeritus of the Biological Sciences Curriculum Study), Samuel N. Namowitz (author of *Earth Science*), Edward Anthony Oppenheimer, M. D. (Chief, Pulmonary Medicine, Kaiser-Permanente Medical Center, Los Angeles, California and Associate Clinical Professor of Medicine, U.C.L.A. School of Medicine), E.M. Reilly (Environmental Consultant), Bea Riemschneider (Editor-in-Chief of *Archaeology*), and Theodore Schultz (IBM's Thomas J. Watson Research Center). And very special thanks to John Wright, without whom this book might never have been published.

Bryan Bunch
Briarcliff Manor, NY

TABLE OF CONTENTS

BIOLOGY   144

ENVIRONMENT 295

THE BIG PICTURE

The history of applied science goes back to before our species, for there is good evidence that our predecessor on this planet, *Homo erectus*, used fire and that erectus's predecessor fashioned the first stone tools. But not much happened for almost 2,000,000 years. Stone hand axes that were made as much as 1,000,000 years apart appear to have been produced by the same methods and used in the same ways. When our own species developed, however, change started to be measured in thousands of years instead of millions.

We know little about scientific ideas before the development of writing, although it is apparent that preliterate peoples had developed a detailed understanding of the apparent movements of the sun, the Moon, and a few major stars. After the development of writing, however, it is possible to chart the course of science from the Sumerians, Greeks, and Chinese to the present day. With some significant setbacks, it is an amazing story of progress in understanding the nature of reality.

Until the end of the eighteenth century, some people were generally familiar with the details of all the scientific knowledge that had accumulated. Furthermore, until the end of the nineteenth century, it was possible for a mathematician to be familiar with all of mathematics, a physicist with all of physics, and a biologist with all of biology. But in the past 80 years or so, scientists have been forced by the sheer amount of available knowledge to specialize more and more. Keeping up with the major developments in biology is difficult even for a biologist, while the ordinary citizen is faced with still greater barriers to understanding—specialized vocabulary, unspoken assumptions, and hidden theoretical frameworks.

The problem of following the broad outline of scientific progress has become more and more difficult to solve since World War II, when science began an explosion that is like nothing else previously seen. *The Science Almanac* is one vehicle for solving that problem. It contains the most important recent developments in all the branches of science. Along with detailed stories about new developments, there are tables of the latest data of scientific interest, lists of resources for further exploration, and a few science stories that would have to be classified as more fascinating than important. *The Science Almanac*, then, is a reference for what is happening now in science and a resource for understanding what will happen tomorrow.

MAJOR DEVELOPMENTS OF THE EARLY 1980S

For a detailed picture of current developments in science, turn to the subject sections of *The Science Almanac*, where new developments are listed by year, ranging from 1980 though 1984. Although there is no way that one can say exactly when a new idea sprang into being, since scientific ideas build on the ideas of the past, it is possible to give a general idea of the accomplishments and trends in science in the early 1980s.

Practical Implications: They are many and most of them are in biology and medicine. In fact, it is hard to tell where theoretical biology at the molecular or cell level stops and the application to medicine begins.

The successes of recombinant DNA techniques, better known as gene splicing, are becoming well known, but the combination of these techniques with a less well-known technology may be what makes the biggest difference. The technology of monoclonal antibodies, cells that produce molecules fine-tuned for extremely specific interactions by a part of the complex immune system, is also revolutionary. Monoclonal antibodies can seek out tiny quantities of a specific protein in a living system. They can be programmed to release a poison when they find the protein, killing a specific cell. They can be programmed to trace the movement of the protein, an ability that can be used either for diagnosis of disease or for research into life processes. When monoclonal antibodies are combined with the use of gene splicing to manufacture specific proteins in quantity, the potential for correction of genetic defects or even cures for cancer seems enormous. While this work is experimental today, progress is so fast that routine clinical applications cannot be far away. The realization that some of our most serious diseases, including rheumatoid arthritis, have a basis in the immune system also suggests the centrality of our new understanding of immunology.

While developments in mathematics are often not accessible to the general public, an important application related to mathematics is changing the way we live and perhaps the way we think: the computer. As computers have moved from their central, air-conditioned, dust-proof rooms onto office desks and into homes and schoolrooms, people have come to appreciate not only their usefulness but also their limitations. So far, computers are not intelligent. However, great progress is being made in providing computers with some forms of artificial intelligence—a development that could mean as much as the invention of the computer itself.

Not all of the practical implications of science are beneficial. Although significant gains have been made in reducing water and air pollution and some endangered species are getting new leases on life, problems persist. Furthermore, we are just learning of the magnitude of such problems as groundwater pollution and depletion, pollution caused by common building materials, and toxic wastes. People were once able to go to the forests to get away from these problems, but today acid rain is killing the fish in forest lakes and killing the trees that form the forests. Pollution used to involve limited areas such as cities and waterways, but now vast regions, including the whole East Coast of the United States, are affected by acid rain, snow, and fog.

It gets even worse on a global scale. The controversies and theories arising out of the probable discovery of the cause of the extinction of the dinosaurs will do much to explain the evolution of life on Earth, but this earth-science discovery also has led to extremely practical considerations. All roads in earth science today, whether from the skies or from sea-floor spreading or ocean currents, seem to lead to climate. Asteroid or comet impacts, or possibly volcanic eruptions, seem to cause the climatic changes that lead to mass extinctions. Changes in the carbon-dioxide level in the atmosphere may have set off the ice ages in the past, and changes caused by modern technology and farming practices today, it is predicted, will cause serious warming of Earth in the near future. Over everything looms the possibility of a sudden "nuclear winter" caused by even a small nuclear war. The nuclear winter would cause untold hardship and the extinction of many familiar species, possibly including the human one.

Theoretical Advances: New understandings that have no immediate practical implications, however, are the backbone of science. Great changes have been made in theory in the early 1980s as well as in application.

No science has shown a greater explosion in understanding in the past twenty years than biology. New discoveries have come at a dizzying pace in the past five years especially, mainly because of our new understanding of genes and how to manipulate them. Understanding genes means more than just the increasingly strong possibility of finding the cause of and effective treatments for cancer. It also gets at the essence of life—growth and development at the molecular level. Knowing what specific genes do may also be the key to understanding learning, memory, and intelligence. Although "whole-organism" biology is still a vital part of the science, molecular biology is increasingly the basis for understanding the behavior and even the evolution of the whole organism.

Physical anthropology of the species that led to modern humans is a little more than a hundred years old. Although precious fossils of these species are exceedingly rare, new discoveries in Africa, Asia, and museum basements in the past five years have considerably changed the basic story of prehuman ancestors during the period from 2,000,000 to 30,000,000 years ago.

The ability to put "eyes" of various kinds in space has changed astronomy from a science dependent almost solely on Earth-based telescopes to one that uses the whole electromagnetic spectrum to study planets closely, as well as stars, galaxies, dust, quasars, and black holes from afar. In the early 1980s the two most productive of these studies were the Voyager surveys of the Saturn system and the Infrared Astronomy Satellite (IRAS) survey of the Milky Way. Voyager's observations revealed mysteries about the rings of Saturn that have yet to be resolved and showed that Titan, one of Saturn's moons, is much like what a primitive Earth is believed to have been. IRAS sent back mountains of data that will take years to work through, but already new phenomena, such as clouds of particles around stars, have been found.

This big picture is certainly mixed. While in some areas of science the gains are purely intellectual, in others there are important practical consequences for good or for evil. In fact, it would appear that if we manage to avoid changing our environment or climate in disastrous ways, a new era of unprecedented health and prosperity is on the way. The choice between these two extremes, or something in between, is largely political, not scientific.

National Medals of Science

The National Medal of Science is the highest science award given by the United States government. Since 1962, it has been presented by the President for achievement in physical, biological, mathematical, engineering, behavioral, or social sciences.

The medal itself is intended to symbolize science. It shows a human being, surrounded by Earth, sea, and sky, contemplating and seeking to understand nature. The scientist holds a crystal in one hand, intended to symbolize the universal order and also the basic unit of living things. With the other hand, the scientist is sketching a formula in the sand.

The nominees are selected by a presidential committee of twelve members acting with the President's science advisor and the President of the National Academy of Sciences.

Year	Winner	Field	Affiliation
1962	Theodore von Karman	Aeronautics	Cal Tech
1963	Luis Walter Alvarez	Physics	University of California, Berkeley
	Vannevar Bush	Electronics	Carnegie Institute
	John Robinson Pierce	Communications	Bell Labs
	Cornelius B. van Niel	Microbiology	Stanford University
	Norbert Wiener	Mathematics	M.I.T.
1964	Roger Adams	Chemistry	University of Illinois
	Othmar H. Ammann	Engineering	Ammann & Whitney
	Theodosius Dobzhansky	Medicine	Rockefeller University
	Charles Stark Draper	Astronautics	M.I.T.
	Solomon Lefschetz	Mathematics	Princeton University
	Neal Elgar Miller	Psychology	Yale University
	Harold Marston Morse	Mathematics	Institute for Advanced Studies
	Marshall W.Nirenberg	Genetics	National Institutes of Health
	Julian Schwinger	Physics	Harvard University
	Harold C. Urey	Chemistry	University of California, Berkeley
	Robert B. Woodward	Chemistry	Harvard University
1965	John Bardeen	Electronics	University of Illinois
	Peter J.W. Debye	Chemistry	Cornell University
	Hugh L. Dryden	Administration	N.A.S.A.
	Clarence L. Johnson	Aeronautics	Lockheed

Year	Winner	Field	Affiliation
	Leon M. Lederman	Physics	Columbia University
	Warren K. Lewis	Chemical Eng.	M.I.T.
	Francis Peyton Rous	Medicine	Rockefeller University
	William W. Rubey	Geophysics	U.C.L.A.
	George G. Simpson	Paleontology	Harvard University
	Donald Van Slyke	Chemistry	Brookhaven Laboratory
	Oscar Zariski	Mathematics	Harvard University
1966	Jacob Bjerknes	Meteorology	U.C.L.A.
	S. Chandrasekhar	Astrophysics	University of Chicago
	Henry Eyring	Administration	University of Utah
	E.F. Knipling	Entomology	U.S. Dept. of Agriculture
	Fritz A. Lipmann	Biochemistry	Rockefeller University
	John W. Milnor	Mathematics	Princeton University
	William C. Rose	Chemistry	University of Illinois
	Claude E. Shannon	Science History	M.I.T.
	J.H. Van Vleck	Physics	Harvard University
	Sewall Wright	Genetics	University of Wisconsin
	Vladimir Zworykin	Physics	R.C.A.
1967	J.W. Beams	Physics	University of Virginia
	A. Francis Birch	Geology	Harvard University
	Gregory Breit	Physics	Yale University
	Paul J. Cohen	Mathematics	Stanford University
	Kenneth S. Cole	Biophysics	National Institutes of Health
	Louis P. Hammett	Chemistry	Columbia University
	Harry F. Harlow	Psychology	University of Wisconsin
	Michael Heidelberger	Immunology	New York University
	G.B. Kistiakowsky	Chemistry	Harvard University
	Edwin H. Land	Photography	Polaroid Corporation
	Igor I. Sikorsky	Aeronautics	United Aircraft
	Alfred H. Sturtevant	Biology	Cal Tech
1968	Horace A. Barker	Biochemistry	University of California, Berkeley
	Paul D. Bartlett	Chemistry	Harvard University
	Bernard B. Brodie	Pharmacology	National Institutes of Health
	Detlev W. Bronk	Medicine	Rockefeller University
	J. Presper Eckert	Computer Science	Sperry-Rand Corporation
	Herbert Friedman	Astrophysics	Naval Research Lab
	Jay L. Lush	Genetics	Iowa State University
	N. M. Newmark	Civil Eng.	University of Illinois
	Jerzy Neyman	Mathematics	University of California, Berkeley

Year	Winner	Field	Affiliation
	Lars Onsager	Chemistry	Yale University
	B. F. Skinner	Psychology	Harvard University
	Eugene P. Wigner	Physics	Princeton University
1969	Herbert C. Brown	Chemistry	Purdue University
	William Feller	Mathematics	Princeton University
	Robert J. Huebner	Medicine	National Institutes of Health
	Jack S. C. Kilby	Computer Science	Texas Instruments
	Ernst Mayr	Zoology	Harvard University
	W. K. H. Panofsky	Physics	Stanford University
1970	Richard D. Brauer	Mathematics	Harvard University
	Robert H. Dicke	Physics	Princeton University
	Barbara McClintock	Genetics	Carnegie Institute
	George E. Mueller	Administration	General Dynamics
	Albert B. Sabin	Medicine	Weizmann Institute
	Allan R. Sandage	Astronomy	Cal Tech
	John C. Slater	Physics	University of Florida
	John A. Wheeler	Physics	Princeton University
	Saul Winstein	Chemistry	U.C.L.A.
1971	No awards	————	————
1972	No awards	————	————
1973	Daniel I. Arnon	Biochemistry	Stanford University
	Carl Djerassi	Chemistry	Stanford University
	Harold E. Edgerton	Mathematics	M.I.T.
	William M. Ewing	Electrical Eng.	University of Texas
	Arie J. Haagen-Smit	Biochemistry	Cal Tech
	Vladimir Haensel	Geology	Universal Oil
	Frederick Seitz	Administration	Rockefeller University
	Earl W. Sutherland	Biochemistry	University of Miami
	John W. Tukey	Statistics	Princeton University
	Richard T. Whitcomb	Aeronautics	Langley Center
	Robert P. Wilson	Administration	Fermilab
1974	Nicolaas Bloembergen	Physics	Harvard University
	Britton Chance	Physics	University of Pennsylvania
	Erwin Chargaff	Biochemistry	Columbia University
	Paul J. Flory	Chemistry	Stanford University
	William A. Fowler	Physics	Cal Tech
	Kurt Gödel	Mathematics	Institute for Advanced Study
	Rudolf Kompfner	Physics	Stanford University
	James V. Neel	Genetics	University of Michigan
	Linus Pauling	Chemistry	Stanford University

Year	Winner	Field	Affiliation
	Ralph B. Peck	Engineering	University of Illinois
	K.S. Pitzer	Chemistry	University of California, Berkeley
	James A. Shannon	Medicine	Rockefeller University
	Abel Wolman	Engineering	Johns Hopkins
1975	John Backus	Computer Science	San Jose Laboratory
	Manson Benedict	Mathematics	M.I.T.
	Hans A. Bethe	Physics	Cornell University
	Shiing-shen Chern	Mathematics	University of California, Berkeley
	George B. Dantzig	Computer Science	Stanford University
	Hallowell Davis	Medicine	Washington University
	Paul Gyorgy	Pediatrics	University of Pennsylvania
	Sterling B. Hendricks	Chemistry	U.S. Dept. of Agriculture
	Joseph O. Hirschfelder	Chemistry	University of Wisconsin
	William H. Pickering	Physics	Jet Propulsion Laboratory
	Lewis H. Sarett	Administration	Merck, Sharp, and Dohme
	Frederick E. Terman	Administration	Stanford University
	Orville A. Vogel	Agronomy	Washington State University
	Wernher von Braun	Administration	NASA
	E. Bright Wilson	Chemistry	Harvard University
	Chien-Shiung Wu	Physics	Columbia University
1976	Morris Cohen	Metallurgy	M.I.T.
	Kurt O. Friedrichs	Mathematics	New York University
	Peter C. Goldmark	Communications	Goldmark Corporation
	Samuel A. Goudsmit	Physics	University of Nevada
	Roger C. Guillemin	Neurology	Salk Institute
	Herbert S. Gutowsky	Chemistry	University of Illinois
	Erwin W. Mueller	Physics	Pennsylvania State University
	Keith R. Porter	Biology	University of Colorado
	Efraim Racker	Biochemistry	Cornell University
	Frederick D. Rossini	Chemistry	Rice University
	Verner E. Suomi	Meteorology	University of Wisconsin
	Henry Taube	Chemistry	Stanford University
	George E. Uhlenbeck	Physics	Rockefeller University
	Hassler Whitney	Physics	Princeton University
	Edward O. Wilson	Zoology	Harvard University
1977	No award	——————	——————
1978	No award	——————	——————

Year	Winner	Field	Affiliation
1979	Robert H. Burris	Biochemistry	University of Wisconsin
	Elizabeth C. Crosby	Anatomy	University of Michigan
	Joseph L. Doob	Mathematics	University of Illinois
	Richard P. Feynman	Physics	Cal Tech
	Donald E. Knuth	Computer Science	Stanford University
	Arthur Kornberg	Biochemistry	Stanford University
	Emmett N. Leith	Electrical Eng.	University of Michigan
	Herman F. Mark	Administration	Brooklyn Polytech
	Raymond D. Mindlin	Civil Eng.	Columbia University
	Robert N. Noyce	Computer Science	Intel Corp.
	Severo Ochoa	Biochemistry	New York University
	Earl R. Parker	Metallurgy	University of California, Berkeley
	Edward M. Purcell	Physics	Harvard University
	Simon Ramo	Administration	Thompson, Ramo, Wooldridge, Inc.
	John H. Sinfelt	Science Advisor	Exxon
	Lyman Spitzer, Jr.	Astronomy	Princeton University
	Earl R. Stadtman	Biochemistry	National Institutes of Health
	G. Ledyard Stebbins	Genetics	University of California, Davis
	Paul A. Weiss		Rockefeller University
	Victor F. Weisskopf	Physics	M.I.T.
1980	No awards	————	————
1981	Philip Handler	Biochemistry	National Academy of Sciences
1982	Philip W. Anderson	Physics	Bell Labs
	Seymour Benzer	Biology	Cal Tech
	Glenn W. Burton	Biology	U.S. Dept. of Agriculture
	Mildred Cohn	Chemistry	University of Pennsylvania
	F. Albert Cotton	Chemistry	Texas A&M
	Edward H. Heinemann	Engineering	General Dynamics
	Donald Katz	Chemistry	University of Michigan
	Yoichiro Nambu	Physics	University of Chicago
	Marshall H. Stone	Mathematics	University of Massachusetts
	Gilbert Stork	Chemistry	Columbia University
	Edward Teller	Physics	Stanford University
	Charles H. Townes	Physics	University of California, Berkeley
1983	No awards	————	————

People Who Deserved the Nobel Prize
(But Did Not Get It)

It has been conjectured that Alfred Nobel tried to assuage his guilty conscience for the wartime use of his invention of dynamite by giving prizes. He specified prizes for physics, chemistry, physiology or medicine, literature, and peace. These prizes have become known as the premier international recognition in those fields. Much of science was omitted, however, including biology, geology, and astronomy. The reasons for these omissions are not known, but for other sciences, such as anthropology, the discipline did not exist as such when Nobel died in 1896. Furthermore, it is rumored that Nobel had feuded with a prominent mathematician, so he did not provide for a prize in mathematics lest it be given to his enemy.

Over the years, the Caroline Institute, which makes the awards in physiology or medicine, and the Swedish Royal Academy, which grants the awards in chemistry and physics, have stretched the terms of Nobel's will in various ways. Although Nobel specified that the awards be for work done that year or just recognized as important that year, in recent years the awards have often gone to people whose principal achievements were many years in the past. The category of physics has been stretched to include astronomy in some cases; physiology or medicine has been stretched to include biology. Despite this elasticity on the part of the prize-givers, over the years since the prizes were initiated, many of the most prominent and important scientists have done their major work without receiving the special recognition of a Nobel. Among them are

Aiken, Howard Hathaway (1900–1973). American mathematician and inventor of the first digital computer, Mark I.

Armstrong, Edwin Howard (1890–1954). American electrical engineer and inventor of FM (frequency modulation) radio.

Baekeland, Leo Hendrik (1863–1944). Belgian chemist who developed the first artificial resin plastic.

Bell, Alexander Graham (1847–1922). American inventor of the first commercial telephone.

Bodenstein, Max (1871–1942). German chemist who first described chain reactions.

Brandenberger, Jacques Edwin (1872–1954). Swiss chemist who developed cellophane for scientific uses.

Bush, Vannevar (1890–1974). American engineer who produced the first analog computer, the differential analyzer.

Carlson, Chester F. (1906–1968). American physicist who invented electrostatic printing, or xerography.

Carothers, Wallace H. (1896–1937). American chemist and inventor of nylon.

Carver, George Washington (1864–1943). American botanist and chemist who developed new agricultural crops.

Cottrell, Frederick Gardener (1877–1948). American chemist who developed a precipitator to reduce pollution.

De Forest, Lee (1873–1961). American electrician and inventor of the triode vacuum tube for television.

De Sitter, Willem (1872–1934). Dutch astronomer who first proposed a theory of the expanding universe.

Dempster, Arthur Jeffrey (1886–1950). American physicist and discoverer of Uranium-235.

Eddington, Arthur Stanley (1882–1944). British astronomer who determined the relationship between star mass and light.

Farnsworth, Philo Taylor (1906–1971). American inventor of the patented image dissector for video screens.

Freud, Sigmund (1856–1939). Austrian neurologist and founder of the science of psychoanalysis.

Geiger, Hans (1882–1947). German physicist and inventor of the geiger counter for measuring radioactivity.

Goddard, Robert Hutchings (1882–1945). American physicist who constructed the first liquid-fuel rocket.

Haldane, J.B.S. (1982–1964). British scientist who applied mathematical principles to genetics.

Hale, George Ellery (1868–1938). American astronomer and inventor of the spectroheliograph for sun study.

Hess, Harry Hammond (1906–1969). American geologist who developed key concepts in the plate-tectonic theory.

Hilbert, David (1862–1943). German mathematician who formulated new axioms for Euclidean geometry.

Hubble, Edwin P. (1889–1953). American astronomer who provided the first evidence for an expanding universe.

Huxley, Julian (1887–1975). British biologist who provided a mathematical analysis of body and organ proportions.

Jansky, Karl Guthe (1905–1950). American engineer who discovered radio waves from outer space.

Kerst, Donald William (1911–   ). American physicist who built the first betatron atomic accelerator.

Meitner, Lise (1878–1968). Austrian physicist who theorized energy release of nuclear fission.

Moseley, Henry Gwyn-Jeffries (1887–1915). English physicist who developed the concept of atomic number.

Nier, Alfred Otto Carl (1911–   ). American physicist who separated isotope U-235 from uranium.

Noyce, Robert N. (1927–   ). American physicist who developed the silicon chip integrated circuit.

Pearson, Karl (1857–1936). British mathematician and founder of the science of statistics.

Russell, Bertrand (1872–1970). British mathematician and developer, with Whitehead, of symbolic logic.

Sabin, Albert (1906–   ). American bacteriologist and developer of oral polio vaccine.

Salk, Jonas (1914–   ). American physician and inventor of the first polio vaccine.

Shapley, Harlow (1885–1972). American astronomer and discoverer of new knowledge about the Milky Way.

Steinmetz, Charles Proteus (1865–1923). German mathematician who developed alternating electrical current.

Strassman, Fritz (1902–   ). German chemist and discoverer of a method to split the uranium atom.

Von Neumann, John (1903–1957). Hungarian mathematician whose research was essential to electronic computers.

Wasserman, August (1866–1925). German bacteriologist who developed a test for syphilis.

Watson-Watt, Robert (1892–1973). British electrician who invented the first successful radar.

Wegener, Alfred L. (1880–1930). German geophysicist who developed the theory of continental drift.

Welsbach, Carl Auer (1858–1929). Austrian chemist who discovered the elements neodymium and praseodymium.

Whitehead, Alfred North (1861–1947). British mathematician and developer, with Russell, of symbolic logic.

Wiener, Norbert (1894–1964). American mathematician who founded the science of cybernetics.

Zworykin, Vladimir (1889–1982). Russian physicist who invented the TV picture tube and electron microscope.

A Universe of Records

As scientists explore the universe, they continually find the extremes—at least the extremes known at a particular time. Here are the most recent extremes recorded by science.

1980

Oldest Fossils Small fossils, dated as 3,500,000,000 years old, have been found in Australia.

1981

Heaviest Object The star R136a has a mass 3,000 times that of the sun.

Most Distant Galaxy The galaxy 3C 427.1 has a red shift of 1.175, implying that it is 10,000,000,000 light years from Earth.

1982

Largest Atom For a tiny fraction of a second, scientists created an atom with 109 protons, 157 neutrons, and 109 electrons.

Most Distant Object The quasar PKS 3000-330 was found to have a red shift of 3.78, implying that it is about 12,000,000,000 light years from Earth.

Largest Known Object A supercluster of superclusters of galaxies is 700,000,000 light years long. It combines the Lynx-Ursa Major supercluster with the Pisces-Perseus supercluster.

Fastest Pulsar The pulsar 1937 +215 is pulsing, and therefore rotating, at 642 times a second.

1983

Oldest Rocks Some zircons found in western Australia were found to be between 4,100,000,000 and 4,200,000,000 years old.

Farthest Man-made Object Until Voyager I passes it, a couple of years from now, Pioneer 10 will be farther away than any other object produced by humans. On June 13, 1983, it left the solar system.

Coldest Recorded Temperature At the Soviet Antarctic station named Vostok it reached –89.4°C (–129°F).

Brightest Object The quasar S5 0014 +81 has an intrinsic magnitude of –33. However, it is so far away that it can only be seen from Earth through the largest telescopes.

Bitterest Pill The chemical denatonium saccharide is so bitter that it tastes bitter even when diluted to one part in 20,000,000.

Ranking the Universities

In 1982 the National Academy of Sciences sponsored an exhaustive survey of doctoral programs in both sciences and humanities. It stressed that no single category of data should be considered a definitive "ranking" of programs in each area. However, most attention has focused on the scoring, by colleagues, of the faculty of each department in the leading universities. What follows is the score, out of a maximum of 5, of the top faculties in the principal areas of doctoral research in the sciences.

Biochemistry

M.I.T.	5.0
Harvard University	4.9
Stanford University	4.9
The Rockefeller Institute	4.6
University of California, Berkeley	4.6
University of Wisconsin/Madison	4.6
Yale University	4.5
Harvard University Medical School	4.4
University of California, San Francisco	4.4
Cornell University	4.3

Chemical Engineering

University of Minnesota, Duluth	4.9
University of Wisconsin, Madison	4.8
California Institute of Technology	4.7
University of California, Berkeley	4.6
Stanford University	4.5
University of Delaware	4.5
M.I.T.	4.3
University of Houston	4.1
Princeton University	4.0
University of Illinois, Urbana	4.0

Chemistry

California Institute of Technology	4.9
Harvard University	4.9
University of California, Berkeley	4.9
M.I.T.	4.8
Columbia University	4.6
Stanford University	4.5
University of Illinois/Urbana	4.5
University of Chicago	4.4
U.C.L.A.	4.4
University of Wisconsin/Madison	4.4

Civil Engineering

University of California, Berkeley	4.8
M.I.T.	4.7
California Institute of Technology	4.5
University of Illinois, Urbana	4.5
University of Texas, Austin	4.2
Cornell University	4.1
Stanford University	4.1

Computer Science

Stanford University	5.0
M.I.T.	4.9
Carnegie-Mellon University	4.8
University of California, Berkeley	4.5
Cornell University	4.3
U.C.L.A.	3.8
University of Illinois/Urbana	3.8
Yale University	3.5

Electrical Engineering

M.I.T.	4.9
Stanford University	4.8
University of California, Berkeley	4.8
University of Illinois, Urbana	4.6
U.C.L.A.	4.1
University of Southern California	4.1
Cornell University	4.0

Geoscience/Geophysics and Geology

California Institute of Technology	4.9
M.I.T.	4.8
U.C.L.A.	4.5
Columbia University	4.3
University of Chicago	4.3
Stanford University	4.2
Harvard University	4.1
Princeton University	4.1
Yale University	4.1
Cornell University	4.0

Mathematics

M.I.T.	4.9
Princeton University	4.9
University of California, Berkeley	4.9
Harvard University	4.8
University of Chicago	4.8
Stanford University	4.6
New York University	4.5
Yale University	4.5
Columbia University	4.4

Mechanical Engineering

M.I.T.	4.8
Stanford University	4.6
University of California, Berkeley	4.6
California Institute of Technology	4.3
University of Minnesota, Duluth	4.1
Princeton University	4.0
University of Michigan, Ann Arbor	4.0

Microbiology

M.I.T.	4.9
The Rockefeller Institute	4.8
Johns Hopkins University	4.3
University of California, San Diego	4.3
University of Washington, Seattle	4.3
Duke University	4.2
University of Pennsylvania	4.2
U.C.L.A.	4.1

Molecular Biology

M.I.T.	4.9
California Institute of Technology	4.8
The Rockefeller Institute	4.8
Yale University	4.7
University of Wisconsin/Madison	4.6
Harvard University	4.3
University of California, San Diego	4.3
University of California, Berkeley	4.2
Columbia University	4.1
University of Colorado	4.1

Physics

California Institute of Technology	4.9
Harvard University	4.9
Princeton University	4.9
University of California, Berkeley	4.9
M.I.T.	4.8
Cornell University	4.7
Stanford University	4.6
University of Chicago	4.6
Columbia University	4.5

Physiology

University of California, San Francisco	4.5
The Rockefeller Institute	4.3
U.C.L.A.	4.3
University of Pennsylvania	4.3
University of Washington, Seattle	4.3
Yale University	4.3
Harvard University	4.1
University of Michigan, Ann Arbor	4.1
Columbia University	4.0
Washington University at St. Louis	4.0

Statistics/Biostatistics

Stanford University	4.9
University of California, Berkeley	4.9
University of Chicago	4.7
University of Wisconsin/Madison	4.3
Cornell University	4.0
Harvard University	4.0
Princeton University	4.0
University of Iowa	4.0
University of North Carolina	4.0

Twenty-five Best Current Science Books

Although the level of difficulty of these books varies, all are recommended for the general reader, not just the specialist. It should be noted, of course, that lists such as this one are necessarily subjective.

THE BIOLOGY OF SPIDERS

Rainer F. Foelix (Harvard University Press, 1982. $30). The author's own English translation of an influential German work that explores in detail the extraordinary complexity of a relatively simple neurological system.

THE BOOK OF WHALES

Richard Ellis, writer and illustrator (Alfred A. Knopf, 1980. $25). The 33 types of great whales are discussed and thirteen of them illustrated in this personal, yet scientifically accurate and captivating coffee-table book.

THE COSMIC CODE: QUANTUM PHYSICS AS THE LANGUAGE OF NATURE

Heinz R. Pagels (Simon and Schuster, 1982. $16.95). The best introduction to quantum physics available to the nonphysicist; even someone familiar with quantum mechanics will learn from the second half of the book.

COSMOLOGY: THE SCIENCE OF THE UNIVERSE

Edward R. Harrison (Cambridge University Press, 1981. $24.95). A survey of the entire universe, past, present, and future, that is a little opinionated and quirky, but always thought-provoking.

THE CREATIVE EXPLOSION

John Pfeiffer (Harper and Row, 1981. $29.95). Human culture and human history explained from the perspective of the meaning of the cave paintings of neolithic times and of the aborigines of today.

DARWIN TO DNA, MOLECULES TO HUMANITY

G. Ledyard Stebbins (W.H. Freeman, 1982. $28.95). The modern synthetic theory of evolution explained with copious illustrations drawn from both the plant and animal kingdoms. The final chapters update the science of evolution with a description of current issues that is never partisan.

A Feeling for the Organism: The Life and Work of Barbara McClintock

Evelyn Fox Keller (W.H. Freeman, 1983. $14.95). The biography of the years of research on mobile genes in corn by a single-minded woman scientist who was rewarded with a 1983 Nobel Prize.

The Fate of the Earth

Jonathan Schell (Alfred A. Knopf, 1982. $13.95). The single most influential forecast of the effects of nuclear holocaust. Considered unscientific and even sentimental by some scientists, it nevertheless continues to provoke debate and to galvanize efforts to limit nuclear armaments.

The Fractal Geometry of Nature

Benoit B. Mandelbrot (W.H. Freeman, 1982. $32.50). Mathematical notions about dimension illustrated with examples drawn from land and plant forms. The book never claims that its theories of line and plane determine every shape in the real world, but its lucid discussions and beautiful illustrations range from the intriguing to the positively fascinating.

Gossamer Odyssey: The Triumph of Human Powered Flight

Morton Grosser (Houghton Mifflin, 1981. $14.95). A discussion of Human Powered Aviation (HPA) from the 1920s through a prize competition in 1980. Grosser's enthusiasm is infectious, and his treatment of the subject is geared to the general public with only hazy notions of the mechanics of aviation.

The Growth of Biological Thought: Diversity, Evolution, and Inheritance

Ernst Mayr (Harvard University Press, 1982. $30). The first of two volumes promised from the director of the Museum of Comparative Zoology at Harvard. It is an opinionated and scholarly history of biological theory, rewarding for those with some previous knowledge of schools of thought within the science.

Inca Architecture

Graziano Gasparini and Luise Margolies (Indiana University Press, 1980. $32.50). The English translation of a monumental reconstruction of the Incan world that places its architecture within political and economic contexts.

Lucy: The Beginnings of Humankind
Donald C. Johanson and Maitland A. Edey (Simon and Schuster, 1981. $17.95). An account of the discovery of a hominid australopithecine, which Johanson claims is more than 3,000,000 years old. The narrative is always engaging, and the attack on rival anthropologists such as Richard Leakey is considered good clean fun.

The Mapmakers: The Story of the Great Pioneers in Cartography from Antiquity to the Space Age
John N. Wilford (Alfred A. Knopf, 1981. $20). An anecdotal survey, filled with curious episodes, from early mariners to Landsat. The general delineations of the growth of measuring distance from an art to a science is convincing, but almost any reader will note arbitrary omissions and shifts of focus.

The Mathematical Experience
Philip J. Davis and Reuben Hersh (Bukhauser, Inc., 1981. $27.95). A collection of brief essays that are designed to convey to the general reader what it is that professional mathematicians actually do, but which can also cause mathematicians to think more seriously about what they are doing.

The Mismeasure of Man
Stephen Jay Gould (W.W. Norton, 1981. $14.95). An illuminating investigation of man's attempts to measure intelligence. Gould emphasizes historical errors and misrepresentations of evidence and convincingly demolishes past efforts to support presumptions about biological determinism.

The Montgolfier Brothers and the Invention of Aviation 1783–1784
Charles Coulston Gillispie (Princeton University Press, 1983. $35). An account of the first manned balloon flights with greater attention to the social history of the times than to the mechanics of invention.

Night Thoughts of a Classical Physicist
Russell McCormmach (Harvard University Press, 1982. $15). A lyrical evocation of the somber imagination of a "composite" physicist living through World War I. This is a fictionalized portrait as interested in politics as it is in science, one that strikes responsive chords in those pessimistic about science.

THE 1980 ERUPTIONS OF MOUNT ST. HELENS, WASHINGTON
Peter W. Lipman and Donal R. Mullineaux, editors (U.S. Government Printing Office, 1982. $35). A large and impressive volume with action photographs, geological analysis, and eyewitness accounts of the recent volcanic "time of darkness."

NUCLEAR ILLUSION AND REALITY
Solly Zuckerman (Viking Press, 1982. $10.95). An insider's report on nuclear weapon proliferation since 1942, when the author, trained as an epidemiologist, first became involved in public arms policy.

REVOLUTION IN TIME: CLOCKS AND THE MAKING OF THE MODERN WORLD
David S. Landes (Harvard University Press, 1983. $20). An historical survey of mechanical chronology from water clocks to digital wrist watches. Lucid descriptions of the craft of clock making are all the more provocative for Landes' underlying assumption that we can measure time but never know what it is.

SCIENCE IN TRADITIONAL CHINA
Joseph Needham (Harvard University Press, 1982. $12.50). A series of five lectures originally delivered in China. In informal but literate fashion Needham contrasts East-West attitudes on topics as diverse as time and biochemistry.

SOUL OF A NEW MACHINE
Tracy Kidder (Atlantic Monthly Press, 1981. $12.95). A journalistic account of round-the-clock efforts to build a minicomputer at the Data General Corporation from 1978 to 1980.

'SUBTLE IS THE LORD...': THE SCIENCE AND THE LIFE OF ALBERT EINSTEIN
Abraham Pais (Oxford University Press, 1982. $25). A biography by a physicist and friend admirably equipped to recount both the work and the personal myth of Albert Einstein.

THE SURFACE OF MARS
Michael H. Carr (Yale University Press, 1982. $45). A richly illustrated book on both the growth and the current state of our knowledge about Mars by a member of the U.S. Geological Survey team studying the planet since the Viking missions.

The Hugo Awards

The World Science Fiction Society Hugo awards for science fiction achievement in the novel form (named after science-fiction pioneer, publisher and author Hugo Gernsback), while not exactly science, suggest what science *could* be. A favorite definition of science fiction is "the world of what-if." This world is presented in the Hugo-award winning novels from 1953's "What if people had true telepathy?" to 1983's "What if the development of society were being directed by a long dead historian?"

Year	Title	Author
1953	The Demolished Man	Alfred Bester
1954	No award	——
1955	They'd Rather Be Right	Mark Clifton and Frank Riley
1956	Double Star	Robert A. Heinlein
1957	No award	——
1958	The Big Time	Fritz Leiber
1959	A Case of Conscience	James Blish
1960	Starship Troopers	Robert A. Heinlein
1961	A Canticle for Liebowitz	Walter M. Miller, Jr.
1962	Stranger in a Strange Land	Robert A. Heinlein
1963	The Man in the High Castle	Philip K. Dick
1964	Way Station	Clifford Simak
1965	The Wanderer	Fritz Leiber
1966	This Immortal	Roger Zelazny
	Dune	Frank Herbert
1967	The Moon is a Harsh Mistress	Robert A. Heinlein
1968	Lord of Light	Roger Zelazny
1969	Stand on Zanzibar	John Brunner
1970	The Left Hand of Darkness	Ursula K. LeGuin
1971	Ringworld	Larry Niven
1972	To Your Scattered Bodies Go	Philip Jose Farmer
1973	The Gods Themselves	Isaac Asimov
1974	Rendezvous with Rama	Arthur C. Clarke
1975	The Dispossessed	Ursula K. LeGuin
1976	The Forever War	Joe Haldeman
1977	Where Late the Sweet Birds Sang	Kate Wilhelm
1978	Gateway	Frederick Pohl
1979	Dreamsnake	Vonda N. McIntyre
1980	The Fountains of Paradise	Arthur C. Clarke
1981	The Snow Queen	Joan Vinge
1982	Downbelow Station	C.J. Cherryh
1983	Foundation's Edge	Isaac Asimov

Recent Losses to the World of Science

The early 1980s saw the deaths of a number of scientists, inventors, and writers about science.

ABELL, GEORGE (Mar. 1, 1927–Oct. 7, 1983). American astronomer, author of the definitive *Abell Catalogue of Clusters*, and discoverer of the Abell Galaxy, once considered the largest object in the universe.

AMOROSO, EMMANUEL CIPRIAN (Sept. 16, 1901–Oct. 30, 1982). British veterinarian who made important discoveries in role of placenta in reproduction.

BARNETT, STEPHEN FRANK (Aug. 10, 1915–Aug. 18, 1981). British veterinarian who studied tick-borne diseases in livestock.

BLOCH, FELIX (Oct. 23, 1905–Sept. 10, 1983). Swiss-born co-winner of 1952 Nobel Prize in Physics for work in nuclear magnetic resonance.

BOK, BART J. (Apr. 28, 1906–Aug. 5, 1983). Dutch-born astronomer, discoverer of "Bok globules" in the Milky Way.

BOYD, WILLIAM C. (Mar. 4, 1903–Feb. 19, 1983). American immunologist, discoverer of the role of lectins in blood chemistry.

BREIT, GREGORY (July 14, 1899–Sept. 13, 1981). Soviet-trained physicist who in U.S. developed resonance theory of nuclear reactions.

BULLARD, SIR EDWARD CRISP (Sept. 21, 1907–Apr. 3, 1980). British geophysicist who developed sciences of seismic measurement and magnetic field statistics.

BURHOP, ERIC HENRY (Jan. 31, 1911–Jan. 22, 1980). Australian physicist, researcher in K-mesons and neutrons, and proponent of nuclear arms control.

BURPEE, DAVID (Apr. 5, 1893–June 24, 1980). American horticulturist who launched W. Atlee Burpee seed business.

CLAUDE, ALBERT (Aug. 24, 1898–May 22, 1983). Belgian founder of modern cell biology and co-winner of 1974 Nobel Prize in Physiology or Medicine for studies of cell structure.

COLE, KENNETH S. (1899–April 18, 1984) American biophysicist who won the U.S. National Medal of Science for his work on how nerve cells transmit messages.

CORNER, GEORGE WASHINGTON (Dec. 12, 1889–Sept. 28, 1981). American embryologist who identified the hormone progesterone in the female reproductive system and synthesized contraceptives.

DARLINGTON, CYRIL DEAN (Dec. 19, 1903–Mar. 26, 1981). British biologist-geneticist who founded the science of cytology.

DELBRUCK, MAX (Sept. 4, 1906–Mar. 9, 1981). German-born molecular biologist, co-winner of 1969 Nobel Prize in Physiology or Medicine for bacteriophage studies.

DENNY-BROWN, DEREK ERNEST (June 1, 1901–April 20, 1981). New Zealand-born neurologist and Harvard authority on neuropathology.

DORNBERGER, WALTER ROBERT (Sept. 6, 1895–June 26, 1980). German rocket engineer who oversaw development of the V-2 rocket.

DUBOS, RENÉ JULES (Feb. 20, 1901–Feb. 20, 1982). French microbiologist, discoverer of antibacterial substances later used in antibiotics, and vocal environmentalist.

FERRANTI, SIR VINCENT (Feb. 16, 1893–May 20, 1980). British electrical engineer and industrialist who pioneered television and radar defense electronics.

FISK, JAMES BROWN (Aug. 30, 1910–Aug. 10, 1981). American physicist who developed microwave high frequency radar.

FLETCHER, HARVEY (Sept. 11, 1884–July 23, 1981). American physicist and head of the team that developed stereophonic sound.

FULLER, ROBERT BUCKMINSTER (July 12, 1895–July 1, 1983). Polymath American inventor and writer best known for design and popularization of the geodesic dome and the prefabricated Dymaxion House.

GIAUQUE, WILLIAM FRANCIS (May 12, 1895–Mar. 28, 1982). American chemist who won 1949 Nobel Prize for research in low-temperature chemistry.

GLUECKAUF, EUGENE (Apr. 9, 1906–Sept. 15, 1981). German chemist who immigrated to England to work in atmospheric gases.

GUTTMANN, SIR LUDWIG (July 3, 1899–Mar. 18, 1980). British neurosurgeon and influential head of the National Spinal Injuries Centre.

HANDLER, PHILIP (Aug. 13, 1901–Dec. 29, 1981). American biochemist who discovered enzymes and studied nutritional diseases.

HARTLINE, H. KEFFER (Dec. 22, 1903–Mar. 17, 1983). American physician who shared 1967 Nobel Prize in Physiology or Medicine for study of the electrical mechanism of the optic nerve.

HASSEL, ODD (May 17, 1897–May 11, 1981). Norwegian chemist who shared 1969 Nobel Prize in Chemistry for study of the molecular structure of organic compounds.

HAVEMANN, ROBERT (Mar. 11, 1910–Apr. 9, 1982). East German scientist, enemy of Nazism, later critic of Communist Party, protected by his international scientific reputation.

HILDEBRAND, JOEL H. (Nov. 16, 1883–Apr. 30, 1983). American chemist, discoverer of means to prevent "bends" in deep-sea divers, and famous teacher at the University of California.

KAMANIN, NIKOLAY PETROVICH (Oct. 18, 1908–Mar. 14, 1982). Soviet aeronautical engineer and head of the cosmonaut training that made Yuri Gagarin the first man in space in 1961.

KAPITSA, PYOTR L. (July 8, 1894–April 8, 1984) Soviet physicist who also worked for thirteen years in Great Britain during the 1920s. His development of methods for liquifying helium and other gases paved the way for modern low-temperature physics. Although a strong advocate of scientific freedom, he was permitted by the Soviet authorities to accept the 1978 Nobel Prize in Physics, awarded for his lifetime of achievement.

KOLLSMAN, PAUL (Feb. 22, 1900–Sept. 26, 1982). American aeronautical engineer who invented the altimeter.

KREBS, SIR HANS ADOLF (Aug. 25, 1900–Nov. 22, 1981). German-born biochemist who shared 1953 Nobel Prize in Physiology or Medicine for work, done in England, on the catalytic chemistry of citric acids.

LANCEFIELD, REBECCA CRAIGHILL (Jan. 5, 1895–Mar. 3, 1981). American bacteriologist who classified 60 types of the common streptococcal bacteria.

LAZAR, MARGULIES (1895?–Mar. 7, 1982). American obstetrician and gynecologist who invented intrauterine coil contraceptive.

LIBBY, WILLARD FRANK (Dec. 17, 1908–Sept. 8, 1980). American chemist who won 1960 Nobel Prize in Chemistry for the radiocarbon dating technique.

MALINA, FRANK JOSEPH (Oct. 2, 1912–Nov. 9, 1981). American aeronautical engineer, founder and first director of the U.S. Jet Propulsion Laboratory.

MOORE, STANFORD (Sept. 4, 1913–Aug. 23, 1982). American biochemist who shared 1972 Nobel Prize in Chemistry for studies in enzyme chemistry.

PIAGET, JEAN (Aug. 9, 1896–Sept. 16, 1980). Swiss child psychologist whose work included new forms of early mathematics training and intelligence testing.

PICCARD, REV. JEANETTE (Jan. 5, 1895–May 17, 1981). American scientist who launched the first balloon into the stratosphere in 1934 and was ordained an Episcopal priest in 1974.

PILYUGIN, NIKOLAY ALEKSEYEVICH (May 18, 1908–Aug. 2, 1982). Soviet electrical engineer who developed electronic controls for space ships and stations.

REICHELDERFER, FRANCIS W. (Aug. 6, 1895–Jan. 26, 1983). American meteorologist who modernized the U.S. Weather Bureau while its chairman from 1938 to 1963.

ROWE, WALLACE P. (Feb. 20, 1926–July 4, 1983). American virologist and director of the National Institute of Allergy and Infectious Diseases.

SELYE, HANS HUGO (Jan. 26, 1907–Oct. 16, 1982). Austrian endocrinologist whose work in Canada documented links between stress and disease.

SLONE, DENNIS (Jan. 9, 1930–May 10, 1982). American epidemiologist who studied health dangers to women of tranquilizers, smoking, and birth-control pills.

SONDHEIMER, FRANZ (May 17, 1926–Feb. 11, 1981). German chemist whose synthesis of steroids and benzene compounds were important advances in organic chemistry.

SPIEGELMAN, SOLOMON (Dec. 14, 1914–Jan. 21, 1983). American microbiologist and geneticist who formulated the molecular basis of cancer.

STEIN, WILLIAM H. (June 25, 1911–Feb. 2, 1980). American biochemist who shared 1972 Nobel Prize in Chemistry for studies in enzyme chemistry.

SZMUMESS, WOLF (Mar. 12, 1919–June 6, 1982). Polish-born epidemiologist who headed the U.S. team that developed a hepatitis-B vaccine.

THEORELL, HUGO (July 6, 1903–Aug. 15, 1982). Swedish biochemist who won 1955 Nobel Prize in Physiology or Medicine for study of enzymes and cell metabolism.

TUVE, MERLE ANTHONY (June 27, 1901–May 20, 1982). American physicist who studied atomic structure and confirmed the existence of neutrons.

UREY, HAROLD C. (Apr. 29, 1893–Jan. 5, 1981). American chemist, winner of 1934 Nobel Prize in chemistry for discovery of "heavy" hydrogen, and researcher in the "birth" of life in amino acids.

VAN VLECK, JOHN H. (Mar. 13, 1899–Oct. 27, 1980). American physicist who shared 1977 Nobel Prize for studies in electromagnetism important to computer technology.

VERNOV, SERGEY NIKOLAYEVICH (July 11, 1910–Sept. 26, 1982). Soviet physicist, researcher in cosmic rays and director of the Nuclear Physics Research Institute, Moscow.

VINOGRADOV, IVAN (Sept. 14, 1891–Mar. 20, 1983). Soviet mathematician, specialist in number theory, and director of the Soviet Institute of Mathematics.

VON EULER, ULF (Feb. 7, 1905–Mar. 10, 1983). Swedish biochemist who shared 1970 Nobel Prize in Physiology or Medicine for study of the nervous system.

WILSON, CARROLL L. (Sept. 21, 1910–Jan. 12, 1983). American physicist and first director of the Atomic Energy Commission.

YUKAWA, HIDEKI (Jan. 23, 1907–Sept. 8, 1981). Japanese physicist, winner of 1949 Nobel Prize in Physics for subatomic research that predicted the existence of mesons.

ZEPLER, ERIC ERNEST (Jan. 27, 1898–May 13, 1980). German physicist and pioneer in radio design.

ZWORYKIN, VLADIMIR (July 30, 1889–July 29, 1982). Russian-born electrical engineer who emigrated to the United States and invented the television camera tube (iconoscope, 1923) and picture tube.

ARCHAEOLOGY AND ANTHROPOLOGY

The dividing line between archaeology and certain aspects of anthropology is hard to discern. Ask an archaeologist about the discovery of a sanctuary in a Spanish cave that is 14,000 years old, and he or she will claim it for archaeology. An anthropologist, however, would also claim the site. In practice, such borderline sites (and, increasingly, many older sites) are investigated both by anthropologists and archaeologists, working in teams.

In archaeology, there are a few continuing trends that are clear. Urban archaeology has become increasingly important, with the excavation of sites under major cities becoming more and more influential. In this section, we examine such sites in Israel, England, Ireland, Mexico, and the United States.

Another archaeological trend that continues to grow is the importance of underwater recovery, especially of ships, but sometimes of drowned buildings or even sunken cities. We include examples for England and Scandinavia.

Aerial and even satellite research has aided in finding new sites and in learning new facts about old ones. A remarkable example from Central America is featured.

One recent trend has been a new understanding of the importance of the Viking era in Western civilization. For 200 years the warriors from Scandinavia dominated in conquest, trade, and exploration. Our feature article for this section tells what we have learned about the Viking era from combining archaeology with historical accounts.

Physical anthropology continues the search for a complete history of the species that led to *Homo sapiens*. Although the picture that results from this search is far more complex than anyone suspected a few decades ago, great progress has been made in filling the gap. Nevertheless, a real gap still remains, almost 4,000,000 years long. New discoveries on each side of the gap occurred in the early 1980s.

1980

Mysteries of the Maya

The Maya formed one of the three great civilizations of what is now Latin America, along with the Aztec and Inca. Unknown to the Western world for several hundred years after early post-Columbian contacts, the Maya were rediscovered in the nineteenth century, when remarkable temple-cities were found in the jungles of southern Mexico and Central America.

Despite the years of archaeological investigation since that rediscovery, the remarkable flowering of the great Maya civilization in Mesoamerica still presents several puzzling problems. How could the inhabitants of these populous cities—some 300 are presently known—support themselves in the inhospitable, swampy jungle environment of the Maya lowlands? The largest of all Maya cities, Tikal in Guatemala, held up to 50,000 people at its peak, and its ruins today cover about 20 square kilometers (8 square miles).

Cities of this size in the middle of swamps are especially difficult to understand because recent research suggests a rapid population growth in the last few hundred years of the Classic period (A.D. 300–900). And what caused the sudden collapse of the Classic Maya civilization at the end of the period?

The problems in understanding the Maya have been complicated by the difficulties involved in trying to read the Maya writing. Although dates and other numbers have been deciphered for a long time, the remainder of the Maya writing system has resisted translation. Thus, we often know that something important happened on a particular day, but we do not know what it was that happened. In part, the difficulty in understanding Mayan writings is that there are so few of them; Spanish priests destroyed all they could find. They argued that if the writings agreed with Scripture, they were unnecessary, but if they disagreed, they were false. As a result, archaeologists have little to go on in trying to solve the mystery of the Maya.

An archaeological research group, trying to find a way to locate additional Maya cities lost in the jungles, inadvertently stumbled on what may well be the answer to the first question. Settling on aerial radar reconnaissance as the most likely method of finding the lost cities, they used a special NASA aircraft carrying a side-scanning radar that had been developed to survey Venus. The first flights were made over the Guatemala and Belize lowlands in 1977–78. By 1980 some 37,000 square kilometers

(14,000 square miles) had been surveyed, and the resulting analyses had been supplemented by additional radar scanning from the SEASAT satellite.

The method proved to be of little help at the time in finding lost cities, but the radar imagery did reveal an unexpectedly widespread pattern of grids and lattices along the rivers and in the larger swamps throughout the whole lowland area. These marked the remains of Maya canals and raised field systems, some of which had already been investigated on the ground. But the extent of the network—some 1,250 to 2,500 square kilometers (480 to 970 square miles) of canals and raised fields—came as a total surprise.

With this discovery the project was quickly switched from lost cities to the further study of this newfound agricultural system. Ground investigation of some of the canals, a difficult undertaking in the steamy, wet jungles, corroborated the aerial evidence. The satellite images had revealed, moreover, what seemed to be a large lake or reservoir in Mexican Campeche near a large Maya city. It had a canal running out of it and concentric rings around it, possibly a water control and storage center. Above all, it was noted that the larger Maya cities were all located near the largest swamps, close to the sites of the newly found canal and raised-field systems.

So here was a plausible answer to the first problem. At the cost of immense labor, this canal and field system had been built to support the rising populations of the cities after A.D. 600, and although the cities and fields were deliberately located in the most inhospitable regions, these regions were potentially the most fruitful ones. Like the famous chinampas of the Aztecs—floating islands—these raised fields could be extremely productive. Not thousands but millions could be fed with the produce from these rich, canal-fed fields.

After the end of the project in 1980, another flight over the Tikal area in a more carefully managed radar scan was able to bring out the city's buildings in reasonable detail. Radar scanning may now, after all, be able to locate those lost cities in the jungle!

The discovery of this sophisticated agricultural system also suggests a possible answer to the second problem. If there had been some political breakdown, possibly a revolution in the highly stratified Maya society at the end of the Classic period, the resulting disorder may well have led to the neglect and deterioration of the canal system, causing the decimation of the population and total breakdown. We know that the population at the time fell disastrously. By A.D. 1000 mighty Tikal was utterly deserted.

1981

Ice Age Shrine Uncovered

A cave in the Cantabrian mountains in northern Spain has yielded the oldest known religious sanctuary, dating from around 14,000 years ago, complete with elaborate offerings and the crude statue of a divinity.

El Juyo lies 5 kilometers (3 miles) inland from the Atlantic coast, with the mountains at its back. Nearby is the famed Altamira cave where Stone Age paintings were first found and identified in the nineteenth century. The whole region is rich in Paleolithic (Old Stone Age) sites, mostly caves. Even in the last Ice Age this was a favored area, relatively warm and sheltered, with abundant game in the mountains and on the coastal strip, fish in the rivers, and edible crustaceans and mollusks for the finding along the shore.

Many of these caves have been excavated, but at El Juyo a team of scientists found an untouched area of deep fill. The cave, which had been rediscovered as early as 1953, was carefully excavated by a team of American (Leslie Freeman and Richard G. Klein) and Spanish (J. Gonzalez Echegaray and I. Barandiaran) anthropologists and archaeologists. Breaking through the remains of ancient rockfalls and covering sheets of limestone, they excavated parts of two untouched layers dating from the Magdalenian period, the high point of Upper Paleolithic culture. These revealed dramatic evidence of intense human activity. The lower layer contained a large, walled, artificial depression filled with the jumbled remains of red deer, for these people dined well on venison, seafood, and foraged plants. Later a small walled pit had been dug here. It was filled with stored valuables such as grindstones, coloring materials, deer antlers, and a lamp.

In the topmost and latest level the sanctuary came to light during the excavations of 1978 and 1979, but the remarkable find was not published until August 1981. Scholars have long inferred some sort of Paleolithic religion from the cave paintings and the practice of burying the dead. Even the graves of Neanderthals reveal special positions and orientations with the compass that suggest religious ideas, with one remarkable Neanderthal burial in which the body was heaped with flowers. Furthermore, the small statues of women with exaggerated breasts and buttocks that are frequently found in Paleolithic and later sites have been assumed to represent some sort of cult. Drawings of part men-part beasts found in cave paintings have also been linked to early shamanism. But here was indisputable evidence for actual worship of a deity, accompanied by elaborate offerings hinting at complex and mysterious rituals. Within a large walled precinct the excavators found three mounded, clay-topped structures covering oblong trenches, the largest topped by a massive 1-ton, 2-meter (6-foot) slab of limestone. Hidden beneath the slab were apparent offering layers of intricate rosettes made of earth and colored clay, interspersed with burnt offerings including brand new bone spear points. The trench itself, lined with red ocher, contained a lump of white clay and an antler tine thrust into the ground. A chunk of red ocher had been buried under the smallest mound, and nearby three small pits held other offerings, including bone sewing needles.

Standing on the smallest mound, facing the cave entrance, stood the statue, a shaped block of stone, its face 33 by 36 centimeters (13 by 14 inches), with a natural fissure that had been used to divide the features, left and right, into the chisled face of a smiling, moustached, bearded man on one side, and a dour, cat-like creature on the other with a snout, a projecting fang, and dotted whiskers—perhaps a lion. Here was an idol with two natures, the good and the evil, a familiar dualism. About 14,000 years ago a final rockfall and layers of stalagmite limestone totally sealed off the cave and these precious remains—until the cave was first entered in modern times.

NOW LASCAUX II—WHAT NEXT?

In 1963 the famous prehistoric cave of Lascaux in France was closed to tourists when it was discovered that the 15,000-year-old Paleolithic paintings were deteriorating. The reason, of course, was that the thousands of tourists trooping through the cave had upset the delicate balance of its closed environment. Another famous cave, Altamira in Spain, was closed for the same reason in 1977. Since then only a few brief visits by scientists have been allowed to either cave while the problem is being studied. "This cave belongs to humanity," said one of the investigators at Altamira. "What good is it if it remains there and we can't see the real thing?"

This concern for reality did not bother the French. In 1983 they opened Lascaux II, sited close to the original, an exact copy in reinforced concrete, with the paintings meticulously reproduced using the original materials. The real cave continues to be open only to specialists, while Lascaux II, of course, is for the tourists.

New Fossils Close the Gap in Our Past (Somewhat)

The exciting search for our ultimate ancestors has been pursued with mounting intensity and increasing expertise for well over a century. In 1981 several important finds began to close the gap between ancestral creatures that were fairly well understood and an earlier assemblage of presumed ancestors.

Cro-Magnon, Neanderthal, and Java discoveries, unearthed in the nineteenth century, with finds in Peking, China, in the 1920s, were judged, after considerable controversy, to be the remains of ancestors who lived many thousands of years ago. Now the paleoanthropologists, those who study the fossil remains of early humans and their ancestors, are investigating the fossils of creatures that lived millions of years ago. These scientists are finding that the picture is far more complex.

A new word had to be invented—*hominids*. Just as the chimpanzee, gorilla, gibbon, and orangutan can be classed as the ape branch of the primates, or *pongoids*, the hominids form a branch closer to the humans. The only extant species of hominid is *Homo sapiens*. Our closest extinct ancestors are also in the genus *Homo*, but other relatives, while clearly nearer to human than the apes, are in another genus, *Australopithecus*. Together, the *Homo* and *Australopithecus* genera form the hominid family as it is known today. The hominids and pongoids together are often called the hominoids.

To date the most famous find of an early hominid is Lucy, the partly complete skeleton of a 20-year old female discovered in Ethiopia in 1974 by Maurice Taieb and Donald Carl Johanson, which they assigned to a new species they called *Australopithecus afarensis*. Lucy was a 1 to 1.2 meter (3.5 to 4 foot) high, small-brained creature who walked upright. She was found lying in a sediment that has been dated as about 3,600,000 years old, although there has been considerable controversy about that date.

To match this sensational find, Paul I. Abell, working with Mary Leakey in 1978, made an equally astounding discovery at Laetoli in Tanzania, where 3,600,000-year-old hominid fossils had already been found. Three of these hominids, possibly two adults and a child, had walked across an area covered by volcanic ash after a nearby eruption, which had then been fossilized, leaving at least 25 meters (75 feet) of clear footprints. These 1.2-meter (4-foot) creatures walked exactly the way humans do today, thus finally disposing of the previous, widely held theory that the ability to walk upright had depended upon the development of a big brain.

As the finds multiplied in recent years, so have the controversies. In 1961 a Leakey son, Jonathan, uncovered a skull that proved to be 1,800,000 years old. It was much like an *Australopithecus* except that it had a much larger brain, placing it firmly in the *Homo* class. The Leakeys dubbed it *Homo habilis* ("handy") because they judged that creatures like it had made the pebble tools previously found in the same area. Here then was another claimant in the *Homo* line, earlier than *Homo erectus*. More skull fragments were found by another son, Richard Leakey, near Lake Turkana, which he also tentatively assigned to the *H. habilis* group. But this "1470" skull was dated to over 2,000,000 years ago. To cap it all, in 1975 the same area yielded an undoubted *Homo erectus* who had lived there 1,500,000 years ago—far earlier than the Java and Peking men (about 700,000 to 500,000 years ago). So *H. habilis*, and possibly even *H. erectus*, had apparently lived contemporaneously with the australopithecines, which to many seemed to rule out *Australopithecus* as a direct ancestor of man.

The anthropologists are moving still further back in time and are now focusing on the Gap, spanning 4,000,000 to approximately 8,000,000 years ago. Few significant fossil remains of hominids dating to this period have been found. This era, it is thought, will eventually produce the evidence for the divergence of the walking hominids, our ancestors, from the larger hominoid group that includes the apes.

A monkey-sized creature called *Ramapithecus*, whose remains have been found in many places from India to Central Europe, has long been considered the most likely candidate for the ancestor of all hominids, largely on the basis of his strong teeth. But we do not know if he walked upright, and new finds of *Ramapithecus* in Pakistan in 1981, dated to 8,000,000 years ago by Harvard's David Pilbeam, indicate that he is more likely the direct ancestor of our cousin, the orangutan, than the first and earliest hominid. Since *Ramapithecus* remains have been dated back as far as 14,000,000 years ago, Pilbeam's elimination of this creature as a hominid (since by 8,000,000 years ago he already showed specializations leading to the orangutan) makes it more likely that the crucial divergence may have taken place much later, perhaps 7,000,000 to 9,000,000 years ago.

Beyond the Gap is *Aegyptopithecus*, found in Africa by Dr. Elwyn Simons. It is the earliest candidate for classification on the hominoid line, dating from 30,000,000 years ago.

Closer to the Gap, Pilbeam also discovered in 1981 in Pakistan another 13,000,000-year-old fossil much like *Ramapithecus*. To date this fossil has not been analyzed. In the Awash River valley in Ethiopia in the same year, J. Desmond Clark and Timothy D. White found two fossil bones from a creature who lived 4,000,000 years ago, definitely walked upright, and had a small brain.

So scientists are now hovering at the edge of the Gap. The requisites for finding more evidence of the latest missing link are sites with a rich and . continuous record of fossil animals, including hominids, and a geological formation that can be securely dated. The Awash Valley is one. Another is in Kenya where Dr. Maeve Leakey, Richard's wife, found a fossil jaw in 1983, tentatively dated between 16 and 18,000,000 years ago. This jaw, and other fossils from the area, have been tentatively identified as *Sivapithecus*, a ramapithecine previously found only in Asia. A third area is in North Africa near the Libyan coast. Here at Sahabi an international team has been searching since 1978 an area rich in fossil remains from 24,000,000 years ago down to about 1,600,000 years ago. Several fragments of a hominoid creature have been recently found in the area, a creature that seems to have had a brain smaller than that of a pygmy chimpanzee as well as a human-like foot. More fossils are needed before any meaningful analysis of this hominoid can be carried out.

WHATEVER BECAME OF—PEKING MAN?

What happened to the original remains of the *Homo erectus* fossils, known to anthropologists as Peking Man, discovered in the 1920s in the famous Zhoukoudian (Choukoutien) cave near Beijing (Peking), China, is still an oft-debated mystery. When the Japanese invaded China in 1937, the excavations ceased and the finds, eventually turned over to the U.S. Marines in 1941, simply disappeared. The loss, however, has since been amply compensated for by the Chinese excavations at the cave since 1949.

Peking Man lived for over 200,000 years in this huge limestone cave, 50 kilometers (30 miles) southwest of Beijing from 460,000 to 230,000 years ago. So far eighteen skulls or skull fragments, 157 teeth, and many bones of more than 40 male and female individuals have been found (including the vanished pre-War fossils, for which plaster casts remain).

From these finds and associated tools, garbage, and remains of the environment surrounding the cave in ancient days, it can be deduced that Peking Man lived and hunted in cooperating bands and brought his prey home to cook and share with others. Anatomical details show that he could talk, and the progress he made in over 200,000 years in shaping his flake tools indicates that these skills were transmitted by teaching.

The Evolution of the Hominoids

No subject in science is more hotly debated than the evolution of human beings, their cousins, and their ancestors. By 1984, the most probable version of this evolutionary history covered 30,000,000 years, with a major gap of from 4,000,000 to 6,000,000 years unexplained.

Years Ago	
30,000,000	*Aegyptopithecus* is tentatively placed as the earliest known ancestor of the hominoids.
20,000,000	*Proconsul africanus* in Africa and Arabia is generally recognized as ancestral to the hominoids, probably directly so of the gibbon.
17,000,000	An African hominoid that may be a form of the known Asian genus *Sivapithecus* is a candidate for the ancestor of the hominids.
10,000,000	*Ramapithecus*, closely related to *Sivapithecus*, is the ancestor of the orangutan. Some would date the split between the gorilla/chimp line and the hominid line around this time, but there is no clear fossil evidence.
8,000,000– 4,000,000	The Gap.
4,000,000	*Australopithecus afarensis* is found, but no one knows what its ancestors were like.
2,000,000	Three or four hominid species coexisted. Two or three australopithecines were evolutionary dead ends, but *Homo habilis* was the first human species. Habilis made simple chipped tools, called pebble tools, and supplemented a diet of roots and other vegetables with carrion. Restricted to Africa.
1,500,000	*Homo habilis* had evolved into *Homo erectus*. Erectus made characteristic hand axes, used fire, and was a hunter. Extended range to Europe and Asia.
150,000– 90,000	*Homo erectus* evolved into *Homo sapiens* during the most recent wave of ice ages. The first *H. sapiens* were apparently of the subspecies known as Neanderthal, who made bifacial stone tools, buried their dead, and probably had an efficient language.
40,000	The modern type of human being, known as *Homo sapiens sapiens* (when contrasted with the Neanderthal subspecies) arose. Extended range to Australia, Polynesia, and the Americas.

1982

The *Mary Rose*: The Raising of Henry VIII's Flagship

"The world's most expensive piece of driftwood," quipped a journalist after the 37-meter-long (120-foot-long) hull of the *Mary Rose* was finally raised from the mud of Portsmouth harbor on October 11, 1982. Seventeen years of work, including 30,000 dives in the murky harbor waters, went into this largest of underwater archaeological projects. The operation cost £4,000,000 (about $6,000,000) and will cost as much again to conserve and display the varied remains—over 17,000 artifacts and the hull itself, much of one side preserved in the protecting silts of the harbor's bottom.

The *Mary Rose*, one of the earliest purpose-built warships, was laid down in 1509 and rebuilt in 1536 to accommodate heavier cannon. On Sunday, July 19, 1545, a French invasion fleet—larger than the Armada—carrying 30,000 men, appeared outside the harbor and sent in some galleys to lure the smaller English fleet out to battle. The Vice-Admiral's flagship, the *Mary Rose*, was overweighted with 91 heavy guns and crammed with 700 men, 285 of them soldiers, mostly archers, high in the fore and aft castles. While the king watched from the shore, she veered about to face the French with gun ports open and cannon at the ready. Suddenly she heeled over, the sea poured into her open ports, and in a matter of seconds she plunged to the bottom. Only about 30 men survived.

The *Mary Rose* was eventually rediscovered by an amateur diving group in 1970–71. After they had found that much was preserved in the deep harbor silt, the prestigious Mary Rose Trust was set up in 1979 with the Prince of Wales as president and an archaeologist in charge. It was decided to remove the hull's top sides and contents and then see if the rest could be raised. A team of archaeologists, volunteer divers, environmentalists, photographers, artists, and ship historians was assembled, and the work pursued with side-scan sonar, underwater magnetometers, computers, and video, backed up by a conservation laboratory on shore. Here the more delicate objects were impregnated with polyethylene glycol and freeze-dried to preserve them.

The hull proved to be a time capsule of Tudor sea life. From it emerged cannon of cast iron and bronze, quantities of rigging and tackle, and such artifacts as pocket sundials; leather boots, shoes, and jerkins; a woolen stocking; table settings of pewter and wood; board games; musical instruments; and fruit pits and peas in the pod. In the barber-surgeon's walnut chest were found syringes, jars of ointment, and a wooden mallet to anesthetize patients! Significant items included the earliest known magnetic compasses, and no less than 139 longbows and 2,500 arrows—for the famous English bowmen were still relied upon. There were many human skeletons, as well as those of rats, insects, a frog, and a pet dog.

The final hazardous lift, recorded by hundreds of media people, was carried out by a huge oil-drilling barge, cables, and a special cradle and lifting frame. The hull is now drying out in drydock under a shed, and it will eventually be partly reconstructed for a Tudor ship museum.

The recovery of the *Mary Rose* was not the only remarkable nautical find during the early 1980s of ships lost in war. In Japan in 1981 an underwater archaeological expedition found the remains of Khubilai Khan's fleet, destroyed by the famous "Divine Wind," or *kamikaze* in 1281. A typhoon had stopped Khubilai Khan's second attempt to invade Japan, and the Japanese people have believed that their protective gods saved them. After 700 years underwater, the archaeological finds are amazing—sword blades, stone anchors, jugs for mixing gunpowder, catapult balls covered with oil and cloth for fire bombs, and ingots of iron ballast. The typhoon had destroyed 4,000 ships of Khubilai Khan's navy. The number of soldiers and sailors killed has been estimated at 130,000. For the past 700 years, remains of the fleet have been washing up on the shore, but now the source has been located. The finds, including those washed ashore, will be housed in a museum near the site.

TIME CAPSULE: A.D. 3982

The British BBC recently marked the 60th year of radio with the burial of a time capsule in Yorkshire on July 21, 1982. An eminent archaeologist and a distinguished panel of 20 experts advised on the contents, which included false teeth, a beer can, seed packets, model cars and planes, radio, music, and TV programs, and a video disk of the battle for the Falklands. The whole was encased in a small steel and concrete chamber filled with argon gas. From the buried capsule a concrete column rises to the surface, bearing instructions on how to open it and when—A.D. 3982. The panel had to consider the likelihood of a future ice age, a nuclear catastrophe, or a Russian takeover. They settled on 2,000 years as a likely period. The archaeologist had his own concerns. Beauty and clever technology, he felt, should come second in the objects chosen, for "the reconstruction of societies and their beliefs and power structures and decision making" would be more interesting to the future finders. There spoke the contemporary archaeologist!

Bonanza of Roman Bones

How tall were the Romans? Were they healthy? What did they eat? To be valid, generalizations of this kind by physical anthropologists must be based on the study of a fair sampling of contemporary skeletons, and this is exactly what has been found at Herculaneum, sister city of Pompeii.

Although both cities were inundated by hot ash and pumice from the eruption of Vesuvius in A.D. 79, Herculaneum, an elegant little seaside resort right under Vesuvius, was spared the first day while 15 centimeters (6 inches) of ash drifted down on Pompeii from the 20-kilometer (12-mile) high volcanic cloud. Until 1982 the world believed that the 4,500 citizens of Herculaneum had learned of the fate of Pompeii and escaped. In fact, Herculaneum was destroyed during the night before Pompeii fell. When the cloud of volcanic ash collapsed onto the city around midnight, searing and burying Herculaneum in a matter of minutes the city was virtually empty. Apparently the citizens had seen glowing clouds approach the city and taken flight. A trenching operation in 1981 designed to drain water that was seeping into the ruins, however, revealed how far they had gotten. Many of the townspeople had gathered on the beach. Pliny the Elder (as reported by his nephew) had succumbed on the beach near Pompeii because strong winds toward Vesuvius, a common phenomenon during a volcanic eruption, prevented his ships from leaving. Perhaps the crowd on the beach was trapped by the same phenomenon.

The victims uncovered on the former town beach are represented by over 150 skeletons so far and more to come—a group of twelve men, women, and children who died in agony huddled under an arcade, a soldier slammed face down on the beach with his sword by his side. Near them is the complete hull of a Roman ship, evidently tossed ashore by the winds or by a tsunami (tidal wave) or merely tossed from a construction site.

And what do the newly found bones reveal about the citizens of Herculaneum, many of whom were wealthy Romans who used the city as a beach resort? Sara Bisel, an anthropologist who was working in Greece at the time of the discovery, was flown quickly to Herculaneum to take over the task of organizing, studying, and protecting the bones. She reports that "The average woman was about 156 centimeters (5 feet 1.5 inches), the average man about 170 centimeters (5 feet 7 inches). And it seems the Romans were a fairly healthy bunch. They had excellent teeth and beautiful mouths." Bones of apparent slaves, however, revealed lives of hard labor and poor diet. The bones, dipped in an acrylic-resin solution, were to have been hardened for future study and displayed in a special on-site museum.

El Templo Mayor

In 1982 an exhibit from the ongoing project of excavating *El Templo Mayor*, or Great Temple, in Mexico City, reached the United States, calling attention to one of the great archaeological projects of our time.

The accidental discovery in 1978 of the great Aztec pyramid in Mexico City and its subsequent excavation has astonished the world and fired Mexico's pride in its Indian heritage.

The Temple of Huitzilopochtli and that of Tlaloc, god of rain and fertility, stood side by side on the pyramid, forming the complex that came to be known as *El Templo Mayor*.

El Templo Mayor has a bloody history. From the top of the 114 steps of the huge, four-tiered pyramid, religious center of the mighty Aztec empire, priests tossed the bodies of sacrificial victims. First the priests tore out their hearts and offered them to the sun god, Huitzilopochtli, to strengthen him in his daily round. Then the bleeding victims tumbled to the bottom and landed on the great circular relief of the dismembered moon goddess, thus symbolizing the daily victory of day over night.

Here the Spanish invaders and the desperate Aztecs fought fiercely in 1520. After his victory Cortes ordered the whole monument destroyed. It was generally assumed that no remains of the temple would ever be found.

The Great Temple, however, had been merely the last of a number of temples built upon the same site, incorporating the earlier ones. Consequently, the destruction ordered by Cortes did not affect the archaeological value of the site.

Actually five to seven major, superimposed temples have been uncovered, each built around its predecessor. It was the discovery of the massive disc of the moon goddess, belonging to temple 4B, that led to the present excavations. The excavations have bared much of the temple area—not under the present cathedral as once thought, but several blocks away—and some of the many subsidiary temples and buildings of the precinct, including the Temple of Serpents with its lively painted frieze, a rack of skulls carved in stone, and many other powerful sculptures. Innumerable precious objects were also found in the numerous offering pits, including rare pieces from earlier Indian cultures, such as an ancient Olmec stone mask—for the Aztecs also took pride in their rich heritage. The offerings also included human skulls and animal remains, ceramics, sacrificial stone knives, and ornaments in gold, silver, and shell. The whole area is to be placed in an archaeological park with a special museum for the finds.

The Ancient History of Jerusalem Revealed

It has long been known that the city of David and Solomon occupied a long narrow rocky spur, often called the Ophel, that descends south from Jerusalem's Temple Mount. It lies entirely outside the much later old walled city of Jerusalem. Partial excavations had been carried out in the area since the mid-nineteenth century, but the largest effort so far, the five-year City of David dig, opened in 1978, has already investigated five well-chosen sites covering four acres. Helped by gangs of enthusiastic volunteers from all over the world, the dig, in the summer of 1980, uncovered a stepped, sloping stone structure, as tall as a five-story building, that has been identified as the underpinnings of the ancient city's acropolis that once supported the palaces of David and Solomon.

The original Canaanite city, which David seized from the Jebusites in 1000 B.C., had been sited on this narrow ridge for defensive purposes and because the Gihon spring, still today an important source of water, gushed out below its northern end. Only 5 hectares (12 acres) were available for the buildings of the ancient city, and as the city expanded under David's Hebrews, it became necessary to build out and down the eastern slope of the ridge. Hence the stepped-stone foundations of the acropolis, and also the discovery of a later residential time, dating from King Hezekiah's time in the 8th century B.C., which had been built up on a series of terraces, the lowest resting against the city wall that ran along the lower slopes of the hill. The excavations uncovered over 37 meters (120 feet) of this wall, which withstood an Assyrian siege under Hezekiah in 701 B.C.

The steep and narrow confines of the Ophel hill also made life difficult for the archaeologists. Nevertheless, the results have been outstanding.

The dig so far has uncovered not only the massive acropolis foundations, the largest monumental work of the period so far found in Israel, but also has illuminated the history of Jerusalem from its beginnings in a Copper Age settlement near the spring of Gihon about 3500 B.C.; its emergence as a town some centuries later; its long history as a Canaanite city, as the capital of Judah and later of the Jewish Hasmonean dynasty; and its existence even into the Herodian, Roman and Byzantine periods.

One conclusion from the dig is that the city of David and Solomon was much smaller than thought and that the city of Hezekiah, perhaps swelled by refugees from the northern state of Israel after it fell to the Assyrians in 721 B.C., was much larger than expected. In the terraced residential area of Hezekiah's city, destroyed in 586 B.C. when Nebuchadnezzar of Babylon took Jerusalem, a series of house foundations were discovered, including a three-roomed mansion with a stone toilet seat and wash basin at the rear, an alleyway with steps, a covered drain carrying rainwater down the hill, and many ovens, cooking pots, oil lamps, wine vessels, weapons, and other finds.

For the inhabitants of the ancient walled city, safe access to water was essential. The City of David had three mostly underground water systems, all emanating from the Gihon spring situated at the northern base of the ridge, outside the city walls. All three were known, but had been badly silted up or blocked by debris. The most complex system, known as Warren's shaft, led 50 meters (160 feet) underground to a point right above the spring, and from there a 12-meter (40-foot) vertical shaft dropped down to the spring, thus frustrating enemy infiltrators. With the aid of skilled alpinists and engineers, the shaft was climbed and the whole system cleaned and recorded, fixing its date—with the other two systems, which were also cleared—securely after the Israelite conquest of the city.

The thundering of the prophets against idolatry and backsliding is familiar to Old Testament readers. Cultic images and fertility statues of Astarte, the Canaanite mother goddess, were found in situ in many of the houses, thus documenting the anger of Jehovah's prophets. Perhaps the most dramatic finds were the numerous human bones discovered amongst the ruins of three burned houses and an alleyway on the eastern slope of Mount Zion. Here refugees fleeing after the bitter Roman siege of A.D. 70 had died, collapsing from starvation or cut down by the relentlessly pursuing Roman troops.

WAVES OVER THE BERING SEA

It has long been assumed that the ancestors of all American Indians came from Siberia across a Bering Sea land bridge about 12,000 years ago, and that they were followed by a second wave about 4,000 years ago, the ancestors of the Eskimos and Aleuts. Now a third wave has been proposed by three anthropologists working independently and together. Joseph Greenberg of Stanford studies the roots of all Indian languages, Steve Zegura of the University of Arizona studies blood groups, and Christy Turner of Arizona State studies Indians' teeth. Using this independent evidence, the three have jointly concluded that a third wave, arriving about 6,000 years ago, were the ancestors of some Northwest Coast Indians and of the Apache and Navajo in the Southwest.

Not all anthropologists agree with this. In fact, Richard S. MacNeish of Boston University offers a range of evidence that puts the first wave of American Indians as early as 20,000 to 30,000 years ago.

1983

The Amazing Tomb of China's Emperor

It is very rare when a major archaeological discovery that encompasses immense wealth and art is found. One reason for the paucity of such rich finds has been the prevalence of grave robbers over the centuries. Among the first traits of humanity is burying the dead with various "grave goods." This was quickly followed by removing the grave goods for resale or reuse. Despite reverence for their ancestors, the Chinese have also been grave robbers. Yet, something prevented them for two thousand years from succeeding in working over what may turn out to be the greatest display of "grave goods" of them all.

The discovery in 1975 of the buried terra-cotta army of China's First Emperor, who died in 210 B.C., inaugurated a major excavation by Chinese experts that promises to become one of the most spectacular archaeological enterprises of our time. To the underground army of more than 7,500 lifesize terra-cotta figures, discovered east of the First Emperor's vast tomb mound, 1983 has added the excavation of the first bronze effigies found in the area—two chariots, half lifesize, with cloaked and hatted drivers, each pulled by four bronze horses.

Ch'in Shih Huang Ti, the self-styled First Emperor, was a tumultuous character, a ruthless tyrant and brilliant innovator who first unified China in 221 B.C. He formed the Great Wall, drew the hatred of the scholars by publicly burning their books, reformed the legal system, abolished feudalism, built roads, canals, and huge palaces, and gave imperial China the shape it would hold for 2,000 years. A recluse who increasingly feared death, he built for himself a vast mausoleum enclosed within a wall nearly 7 kilometers (4 miles) long. According to legends and later chronicles, the tomb was constructed by over 700,000 conscripts from all over China and contained a model of the Empire with the heavens above, the earth below, and the ocean and great rivers of China simulated in flowing mercury. After he was buried, the legends continue, his coffin was sealed in a giant ingot of copper. Then, the whole tomb was covered by a colossal mound. Although various attempts have been made to rob the Emperor's grave in the past, nothing of value was found in these efforts until 1975. The legend that the great mound in Shensi Province, near Sian, was the Emperor's tomb was therefore assumed to be just that—a legend.

Farmers digging a well in 1975 found the buried army east of the mound, in three pits or vaults once covered with a timber roof (a fourth pit was never used). They were drawn up in battle formation to defend their Emperor, a lightly armed vanguard protecting the armed infantry, led by groups of officers who had the use of chariots and horses. The clothes, the hair styles, the belts, caps, and armor of these lifesize figures, once brilliantly painted and provided with real weapons, long vanished, are minutely rendered. Despite a certain formalism, the faces are individual portraits, reflecting the many races of the empire. As they emerged, the figures were reconstructed and now again stand in ranks, the bulk of them under a huge shed the size of two football fields covering pit number one.

Work continues on this mammoth archaeological project. Recently the underground stables of the tomb complex have been found, 93 pits containing buried horses, fodder, and farm equipment. In 1980, lying between the tomb and the buried army, seventeen graves were located, of which eight have been excavated. Bronze seals attest that the people buried here were the brothers and sisters of the First Emperor's youngest son and successor, along with some court ministers. All were killed, a historian tells us, by the young Emperor as possible rivals. All had been dismembered, some hanged and others shot with bronze arrowheads found behind the ears.

The project, one of the archaeologists said, may take "one or two generations" to complete. This may be an underestimate, since Pompeii and Herculaneum have been under excavation for over 235 years, and the work is not complete.

COLUMBUS REALLY DID GET HERE

Archaeology has confirmed in 1983 what was long suspected, that Columbus made his first New World landing in 1492 on the tiny Bahamian island of San Salvador. After years of searching, a cache of Spanish beads, buckles, spikes, and a fragment of crockery was found here just below the surface, mixed with Indian artifacts. The beads, of a type that Columbus in his journals said he traded with the Indians on landing, have been dated between 1490 and 1560—and there were no Indians on the island after 1520. Also in 1983 Dominican archaeologists, probing at the site of the New World's first colony, La Isabela, established by Columbus on Hispaniola in 1493 and long abandoned and forgotten, discovered the foundation of five stone buildings, one of which they identified as Columbus's own house. A cemetery yielded both Indian and Spanish burials, the latter identified by their Christian-style interment facing west, with folded hands. The church, fort, and a warehouse have also been traced. Columbus soon returned to Spain, and the colony lasted only five years. The starving inhabitants decamped elsewhere in search of gold.

New Discoveries at Roman Bath

In recent years the Roman baths in the lovely city of Bath have become one of the most popular ancient monuments in England. This is largely due to the arduous underground excavations of the inner precinct of the Temple of Sulis Minerva, which was completed in January 1983. The precinct lay next to the baths proper, and now forms an extension of the previously existing Roman Baths museum. The new display area lies directly under the famous eighteenth century Pump Room, which had to be shored up to make the excavations possible. The work began in 1978 with the temporary removal of the floor of the old King's Bath, which was showing signs of instability, to uncover the spring and Roman reservoir beneath it. The hot mineral waters boiling from this spring—more than a quarter million gallons a day at 48°C (120°F)—was the essential reason behind the two great periods of Bath's popularity. As *Aquae Sulis* the town was well known in the Roman Empire for its waters and luxurious bathing establishment, and as the city of Jane Austen and Beau Nash, it once again became famous as an elegant watering spa.

In fact, Bath remains one of the most popular of English cities today, as well as being an archaeological treasure. After London, Bath is in many ways the most urban of English cities, no longer dependent mainly upon the resources of its famous spring.

The Romans contained the ancient spring in the first century A.D. They built a polygonal, lead-lined reservoir, which acted as a settling tank for the sediments periodically cleared out through a massive drain; it was a masterpiece of hydraulic engineering. So, too, was the spacious bathing establishment next to it with its Great Bath, once vaulted but now open to the air, and its ancillary hot and cold pools and basins, its lead pipes, drains, flues, and underwater heating systems. But all of this, cleared in earlier years, has long been a popular tourist center. During an earlier uncovering of the Roman reservoir and the recent probes beneath the tumbled masonry of its once-vaulted roof, an interesting collection of offerings to the gods has been found, including pewter objects engraved with curses—many displaying Celtic names—metal vessels, some 15,000 coins covering the entire Roman period, and much else.

The excavations under the Pump Room began in 1981 and uncovered the entire paved precinct around the great sacrificial altar outside the temple, with its ornamented corners and a statue base nearby donated by an augurer—an important functionary who foretold the future by reading the entrails of slaughtered animals. The diggers found, too, the great buttresses and ceremonial entrance to the spring, built up against its vault in the third century to stabilize it against collapse. This entrance, its pediment adorned with the head of the sun god Sol, was apparently and appropriately balanced across the inner precinct by an ornamental facade called the Four Seasons for its sculpture, which featured the moon goddess Luna. Fragments of the facade, its site now inaccessible, had turned up earlier, but its position had not been known.

The temple itself, a simple classical structure with a steep pediment above four front columns, now lies largely beneath the main street outside the Pump Room and is inaccessible. Cellar probes have found that it was later flanked by two small shrines. But much is known about it. A corner of its concrete podium was twice glimpsed during construction work, and part of its steps, much worn from over 300 years of constant use, have been exposed and are on view in the new museum. The lovely gilt bronze head of the cult statue of Minerva was found in 1727, and some of the pediment sculptures have also turned up, including the great Gorgon's head, a magnificent piece, very Celtic in feeling, which was flanked on the pediment by purely Roman winged victories.

After the departure of the Romans in A.D. 410, the reservoir's drainage system gradually clogged up, as the excavations have shown, immersing the buildings deep in marshy sediments into which they eventually fell. But the main structures stood until at least the eighth century, when an Anglo-Saxon poet lamented: "Strange to behold...the work of giants decaying, the roofs are fallen...time scarred, tempest marred...There stood arcades of stone...There the baths were hot with inward heat...."

JEFFERSONIAN ARCHAEOLOGY

At Monticello, Thomas Jefferson had tried to create a mix of working farm with private gardens, as well as a self-sufficient plantation with its own home industries. Excavations, using the latest archaeological techniques and Jefferson's own surveys, began in 1979. Stains found in the soil recovered the positions of 57 orchard trees as well as the stakes of the vineyard and the line of a fence. The smokehouse/dairy, nail-making building, and other dependencies were located and their building history determined by the soil stratigraphy within them. A lost pavilion is being reconstructed on its own foundations.

Among artifacts recovered here and there were five wine bottles filled with a preserved mixture of cherries and cranberries in good condition, a fine Madeira decanter, the seal of a French wine bottle, and enough broken china and glassware in one cache to make a complete setting for the great man's table. Jefferson would be pleased.

Rescuing the Hohokam

Centuries ago up to 1,000,000 Hohokam Indians lived in scattered cities in a huge area of southern Arizona, perhaps 25,000 square kilometers (10,000 square miles), centering on modern Phoenix. Today well over 1,000,000 people live in Phoenix alone, a city that will not stop growing. There is reason to believe that the last unexplored Hohokam remains lie just under the surface of Phoenix's streets, so far undisturbed because the houses there were built on shallow concrete foundations. But now the developers are moving in, digging deep for new housing, highways, factories, and high rise buildings. Despite legal safeguards, the developers and government agencies manage every day to destroy more of the Hohokam heritage. In the language of the Pima Indians, *Hohokam* means the people who are "all used up." Soon the Hohokam will be all used up indeed. By 1983, the Hohokam had become an explosive political issue in Phoenix.

Not enough is known about this remarkable civilization, one of the three great prehistoric civilizations of the Southwest, along with the Anasazi and Mimbres/Mogollon. The three civilizations all flourished roughly from A.D. 1100 to the 1400s, and then all three mysteriously disappeared.

The disappearance of many ancient peoples who flourished in what are now inhospitable regions has been one of the great puzzles of archaeology. Among the more famous instances have been the Maya and ancient inhabitants of Babylonia. In some cases, we know why the civilization ceased—usually because the natural resources were used up by the methods of cultivation practiced by the civilization. In other cases, however, little is known about the way the people lived and less still about the way that their civilization vanished. Furthermore, while there are often remnant populations from the once great civilization, in other cases not only has the civilization vanished, but also the people themselves. This last mystery applies to the three great American Indian civilizations of the Southwest. The present-day populations do not seem to be the remnants of the civilizations of half a millennium ago. Instead, they appear to be groups that occupied the region after the more advanced civilizations vanished. They only have vague tribal memories of the earlier great peoples, such as the people who became "all used up"—the Hohokam.

The secret of the Hohokam during the years of their greatness apparently was their irrigation system, nearly 15,000 kilometers (10,000 miles) of canals watering fields that produced two crops a year of corn, beans, squash, and cotton in a desert environment. The canals trapped the spring runoff and channeled it into the fields. With this system the Hohokam managed to support about the same population as Phoenix without importing food. Imported food is the main source of nourishment to Phoenix's population today, and little is grown in the desert.

With the wealth produced by their agricultural system, the Hohokam built sprawling cities, some with huge enigmatic structures like the surviving four-story Casa Grande ruin near Phoenix. Most of their cities, however, were within the area now covered by Phoenix.

Hohokam pottery was superb, as was their jewelry and other objects. Among the remains of the civilization are shells imported from the Gulf of Mexico and etched with designs produced with acid. The Hohokam dug huge oblong pits, perhaps ceremonial ball courts like those of the Maya, and built flat-topped pyramids, both suggesting influences from nearby Mexico. But all this disappeared in the 1400s. Was it drought, or overexploitation of local resources? Canal systems have a way of eventually destroying the soil they are watering. It is believed that a similar fate caused Mesopotamia to become the desert it is today. No one really knows, however, if the Hohokam canals, which had led to their prosperity, also led to their downfall. And we may not know unless the archaeologists in Phoenix keep ahead of the developers.

EARLY MAN IN MAINE

An abnormally low water level in Aziscohos Lake in Maine's wilderness, a handful of unmistakable Paleoindian artifacts spotted by some amateurs, and a hurried five-months' excavation of the site before the water level rose again, all added up to an important discovery in American anthropology.

Aziscohos Lake is formed by a dam on a river. The lowered water level revealed a killing ground where game animals, mostly caribou, had to cross the river. As they crossed, they were ambushed by the Indians. Not far off was the rare find of an early habitation site, seasonally occupied by the hunters and radiocarbon-dated to 11,000 years ago. In other words, these early hunters occupied Maine less than a thousand years after the northern retreat of the last ice sheet.

Scatterings of stone tools and points (arrow and spear heads) marked the positions of eight flimsy huts, only 225 meters (725 feet) downwind from the killing ground where ten fluted spear-points were found. The finds indicated that the site was successively occupied for two or three generations by a small Indian band. They apparently chose the spot to command the river crossing of the caribou and because to the north were abundant outcroppings of fine quality chert, the rock used in making their delicately crafted, fluted points. Undoubtedly other such sites at river crossing points in the north woods will soon be discovered.

A Selection of 50 of the Most Significant Archaeological Sites Around the World

Any selection of 50 sites is bound to leave out some that are quite important in one way or another. The criteria used here include scientific importance, degree of preservation, intrinsic interest, and typicality. These sites are either places in which continuing excavation is revealing new information or in which the display of previous excavations helps make the story of civilization clear to the visitor.

Africa

Algeria Tassili N'Ajjer	Huge sandstone plateau in the Sahara. Its fantastic rock formations show a wealth of paintings, engravings and carvings of animals and people, from Neolithic times.
Egypt Abydos	An early center of Osiris worship in Upper Egypt, with First Dynasty tombs. The nineteenth dynasty temple of Seti I (about 1300 B.C.) here has superb colored reliefs.
Egypt Luxor and Karnak	These two New Kingdom temple complexes of the god Amun lie close together near Thebes, the capital. Karnak is immense, but the earlier Luxor (eighteenth dynasty) is by far the finest.
Egypt Sakkara	Necropolis of the Old Kingdom, near its capital Memphis, Lower Egypt. Outstanding here is the first true pyramid, the Step Pyramid of Djoser (about 2,630 B.C.).
Libya Cyrene	Greek colony, founded by Aegean Thera in 631 B.C., an extensive and beautiful site still under excavation, its varied remains mostly Roman or Hellenistic.
Libya Leptis Magna	Finest of the opulent Roman cities of North Africa, birthplace of Septimius Severus. A theater, huge baths, a market, forums, arches, a basilica, and temples adorn the vast site.
Mali Jenne-Jeno	Excavated remains of a prosperous and cosmopolitan city of 10,000, founded in the third century B.C., which by A.D. 800 was engaged in widespread trade.
Tanzania Olduvai Gorge	Important Paleolithic site, dug by Louis Leakey and his successors from the 1930s. Its strata contains tool assemblages and fossil hominid bones up to 1,900,000 years old.

| *Zimbabwe*
Great Zimbabwe | A group of imposing stone constructions, built from the twelfth to the fifteenth centuries for the court of the Shona king, grown rich on the Indian Ocean trade in gold and ivory. |

The Americas

Newfoundland L'Anse aux Meadows	Undoubted Norse seasonal settlement of about A.D. 1000, proving that Vikings discovered America.
Illinois Cahokia	A large prehistoric city (A.D. 900–1300), about 15–20 square kilometers (6–9 square miles) in area, was the only one in the eastern United States.
New Mexico Chaco Canyon	Hundreds of multi-room stone and timber pueblos in this area, dominated by the 650-room Pueblo Bonito, once housed over 10,000 Anasazi Indians (A.D. 1100–1200).
Mexico Teotihuacan	Great urban center near Mexico City, at its height A.D. 300–650. It is dominated by the huge pyramids of the Sun and Moon.
Guatemala Tikal	Magnificent ruins of the largest of all Maya sites. It flourished in the Late Classic period, A.D. 600–900.
Peru Chan Chan	Capital city of the Chimu Empire, conquered by the Incas about A.D. 1470.
Peru Machu Picchu	Fabulous "lost" city of the Incas, dramatically sited high in the Andean peaks (fifteenth century A.D.). Its ruins, beautifully preserved, were rediscovered in 1911.
Bolivia Tiahuanaco	Once widely influential urban and ceremonial center (200 B.C.–A.D. 600), in the Andes near Lake Titicaca. Impressive stone-built ruins include the huge gateway of the Sun.

Asia

| *Turkey*
Pergamum | Best preserved of the Hellenistic cities of Turkey, a center of arts and sciences under the Attalids and Rome. |
| *Turkey*
Troy | Famous Homeric site, identified and first excavated by Schliemann after 1871. Of the nine successive cities, from the Neolithic, scholars now propose VI or VII as Homer's Troy. |

Israel Masada	Dramatic Dead Sea fortress, topped by Herod's palace. Here the remnants of the Jewish revolt of A.D. 70 held out against the Romans, whose siege works can still be seen.
Israel Megiddo	Great fortress city (Armageddon), massively excavated, dating from 3000 B.C. to 350 B.C. Noteworthy are the Canaanite water system, and the Israelite palace, shrine and stables.
Israel/Jordan Jericho	Important Jordan Valley site, extensively excavated, revealing one of earliest walled settlements (about 7000 B.C.). No trace of Joshua's walls survive from the Hebrew siege.
Iraq Ur	Major city and capital of Sumer about 2100 B.C., excavated in the 1920s. Its early royal tombs (2800 B.C.) yielded fabulous treasures.
Iraq Samarra	Ruins of a magnificent city, briefly Abbasid capital (A.D. 836–882), stretching 22 miles along the Tigris.
Iran Persepolis	Magnificent ruins of the ceremonial capital of Achaemenid Persia (sixth to fifth centuries B.C.), with massive palaces and lively reliefs on a huge platform. It was burned by Alexander.
U.S.S.R. Samarkand	Tamerlane's capital in Central Asia, an ancient city adorned by him around A.D. 1400 with massive mosques, caravanserais, madrasahs, and his own mausoleum, all decorated with tiles.
Pakistan Mohenjo-Daro	One of the twin capitals of the early Indus Civilization, a large planned city of baked bricks with monumental buildings, a street grid and drains; abandoned about 1750 B.C.
Cambodia Angkor	Vast capital city of the Khmer Empire. Its Hindu-inspired buildings, mostly from the twelfth and thirteenth centuries, include the huge Angkor Wat Temple complex and the city of Angkor Thom.
China Choukoutien	Famous cave near Beijing (Peking), residence of *Homo erectus* from 430,000 to 230,000 years ago. The creature—bones of over 40 have been found—hunted, used fire, made crude tools.
China Sian	The enormous burial mound of the First Emperor (died 210 B.C.) near Sian was guarded by a terracotta army of over 7,500 figures.

Europe

England Avebury	Impressive Neolithic monument in Wiltshire, larger than Stonehenge, with 98 huge stones within a circular ditch and bank, and two smaller circles inside it.
England Bath	Famous Roman hot springs spa of Aquae Sulis, with monumental baths and temple. Much survives, and more has recently been excavated underground, including the great altar.
England Fishbourne	Large Roman Palace in Sussex, built in the first century A.D. for a British king. Excavations revealed four wings with colonnades, an audience hall, gardens, and mosaics.
England Hadrian's Wall	Stone-built wall defending Roman Britain's northern frontier, built by Hadrian A.D. 122–28. About 117 kilometers (73 miles) long, it had sixteen major forts and many small forts and signal towers.
Denmark Jelling	Royal site of tenth century Viking Danish kings, in Jutland. Two large burial mounds and a number of standing stones, including two famous decorated rune stones, are here.
Germany Heuneburg	Great Iron Age hillfort, Hallstatt period, on the upper Danube, with five building periods. The second phase showed Greek imports and Greek-style ramparts.
France Carnac	Megalithic site in Brittany with nearly 3,000 standing stones in three groups, each with ten to thirteen rows; also large stone tombs. It was obviously a Neolithic ritual center.
France Lascaux	Famous painted cave in the Dordogne, with Upper Paleolithic paintings and engravings. Discovered in 1940, it is now closed for protection and a replica can be seen nearby.
France Nîmes	Important city of the Roman Empire with many fine surviving monuments. These include the amphitheater, the Maison Carree, a temple, and the aqueduct Pond du Gard nearby.
Spain Merida	Augusta Emerita, a major Roman center, founded by Augustus in 25 B.C. as a veterans' colony. Remains include a theater, arch, temples, a bridge, and luxurious houses.

Italy (Sardinia) Barumini	Su Nuraxi here is the best known of Sardinia's Bronze Age Nuraghe—complex towered structures of heavy masonry with outer walls and internal passages.
Italy Herculaneum	Though smaller than nearby Pompeii, the site is more rewarding because the remains of this wealthy resort are better preserved in the ash from Vesuvius's eruption in A.D. 79.
Italy (Sicily) Syracuse	The most splendid Greek city of the West, home of Archimedes, visited by Plato. Its remains include three temples, a Greek theater, and massive late Greek fortifications.
Italy Tarquinia	Major Etruscan city, famous for its painted tombs and the Tarquin kings of early Rome. The foundations of an Etruscan temple, parts of its walls, and 3 miles of tombs survive.
Poland Biskupin	Iron Age, pre-Slavic settlement on an island, defended by elaborate timber breakwaters and ramparts. Some of the 100 timber houses, laid out on regular streets, have been restored.
Greece Athens	The most splendid city of the Greek homeland. The monuments on the Acropolis and elsewhere are well known; less known is the Agora excavation, extensive and meticulous.
Greece Delphi	The oracle and temple of Apollo, dramatically sited. It was a powerful unifying force in the quarrelsome Greek world, both at home and abroad. Magnificent monuments.
Greece (Crete) Knossos	Sprawling palace and epitome of the Minoan, Greece's first civilization, excavated and restored by Sir Arthur Evans after 1899. Much fruitful digging still goes on in the area.
Greece Mistra	Dramatic deserted hill town, successor to Sparta below it. Its ruined palace, monasteries, and many churches, with superb paintings, form a monument to late Byzantine life.
Greece Thera (Santorini)	The explosion of this Aegean island about 1500 B.C. probably destroyed Minoan civilization. It also buried the island town of Akrotiri, now excavated with dramatic finds.

1984

Update 1984: Archaeology and Anthropology

NEW EGYPTIAN TOMBS FOUND

Although the ancient Egyptian pharaohs were generally buried in the Valley of the Kings at Thebes, other high-ranking Egyptians had impressive tombs in other sites. The Necropolis at Sakkara, 24 kilometers (15 miles) south of Cairo was one of them. Although Sakkara had produced little in the way of new finds for the last hundred years, it was reported in February 1984 that new excavations are turning up many previously unknown tombs.

The new finds at Sakkara consist of tombs of high officials of the reign of Ramses II, perhaps the most powerful pharaoh of them all. Ramses the Great, as he was also known, ruled Egypt and much of the eastern Mediterranean area from 1304 to 1237 B.C. Among the tombs found are that of the army commander, treasury minister, and royal scribe of the Ramses regime. These officials were not necessarily buried in these tombs; tombs were built during a person's lifetime; if the person attained greater status or wealth, the original tomb would be abandoned and another, grander one, built. Nevertheless, the team, led by Sayed Tewfik, had hopes of finding mummies. Whether there are mummies or not, the archaeologists have found various reliefs showing life in the New Kingdom, a treasure in themselves.

MASSIVE NEW DICTIONARY STARTED

Even when you know the basic principles of a foreign language, reading a long passage is difficult without a dictionary. This has been a serious problem for scholars and archaeologists specializing in the world's oldest civilization—Sumeria. Although the baked clay tablets of the Sumerians have survived 5,000 years relatively intact—and there are hundreds of thousands of them—almost no one can read them. Sumerian is not related to any other known language and, until 1984, there was no dictionary to provide aid.

While many of the tablets involve simple commercial transactions, others include some of the world's earliest known poems and tales. There are so many tablets preserved that the documentation of the Sumerian civilization is better than that of the Roman civilization, since the Roman records were much more perishable than baked clay.

The new dictionary will take many years to be published and is expected to run about 16,000 entries covering more than 22 volumes. (The Assyrian dictionary, a comparable project, was started in 1921 and is still not complete.) The editor of the *Pennsylvania Sumerian Dictionary* is Ake R. Sjoberg from the University of Pennsylvania.

First Unlooted Maya Tomb in Twenty Years

On May 15, 1984, Richard E.W. Adams uncovered in a remote jungle in Guatemala the first unlooted Maya tomb found anywhere since the early 1960s. Although no gold or jewelry was found, this archaeological discovery made front-page news in New York City for several days. The reason was that an enterprising reporter for the *New York Times* learned of the discovery while in Guatemala City, before its official announcement. She and a photographer made a strenuous five-and-a-half hour jeep journey though the jungle to the site of the tomb. When she arrived, Adams allowed her to cover the story only if her companion and driver left, which stranded the reporter for three days at the camp near the site. As a result, she collected a great deal of information about the discovery .

The principal value of the tomb is its untouched nature and the fact that it had been sealed effectively in the fifth century A.D., so the contents are well preserved. Especially important finds were wall paintings and the only jar with a screw top ever found in Mayan ruins. The skeleton of a man was clearly visible. He is thought to have been a relative of the ruler of the site.

Because the first archaeologist to see the tomb was Grant Hall, the associate director of the dig, the tomb is known informally as "Grant's Tomb," but it will enter the archaeological record as Tomb 19, Río Azul.

Chipping Away at the Gap

In April, the National Science Foundation announced that a part of a fossilized jawbone found in Kenya by an expedition led by Andrew Hill was the oldest-known representative of the hominid line. Hill is from Harvard's Peabody Museum of Archaeology and Ethnology, but the expedition also included such notables as David Pilbeam, also of Harvard, and Richard Leakey, from the National Museums of Kenya.

The fossil, about 5 million years old according to preliminary dating, appears to be a representative of *Australopithecus afarensis*, best known from the fossil Lucy. Lucy, however, lived about a million years after the newly discovered creature.

New View of the Vikings

Neglected in many ways for years, the archaeology of the amazing Viking expansion of the ninth, tenth, and eleventh centuries became a major thrust during the early 1980s. As a result, it is now possible to get an overall view of the Viking world, a view that was not clear only a few years ago.

The word *Viking*, which means "warrior," is used to represent a specific maritime culture of Scandinavia, which included several tribes, who have among their descendents not only the Norwegians, Swedes, and Danes in their native homelands, but also Icelanders, French, English, Germans, Russians, Irish, and others throughout the Western world. In fact, the best place to begin to understand the Vikings is not, as you might expect, in Scandinavia or Iceland, but in England.

One of England's largest urban excavations, that of Viking York, or Jorvik, which closed in 1981 after more than five years of intensive work, has been a major factor in completely revolutionizing our knowledge of the Vikings. Those fierce and intrepid mariners with their beautiful ships, their far-flung raids and explorations, have long fascinated the public, although their actual life style was little known. Although archaeologists know better, most people still think of the Vikings as brutal raiders or hit-and-run destroyers. The picture is of fierce warriors who would appear off the coasts of settled Europe in their sleek ships, then flee, leaving death and destruction behind them. There is some truth in this point of view, especially with regard to the early Vikings. For example, of the early sack of Lindisfarne, the English scholar Alcuin wrote from the court of Charlemagne in the eighth century: "...never before has such terror appeared in Britain as we have now suffered from a pagan race, nor was it thought that such an inroad from the sea could be made."

Recent excavations of York, as well as those at Lincoln, at Dublin, Ireland, and at the Viking commercial towns in Scandinavia itself have proved conclusively that, though they started as raiders, the Vikings quickly became the greatest maritime traders and colonizers of their day. In the tenth and eleventh centuries Viking York, a Scandinavian city on English soil, was one of the wealthiest cities of the West, a center of manufacturing and far-flung commerce; after its fall to the English in A.D. 954, it was succeeded by Viking Dublin, which became the commercial center of the Norwegian Vikings in the West. The worldwide Viking trade routes ran from Baghdad and Samarkand in the east to Iceland and Greenland,

and even briefly to America in the west, and from North Africa and Constantinople to Ireland and the northern Scottish islands. It was the Swedes who explored the trade routes down the Dnieper to Constantinople, and down the Volga, selling slaves and furs, to tap the riches of the Islamic empire at Baghdad, and even as far as India. The Norwegians sailed west to the Scottish islands, Ireland, western England, Iceland, and Greenland. The Danes, in between, moved down throughout Europe and heavily colonized both Normandy and the northeastern half of England.

THE VIKING EXPANSION

The Viking drive, which began in the late eighth century, seems to have been impelled by overpopulation (and widespread polygamy that contributed to population growth), as well as the custom of passing inheritances to the eldest son of a family, leaving many younger sons to make their own fortunes elsewhere. The Vikings, however, were pragmatic warriors and would just as soon trade as raid to gain the wealth they coveted. Though the great expansion began with raids, which never wholly ceased for 200 years, the alternate of trading led to colonization, as it almost always does, and the resulting interchange of goods and ideas between Western Europe and Scandinavia led gradually to the "civilizing" as well as Christianizing of the Viking homeland. In fact, although it was not Christian to begin with, the Vikings had a well-developed civilization before the raids started, with their own system of writing (called *runic*) well developed. Finally, the development of strong monarchies at home in Norway, Denmark, and Sweden marked the end of the Viking expansion and the beginning of the medieval world, both in Scandinavia and Europe.

It all started, however, with the raids. Viking raids began in the summers. When the Vikings had established a presence with several raids, they wintered over for the first time. Wintering often led to a continued presence. Wives and families began to join them. Soon, a Viking outpost became a Viking nation.

In England the Viking harassment began inconspicuously. In 789, according to the Anglo-Saxon Chronicle, "came three ships of the Northmen from the land of robbers" off the south coast at Portland, "whereupon the royal official rode to them and tried to force them to go to the royal manor, for he thought they were traders; and then they slew him." Brutal surprise was one of the Vikings' canny practices, and at first the wealthy and unprotected monasteries were their principal targets. Lindisfarne, or Holy Island, off the northeast coast was next. In 793, reports the Chronicle, "the harrowing inroads of heathen men made lamentable havoc in the church of God in Holy Island, by rapine and slaughter." This was the first raid; many more were to follow.

In 850 a Danish army wintered over for the first time in the Isle of Sheppey. In 865, the first year that the Vikings in England were bought off by the payment of "Danegeld" (Dane gold), a great army wintered in East Anglia. Once the Vikings began wintering over, they generally stayed around for quite a time. Held off by the southern kingdom of Wessex, Ivar the Boneless turned his Viking army north and in 866 took York, ensconcing himself behind the Roman walls of that ancient city. For York, known by the Roman name of Eboracum, had been the principal military center of Roman England in the north and a legionary center.

What else was happening in the Viking age in the 860s, when York fell to Ivar the Boneless? In southern England the Danish chieftain Rorik had plundered both London and Canterbury from his base on Thanet. The Danes in the south, soon to be joined by the army from York, were preparing for the great confrontation with King Ethelred of Wessex and his brother, soon to become Alfred the Great.

Across the English Channel, Rorik the Dane had settled in Frisia, the Lowlands. By decree of Lothar, one of the Carolingian rulers, the Frisian island of Walcheren had been granted to Rorik and his brother Harald in the 860s. Other Viking bands had been running up the Seine, repeatedly sacking its towns, including Paris. Finally, in desperation, Charles the Simple, taking his cue from the Walcheren episode, granted Normandy to the Viking Rollo in 911, hoping thus to set Viking against Viking for his own protection. Thus began the great dukedom of Normandy.

Farther south the Vikings, chiefly Norwegians, had been raiding deep into France and even into Moorish Spain. They established a base on the island of Noirmoutier, off the mouth of the Loire, chasing the monks from a monastery there. One famous expedition left the Loire in 859 with 62 ships, skirted Spain, sacked Algeciras, then raided the coasts of North Africa and southern Spain before setting up a camp in the French Camargue, whence they raided up the Rhone as far as Nîmes and Arles. Finally beaten off by the Franks, the Vikings turned to Italy and sacked Pisa. Then they moved on to a walled town called Luna (since disappeared), which they thought was Rome, and entered by a ruse. They may have gone farther east; no one knows. In 861 they fought their way back through Gibraltar against the Moorish fleets, raided Navarre on their way, and captured its prince for a fat ransom. The survivors reached the Loire again in 862.

In the meantime the Swedish Vikings, moving across the Baltic, had by the 860s already opened up the great trading routes down the Dnieper and Volga through Russia to trade with the other end of the Islamic empire and with the Byzantines. In 860 the foolhardy Swedes actually attacked the mighty walls of Constantinople itself, with no success, however. These Rus, as they were called in Russia, set up states along the great rivers, notably at Novgorod and Kiev, ruled by Viking princes. They thus created the foundations of the Russian state, though the Soviets today tend to play down the Viking contribution to their origins for obvious reasons. Vikings later formed the core of the Byzantine imperial guard, which policed the Mediterranean for the emperor.

In the western area, the Norwegian Vikings had raided and, after the 860s, started to settle, first on the northern islands, the Orkneys and Shetlands, then the Hebrides and the Isle of Man and northern Scotland, then its east coast. Some of the outer islands in the Shetlands still celebrate Viking festivals to this day.

They had been raiding Ireland, too, after the 830s, and established winter bases, the first one at the site of modern Dublin in 841. Later they made Wexford,

Waterford, and Cork into Viking bases. Norwegians began to settle along the nearby western coasts of England, where they were to come into conflict with the Danes at York and eventually take over that city.

In the 860s, too, Norwegians were sailing from their bases in the British Isles on the first exploratory voyages to Iceland, where they were to set up a remarkable independent republic and the earliest popular general assembly in Europe, the Althing, which traditionally first met in 930. It was in this assembly that Iceland, by general assent, was converted en masse in A.D. 1000 to Christianity. Already Greenland had been colonized by Erik the Red in about 985, and in the year 1000 America was discovered by these hardy explorers, a fact confirmed in the 1960s by the identification and excavation of an undoubted seasonal Norse settlement at L'Anse aux Meadows in northern Newfoundland.

SHIPS OF THE VIKINGS

And so in pursuit of the Vikings, we have moved from the area of the Caspian Sea in Asia to the shores of North America. It was their ships, primarily, that made it all possible. Vikings always preferred to raid and explore and even to settle with their ships, buying, seizing, or sometimes carrying along the horses which they rode on their pillaging raids or to battle. These superb ships were flexible enough to ride out ocean storms but had a shallow draught so that they could be pulled up on beaches or penetrate the rivers of Europe and Russia. Such ships had a long history of development, dating back to cruder examples, such as the Nydam ship of about A.D. 400 of the type that carried the Anglo-Saxons to England, or the Sutton Hoo ship of the seventh century, discovered in a ship burial in England.

A number of fine examples of later Viking ships have been found and reconstructed, mostly from ship burials in Scandinavia from the ninth and tenth centuries—the early and exquisitely carved Oseberg ship, for instance, and the later Gokstad and Tune ships. Most were propelled by oars and a square sail, and the longships and dragon warships could be up to 90 feet in length. The crews rowed into battle and navigated the oceans by a sophisticated system of dead reckoning. When the sailors and rowers piled onto the shore, they became the fierce warriors, armed with spear, shield, sword, and lance, so much feared by their often helpless victims.

The five Skuldelev ships, first found by a diver in 1956 where they had been sunk to block Denmark's Roskilde fjord, have shown that by the tenth and eleventh centuries specialized types of Viking ships had been developed. Two were warships, one a rare dragon ship of the type used as the flagship of royal fleets. Two others were capacious cargo vessels that relied on sail more than oars, and the fifth was a small fishing boat or ferry. But all had the elegant, long lines of the Viking prototype.

A typical Viking warship, such as the 21-meter- (70-foot-) long Gokstad ship, must have made a brave sight as it rode in harbor in preparation for sailing. Along its long, low sides from prow to stern there were displayed 33 shields on each rail, two for each oar-hole, overlapping and painted alternately yellow and black, while above them on the 12-meter (40-foot) mast the great square sail might be poised, red in color or perhaps striped red and blue. The narrow gangway, about 7 meters (22 feet) long, might still be up though the helmsman would already be standing by the steering oar, some 10 feet long, fastened to the starboard side of the ship.

The discovery of these ships revolutionized our understanding of the technology of the Vikings. Although ancient chroniclers told of Viking ships with 60 rowers, modern archaeologists had doubted that the ships could have been this long, guessing that numbers such as 20 or 40 rowers might be more likely. The ancients, however, were not exaggerating after all.

VIKING YORK

In such ships as these the Danes sailed to take York in 866. Ten years later the army, now under Halfdan, decided to settle down. They "shared out the land of the Northumbrians, and they proceeded to plough and to support themselves." In other words, there was a deliberate decision to colonize, and the colonists were reinforced by a new group of Danish settlers in 896. Farther south, other parts of the great army had also settled down, so that the Danes were now in control of most of eastern England, from London to York. By the 870s, Alfred the Great of Wessex had been forced to take refuge in the marshes and forests of the west. But this remarkable man regrouped his forces and fought back, defeating the Danes and retaking London. Eastern England, called the Danelaw, extended from East Anglia to the borders of Scotland, and here the Danes created a society of peasants, freemen with their own language, customs and laws—some of which still survive, particularly in the numerous places with names of Scandinavian origin. At the north of the Danelaw was the kingdom of York, to the south the kingdom of East Anglia, and in between a number of fortified boroughs, including Lincoln.

York in the tenth century could have held over 10,000 people. It had become one of the largest European cities of the period. It must have been a crowded rabbit warren of craft workshops, merchant shops and houses, importers and exporters. In the areas uncovered by the excavations, the lines of the streets and even the boundaries of each individual plot remained exactly the same until the recent demolition of nineteenth century buildings to make way for the excavation and redevelopment. Most of the street names in the inner city are still Scandinavian in origin.

York indeed had become a key mercantile center for the worldwide network of Viking trade, as many of the finds attest. The finds include wine jars from the German Rhineland, Irish metalwork, soapstone bowls from the Shetlands, fine-grained stone from Norway for honing knives, items from the Gulf of Aden or Samarkand, pottery from nearby Lincolnshire, and fine silks from the Near East or Asia. For example, a lady's gold-colored silk cap, probably from Constantinople, was almost perfectly preserved in the wet earth. York not only imported such luxury goods, but also produced many other items in its workshops, as we shall see.

The Viking kingdom of Jorvik was ruled by a succession of Danish kings until A.D. 919, when a Norwegian dynasty took over. Vikings continued to rule York, with the exception of the years 927 to 939, until 954. Then the English successors of Alfred the Great finally expelled the last Norwegian king, Erik Bloodaxe. York itself remained predominantly Scandinavian, however, right up to the Norman Conquest and up to and even beyond the subjugation of the city in 1067. The Normans were, of course, Vikings themselves by origin, but generally are viewed as having evolved into a different culture, perhaps because they spoke French. Moreover, between the English reconquest of York and the Norman conquest, Danish kings, particularly Sven Forkbeard and Knut the Great, his son, ruled all of England, including York, from 1013 to 1042. For this reason the archaeologists investigating York have considered the whole period of 200 years, from the Danish capture of the city to the Norman conquest, as a single, Anglo-Scandinavian period with its own distinctive characteristics.

After the English earls took over in York in 954, much of the direct Scandinavian trade was lost. However, the city continued to flourish without interruption as a wealthy trading center. Around the year 1000, York was described as "a city enriched by the treasures of merchants who come from all quarters, particularly from the Danish people."

REBUILDING THE VIKING PAST

Urban archaeology, the salvaging of the remains of the past in advance of the bulldozer, as the crowded modern cities of the West build and rebuild, has become a highly specialized discipline. The recent excavations of both Dublin and York are shining examples of the success of the new techniques. Both took place in water-logged conditions that preserved the wood, the fabrics, the leather, and metal objects in remarkably fine condition. Modern methods of conservation are then applied to keep these objects from disintegrating in the open air.

As in Dublin, much rescue archaeology in advance of redevelopment took place in York in the 1960s. The results were so spectacular that, as the city entered a period of even more intense redevelopment, a full-time organization, the York Archaeological Trust, was set up in 1972 to handle rescue operations as they arose, in order to fill in the archaeological history of the city, whether Roman, Viking, or medieval. The digging of new vaults under Lloyds Bank led to an eerie under-ground excavation in 1972-73 that revealed deep deposits from the Anglo-Scandinavian period, the remains of ten superimposed buildings that had housed the leatherworkers and cobblers for many years. Perfectly preserved footgear of many types and skins were found, as well as leather being prepared for the footwear, clothes, and tools of the workmen. Other rescue operations in a sewer trench, an elevator shaft, and other places brought up more finds.

Finally, in 1976, the city purchased four long, narrow properties running downhill to one of York's two rivers in the Coppergate area—Scandinavian for the street of the coopers, or woodworkers—and demolished the nineteenth century buildings in preparation for redevelopment. In the meantime the Trust was invited to excavate the entire site, an immense task that took over five years to accomplish.

At the height of the dig, in the early 1980s, more than 50 volunteers were working down through the remains of debris from over 50 generations of Anglo-Scandinavian inhabitants, supervised by the 30-odd permanent members of the Trust. Five tons of animal bones and roughly a million pieces of pottery came to light, and almost 15,000 other small finds of value, including spindle whorls for spinning, bits of textiles, a set of pan pipes—still playable—dice, and gaming boards. There was a workshop devoted to turning wooden bowls and cups. A jeweler nearby worked with imported amber and native Yorkshire jet, while elsewhere combs, needles, and the like were fashioned from bone and antler. Metalworkers wrought many objects in gold, silver, copper, lead, and bronze.

There also seems to have been a mint at York for the issuing of coins, which first appeared about 900 and continued right through the later, English period. Two coin dies were found, one of them used to mint coins of the English king Aethelstan (924-939) when he was briefly in control of York. A later English king, Aethelred the Unready (978-1016), was forced to issue enormous quantities of coins, some of them at York, to pay the vast Danegelds extorted by the Viking armies.

Other details were learned about Viking York from a special team of environmental archaeologists who studied the animal, fish, and bird bones found, as well as the tiny remains of grain, pollen, seeds, and insects. They discovered, for instance, that the inhabitants ate enormous quantities of seafood and ocean fish, as well as many varieties of birds, and that most of the inhabitants suffered from intestinal parasites.

The last chapter of the Jorvik Story was written in early 1984 when the ambitious Jorvik Viking Centre opened. An innovative project combining the latest museum techniques with more than a dash of Disneyland know-how, the center recreates Viking York, specifically the Coppergate site, in the basement of the large shopping complex above it. Tourists, who already contribute a major share of York's wealth, enter an orientation section where the Vikings are explained. Then they step into electronically controlled "time-cars," equipped with concealed speakers, which take them through a time tunnel. Here special effects such as stereophonic sound and back-projected films take them back through the centuries to the Viking Age. Next they pass through a Viking street, the Coppergate, exactly as it was (and was found to be through the excavations) 1,000 years ago, with its buildings, animals, workshops, and even latrines meticulously restored. At the end of the street there is a Viking wharf with Viking ship prows looming above, and here the visitor is told in detail about the far-flung Viking trade and its monetary system. A second time tunnel brings the visitor back to 1979 and an exact reproduction of the Coppergate dig as it looked in that year, with the actual timbers, fences, hearths, and wood-lined wells put back into place again, while the sound track discourses on the techniques of modern archaeology—conservation, photography, dating methods, and the importance of pottery. Climbing out of a time car, the visitor next enters a hall where the conserved small objects from the dig are displayed.

ASTRONOMY AND SPACE

In the early 1980s attention in astronomy and space was focused on space near Earth, the Milky Way and its environs, and the farthest reaches of time and space.

Nearest Earth, space travel was dominated by the American space shuttle, and the Soviet's steps toward building a space station. The emergence of the shuttle as a workable way to get in and out of orbit was the dawning of a new age in space travel. Meanwhile, the Soviet Union continued to set new records for long space missions as they slowly put together the information and techniques needed to build a permanent space station. By 1984 the United States had also made the commitment to a space station.

Emphasis on the shuttle and space stations, especially in the United States, was bad news for planetary astronomers. Money was taken from projects designed to study the solar system and applied to development of the shuttle and space station. Planetary astronomers, however, had their hands full. The Voyager missions to Jupiter and Saturn, with Uranus and Neptune coming up later in the decade, provided a wealth of new data, much of it very difficult to interpret. The picture was somewhat clearer for Venus, which was visited by Soviet spaceprobes, and Mars, where Viking 1 continued to relay data until the end of 1982. As a result of these spaceprobes and continued observations from Earth and from Earth-orbiting satellites, a picture of the solar system emerged that was significantly different in detail from the one held just a few years ago.

One notorious example of the drift of money away from exploration of the solar system to the shuttle program was the decision by the United States not to send a spaceprobe to Halley's Comet, which is already in view in the largest telescopes on its periodic visit to our neighborhood. There will be spaceprobes to the comet from other countries, including Japan and the Soviet Union. However, a sun-orbiting U.S. satellite was redirected to another small comet.

In the meantime, unexpected new insights may come from spaceprobes launched *from* space—meteorites. Analyses of meteorites established that some come from the Moon, some probably come from Mars, and some may even come, at least in part, from outside the solar system entirely.

While spaceprobes provided new data about the solar system, satellites continued to expand the spectrum (literally) of our observation of the rest of the universe as well. While the 1970s was the decade of X-ray astronomy, based on the Uruhu and Einstein X-ray satellite telescopes, the 1980s continued the exploration of new wavelengths with IRAS, an orbiting infrared scanner. The X-ray sky had revealed many new and unexpected phenomena, and so did the infrared spectrum. While analyzing all of the data from IRAS will take the rest of the decade, early results included small particles orbiting stars, new regions of star formation, and new comets.

Beyond the solar system and, for the most part, beyond our Milky Way galaxy, astronomers continue to find truly spectacular phenomena of types that are far removed from any experience we have on Earth. The early 1980s cataloged the second and possibly the third black hole, the fastest pulsar (first of a new class of pulsars), and the farthest quasar. What all these phenomena have in common is that they are composed of collapsed matter. Although this is not certain for quasars, most explanations of the quasar phenomena assume that they involve black holes. Gravity powers the stars but also leads to the collapsed state, which, in its most extreme form—the black hole—is so dense that light itself cannot escape from the gravitational attraction of the former star.

One of the major questions for cosmology, that branch of astronomy and physics that deals with the origins, structure, and fate of the universe as a whole, is how the universe will end. While the ending seems to be at least 20,000,000,000 years away or much farther, human beings want to know how any story comes out. For the universe, the answer most likely depends on how much mass there is. If there is not very much mass (compared with the size of the universe), the universe will continue to expand forever, with the stars gradually dying by becoming cinders of themselves or black holes. If there is rather more mass, the universe will collapse itself. Although new mass was found surrounding galaxies and in interstellar space during the early 1980s, astronomers say that they still have not found enough mass to guarantee collapse. Gravitational collapse is the preferred alternative philosophically, because from such a collapse, the universe could be reborn. In fact, new theories about the origin of the universe that gained quick acceptance in the early 1980s almost demand that a collapsed universe give birth to a new expansion.

1980

Solar Maximum Mission

The United States Solar Maximum Mission satellite, informally known as Solar Max, was launched on Valentine's Day 1980 as part of a major effort by scientists to study the solar maximum. The solar maximum is a peak of sunspot and solar flare activity on the sun that occurs approximately every eleven years. The launch was part of the Solar Maximum Year, a nineteen-month international project. But failure in the satellite's attitude control system began to appear by November 1980, and Solar Max was ultimately unable to complete its mission, although its ten months of data have been extremely useful.

Solar Max is a 2,313-kilogram (5,100-pound) satellite designed to study solar flares, and it carries sensors for ultraviolet, X-ray, and gamma-ray radiation. For observing solar flares, Solar Max can use one X-ray sensor to point the others in the correct direction. This is essential because most solar flares, a prime object of interest, last less than half an hour, which is not a long enough time to aim the sensor from Earth.

Another part of its original mission included measuring the sun's total output of radiation for one year. This was part of an effort to see if changes in the sun's heat could affect Earth's weather. Although it missed the period for which it was designed, all these measurements would be useful.

For example, Solar Max carefully monitored the solar constant, which is the total energy received from the sun in one second on a square meter (1.2 square yards) of Earth. An almost incredibly accurate measurement from space by Solar Max proved for the first time that the solar constant varies as much as 0.02 percent from day to day and week to week. This is enough, perhaps, to affect weather on Earth.

Considering its failure, it is ironic that Solar Max was the first satellite designed to be serviced in space by a crew from the space shuttle. It was built with special "handles" that the shuttle could grasp and standardized parts for easy replacement. The plan originally was to change the instrument package after Solar Max had observed the sun at its peak, not to repair an abortive mission. As a result, when the mission failed after only ten months, the shuttle craft was far from operational. After the shuttle became operational in 1982, however, Solar Max was slated to become the first satellite to be repaired—which was accomplished in 1984. Of course, the solar maximum was long since past by 1984 and the solar minimum had just passed in 1983. Perhaps the satellite should be renamed Solar Miss!

Very Large Arrays and Long Baselines

Many of the most remarkable discoveries in astronomy since World War II have come as a result of radio astronomy. Radio astronomy relies on electromagnetic waves that are from a few centimeters to a few meters long, as compared with optical astronomy which uses electromagnetic waves that are millionths of a centimeter long. As a result, images from single radio telescopes cannot tell the difference between two nearby sources of radio waves nearly as well as optical telescopes can separate nearby sources of light. This ability to separate nearby sources is called the resolution of a telescope. The better the resolution, the better the telescope. The amount of magnification produced by the telescope is not nearly so important as its resolution.

Completion of the VLA (Very Large Array) radio telescope in 1980 at Socorro, New Mexico, gave astronomers an important new tool for peering deep into space. Images produced by the VLA are of far greater resolution than any single-dish radio telescope.

The new VLA radio telescope, built at a cost of $28,000,000, consists of 27 separate dish-like radio antennae. The antennae are 25 meters (82 feet) in diameter and are arranged in a Y-shaped pattern that covers 61 kilometers (38 miles). Because all the antennae are linked by cable to a central computer that processes the incoming signals, the VLA system produces an image that has the resolution equivalent to a single-dish radio telescope 27.4 kilometers (17 miles) in diameter.

Improving image resolution has been a major stumbling block in the field of radio-telescope technology. The physics of both optical and radio telescopes is essentially the same—resolution is a ratio of the wavelength and the diameter of the collector, whether the collector is a lens, a mirror, or a radio dish. But the wavelength of radio emissions is so much greater than that of light that, even to approach the resolving power of an optical telescope, the radio dish must be very large. One way to do this is with a single dish, of which the largest currently in existence is 305 meters (1,000 feet) in diameter. However, to reach the resolving power of a typical large astronomical light telescope, a single-dish radio telescope would have to be 240 kilometers (150 miles) in diameter! There is, however, a clever alternative. A large array of smaller antennae can be used. The new radio telescope at Socorro, which uses the latter technique, has about the same resolution (1 arc-second) as that of a large optical telescope.

The basic principle involved in the new radio telescope is called radio interferometry. As in optical interferometry, the comparison of the patterns made by two versions of the same electromagnetic wave is used. If the patterns are in phase, the resolution can be improved. Interferometry has long been a powerful tool. For example, it can be used to detect motions as slow as a centimeter in over a hundred years. Interferometry was the basis of the famous Michelson-Morley experiment in 1881 in which the same principles were used to find a very good value for the speed of light.

The application of interferometry to extending the resolution of radio telescopes has been made possible by development of powerful computers over the past decades. Through a complex process, the computers analyze and compare incoming signals received simultaneously by the various antennae in the array, which are all connected. The resulting image produced by the computer has the same resolution that a single dish as large as the whole array of small dishes would have.

Since the late 1960s, when accurate atomic clocks became available, it has been possible to take simultaneous readings at radio telescopes thousands of kilometers apart, without any direct electronic link between the dishes. This technique, called Very-Long-Baseline Interferometry (VLBI), has been used with increasing success to achieve high-resolution images. By this system, signals received at each of the several remote antennae are recorded on a videotape along with timing signals from the atomic clock. The tapes are then processed at a central computer, and an image is produced in the same manner as at the VLA radio telescope. Obviously, the resolution is even greater, since the radio telescopes are much farther apart.

An important factor in this technique is that differences in wave phase caused by atmospheric conditions or other variations can be eliminated by using at least three different radio telescopes. Very-Long-Baseline Interferometry was at first hampered, however, by differences in atmospheric turbulence, clouds, and so forth, between the different remote antennae. A new type of system, using data from a closed loop of four radio telescopes, was tested in 1975 and produced images of very high resolution (0.01 arc-second), a thousand times better than the typical large optical telescope. This system of VLBI, called hybrid mapping, has been used with existing radio telescopes with great success since then. No one would have ever dreamed when the first radio telescopes were invented that they would ever equal, much less surpass, optical telescopes.

Even this resolution could be improved, however, by the construction of special radio telescopes designed specifically to use the new system. As yet, such telescopes have not been built, although, naturally, astronomers are making plans.

Magsat Maps Earth's Magnetism

The United States scientific satellite called Magsat (for magnetic field satellite) crashed to Earth in June 1980, after completing its eight-month mission to survey Earth's magnetic fields. The huge amount of information sent back is being used to compile a detailed map of the variations in Earth's magnetic field.

When artificial satellites were first launched in 1957, no one had foreseen how useful they would be in surveying the Earth in detail. What we currently know about the shape of the Earth and the radiation from it has almost all been learned from satellite studies.

The scope of the project is enormous, and scientists will spend years analyzing and interpreting the data. But the information is invaluable. The magnetic maps record areas of high and low magnetic intensity in the same way contour maps show changes of elevation, and, for one thing, the maps are expected to reveal the geology of the deeper parts of Earth's crust. They could also provide new information about the crustal plates and their movements over the past millennia.

The magnetic maps are likely to have great economic significance as well. Different types of rock have different magnetic signatures, although geologists have yet to clearly define which signature goes with which type of rock. Evidence of the signatures on magnetic maps can be used to identify areas where valuable mineral resources may be found.

The concept of magnetic mapping is not new. Surveys of small areas have been made in the past for mining and oil companies, using nothing more than airplanes equipped with magnetometers. In addition, data on magnetic fields in specific areas has been gathered by other United States satellites. But the Magsat project was the first all-out effort to map the entire surface of the Earth.

During its eight months in orbit, Magsat made over 1,000,000,000 readings. The readings take into account both the Earth's main magnetic field and magnetic anomalies, which are weak, irregular magnetic fields caused by localized phenomena deep inside the Earth. With this information and computers programmed to sift through the mountain of data, geologists are developing maps that show how subsurface rock formations affect the strength of the main field—in some areas such formations intensify the field slightly, while in others the subsurface rocks either dampen the field or leave it unchanged. This, the scientists believe, is tied to variations in the composition, thickness, and temperature of the Earth's crust at any given point.

Saturn: The Spectacular Voyager 1 Flyby and the Voyager 2 Encore

When the United States's Voyager 1 spaceprobe flew by Saturn on November 12, 1980, it amassed a huge amount of data that both answered scientists' questions about the huge planet and raised a host of new ones. Voyager 1 was not the first United States spaceprobe to reach Saturn—Pioneer 11 had passed by in 1979—but Voyager's equipment was far more sophisticated and provided much more detailed information on Saturn, its rings, and its moons than Pioneer had. Almost a year later Voyager 2 swept past the planet on August 27, 1981, and added still more detail to the emerging picture of Saturn. The two Voyager flybys succeeded not only in providing a wealth of scientific data about Saturn, but they also renewed the interest of the average citizen in the United States space program.

Voyagers 1 and 2 were launched only a few days apart in 1977. They were identical spacecraft, weighing a ton each, and were loaded with scientific equipment. The plan of having two identical spacecraft on the same mission might seem strange, but there were good reasons to do it this way. If one spacecraft malfunctioned, the other could take over. If both were working, the first one to arrive could make a broad survey. Then, the second one could be programmed to focus more directly on objects of special interest. And that was the way it worked.

Saturn is about ten times as far from the sun as Earth is. Because of the immense distance involved, the in-space guidance systems on the Voyagers were designed to take over many functions normally carried out by ground stations on Earth. When the two spaceprobes were in the vicinity of Saturn, they would each travel 80,500 kilometers (50,000 miles) in the three hours required for radio waves to reach the vicinity of Saturn. So the Voyager spacecrafts received instructions from Earth on a periodic basis only and otherwise automatically monitored themselves.

Both Voyager 1 and 2 flew by Jupiter in 1979 on their way to Saturn and sent back spectacular pictures from that planet. In 1980, analysis of Voyager data revealed two new moons of Jupiter. The overall courses of the two Voyagers toward Saturn from Jupiter were different, though, which accounts for the long difference in their arrival times at Saturn. After leaving Saturn, the paths were even more different. Voyager 1 headed out into deep space toward the edge of the solar system. Voyager 2 was directed to make the closest pass by Saturn, 101,400 kilometers (63,000 miles), and to use Saturn's gravity to redirect its course toward the next planet in the solar system, Uranus.

New Understanding of the Milky Way

The Milky Way galaxy, home to our own solar system, is a spiral galaxy; that is, a vast disk of millions of stars that gravitational forces have formed into a flat spiral. Until recently, the Milky Way galaxy was believed to have a radius of about 65,000 light-years. (A light-year is the distance that light travels in a vacuum in one year—9,460,000,000,000 kilometers, or 5,878,000,000,000 miles.) But studies of the velocity of rotation of objects around the center of the galaxy have consistently found rotational speeds that are too high for the amount of mass calculated to be within the 65,000-light-year radius.

These findings were confirmed in a new study in 1980, providing further support for theories that the Milky Way could be much, much larger. One such theory postulates the existence of a huge spherical corona that surrounds the known, spiral configuration of the galaxy. By this theory, the Milky Way has an actual radius of 325,000 light-years. Its total mass is estimated to be 2,100,000,000,000 solar masses, which is a sevenfold increase over earlier calculations.

While astronomers have been adjusting their notions about the size of our galaxy, they have also been filling in what is an intriguing, if speculative, picture of the structure of the Milky Way. In the middle of the galaxy is the large "central bulge." It has a radius of about 16,000 light-years and contains clusters of old stars, clouds of gas, and dust. Infrared pictures show an intense infrared source inside the central bulge, at the very center of the galaxy, 28,000 light-years from our sun. Many astronomers think the infrared source may be a giant black hole—a "stellar graveyard" that has sucked in hundreds of thousands of stars because of its powerful gravitational attraction.

Surrounding the black hole out to a radius of about ten light-years is an incredibly dense conglomeration of stars—perhaps several million—and clouds of gas that orbit around the center. Gas clouds have been calculated to complete the orbit in a scant 10,000 years. Between 30 and 10,000 light-years from the center are four distinct rings. The first (at 30 light-years) contains ionized atoms. It is believed to be dense, hot (about 5,000° Kelvin, or 8500°F), and rotating. At 1000 light-years there is another ring. It is much hotter (10,000°K, or 17,500°F) and contains atomic and molecular hydrogen, other molecular substances, and clouds of dust. The ring is studded with clusters of new, blue-white supergiant stars. Beyond this, at 5,000 light-years is a ring of atomic and molecular hydrogen that is rotating and expanding in diameter. The fourth ring, at 10,000 light-years, is also rotating and expanding rapidly. It is composed of neutral (not ionized) hydrogen.

From the outer edge of the central bulge out to about 50,000 light-years is the outer disk of the Milky Way. It contains many younger stars, gas clouds, and dust and is generally described as having a three-armed, spiral configuration (the arms are the inner, Sagittarius arm; the Orion arm; and the outer, Perseus arm). The spiral arms are regions of active star formation, and some stars in the arms are only about 10,000,000 years old. The arms rotate about the galaxy center. The sun, which is located in the Orion arm, has a rotational speed now estimated at 230 kilometers (143 miles) per second, or one complete orbit in 200,000,000 years.

The galactic "halo" of the Milky Way is a flattened sphere that surrounds the central bulge and disk. It reaches out to about 65,000 light-years and in the past was thought to mark the outer boundary of the galaxy. It is made up of old stars and about half of the globular clusters (groups of old stars).

Beyond the halo is the hypothetical galactic corona. In this zone, from 65,000 to 300,000 light-years, a few globular clusters, dwarf spheroidal galaxies, and the Magellanic Clouds are known to exist. But it is believed that most of the mass present in the corona is "dark matter," such as burned-out stars, which cannot be detected by optical telescopes.

INTERGALACTIC HITCHHIKERS: DON'T DRINK THE VODKA!

Studies of interstellar gases have revealed that, in addition to the most abundant element, hydrogen, there are vast quantities of molecules similar to those found on Earth. The gases floating around in space are, of course, very thin, but they do contain molecules of such familiar compounds as water (H_2O), carbon monoxide (CO), and ammonia (NH_3), as well as some very complex compounds. One of the more exotic substances discovered might well appeal to a weary intergalactic traveler of the future. It is CH_3CH_2OH, specifically ethyl alcohol, or just plain vodka!

Scientists have actually estimated the amount of the interstellar vodka that can be found at the center of the Milky Way galaxy. Apparently there is enough to make about 10,000 very dry vodka martinis in glasses that have a capacity somewhat larger than the total volume of Earth. Apart from the fact that olives of the proper size are presumably unavailable anywhere in the universe, studies have shown that the vodka in its natural state is very dilute. Water content in clouds where intergalactic vodka can be found tends to be high, and the vodka is believed to be only 0.002 proof. What is worse, however, is the fact that the clouds are also spiked with hydrogen cyanide (HCN) and other undrinkable compounds.

GALAXIES: AS ANDROMEDA TURNS

When an artist wants to show a galaxy, something resembling Andromeda—a galaxy in the shape of a spiral—is usually drawn. The Andromeda galaxy, a spiral of billions of stars a scant 2,000,000 light-years away from Earth, is the closest neighbor to our own galaxy, the Milky Way, that is also a spiral galaxy. As such it has for decades supplied scientists with many opportunities to learn about the structure and evolution of galaxies and has formed the public concept of what a galaxy is. But just how spiral is Andromeda? And which way is the spiral turning?

On the one hand, there is a more traditional view that Andromeda has two spiral arms whose ends trail the direction of rotation. A newer theory holds that Andromeda has only one spiral arm, and that its end leads the direction of rotation.

At the very least, the Andromeda galaxy is a disk-like array with a bulge at the center. It is roughly twice as big as the Milky Way, is thought to be 125,000 light-years across, and contains some 400,000,000,000 stars.

Clouds of hot hydrogen gas were also located in the galaxy as well as stars. These clouds were found to be heaviest in the middle section of the disk, where most new stars were being formed. But the cold, thin clouds of neutral (not ionized) hydrogen were harder to trace because there was no way to locate them. This only became possible when it was discovered that neutral hydrogen emits radio waves at a wavelength of 21 centimeters (8.27 inches). Radio telescopes were then used to map these clouds, and it was found that while neutral hydrogen clouds were concentrated in the middle of the disk, they also extended outward far beyond the hot hydrogen clouds.

Of even more importance to the controversy over the spiral structure of the galaxy, however, was the motion of these gas clouds. It was expected they would rotate with a near circular orbit around the center of the galaxy (along with stars, dust, and so forth in the disk). The clouds do rotate but the innermost of three great concentrations of these gas clouds is also falling in toward the galaxy center at some 100 kilometers (62 miles) per second.

Halton Arp of the Hale Observatories first defined the shape of the spiral arms using data on hydrogen gas cloud distribution, and he developed the two-armed spiral theory. In this view, two separate arms begin at the center and spiral outward in the same direction, one inside the other. The configuration is only a rough fit, however, and deviations are explained as gravitational distortion caused by M32, a small companion galaxy.

Agris Kalnajs of the Mount Stromlo Observatory (Australia) challenged this theory and, again using the distribution of hydrogen clouds, devised the one-armed spiral configuration. His spiral also begins at the galaxy center, but it spirals outward in the opposite direction. Thus the leading, outside arm of the spiral is open, something which cannot be found in any other spiral galaxy. The companion galaxy, M32, is invoked here, too, to explain the anomaly.

Both theories have gained support from additional studies, and it does not appear that the controversy will be resolved in the near future. A study of positions of hot, young stars by a group of European scientists supported the single spiral-leading arm theory. But a later study reported early in 1981—based on locations of over 400 recently discovered open clusters (groups of young stars)—favored the two-armed spiral theory.

1981

The Emerging Frontier of Neutrino Astronomy

Neutrinos are subatomic particles with little or no mass (whether it is "little" or "no" is still being debated), no electrical charge, and they travel at the speed of light. Neutrinos are unaffected by magnetic fields, dust clouds, the atmosphere, or any other disturbances that plague electromagnetic radiation. In fact, neutrinos can pass straight through vast quantities of matter, regularly zipping right through the Earth.

Neutrinos are formed during collisions of subatomic particles, so they are produced in vast quantities by the nuclear fusion reactions going on inside the sun and all other stars of the universe.

These qualities could make neutrinos ideal tracers that could help astronomers pinpoint far-off galaxies and learn more about quasars, pulsars, supernovas, and so forth—except for one thing. Unfortunately, the neutrino's most attractive qualities also make it very hard to detect, and detection of the particles is, of course, essential.

A crude neutrino "collector," designed to capture high-energy neutrinos emitted by the sun, has been in operation for over fifteen years. It is located a mile below the Earth's surface in a deep gold mine in Lead, South Dakota, and consists of a huge tank filled with 4,400 kiloliters (100,000 gallons) of cleaning fluid. When a neutrino collides with a chlorine atom (in the cleaning fluid molecule), an atom of argon-37 is created.

The collector does show evidence of neutrino collisions, but it has consistently registered just about a fourth of the anticipated number of hits (only about two a day instead of eight a day). Hundreds of ideas have been put forth to explain this deviation from theory, but none has won universal acceptance.

The first steps toward building a more ambitious experimental neutrino telescope were taken in 1981. The device will be located in the ocean, 5 kilometers (3.1 miles) below the surface off the coast of Hawaii. Called DUMAND, for deep underwater muon and neutrino detector, the device consists of strings of about 1,300 highly sophisticated light sensors set to monitor an area of about a cubic kilometer (0.2 cubic mile) of seawater.

When a neutrino strikes an atom in the seawater, it creates a shower of charged particles. If one of these particles is moving faster than the speed of light in water, a flash of light, called Cerenkov radiation, will be produced. The Cerenkov radiation then triggers the light sensor (a photomultiplier), which in turn sends out an electronic signal, recording the position of the collision.

Stars Show "Sunspot" Cycles

Following a fourteen-year study of 91 nearby stars, scientists reported in 1981 that many of these stars exhibit signs of cyclic "sunspot" activity (that is, activity that is like sunspots, but on a star other than the sun; call them "starspots"). Moreover, they estimate that about half of all sun-like stars (small, middle-aged stars of the type known as the main sequence) will show cyclic starspot activity, though they believe the proportion could be much higher. Where cyclic activity was observed, starspot cycles ranged from seven to fourteen years, as compared to an eleven-year sunspot cycle for our own sun.

Studies of our sun have led scientists to believe that sunspots (and starspots) are caused by differences in the speed of rotation of parts of the sun's outer surface (faster at the equator than at the poles). According to one theory, sunspots are really "kinks" in the sun's magnetic field that develop because of these differences in motion. A sunspot, then, would be the point at which magnetic lines break through the surface and then dive back in again.

The high magnetic fields associated with sunspots provided scientists with the key to investigating such activity on other stars similar to the sun. Spectrum analysis of the sun showed that magnetic fields caused an enhancement of the H and K lines of the spectrum for calcium (wavelengths of 3,968.470 and 3,933.664 angstrom units). Using special equipment, scientists in 1966 began examining nearby stars on a regular basis for variations in the H and K emission lines. The findings were then cataloged and charted to establish a profile of the rise and fall of starspot activity on each of the stars studied.

Cyclic activity was observed (or fairly well indicated) on fifteen of the stars in the sample that were similar to the sun. But, for a variety of reasons, scientists believe the actual number of stars exhibiting cyclical activity is much higher. For one, the period of actual observation of a single star at any one time was brief. A substantial variation in the H-K emission lines could have easily occurred between observations, which may explain why some stars did not show cyclic behavior.

A second major reason is the relatively short span of the study. Our sun is known to have undergone a period of about 70 years in which sunspot activity came to a near standstill (known as the Maunder minimum). Some of the stars observed may be undergoing a similar period of inactivity. Other stars may simply have a cycle that is longer than the fourteen years of the study.

The Age of the Shuttle: Year One

The first flight of the United States space shuttle *Columbia* (April 12-14, 1981) was a resounding success. It marked not only the first round-trip flight of a reusable spacecraft, but also the beginning of what is hoped will be a new age of routine—and relatively inexpensive—flights into space. In that vision of the future, the shuttle will be the space truck that hauls people, satellites, and material into space; it may eventually make possible a permanently staffed space station.

Piloted by Commander John Young (a veteran of numerous spaceflights) and his copilot Robert Crippen, the *Columbia* took off from Cape Canaveral at 7 A.M. on April 12, 1981, and blasted up to a circular orbit 272 kilometers (169 miles) in space. After two days in space, *Columbia*, with Young at the controls, glided to a perfect landing at Edwards Air Force Base in California. That landing marked the end of two and one-half years of delays, doubts, and a total of about $9,000,000,000 in expenses. The success gave the United States space program a much-needed shot in the arm.

The space shuttle *Columbia* (one of several shuttles constructed for the Space Transportation System, or STS, as the shuttle program is officially known) is about the size of a DC-9. At launch, however, the shuttle itself is attached to a huge fuel tank and two booster rockets, which vastly increase the size of the assembly on the launch pad. The shuttle's three main engines develop about 1,100,000 pounds of thrust at blast-off, while the boosters add another 5,000,000 pounds of thrust. The two booster rockets are designed to be reusable. They were retrieved from the ocean after separation from the shuttle during *Columbia's* launch. The fuel tank was released without parachutes and fell into the Indian Ocean.

The shuttle was designed to take off like a rocket and return like a glider. A crucial facet in this concept is the heat shield. An effective shield is needed during reentry to prevent the spacecraft from burning up. The heat shields used on the space capsules previously used for manned flight were good for only one trip; so NASA created a reusable heat shield from special tiles made of glass fibers and coated with a ceramic glaze. Before the shuttle's flight, there were many demonstrations on television showing that a tile could be heated with a blowtorch on one side with none of the heat penetrating to the other side. Some 27,000 such tiles were glued to the space shuttle, and they proved to be a continuing headache up until the flight. But only about fifteen tiles were knocked loose during the launch, and the lost tiles were in a noncritical area.

Though about 16 tons of payload can be crammed into the shuttle's 4.6-by 18.3-meter (15- by 60-foot) cargo hold, the primary objective of the first trip was to test out the shuttle itself. So monitoring equipment filled the cargo bay and took readings for stress on critical parts, changes in temperature, and so forth.

As it turned out, the first shuttle flight was nearly trouble free. The cargo bay doors, which engineers had thought might warp and become stuck in the open position in the cold vacuum of space, were tested and worked perfectly. Only one of the flight data tape recorders and the zero gravity space toilet suffered any malfunctions.

A half year after its first successful flight, the United States space shuttle *Columbia* returned to orbit on November 12, 1981, proving that the shuttle was indeed the world's first reusable spacecraft. Mechanical problems had delayed the launch for over five weeks. Other mechanical problems in space forced Mission Control to shorten the original five-day flight to two and one-half days. But the second voyage of the *Columbia*, piloted by astronauts Joe Engle and Richard Truly, was nevertheless considered a success.

For one thing, the 15.25-meter (50-foot) mechanical arm in the shuttle's cargo bay was tested for the first time. Operated by remote control from inside the *Columbia*, the arm is crucial for unloading cargo, such as satellites, in space. Other important aspects of the second flight were tests of *Columbia's* maneuverability and stability during reentry. Also, several scientific experiments scheduled for this mission were either finished or largely completed despite the shortened flight time.

The mechanical problem that forced an end to the flight—a failed fuel cell—followed a string of minor problems that had held up the shuttle's launch. The trouble-prone tiles on *Columbia's* heat shield were among the prelaunch problems, along with failures in the shuttle's hydraulic system and in communications gear. But the loss of one of *Columbia's* three fuel cells, which provide electric power for the spacecraft, was the only serious in-flight problem.

Of the scientific work accomplished on the mission, data gathered by the imaging radar equipment was perhaps the most important. Some 7,750,000 square kilometers (3,000,000 square miles) of Earth's surface were mapped by the equipment, which scans below vegetation to record subsurface features. Infrared sensors were also aboard, and they were used variously in experiments to scan for mineral deposits; for differences in readings from clouds, the ocean, vegetation, and so forth; and also for infrared emissions from carbon monoxide in the atmosphere.

American Manned Space Flight

Craft	Rocket	Date	Duration	Crew	Remarks
Mercury 3	Redstone	5/5/61	15 min.	Alan B. Shepard, Jr.	*Freedom 7*; suborbital.
Mercury 4	Redstone	7/21/61	16 min.	Virgil I. Grissom	*Liberty Bell 7*; suborbital.
Mercury 6	Atlas	2/20/62	4 hr. 55 min.	John H. Glenn, Jr.	*Friendship 7*; first orbital flight (3).
Mercury 7	Atlas	5/24/62	4 hr. 56 min.	M. Scott Carpenter	*Aurora 7*; 3 orbits.
Mercury 8	Atlas	10/3/62	9 hr. 13 min.	Walter M. Schirra	*Sigma 7*; 6 orbits.
Mercury 9	Atlas	5/15/63	34 hr. 20 min.	L. Gordon Cooper	*Faith 7*; 22 orbits.
Gemini 3	Titan	3/23/65	4 hr. 53 min.	Virgil I. Grissom John W. Young	First 2-man crew; 3 orbits.
Gemini 4	Titan II	6/3/65	97 hr. 56 min.	James A. McDivitt Edward H. White	62 orbits; first extravehicular activity; first use of personal propulsion unit.
Gemini 5	Titan II	8/21/65	190 hr. 56 min.	L. Gordon Cooper Charles Conrad, Jr.	120 revolutions; demonstrated feasibility of lunar mission; simulated rendezvous.
Gemini 7	Titan II	12/4/65	330 hr. 35 min.	Frank Borman James A. Lovell, Jr.	206 revolutions; extensions of testing and performance; target for 1st rendezvous.
Gemini 6A	Titan II	12/15/65	25 hr. 51 min.	Walter M. Schirra Thomas P. Stafford	15 revolutions; accomplished first rendezvous (with Gemini 7).

75

Craft	Rocket	Date	Duration	Crew	Remarks
Gemini 8	Titan II	3/16/66	10 hr. 42 min.	Neil A. Armstrong David R. Scott	7 revolutions; first dual launch and docking with Agena; first Pacific landing.
Gemini 9A	Titan II	6/3/66	72 hr. 21 min.	Thomas P. Stafford Eugene A. Cernan	44 revolutions; unable to dock with target vehicle; 2 hr. 7 min. of EVA.
Gemini 10	Titan II	7/18/66	70 hr. 47 min.	John W. Young Michael Collins	43 revolutions; first dual rendezvous, docked vehicle maneuvers; umbilical EVA.
Gemini 11	Titan II	9/12/66	71 hr. 17 min.	Charles Conrad, Jr. Richard F. Gordon, Jr.	44 revolutions; rendezvous and docking.
Gemini 12	Titan II	11/11/66	94 hr. 34 min.	James A. Lovell, Jr. Edwin E. Aldrin, Jr.	59 revolutions; final Gemini mission; 5 hrs. of EVA.
Apollo 7	Saturn IV	10/11/68	260 hr. 8 min.	Walter M. Schirra Donn F. Eisele Walter Cunningham	8 service propulsion firings; 7 live TV sessions with crew; rendezvous with S-IVB stage performed.
Apollo 8	Saturn V	12/21/68	147 hr.	Frank Borman James A. Lovell, Jr. William A. Anders	First manned Saturn V flight; first lunar orbital mission (10 orbits); returned good lunar photography.
Apollo 9	Saturn V	3/3/69	241 hr. 1 min.	James A. McDivitt David Scott Russell Schweickart	First manned flight of all manned lunar hardware in Earth orbit, including lunar module.
Apollo 10	Saturn V	5/18/69	192 hr. 3 min.	Eugene A. Cernan John W. Young Thomas P. Stafford	Lunar mission development flight to evaluate LM performance in lunar environment; descent to within 50,000 feet.

Craft	Rocket	Date	Duration	Crew	Remarks
Apollo 11	Saturn V	7/16/69	165 hr. 18 min.	Neil A. Armstrong Michael Collins Edwin E. Aldrin, Jr.	First lunar landing; limited inspection, photography, evaluation and sampling of lunar soil. Touchdown: July 20.
Apollo 12	Saturn V	11/14/69	244 hr. 36 min.	Charles Conrad, Jr. Richard F. Gordon, Jr. Alan L. Bean	Second lunar landing; demonstrated point landing capability; sampled more area; total EVA time: 15 hr. 32 min.
Apollo 13	Saturn V	4/11/70	142 hr. 55 min.	James A. Lovell, Jr. Fred W. Haise, Jr. John L. Swigert, Jr.	Third lunar landing attempt aborted due to loss of pressure in liquid oxygen in service module and fuel cell failure.
Apollo 14	Saturn V	1/31/71	216 hr. 42 min.	Alan B. Shepard, Jr. Stuart A. Roosa Edgar D. Mitchell	Third lunar landing; splashed down in the Pacific Ocean; returned 98 pounds of lunar material.
Apollo 15	Saturn V	7/26/71	295 hr. 12 min.	David R. Scott Alfred M. Worden James B. Irwin	Fourth lunar landing; first to carry Lunar Roving Vehicle; total EVA time: 18 hr. 46 min.; returned 173 lb. of material.
Apollo 16	Saturn V	4/16/72	265 hr. 51 min.	John W. Young Thomas Mattingly II Charles M. Duke	Fifth lunar landing; second to carry LRV; total EVA time: 20 hr. 14 min.; returned 213 lb. of material.
Apollo 17	Saturn V	12/7/72	301 hr. 52 min.	Eugene A. Cernan Ronald E. Evans Harrison Schmitt	Last manned lunar landing; third with LRV; total EVA time: 44 hr. 8 min.; returned 243 lb. of material.
Skylab 2	Saturn IB	5/25/73	28 days 49 min.	Charles Conrad, Jr. Joseph P. Kerwin Paul J. Weitz	First Skylab launch; established Skylab Orbital Assembly in Earth orbit; conducted medical and other experiments.

Craft	Rocket	Date	Duration	Crew	Remarks
Skylab 3	Saturn IB	7/29/73	59 days 11 hr.	Alan L. Bean Owen K. Garriott Jack R. Lousma	Second *Skylab*; Crew performed systems and operational tests, experiments and thermal shield deployment.
Skylab 4	Saturn IB	11/16/73	84 days 1 hr.	Gerald P. Carr Edward G. Gibson William R. Pogue	Third *Skylab*; crew performed unmanned Saturn workshop operations; obtained medical data for extending space flights.
ASTP	Saturn IB	7/15/75	9 days 1 hr.	Thomas P. Stafford Vance D. Brand Donald K. Slayton	Apollo Soyuz Test Project, cooperative U.S.-Soviet mission. Soviet crew: Aleksey Leonov and Valeriy Kubasov; docked with *Soyuz 19*.
STS-1		4/12/81	2 days 6 hr.	John W. Young Robert L. Crippen	First flight of reusable Space Shuttle *Columbia*; proved concept.
STS-2		11/12/81	2 days 6 hr.	Joe H. Engle Richard H. Truly	First reuse of spacecraft *Columbia*; ended early due to loss of fuel cell.
STS-3		3/22/82	8 days	Jack R. Lousma C. Gordon Fullerton	Third flight of orbiter *Columbia*; payload included space science experiments.
STS-4		6/27/82	7 days 1 hr.	Thomas Mattingly II Henry Hartsfield, Jr.	Fourth space shuttle mission; final development.
STS-5		11/11/82	5 days 2 hr.	Vance D. Brand Robert F. Overmyer Joseph P. Allen William B. Lenoir	Fifth flight of *Columbia*; first operational mission; first 4-man crew; first deployment of satellites from shuttle.
STS-6		4/4/83	5 days	Paul J. Weitz Karol J. Bobko Donald H. Peterson Story Musgrave	First flight of space shuttle orbiter *Challenger*; deployed TDRS tracking satellite; first shuttle EVA.

Craft	Date	Rocket	Duration	Crew	Remarks
STS-7	6/18/83		6 days 2 hr.	Robert L. Crippen Frederick H. Hauck John M. Fabian Sally K. Ride Norman E. Thagard	Second flight of *Challenger*; first 5-person crew; first American woman in space; first use of Remote Manipulator Structure to deploy and retrieve a satellite.
STS-8	8/30/83		6 days	Richard Truly Daniel Brandenstein William Thornton Guion Bluford, Jr. Dale Gardner	Third *Challenger* flight; first night launch; first Black American (Guion Bluford, Jr.); launched weather/com. satellite for India; experiments with pharmaceuticals in zero gravity.
STS-9	11/28/83		10 days	John W. Young Brewster Shaw, Jr. Robert Parker Owen K. Garriott Byron Lichtenberg Ulf Merbold	*Columbia* launched Spacelab; 6-man crew performed numerous experiments in astronomy and medicine, including tests on human ear; computer malfunctions caused concern during landing.
STS-10	2/3/84		8 days	Vance D. Brand Bruce McCandless Robert Stewart Ronald McNair Robert Gibson	Fourth *Challenger* flight; jet-propelled backpacks carried 2 astronauts on first untethered spacewalks; 2 satellites (Western Union and Indonesia) lost; first landing at Kennedy Space Center.
STS-11	4/7/84		8 days	Robert L. Crippen Richard Scobee Terry Hart George Nelson James van Hoften	Fifth *Challenger* flight; deployed Long Duration Exposure Facility for experiments in space durability; snared and repaired attitude control system of Solar Max satellite.

Source: NASA

Masers: A Better Yardstick

Determining the distance to far-off objects in our galaxy has always presented astronomers with a thorny problem. In the past they have often had to rely on indirect and very complex methods, such as measuring the luminosity of a star of a given type and extrapolating the distance from that. But now maser radiation, a highly intense radio signal emitted from gas clouds surrounding newly formed stars, may give scientists the means to calculate distances directly. Masers provide benchmark distances that can be used to make calculations for other nearby celestial bodies and will make possible a more accurate map of our galaxy, the Milky Way.

The maser is a transient phenomenon that follows the birth of a large star. Once the star has been born—that is, once it has condensed sufficiently inside a cloud of interstellar gas and dust for thermonuclear reactions to begin—the intense radiation from the star's surface, known as the stellar wind, blows against the surrounding gas and dust clouds that still remain at hundreds of kilometers a second. As the gases are pushed outward, huge gaseous "clumps" condense within the expanding cloud. Energy absorbed from the newborn sun (far larger than our own) is then reradiated by these clumps as intense microwave emissions.

The process is not fully understood in stars, but it is believed to be essentially the same as that for earthbound masers. Masers were the predecessors of the more famous lasers, except that radio waves and not light waves are produced in masers. In other words, vast numbers of molecules or other clusters of atoms have their electrons pushed into higher energy states. As they return to their more normal, lower states, they emit coherent beams of radio energy. The maser phenomenon, which may be short lived in any individual cloud, continues in one cloud after another for tens of thousands of years after the star has been born.

The new technique was tested in 1981 on masers in the Orion Nebula, which at 1,600 light-years away is fairly close, and the findings agreed with calculations of the distance by other methods. Basically, the technique involves measuring the Doppler effect on the frequency of the maser radio emissions. (The Doppler effect raises the frequency of light for approaching objects and lowers it for objects that are moving away from the observer.) The changing position of the gas cloud is also tracked over a several-month period. These two factors are then used to find the distance of the maser from the Earth by elementary geometry. As astronomers apply the maser method to other sources, they are developing the first direct measurements of the universe beyond the immediate neighborhood.

1982

The News from Venus

Two Soviet spacecraft, *Venera 13* and *Venera 14*, made successful soft landings on Venus within four days of each other in March 1982. The two probes sent back pictures and data from the planet's surface, making a significant contribution to the growing body of knowledge of Venus.

The two *Venera* probes completed the 298,000,000-kilometer (185,000,000-mile) journey to Venus in four months and parachuted to the planet's surface. *Venera 13* landed on a rolling plain, while *Venera 14* set down about 965 kilometers (600 miles) away in a lowland region. The probes, bringing the total of Soviet probes to reach Venus to seven, completed their data-gathering missions before the searing 460°C heat (860°F) and tremendous atmospheric pressure (90 times that on Earth) overcame them.

From data sent back, scientists have discovered that Venusian geology is somewhat similar to Earth's. X-ray fluorescence equipment revealed that rocks at the two landing sites were all varieties of basalt. Other tests showed that the surface matter is very porous (up to 50 percent) and that it has a low bearing strength, only a few kilograms per square centimeter. Its density is under 1.5 grams per cubic centimeter (0.09 ounces per square inch).

Photographs of the Venusian surface provided tantalizing shreds of information about rock formations on the planet. *Venera 13* showed a mixture of large slabs and sharp fragments in the rolling plains region, while *Venera 14* pictures of the lowland showed a vast expanse of large slabs. Scientists speculate that the rock may well be sedimentary in origin, but they cannot be sure what the process on Venus was. The absence of water on the surface rules out any process similar to that on Earth, and it may be that a form of surface metamorphosis took place as a result of chemical interaction between the rock and the harsh Venusian atmosphere.

The Venusian atmosphere, thick with clouds that have shrouded the planet's surface from view, has been studied extensively by a series of United States spaceprobes. The atmosphere is made up mainly of carbon dioxide (96 percent) and nitrogen, with small amounts of other substances. The dense clouds that completely encircle Venus range from about 45 to 60 kilometers (28 to 37 miles) above the planet's surface and are differentiated into three layers. Droplets of sulfuric acid and water have been identified in the clouds.

The Age of the Shuttle: Year Two

After a successful first year of flight tests, the second year was largely devoted to more flight tests—but the shuttle went "operational" by the end of its second year. Here is an account of that year.

The United States space shuttle *Columbia* returned to Earth on March 31, 1982, to complete its third successful mission. The eight-day flight was marred by minor difficulties, but it nevertheless provided further proof for the basic concept of a spacecraft that could be reused.

In fact, the most worrisome aspect of the shuttle program had become the cost. Flights average out to $28,000,000 (in 1975 dollars) each, as opposed to the planned $18,000,000.

This mission of the *Columbia*, flown by Jack Lousma and Gordon Fullerton, included additional testing on the shuttle itself and some scientific experiments as well. The mechanical arm in the shuttle's cargo bay, first tested on the previous flight, was given a second workout. This time it was used to move a sensor around to measure electric fields and particles outside the shuttle. The shuttle was also positioned to test the ability of various parts to withstand the extremes of heat and cold that result when surfaces are exposed to direct sunlight or are hidden in a shadow in the vacuum of space.

Experiments aimed at showing how other living things react to the weightlessness of space turned up some interesting results. Small plants, brought along to see how they would grow, grew up toward the light as they do on Earth, but some of their roots also grew upward. Insects, at least those brought along on this flight, seem to be ill-suited to space flight. Flies apparently did not try to fly at all (they preferred walking) and bees did pinwheels whenever they tried flying (they died before the shuttle landed). Moths, on the other hand, put on a wild and erratic display.

Columbia completed its fourth and final test flight on Independence Day—July 4, 1982. It then went on to make the first official "operational" flight on November 11-16, the fifth flight in the shuttle program.

The fourth shuttle mission included tests of the cargo bay doors, experiments relating to pharmaceutical manufacturing in space, and the ferrying of 10,000 kilograms (22,000 pounds) of cargo up into space. A secret military satellite was included in the cargo.

The fifth mission, being operational, included the launching of two commercial satellites. The flight was generally successful, though space suit malfunctions prevented a space walk and problems with the computers prevented a landing controlled completely by automatic pilot.

Corporate Rocket in a Texas Cow Pasture

Space Services Incorporated of America (SSI) became the first privately owned American company to enter the space business on September 9, 1982, when it launched its own rocket from an impromptu pad in a Texas cow pasture. The rocket, dubbed Conestoga I, performed perfectly on this test flight. It blasted 315 kilometers (196 miles) up into space and then splashed down on target in the Gulf of Mexico, 517 kilometers (321 miles) from the cow pasture. The flight lasted 10.5 minutes.

SSI, formed by a Texas millionaire and staffed by former NASA officials and others, hopes to get into the emerging commercial satellite launching business. While they are among the front runners, competition for the multimillion dollar launch fees is already lining up. Heavyweights are, of course, the United States space-shuttle program and the European Space Agency, both of which are strongly promoting their satellite-launching services. (However, the same week as the successful launch of the Conestoga 1, the European Space Agency's rocket Ariane failed its first operational flight, and landed the satellite it was carrying at the bottom of the Atlantic Ocean. And, in the next two years, the shuttle also had its problems with satellite launches.) Governments of other countries also appear to be moving into the space business, and a West German company successfully launched its own rocket last year in Libya.

SSI's rocket was a $2,500,000 surplus special. The firm bought an old second stage of a surplus Minuteman rocket (cost: $365,000), which is powered by solid fuel. Mock-ups of two additional rocket stages and a payload were then fabricated by SSI. The rocket developed 45,000 pounds of thrust and burned for 61 seconds, giving SSI's test rocket a maximum speed of 8,800 kilometers (5,450 miles) per hour.

The Minuteman rocket, noted for its reliability, performed flawlessly during the test flight. Separation of the payload stage and orientation for the final boost into orbit were also completed, showing that SSI was indeed capable of getting a satellite into outer space. The confidence of SSI's investors ran high following the successful launch.

SSI's next Conestoga rocket will be a four-stage fully operational rocket that may cost as much as $20,000,000. It will deliver a 227 kilogram (500 pound) satellite to a 805-kilometer (500-mile) high orbit. However, SSI eventually hopes to be able to put operational satellites in orbit for $5,000,000 or less, a price they believe will be affordable by oil companies and other corporations that need small satellites but which could not afford the prices of NASA's shuttle launches.

The Universe: New Interpretations

The Big Bang theory was first proposed in 1948 and for some years has been the prevailing view among cosmologists on the origin of the universe. (Cosmologists are astronomers or physicists who try to explain the origin, shape, and fate of the universe as a whole.) But in an expanding intellectual universe, at least, no theory as important as the Big Bang is safe for long.

The Big Bang theory was given its name in mild mocking by the opposition, Fred Hoyle, who, along with Thomas Gold, had developed what was then the alternative Steady-State (Continuous Creation) theory of the origin of the universe. The density of the expanding universe in the Hoyle-Gold theory remained constant despite the expansion of the universe, Steady-State theorists contended, because new matter in the form of hydrogen atoms was being created continuously out of nothing.

Astronomers like to believe in a principle of uniformity that says, in effect, we are not in a privileged place in the universe; what we see in the stars and galaxies around us is like what any other astronomer at any other place in the universe would see. The Steady-State theory went this principle one better. It also assumed that we are not in any special time in the history of the universe. What we see would also be seen at any other time.

The Big Bang theory was developed by George Gamow and Ralph Alpher, who, in a famous jest, published their paper on the subject with the name of Hans Bethe added, so the authorship would read "Alpher, Bethe, Gamow" in imitation of "alpha, beta, gamma," the beginning of the Greek alphabet. Gamow and Alpher postulated a horrendous explosion at the beginning of the universe. At first there was only high-energy radiation, and everything else was formed from that. The universe was continuously expanding (and cooling) because of the explosion, thus explaining why all the galaxies are receding from each other. The Big Bang theory did not have to depend upon unknown forces spontaneously creating hydrogen atoms, as the Steady-State theory did.

The Gamow-Alpher theory also predicted that some of the primordial radiation would survive in our own time, though it was expected that this residual radiation would have degraded from gamma to microwave radiation as a result of the expansion and cooling of the universe since the explosion. That prediction was overlooked until 1965, when it was resurrected just in time for the predicted radiation to be discovered by accident. The discovery of microwave background radiation from space in 1965 by Arno Penzias and Robert Wilson provided what was considered conclusive evidence in favor of Big Bang. Although accidental, they managed to garner a Nobel Prize for the discovery. Since the background radiation was found, the Big Bang theory has been the predominant view among cosmologists.

But the remarkable evenness of the background microwave radiation has led to yet another stab at explaining how the universe around us got where it is today. On December 6, 1979, Alan Guth wrote out the first account of what has since become a popular modification of the Big Bang, the inflationary universe. The inflationary theory attempts to take into account recent developments and solve such problems as the evenness of the background radiation. In the early universe, according to Guth, any number of separate universes supposedly form as low density bubbles in the "primordial fireball" that existed before the Big Bang. Our universe in effect bubbled up out of the fireball and began expanding. It inflated very rapidly, like a bubble. Communication with other universes formed at the same time from other bubbles is blocked by complex physical processes. Guth's theory was a hit in its first version, but other cosmologists provided various modifications in the early 1980s to make it fit the facts better. By 1982 the inflationary version of the Big Bang, in one modification or another, was viewed as the most likely explanation of the beginning of the universe by most specialists.

Another Big Bang variation in 1982 explained the formation of matter from a sea of quarks on the basis that the force between quarks is stronger the farther apart the quarks are. As the universe expanded, the quarks became far enough apart to be permanently bound into protons and other heavy particles.

The universe had a few more surprises that year, as well. Also in 1982 the British astronomer P. Birch reported finding evidence that indicated that the universe was rotating.

Though far from conclusive, the rotation thesis was advanced after a study of 94 distant galaxies by radio telescope showed that shape of galaxies appeared to depend on their position in the universe. Looking out from Earth, the galaxies in one part of the sky are bent in an S-shape. Arms of galaxies in the other part of the sky bend in the opposite direction and form a Z-shape.

In fact, it is believed, all the galaxies are bent in the same direction. But, because from Earth we see galaxies from the top in one part of the sky and from the bottom in the other, the arms look as though they bend in opposite directions. The cause of the bent arms is assumed to be the rotation of the universe itself. One complete rotation is calculated to take about 60,000,000,000,000 years.

Quasars: Giants of the Universe

Quasars, or quasi-stellar radio sources, are the most distant objects in the universe known to man. Until 1982 the farthest out quasar was OQ 172, discovered in 1973 at just under 12,000,000,000 light-years away (a light-year is about 6,000,000,000,000 miles). In 1982 a new record holder, PKS 3000-330, was found by Australian astronomers at 12,000,000,000 light-years, slightly farther than OQ 172.

A quasar is a starlike object that is believed to have enormous power. Its light can be up to 1,000 times brighter than a whole galaxy, and it is often a strong radio wave source as well. The most telling feature of distant quasars is the distinctive "red shift" in their spectrums. This sharp displacement of light toward the red part of the spectrum is believed by most scientists to indicate the quasar is moving away at speeds close to that of light (OQ 172 was calculated to be receding at 91 percent of the speed of light). That in turn means the quasars are probably very distant, and the larger the red shift the more distant the quasar.

The reason that the red shift is correlated with distance is that the universe is expanding. In an expanding universe, the more distant an object is, the faster it will be going away from every other point in the universe. If you picture the stars, galaxies, and quasars as points on the surface of a balloon that is being blown up, you can visualize how this happens. While all the points on the balloon will move apart as it is inflated, the relative distance between two points that are far apart will increase faster than the distance between two points that are close together.

The inner workings of quasars are a mystery and the object of wide speculation. Even the well-defined red shift associated with quasars is thought by some astronomers to be the result of an as yet unknown physical process associated with the quasar itself—not its rapid motion. Thus the quasars could be much closer than now believed.

But these disagreements aside, the discovery of PKS 3000-330 had some important cosmological implications. Scientists have found that quasars are more numerous at greater distances than they are in the immediate area of our galaxy. Because they are so distant and receding from us at tremendous speeds, quasars are believed to have been formed by some process soon after the Big Bang, a process that has since ceased to exist. Furthermore, the light emitted by the most distant quasars, which astronomers see today, is really a part of the image of the universe when it was only 4,000,000,000 years old. (Most cosmologists believe that the universe is now about 18,000,000,000 years old.)

Since 1973 astronomers have actively searched for some stellar object beyond the 11 to 12,000,000,000-light-year range but without any success. While there is considerable speculation as to why, some scientists believe that quasars mark the outer limits, the edge, of the universe. Others disagree with the notion that there is nothing more to see beyond 12,000,000,000 light-years and contend that limitations of instruments or an obscuring dust cloud could easily be responsible.

As if the quasars' position as guardians of the edge of the universe were not enough, some astronomers believe that quasars can be found at the center of every galaxy and that these strange stellar objects are in fact responsible for holding the galaxies together. This is, of course, a highly speculative proposition, but evidence tends to point in this direction. In the case of very distant quasars, the rest of the galaxy around the quasar is simply too far away to detect as more than a faint fuzz.

In the early 1980s the suspicion that quasars are surrounded by galaxies was confirmed. One group of astronomers at Arizona State University at Tempe, led by Susan Wyckoff, was able to show that the fuzz around some quasars had the shape and brightness patterns of large galaxies. Similarly, astronomers working at Mount Palomar found the fuzz had the same spectrum as galaxies. Finally, astronomers working with the unusually clear conditions at Mauna Kea in Hawaii were able to demonstrate by 1983 that all the nearby quasars they surveyed had the fuzz.

So, what is the quasar itself? Here the speculation becomes even more intense. One common view is that inside each quasar is a black hole. Black holes are created by the collapse of gigantic stars that have become so dense that they distort space and disappear into themselves. Only the tremendous gravitational force of the collapsed star remains, so the theory goes, and all matter in the galaxy surrounding this black hole is being drawn inexorably toward it. Massive stars, dust clouds, and anything else that happens to be around are then swallowed up in huge quantities by the black hole, which has such tremendous gravitational force that even light cannot escape it.

This is where the black-hole explanation of quasars gets tricky. Quasars are light-emitting bodies, and the black holes that are supposed to be inside them are light-absorbing bodies. The difference, theorists contend, can be explained by what happens as matter approaches the fringes of the black hole. The black hole, they say, is surrounded by a huge accretion disk, in which particles are accelerated to very high speeds and, swirling around the black hole before being drawn in, generate enormous amounts of heat and radiation. This could explain the tremendous power (that of 10,000,000,000,000 suns)quasars need to become as luminous as they are.

The concept of the accretion disk is also used to explain another strange aspect of some quasars—the gas jets up to millions of light-years long that literally shoot up out of them. Here, the disk could be supplying gases faster than they can be consumed by the black hole, thus forcing the excess up and outward away from the hole. In another explanation of this phenomenon, the spinning accretion disk builds up tremendous magnetic forces. These in turn create magnetically induced jets of charged particles directed outward from the black hole.

In June of 1982 the black-hole theory of quasar energy got a new twist. Nearby quasars were found by Alan Stockton to result from the collisions or near collisions of galaxies. Throughout the remainder of 1982, both radio and optical telescopes were trained on nearby quasars in an effort to confirm or reject the idea that Stockton's colliding galaxies were common. All such studies found evidence of quasars fueled by collisions or near collisions of galaxies.

The explanation for this phenomenon is that galaxies originally form around or produce a black hole at their centers. The black holes consume stars and gases at a prodigious rate. Eventually, all the nearby stars and gases are gone, and the quasar dies down. But the black hole is still there. When another galaxy comes close, the black hole draws matter from the second galaxy, refueling the quasar.

At the edge of the universe (or 12,000,000,000 years ago) there may have been more quasars than today for one of two reasons. Either the original source of energy for the black holes may not have been consumed; or, in the universe's earlier days, galaxies may have been so close that collisions were much more frequent.

SUNSPOTS MEAN LESS HEAT REACHES EARTH

Spectrum analyses of the sun conducted since 1975 have revealed a relationship between sunspots and the apparent temperature of the sun. During periods of high sunspot activity, the sun "cools off" by about 5°C (2.8°F). At first this was thought to be part of a permanent decline in the sun's temperature, but when sunspot activity subsided in 1981, scientists noticed that the apparent temperature began to increase.

Sunspots are areas on the sun's surface that have lower temperature, and it is believed that the relatively large number of these cool spots accounted for the drop in the sun's temperature, as perceived from Earth (the "apparent" temperature). Scientists speculate that strong magnetic fields associated with sunspots actually retard the outward flow of the sun's radiation, causing the lower apparent temperature.

Halley's Comet and the Origin of Life

A team of Caltech astronomers led by David C. Jewitt and G. Edward Danielson using the 508-centimeter (200-inch) Mount Palomar telescope became the first Earthlings to catch sight of Comet Halley on its return journey toward the sun. The comet, which completes one of its long, elliptical orbits around the sun every 76 years, will pass by Earth in 1986. It last appeared in 1910, when it became familiar as Halley's Comet.

The sighting was made on October 16, 1982. The comet appeared as a tiny speck of light in the constellation Canis Minor, and at that point it was still over 1,600,000,000 kilometers (1,000,000,000 miles) away. The comet's characteristic glowing tail, or "hair," is expected to begin growing once the comet arrives in the vicinity of Jupiter and Mars in 1985. At that point the snowball of frozen gases and dust particles will have come close enough to the sun to cause some of its icy outer surface to evaporate. Solar winds will then blow the gases and dust outward to form the tail. The closer the comet gets to the sun, the longer and more spectacular the tail gets, finally growing up to 80,000,000 kilometers (50,000,000 miles) long. Although it will pass closest to Earth on April 11, 1986, this particular pass of the comet is not expected to be very spectacular to the naked eye.

Comet Halley is perhaps the most famous of the comets that pass by Earth at intervals, though there is precious little information about any of them. It gained its fame in 1705 when Edmund Halley claimed that the comet that had been sighted in 1682 was the same as the one seen in 1456, 1531, and 1607. He therefore predicted that it would return in 1758. No one had ever predicted a comet before. Although Halley did not live to see it, his prediction (slightly modified by later astronomers) came true. Since then, Comet Halley has been watched for eagerly on every return.

Because little is known about comets a number of spaceprobes have been scheduled to rendezvous with Halley's as it speeds through our solar system. Two joint Soviet-French spaceprobes will fly through the comet's tail, crossing about 9,650 kilometers (6,000 miles) away from the comet itself, in early 1986. They will take pictures and sample gas and dust in the tail. A European Space Agency spaceprobe (the agency's first) will close to within 965 kilometers (600 miles) of the comet in March 1986, after it has passed around the sun and begun its journey back out into deep space. The spaceprobe will take close-up pictures of the comet and sample gas and dust from the comet's surface. Two Japanese spacecraft, their first, will also be launched to study Halley's Comet. The United States scrapped plans to send out probes because of budget cuts.

Though the structure of the body of a comet remains a mystery, most scientists believe comets are balls of frozen gases and dust particles. They could well be composed of water, methane, ammonia, carbon dioxide, and other substances left over after formation of the solar system.

There are various theories about where comets come from. Perhaps the most widely accepted one postulates that billions of comets can be found at the farthest fringes of the solar system. Trillions of kilometers from the sun, this region is certainly cold enough to freeze clumps of gas and dust and form the comet body. The comet can remain in this region of deep space for millions of years. Then a change in the gravitational forces acting on it sends it careening in toward the sun.

That journey can take up to 2,000,000 years, for those that make it to the center of our solar system. But few ever pass beyond the outer planets. Many are trapped by the gravity of Jupiter and Saturn. Some comets collide with planets, and still others crash into the sun itself (a spectacular collision with the sun was photographed in August 1979).

Thus only a small number of comets survive the first pass through the solar system and go into a periodic orbit such as that of Comet Halley. About 600 are known and just a small fraction of these can ever be seen with the naked eye.

The possibility that comets could have rained down on Earth in large numbers long ago has stirred considerable speculation on their role as bearers of the "seed" of life. Comets could have brought to Earth millions and millions of tons of the compounds needed to form the first amino acids through outright collisions and near misses. Others think it possible that the comets contained amino acids or even living organisms, which were locked in a deep freeze until a collision with Earth. Studies of interstellar gases in the Milky Way galaxy have turned up about 60 compounds, many of them organic.

In an attempt to prove that the building blocks of life could be formed in space, scientists recently conducted an experiment in which they duplicated conditions in deep space. A vacuum chamber was filled with interstellar gases (water, methane, carbon monoxide, ammonia) and chilled to $-263°C$ $(-441°F)$. It was warmed slightly on a periodic basis with ultraviolet radiation, (simulating starlight) and then allowed to cool again. After a period of time, scientists found that through molecular action many of the complex chemical compounds found in space were formed. In addition, an unidentified yellow substance was formed on the surface of a glass rod inside the chamber. This substance, or something like it, could also be formed in space around dust particles.

Expected Return Dates of Periodic Comets

Comet	Expected Return	Period (Years)	Certainty (Code)
Tsuchinshan I	1985 Jan. 2	6.67	A3
Schwassmann-Wachmann 3	1985 Jan. 11	5.35	A4
Honda-Mrkos-Pajdusakova	1985 May 23	5.30	A1
Schuster	1985 June 2	7.25	B3
Russell 1	1985 July 5	6.10	B3
Gehrels 3	1985 July 7	8.34	B3
Kowal 2	1985 July 10	6.47	B3
Tsuchinshan 2	1985 July 21	6.85	A3
Daniel	1985 Aug. 3	7.07	A3
Giacobini-Zinner	1985 Sept. 5	6.59	A1
Giclas	1985 Oct. 3	6.94	B3
Boethin	1986 Jan. 23	11.2	B3
Ashbrook-Jackson	1986 Jan. 24	7.46	A3
Halley	1986 Feb. 9	76.0	A2
Holmes	1986 Mar. 14	7.08	A2
Wirtanen	1986 Mar. 19	5.50	A2
Kojima	1986 Apr. 5	7.89	A3
Shajn-Schaldach	1986 May 27	7.46	A2
Whipple	1986 June 25	8.49	A2
Wild 1	1986 Oct. 1	13.3	A2
Forbes	1987 Jan. 2	6.26	A3
Neujmin 2	1987 Apr. 2	5.39	C3
Jackson Neujmin	1987 May 24	8.42	A3
Grigg-Skjellerup	1987 June 20	5.10	A4
Russell 2	1987 July 4	7.10	B3
Encke	1987 July 17	3.29	A1
Klemola	1987 July 22	10.9	A1
West-Kohoutek-Ikemura	1987 July 27	6.40	A3
Denning-Fujikawa	1987 Aug. 5	8.85	A3
Gehrels 1	1987 Aug. 13	15.1	B3
Comas Sola	1987 Aug. 18	8.78	A2
Schwassmann-Wachmann 2	1987 Aug. 30	6.39	A3
Wild 3	1987 Aug. 31	6.90	B3
Brooks 2	1987 Oct. 16	6.89	A3
Reinmuth 2	1987 Oct. 25	6.72	A1
Kohoutek	1987 Oct. 29	6.65	A3
Harrington	1987 Oct. 30	6.84	A2
de Vico-Swift	1987 Dec. 7	7.40	A3
Bus	1987 Dec. 11	6.52	B3
Borrelly	1987 Dec. 18	6.86	A2

Code: A1—sure; A2:A4—likely; B—doubtful; C—unlikely.

From Comets, edited by Laurel L. Wilkening. By permission from The University of Arizona Press, copyright © 1982.

Solar System and Milky Way Made of Different Stuff

Astronomers regularly have to resort to indirect, sometimes downright devious, means to learn more about the distant reaches of the universe. Such was the case when they wanted to know if the distribution of elements in our solar system was the same as in our galaxy, the Milky Way. After four years of satellite observations, the answer—very tentatively—appears to be that the solar system is not the same as the rest of the galaxy.

Since the relative abundance of elements in our solar system is fairly well known, scientists sought a reliable way to measure their distribution in stars of the far reaches of the galaxy. Cosmic rays, the nuclei of atoms blasted toward us by stars that have exploded as supernovas, provided the necessary key to the composition of these stars. Counting the number of the various types of cosmic ray particles produces a fairly accurate picture of substances that made up the original star.

Earlier studies had shown the expected—only 1 percent of cosmic rays were nuclei of elements other than hydrogen and helium, the two main components of the galaxy and our solar system. But equipment was not sensitive enough to measure accurately proportions of the heavier elements present in gases when a star is born or that are synthesized during the thermonuclear fusion reaction by which the star "burns" hydrogen and helium.

In 1978 the United States launched the satellite ISEE-3 (International Sun-Earth Explorer) into orbit some 1,500,000 kilometers (900,000 miles) from the Earth. The orbit was the most peculiar ever for a satellite. ISEE-3 orbited the sun, keeping continually at the point where the gravitational pull from the sun and the Earth are equal. Aboard the satellite was a highly advanced, miniaturized mass spectrometer that could detect the cosmic ray nuclei of a wide range of elements and many isotopes as well. An element is determined by the number of protons in a nucleus, but different isotopes of the element, while sharing the same number of protons, have differing numbers of neutrons. The neutrons change the mass of the nucleus. The satellite's sensor was capable of determining the particle mass to an accuracy of better than 1 percent, thus determining the isotope represented by the particles it monitored.

After four years of gathering data, the satellite produced the most interesting—and controversial—results for rare isotopes of neon, magnesium, and silicon. After correcting for "secondaries"—nuclei produced by collisions of heavier cosmic ray nuclei with particles of interstellar gas— scientists discovered there was a 60 percent greater concentration of rare isotopes of magnesium (Mg-25 and Mg-26) and of silicon (Si-29 and Si-30) than in solar system samples. Even more startling was a 200 percent higher concentration of the isotope neon-22.

Though further studies of isotopes of other elements are needed, these results have aroused speculation about the distribution of elements—or at least of isotopes of elements—in our solar system as opposed to other parts of the galaxy. One possible explanation is that there was a different proportion of carbon, nitrogen, and oxygen present in the gas clouds that gave birth to the distant stars (these elements are important catalysts in part of a star's fusion reaction, the CNO Cycle). Another explanation is that there is an as yet unknown nuclear transformation involved in the supernova explosion, though evidence does not seem to point in this direction.

A more radical hypothesis contends that the composition of the gas cloud from which our solar system was formed was altered by the explosion of a nearby supernova. The resulting blast would, among other things, inject huge amounts of new material into the gas cloud. Our solar system, then, would be a local anomaly, with a composition slightly different from that of the rest of the galaxy.

After completing its work, ISEE-3 was diverted to another mission. For the surprising twist in its fate, see page 103.

THIS CLOCK WILL DIE BEFORE IT LOSES A SECOND

Scientists at the Harvard-Smithsonian Center for Astrophysics in Boston build the world's most accurate atomic clocks, called hydrogen maser clocks. Accurate to 50/1,000,000,000,000 of a second a day, the clocks will lose only one second in 50,000,000 years. Scientists admit, however, the clock itself would never last that long.

A hydrogen maser is much like a laser, except that intense microwave radiation, not light, is produced. This particular maser, first applied to timekeeping in 1960, happens to produce a very stable signal that oscillates in excess of 1,000,000,000 times a second (1,420,405,751.68, to be exact). This oscillation is, in effect, the balance wheel.

The phenomenal accuracy of these clocks is crucial for guiding space probes on missions beyond the moon, for making extremely accurate observations of interstellar objects in deep space, and for other research, such as measuring the movement of the Earth's crust.

If you don't happen to be engaged in any of these endeavors, you probably won't want to run right out and get a maser clock. For one thing, they cost over $350,000, and the newest version is about the size of a small refrigerator. Of course, there are no moving parts to speak of, and just a quart of hydrogen will keep this one ticking for ten years.

Pulsar with Something Extra

Late in 1982 astronomers using the 305-meter (1,000-foot) radio telescope at Arecibo, Puerto Rico, discovered a new breed of pulsar in the constellation Vulpecula. Designated 1937 + 215, the pulsar is emitting a characteristic pulsing signal, but at an incredibly fast rate. From the signal, scientists calculate the pulsar is spinning at 642 revolutions per second, or once each 1.6 milliseconds. This is about two-thirds less than the speed at which centrifugal force would break up the pulsar (and twenty times faster than other known pulsars).

Pulsars are believed to be neutron stars, incredibly dense clumps of matter about 16 kilometers (10 miles) in diameter. A neutron star is the remnant of an old star that has exploded into a supernova. The force of the explosion actually compresses more matter than is found in our own sun into a 16-kilometer-wide ball. The resulting pulsar is composed mainly of neutrons.

Characteristically, neutron stars spin at high rates of speed. The rotational force of the original star speeds up because matter is compressed into a smaller diameter of rotation. The same conservation of momentum is observed when a spinning figure skater pulls his or her arms close to the body to speed up the rate of spinning.

The pulsing radio signal given off by pulsars is believed to be the result of streams of high energy particles being emitted at the magnetic poles. Each time the neutron star rotates, it sends out a signal that sweeps across the skies in the same way a rotating beacon of light does.

The new pulsar rotates far faster than the 300 or so other pulsars that have been discovered so far. In fact, its rate of rotation is too great for it to be simply the remains of a supernova explosion. One possible explanation for the high rotational velocity of 1937 + 215 is that it has been "spun up," that is, the neutron star has drawn in huge amounts of matter from a companion star. The ongoing collision of this stream of matter, falling into the neutron star, is the source of the extra "push" needed to make it spin faster than normal. While a standard supernova explosion would be expected to blow apart a binary system, some types of explosions would blow away the companion slowly enough for the pulsar to form first. Although 1937 + 215 is not part of such a binary system, there is good reason to believe that at one time it had been. In 1983 another fast pulsar, almost as fast as 1937 + 215, was found. It seemed to fit the theory of fast pulsars being parts of binary systems. Astronomers are now searching the skies for other examples of this new class of pulsar.

New Space Endurance Record Set

Soviet cosmonauts Anatoli Berezovoy and Valentin Lebedev set a new record for the longest stay in space on December 11, 1982, when they completed a 211-day space flight.

Soviet cosmonauts have been regularly setting new records in this kind of space endeavor, apparently as part of an ongoing effort to build the first permanent manned space station by the end of this decade. For example, Leonid Popov and Valery Ryumin held the previous record of 185 days in space aboard the orbiting spacecraft *Salyut 6*, set in 1980.

Perhaps the most important aspect of these missions was to show that humans can adapt to long periods of space travel. Although little is known about the causes of problems that people in space have, it is clear that the very low gravity of space flight causes physiological changes. If a permanent space station is to become operational, these changes have to become well understood. But during their flights the cosmonauts also conducted experiments on the effects of zero gravity on formation of metal alloys and crystals and made observations of Earth.

Berezovoy and Lebedev completed their entire mission in the Soviet spacelab *Salyut 7*, considered the prototype for the larger, permanent space station. *Salyut* was put into orbit about a month before the cosmonauts started their mission. The cosmonauts were sent up by a second rocket on May 13, which then linked up with *Salyut*.

Reports in October 1983 of malfunctions and other mishaps surrounding *Salyut 7* indicated that the Soviet program had run into some snags. In August the Cosmos 1443 module, which had been attached to *Salyut 7* to enlarge the crew's quarters, malfunctioned and was jettisoned. There was also a problem with the environmental control system on the *Salyut 7*. The cosmonauts were nearly forced to abandon the spacelab.

Then on September 9, 1983, cosmonauts again came close to abandoning ship when a leak in a fuel line spilled nitrogen tetroxide (propellant oxidizer) from two of the spacelab's three tanks.

Finally, a rocket carrying replacement crewmen to *Salyut 7* exploded on its launch pad on September 27, 1983. The crewmen escaped with their lives when their escape module blasted free of the rocket.

Despite these difficulties, however, the cosmonauts were evidently in no serious danger. They returned to Earth in December 1983, using the same spaceship that had carried them into orbit five months earlier, a potentially dangerous maneuver, since the spaceship was not designed originally for reuse.

Soviet Manned Space Flights

While the Soviet Union was the first to put human beings in space, its space flights with humans aboard have included several serious disasters and some near misses. The Soviet space program has specialized in long-duration flights.

Craft	Date	Remarks
Vostok 1	Apr. 12, 1961	First manned space flight; 1 orbit; 1.8 hr.; Cosmonaut–Yuri Gagarin
Vostok 2	Aug. 6, 1961	17 orbits; 25.6 hr.; Cosmonaut–G. Titov
Vostok 3	Aug. 11, 1962	64 orbits; 94.4 hr.; A. Nikolayev landed by parachute
Vostok 4	Aug. 12, 1962	Dual launch with Vostok 3; P. Popovich made 48 orbits in 71 hr.
Vostok 5	June 14, 1963	81 orbits; 119 hr.; V. Bikovsky landed by parachute
Vostok 6	June 16, 1963	Second dual launch; V. Tereshkova, first woman, made 48 orbits in 78 hr.
Voskhod 1	Oct. 12, 1964	First 3-man crew; made 16 orbits in 24.3 hr.
Voskhod 2	Mar. 18, 1965	17 orbits; 26 hr.; first Cosmonaut to leave spacecraft–A. Leonov (20 min.)
Soyuz 1	Apr. 23, 1967	18 orbits; 26.8 hr.; Cosmonaut V. Komarov killed when parachute failed
Soyuz 3	Oct. 26, 1968	64 orbits; 94.9 hr.; maneuvered to 650 ft. from unmanned Soyuz 2
Soyuz 4	Jan. 14, 1969	Docked with Soyuz 5 in first space link-up of 2 manned vehicles; 48 orbits; 71.2 hr.
Soyuz 5	Jan. 15, 1969	3 cosmonauts performed EVA, transferred to Soyuz 4 in rescue rehearsal
Soyuz 6	Oct. 11, 1969	First triple launch (with Soyuz 7 & 8); 7 cosmonauts tested controls of 3 flights
Soyuz 7	Oct. 12, 1969	With Soyuz 6 & 8, conducted experiments in navigation, photography
Soyuz 8	Oct. 13, 1969	With Soyuz 6 & 7, orbited 80 times in 118.7 hr.; soft landed

Soyuz 9	June 2, 1970	Longest flight to date—17 days, 16 hr.
Soyuz 10	Apr. 23, 1971	3 crew members docked with Salyut 1, first space station
Soyuz 11	June 6, 1971	Docked with Salyut 1; 3 cosmonauts killed during reentry
Soyuz 12	Sept. 27, 1973	2-day flight; first manned flight since 1971 Soyuz 11 tragedy
Soyuz 13	Dec. 18, 1973	8-day flight; performed astrophysical and biological experiments
Soyuz 14	July 3, 1974	Occupied Salyut 3 space station; performed earth resources work
Soyuz 15	Aug. 26, 1974	Made unsuccessful attempt to dock with Salyut 3
Soyuz 16	Dec. 2, 1974	Precursor flight to check modified Soyuz systems
Soyuz 17	Jan. 10, 1975	Docked with Salyut 4, space station; set endurance record of 30 days
Soyuz 18	May 24, 1975	Docked with Salyut 4, space station; crew stayed for 63 days
Soyuz 19	July 15, 1975	Joint U.S.–Soviet mission; docked with Apollo
Soyuz 20	Nov. 17, 1975	Biological mission; docked with Salyut 4
Soyuz 21	July 6, 1976	Docked with Salyut 5; 49-day mission; performed earth resource work
Soyuz 22	Sept. 15, 1976	Took earth resources photographs
Soyuz 23	Oct. 14, 1976	Unsuccessful attempt to dock with Salyut 5
Soyuz 24	Feb. 7, 1977	Docked with Salyut 5 for 18 days of experiments
Soyuz 25	Oct. 9, 1977	Unsuccessful attempt to dock with Salyut 6
Soyuz 26	Dec. 10, 1977	Docked with Salyut 6; Romanenko & Grechko set endurance record of 96 days
Soyuz 27	Jan. 10, 1978	Carried second crew to dock with Salyut 6 space station
Soyuz 28	Mar. 2, 1978	Carried third crew to board Salyut 6; first non-Russian or American in space (Czech)
Soyuz 29	June 15, 1978	Docked with Salyut 6 crew; crew had new space record of 139 days

Soyuz 30	June 27, 1978	Carried second international crew to Salyut 6; first Polish cosmonaut
Soyuz 31	Aug. 25, 1978	Carried third international crew to Salyut 6; first East German
Soyuz 32	Feb. 25, 1979	Carried crew to Salyut 6; new record–175 days
Soyuz 33	Apr. 10, 1979	Engine failure prior to docking forced early termination; first Bulgarian
Soyuz 34	June 6, 1979	Launched unmanned; returned with crew from Salyut 6
Soyuz 35	Apr. 9, 1980	Carried 2 crewmembers to Salyut 6; first Hungarian
Soyuz 36	May 26, 1980	Carried 2 crewmembers to Salyut 6; crew returned in Soyuz 35
Soyuz T 2	June 5, 1980	Test of modified Soyuz craft; docked with Salyut 6
Soyuz 37	July 23, 1980	Exchanged cosmonauts in Salyut 6; returned Soyuz 35 crew after 185 days in orbit
Soyuz 38	Sept. 18, 1980	Ferry to Salyut 6; first Cuban in space
Soyuz T 3	Nov. 27, 1980	Ferry to Salyut 6; first 3-man crew since Soyuz 11
Soyuz T 4	Mar. 12, 1981	Mission to Salyut 6
Soyuz 39	Mar. 22, 1981	Docked with Salyut 6; first Mongolian
Soyuz T 5	May 13, 1982	First flight to Salyut 7; space station equipped to monitor body functions
Soyuz T 6	June 24, 1982	Mission to Salyut 7; Soviet/French team
Soyuz T 7	Aug. 16, 1982	Mission to Salyut 7; second Russian woman in space
Soyuz T 8	Apr. 20, 1983	3 cosmonauts failed in planned rendezvous with Salyut 7
Soyuz T 9	June 27, 1983	Crew spent 149 days in Salyut 7 after Soyuz T 10 failed in relief mission
Soyuz T 10	Feb. 8, 1984	Mission to Salyut 7 to repair propulsion system
Soyuz T 11	Apr. 2, 1984	Docked with Salyut 7; first Indian cosmonaut

Source: NASA

1983

Infrared Telescope in Orbit

IRAS, the Infrared Astronomy Satellite, was launched into orbit around the Earth on January 25, 1983, and on February 1 began the massive task of creating an infrared map of the heavens. By mid-1983 IRAS had already completed the first of two planned sweeps across the sky and made numerous important discoveries about specific stellar objects as well. Ultimately, it made nearly six complete such sweeps before it literally ran out of gas on November 21, 1983. The "gas" involved, however, was not fuel, but liquid helium used to cool the telescope to only two degrees above absolute zero. A joint project of the United States, Britain, and the Netherlands, IRAS was the first orbiting infrared telescope, and it was 1,000 times more sensitive than any other infrared telescope. It was able to "see" many previously unknown celestial objects, especially cold, nonluminous objects such as asteroids, planets, interstellar dust clouds, and protostars.

IRAS could see "cold" objects because everything in the universe is warmer than absolute zero and therefore gives off *some* infrared radiation, even when the object cannot be seen through an optical telescope. The hotter the object, the stronger the infrared radiation, but IRAS is so sensitive it can detect the heat from a small light bulb at a distance of 6,500,000,000 kilometers (4,000,000,000 miles). However, IRAS was not designed to detect high-temperature objects, so, for example, stars themselves register as empty regions, rather than as bright points.

One of the main reasons for this great sensitivity is that IRAS is orbiting high above the Earth's atmosphere. Infrared telescopes on the Earth are hampered because much of the incoming infrared radiation is absorbed by water vapor in the atmosphere. There is also considerable background infrared radiation from the atmosphere itself.

The ability to place telescopes of various kinds above the Earth's atmosphere has resulted in a great explosion of astronomical knowledge. The atmosphere is relatively transparent to visible light and to the long waves of electromagnetic radiation that we use for radio. However, it is relatively opaque to most of the electromagnetic spectrum. Often the most interesting phenomena are observable only in the X-ray, ultraviolet, or infrared portion of the spectrum (with high-energy phenomena producing mostly high energy radiation, such as X-rays or gamma rays, and low-energy phenomena producing infrared and longer waves). "Seeing" the universe with the whole electromagnetic spectrum with the use of satellites in space is revolutionizing our knowledge of the universe.

The IRAS satellite weighed 0.9 metric ton (1 ton) and cost $180,000,000. It consisted of a 57-centimeter (22.4-inch) telescope that focused infrared radiation on solid-state sensors for wavelengths ranging from 8 to 120 micrometers. The whole telescope was chilled to just two degrees above absolute zero by means of the liquid helium stored aboard the satellite. This refrigeration system eliminated any background infrared radiation from the equipment. As long as a wisp of liquid helium remained, it kept the satellite properly chilled. When the last helium evaporated, as a result of the infrared emissions from Earth and the molecular motions of the material in the satellite itself, the telescope went dead with no warning at all. The helium, however, had been effective for about three months longer than had been planned, giving the satellite nearly 50 percent longer life than scientists expected.

IRAS generated a huge amount of data—some 700,000,000 bits of data per day—that was dumped directly into storage at the Jet Propulsion Laboratory in Pasadena, California. Computers then analyzed the 100,000 or so detections recorded by IRAS each day, eliminating about 80 percent of the sightings as false images (caused by cosmic rays, dust particles, and so forth). The rest were retained for inclusion on the infrared map.

The job of studying IRAS data will go on for years, and even now computers are months behind on processing the raw data. But by November of 1983, as the telescope went dead, the preliminary findings had more than lived up to scientists' expectations. Here is a summary of what is known so far about the first set of discoveries.

The most dramatic discovery, at least for the general public, was that solid, but very small, particles were found orbiting around the star Vega. This was the first direct observation of solid matter around a star other than our own sun. Later analyses of data from IRAS revealed similar particles around Fomalhaut and some other stars. While these particles are not planets, they may well be a precursor of planets. In fact, if IRAS had been orbiting one of the stars nearest the sun, Alpha Centauri, it could not detect planets in the solar system. There is some reason to suspect that the particles are caused by comets around Vega and Fomalhaut.

Many interstellar clouds of gases and dust particles were located in our Milky Way galaxy and in other galaxies. Also, previously unknown regions of star formation were found within the clouds. In one nearby cloud, Barnard 5, astronomers found a newly forming star that is about the same size as the sun. Previously, astronomers were only aware of giant stars forming in clouds far away. IRAS showed that stars are forming much nearer than anyone knew. In fact, the Milky Way seems to be forming about the equivalent of one solar mass each year. The exact causes of star formation are poorly understood, so the IRAS data will give astronomers much more information to help solve the puzzle.

IRAS also discovered five comets (including IRAS-Araki-Alcock, see page 109) and five asteroid-like "minor planets" within our solar system. One of them, 1983TB, is believed to be the parent body for space debris that periodically gives rise to the Geminid meteor shower on Earth. In this case, 1983TB may actually be a comet that has lost most of its gases from traveling too close to the sun on a highly elliptical orbit that crosses both the orbits of Mars and Mercury. This discovery suggests that several of the 58 known Earth-orbit-crossing asteroids may be burnt-out comets, since comets are likely to cause meteor showers, but ordinary asteroids are not expected to do so.

A vast ring, or rings, of dust around the sun was located between the orbits of Mars and Jupiter. This ring may be a source of the faint zodiacal light that is visible to the naked eye at dawn and dusk. Dust of the size found would not be able to last for more than a few tens of thousand of years, so it must be created continuously. Astronomers believe that the dust is caused by comets colliding with asteroids as the comets travel around the sun.

IRAS also found a mysterious background emission in the Milky Way as a whole that may account for some of the "missing" matter in the galaxy.

The telescope also uncovered a series of dust shells around the northern portion of Betelgeuse, suggesting both that a number of eruptions had occurred and that Betelgeuse was, as a result, emitting dust clouds the way an old fashioned coal-burning steam engine emitted puffs of smoke as it traveled.

IRAS also discovered a large number of galaxies that emitted a lot of infrared radiation and very little light. The nature of these galaxies is still largely unknown. No one had predicted that the ratio of infrared radiation to light would be as high as it is.

Finally, about 10 percent of the plates showed nothing, which is odd in itself. No one has offered a convincing explanation of these blank plates.

Because infrared rays can penetrate dust and gas clouds, IRAS data should be able to shed new light on phenomena at both the center of our galaxy and the far side as well. It should also provide new findings on such distant celestial objects as galaxies, quasars, and so forth, as the data are analyzed over the next several years.

Another Black Hole Found

Black holes are probably the strangest phenomena predicted by relativity theory, since they have gravity so strong that light cannot escape from them.

The discovery of what astronomers believe is a black hole was reported early in 1983. Identified as LMC X-3, the black hole is located in the Milky Way companion galaxy called the Large Magellanic Cloud. It was just the second black hole ever found, the first being Cygnus X-1, which was identified ten years earlier. (Some astronomers also believe that Cir X-1 and GX339-4 are black holes, but these X-ray sources cannot be shown to have the mass required to be a definite black hole.)

Puzzled by what they first noticed as a strong X-ray source in the Large Magellanic Cloud, astronomers traveled to Chile in November 1982, to make further studies. The team included Anne Cowley from the United States and David Crampton and John Hutchings from Canada.

The Magellanic Clouds are companion galaxies, traveling along with the Milky Way, that can only be observed from the Southern Hemisphere. By training the 401-centimeter (158-inch) Inter-American Observatory telescope on a visible companion star, which orbited a common center with the X-ray source, the scientists were able to get the data they needed to calculate the mass of the X-ray source. Its mass turned out to be about ten times that of the sun, thus confirming that the X-ray source was indeed a black hole and not a neutron star.

Both black holes and neutron stars are thought to be formed during supernova explosions of old stars. They are a remnant of the core of the exploded star, created when the tremendous force of the explosion squeezed the mass of several of our suns into a dense ball of matter only a few miles across. When the compacted mass contained in this ball is greater than two to three times our sun, the object is classed as a black hole; when it is less than that, it is a neutron star.

In the case of LMC X-3, the companion star should be distorted into the shape of an egg by the gravitational attraction of the black hole. As the visible companion orbits, the area emitting light as viewed from Earth should vary, thus changing the brightness. Dutch astronomers claim to have observed such changes, tending to confirm that LMC X-3 is indeed a black hole.

The gravitational force of the compressed matter in a black hole is so strong that even light cannot escape it. Consequently a black hole is always identified indirectly, such as by examining the orbit of a companion star or by high energy emissions of matter being drawn into the object.

Intergalactic Gas Cloud Discovered

Astronomers from Cornell University reported in March of 1983 that they have found the first cloud of neutral (not ionized) hydrogen in the vast reaches between the galaxies. Intergalactic space had been thought to be virtually empty, and the cloud was accidentally discovered by Stephen Schneider while the team of astronomers was calibrating the 305-meter (1,000-foot) radio telescope at Arecibo, Puerto Rico, against a known void between two galaxies in the constellation Leo.

In the supposedly empty void, the Cornell team found a huge cloud of cold hydrogen gas about 300,000 light-years long and containing 1,000,000,000 times the mass of the sun. Located about 30,000,000 light-years away, the cloud is rotating very rapidly although not uniformly. At the extremities, the rotation rate is about 80 kilometers (50 miles) per second. Astronomers believe that there is some invisible mass at the center of the cloud that holds it together, perhaps a black hole (everyone's favorite candidate for any unexplained phenomenon), hidden stars, or swarms of elementary particles. Should the invisible mass turn out to be hidden stars, the cloud may just be an unusual galaxy.

It is also possible that this huge cloud of hydrogen is a tidal fragment, a mass of hydrogen ripped away from the edge of a galaxy when another galaxy passed close by.

The most intriguing possibility, however, is that the cloud is a protogalaxy that for some reason never formed stars. If this were so, the cloud would be a "living fossil" of the early universe, and could provide important information about how galaxies form.

VISIT TO A SMALL COMET

When it became apparent that the United States was not going to join the Soviets, Europeans, and Japanese in sending probes to Comet Halley, Robert Farquhar suggested a cheap way to beat the competition. The ISEE-3, which had been monitoring cosmic rays from an orbit around the sun, could be redirected to visit Comet Giacobini-Zinner, which comes around every 6.5 years. NASA agreed, and in 1982, ISEE-3 was shifted out of its orbit and headed toward the Moon. Using the Moon's gravitational pull to redirect its path and to provide more energy, ISEE-3 zipped around the Moon several times in 1983, passing within 116 kilometers (72 miles) of the surface on December 22, 1983. Last seen, the former satellite was zooming toward a September 1985 rendezvous with Comet Giacobini-Zinner, a few months *before* the Soviet probe will reach Comet Halley!

SETI: Search for Extraterrestrial Life in the Cosmic Haystack

Radio telescopes have proved to be powerful tools for studying distant and strange objects in our galaxy and beyond; now astronomers are using them in a search for radio messages from other civilizations that they feel certain must exist.

Even before radio waves were discovered, some scientists had proposed that there might be other intelligent beings in the universe. Before radio, however, the best that people could offer were such schemes as marking a giant right triangle in the desert in hopes that it could be seen from the Moon or Mars. Radio offered more practical opportunities—and the stars.

SETI (the search for extraterrestrial intelligence) finally began in earnest in 1983 on two fronts. On the first, a privately financed search will be conducted full-time for several years using Harvard's Oak Ridge 26-meter (85-foot) radio telescope and a special receiver capable of searching and analyzing 128,000 channels. The second front involves a more ambitious project, sponsored by NASA, at the Deep Space Network facility's 64-meter (210-foot) telescopes in California, Spain, and Australia. Receiving equipment will be continually upgraded until it is capable of searching and analyzing some 8,000,000 different channels.

Even with this powerful equipment, the odds of finding a message from some far-off civilization are quite slim. Assuming that there are 1,000,000 other civilizations in our own Milky Way galaxy, which consists of about 100,000,000,000 stars, astronomers would have to search through 200,000 stars before finding a signal from extraterrestrial life. One SETI study, conducted over ten years ago, said that to do the job properly, 1,000 or so radio telescopes of 91 meters (300 feet) each would be required. The total cost: $10,000,000,000. Even the use of existing equipment poses problems. The search for messages from intelligent life requires the continuous use of expensive radio telescope facilities that might well be dedicated to research that has a better chance of success.

Then there is the problem of what frequency to search. Scientists believe that any message they will receive from intelligent life will be over a narrow band, perhaps as narrow as .01 hertz. While that makes it fairly easy to distinguish a message from natural radio emissions that are all wide-band emissions, it also means that messages could come in on any of billions of possible channels. There is, however, a "window" from one to ten gigahertz in the microwave band that scientists have identified as the best place to search.

The microwave window is relatively free of interference; at these frequencies natural background signals from distant civilizations would have a much better chance of reaching Earth. Presumably, other intelligent life forms would also know this.

To further reduce the problem of sifting through various frequencies, and to make a concerted SETI project possible, scientists have developed what is called a multichannel analyzer. It is a highly sophisticated receiver that splits a radio signal into 8,000,000 one-hertz-wide channels by means of a high-speed computer. A second computer then analyzes the data, eliminating known sources and searching for messages.

The two SETI programs begun in 1983 represent the largest scale searches to date, but they are by no means the first nor the only projects involving the quest for life on other planets. The first organized search using a radio telescope was project Ozma, conducted in 1960 at the National Radio Observatory in West Virginia. No evidence of intelligent life was discovered. Since then, many other small scale searches have been mounted with similar results. But the effort goes on. In 1982 Soviet astronomers began using an 80-meter (230-foot) radio telescope to try to find some sign of life in other galaxies. The Japanese, meanwhile, are searching for messages with a 52-meter (170-foot) radio telescope.

The supposition that extraterrestrial life exists, which is not without significant opposition in the scientific community, has spawned considerable interest in how life might have begun on other planets. Thirty years ago, two University of Chicago scientists, Harold Urey and Stanley Miller, performed a pioneering experiment: they synthesized life's building blocks—amino acids—by sending electric sparks through a mixture of water vapor, methane, ammonia, and hydrogen. Since then, biologists have synthesized "protocells" that are similar to very simple organisms.

Of even greater interest has been the discovery that interstellar dust clouds contain a wide range of organic molecules, most notably those needed to construct proteins and the all-important RNA and DNA, which form the genetic code for life on Earth. Some scientists speculate that meteors, or even comets, could have seeded Earth with the building blocks of life.

It seems likely to many that life in some form exists beyond our own planet. Given the amount of time that has passed since the universe came into being, it is possible that highly advanced civilizations have developed in the far reaches of the galaxy. Such civilizations might already be aware that there is intelligent life on Earth. The big question is, Do they really want to talk to us?

Star Wars Becomes Policy

On March 23, 1983, U.S. President Ronald Reagan committed the nation to developing a defense against ballistic missiles. His speech on national television became widely known as "the Star Wars speech," after the popular movie in which the good guys blow up the bad guys with lasers. It was widely assumed that what the President had in mind was the development of laser weapons to be used from space.

Laser technology has progressed to the point where the argument in United States defense circles is whether laser weapons in space are cost effective, not whether they can be built at all, although many scientists still disagree about their viability. Military planners, however, believe that a laser satellite could be put into orbit in a few years, and that, if the alarmists are correct, the Russians will do it first.

The idea of an orbiting laser weapon is attractive to the military. In the event of an attack, the satellite could locate and destroy ballistic missiles as they arc up into space on the way to their targets. And the laser beam is ideally suited to the task. It travels at the speed of light in a perfect, straight line and can deliver sufficient force over long distances.

But to accomplish this requires an exceptionally sophisticated weapons system. First, a five megawatt chemical laser is needed, along with equipment for focusing the laser beam. Then there must be a means of locating and tracking enemy missiles that are traveling at 14,500 kilometers (9,000 miles) per hour. The aiming device must be accurate to one one-thousandth of a degree. The weapon must have sensors to indicate when a target is destroyed, so that the beam can be aimed at the next incoming missile. A ground communications system must be able to function despite enemy jamming devices, and finally there is the matter of defending the satellite itself. A satellite that costs $1,000,000,000 or more is a tempting target, and some system for defense against attack is needed. Not only could the satellite be destroyed by ground-based weapons, there is even talk of "space mines," small killer satellites that could orbit next to their targets and be set off when needed.

While these problems are difficult, solving them is not technologically impossible. The real problem with orbiting laser weapons is the same as that for the now defunct ABM (antiballistic missile system): for every laser weapon launched into space, an enemy can attack with hundreds of missiles, decoys, and other distractions at far less cost. Defense experts agree that under this kind of barrage, some nuclear missiles will always get through to their targets.

The Age of the Shuttle: Year Three

The Space Transportation System's year began with the first flight of the second United States space shuttle—called the *Challenger*—and from the launch on April 4, 1983, to touchdown at Edwards Air Force Base five days later, the shuttle's performance was near perfect. The success of the shakedown flight for the new shuttle (the sixth flight in the shuttle program) was important because NASA officials are trying to establish a fleet of reusable shuttles that can be used in rotation.

The *Challenger* is more powerful than the first space shuttle, the *Columbia*, but its huge fuel tank and booster rockets are about nine tons lighter than *Columbia's*. As a result of this combination of power and less weight, the *Challenger* is capable of carrying aloft 40 percent more payload.

During the five-day flight, astronauts F. Story Musgrave and Donald H. Peterson became the first United States astronauts to go on a spacewalk in nine years. In addition, the *Challenger* released its payload, a $100,000,000 communications relay satellite (TDRS). The first of three TDRS satellites designed to replace ground control stations, the satellite was to be boosted into a geosynchronous orbit (35,776 kilometers, or 22,235 miles) by onboard rockets controlled from the ground. A malfunction, however, left the satellite in a lower orbit. Eventually, scientists from the ground were able to maneuver the TDRS into a usable orbit using the satellite's positioning rockets.

The seventh shuttle program flight, and the second mission for the shuttle *Challenger*, was also marked by some important firsts. Launched on June 18, 1983, for a six-day flight, the *Challenger* carried aloft five crew members, the largest crew to date in the shuttle program. Among them was the first United States woman astronaut, Sally Ride, a physicist. The *Challenger* was commanded by Robert Crippen, the only astronaut at that point to have made two shuttle flights.

The third year of the Space Transportation System saw the beginnings of commercial success for the U.S. approach toward space flight. Astronaut Ride and another crew member, John Fabian, launched two commercial satellites into orbit from the shuttle's cargo bay, earning the shuttle program a much welcomed $22,000,000.

While the satellite-launching part of the mission underscored the usefulness of the shuttle as a space transport, another phase was designed to show off the shuttle's ability to recover, fix, and relaunch malfunctioning satellites. Practicing with a reusable space laboratory stowed in the shuttle's cargo bay, astronauts used the shuttle's mechanical arm to deploy and then retrieve the laboratory several times. There were other scientific tests on the mission as well, including tests of manufacturing processes in the weightlessness of space.

The *Challenger's* successful third flight, from August 30 to September 5, 1983, marked the eighth flight in the shuttle program and the first time that the shuttle lifted off and landed at night. Among the *Challenger's* crew were the first black United States astronaut and the oldest astronaut ever to fly in space. Guion S. Bluford, Jr., an Air Force colonel and an aerospace engineer, earned the distinction of becoming the first black on a United States space flight (a Cuban became the first black in space during a Soviet flight in 1980). Dr. William Thornton, a physician and expert on space sickness, was, at age 54, the oldest astronaut to enter space.

The shuttle *Challenger* performed without mishap on this, its third flight, and the crew completed the mission program. This program included launching a weather/communications satellite built for India, practice sessions with the mechanical arm in the shuttle's cargo bay, communications tests with the TDRS satellite launched on the sixth mission, and experiments concerned with the manufacture of pharmaceuticals in zero gravity.

The eighth flight saw the return of *Columbia* after some refurbishing and upgrading. *Columbia* took the European-built Spacelab aloft from November 28 though December 8. Spacelab was planned as the major way to get European scientists involved in the shuttle program.

Spacelab, a 17-ton portable laboratory, was carried inside the shuttle's cargo bay. During the flight, the six-man crew performed about 70 experiments, including a number in the field of materials science, as well as others in medicine and astronomy.

While the overall mission was pronounced a success, there were a number of technical problems. A premature shutdown of two auxiliary power units was apparently caused by an explosion of some fuel. Then, a few hours before the scheduled landing, two of the shuttle's five main computers shut down. The system is planned to work, however, with fewer than five computers. Some are merely used as backups.

The landing was briefly postponed, and then mission control gave the go-ahead for landing even though the cause of the computer malfunction was not known. The computers worked without a hitch through re-entry and the approach to Edwards Air Force Base in California. But as soon as *Columbia* made contact with the ground, they shut down for the second time.

Comet Made Closest Call

On May 11, 1983, the newly discovered Comet IRAS-Araki-Alcock passed by Earth at a scant 5,000,000 kilometers (3,100,000 miles), closer than any other known comet over the past 200 years. The comet was discovered by the orbiting infrared telescope IRAS on April 25 and was subsequently sighted visually by amateur astronomers George Alcock and Genichi Araki. Following the convention for naming comets, the three names of its discoverers were used as the comet's name—the first time that a comet ever included a satellite's name in its official name.

IRAS had, as one of its tasks, to compile a catalog of the 15,000 to 20,000 suspected asteroids, of which only about 3,000 are presently identified. When the satellite spotted Comet IRAS-Araki-Alcock, astronomers at first believed that the comet was just another previously undiscovered asteroid, and so the IRAS team routinely asked astronomers to photograph it. Although Swedish astronomers calculated the orbit, which identified the object as a comet, the news failed to reach the International Astronomical Union, the group charged with alerting astronomers to unusual phenomena. Word did not get to the International Astronomical Union until the amateurs spotted the comet on May 3.

Comets are thought to be "dirty snowballs" of water ice, gas, and dust. As they near the sun, the sun turns some of the material into a cloud around the comet, called a *coma*. Then the solar wind "blows" the coma away from the direction of the sun, forming a tail. By the time Comet IRAS-Araki-Alcock passed by Earth, its coma had grown to 161,000 kilometers (100,000 miles) in diameter.

The comet had also produced a mountain of data for curious scientists to sift through.

For one thing, radar waves were bounced off the comet's nucleus as it approached Earth and thereby proved it had a solid core. Results of the radar scan were announced in October 1983. According to the radar astronomers, the comet had an irregular shape and apparently either rotated very slowly or did not rotate at all. The nucleus of the comet was surrounded by a cloud of particles that were much larger (an inch or more) than had been expected.

IRAS took infrared readings during the flyby, and scientists gathered ultraviolet data from the comet by using the International Ultraviolet Explorer (IUE) satellite. These readings indicated the presence of diatomic sulfur, two sulfur atoms joined to form a molecule. This was the first time that the two-sulfur molecule had ever been observed in a celestial body.

Viking Sign-Off

Almost as spectacular as the first manned landings on the Moon at the time, the Viking unmanned missions to Mars gradually faded from the consciousness of the general public. Even NASA at one time almost shut down its monitoring of reports from the still-functioning Viking lander as part of a budget cut; but donations from people interested in solar-system exploration kept the project going. Although Viking 1 outlived all expectations for performance, eventually no efforts from Earth could keep it in operation.

Any last hope for Viking 1, which had landed on Mars in 1976, was abandoned in May, 1983, when the spaceprobe failed to radio data back to Earth as scheduled. The Mars probe at first failed to respond to radio commands in November 1982, after spending 2245 Mars days, known to astronomers as Mars *sols*, on the red planet. The 2245 sols amounted to 3.3 Martian years (7.3 Earth years). Although all efforts to get a response after November failed, astronomers waited until May for a routine data transmission that never came before giving up hope.

Viking 1 had outlived Viking 2, a second probe which landed on Mars in 1976. During its unexpected long lifetime, Viking 1 had transmitted a tremendous amount of information on the planet. Most spectacular were the two dust storms, each of which lasted about 100 sols, but much basic information about Martian soil and weather was also returned to Earth.

On one important mission, Viking 1 gave ambiguous results. Although experiments were conducted to determine whether or not life exists on Mars, the results—while probably negative—were not definite.

Viking 1 was supposed to have lasted for three Earth months.

THE ROCKY LEGACY OF APOLLO

At a special facility in the Johnson Space Center in Houston, Texas, there are the Earth's only examples of such truly exotic minerals as tranquillityite and armalcolite. Tranquillityite is named after the Sea of Tranquillity, where Apollo 11 landed, while armalcolite is formed from the names Armstrong, Aldrin, and Collins, the crew of Apollo 11. Of course, these are among the "Moon rocks," which are kept at the Lunar Curatorial Facility.

All in all, there are over 360 kilograms (790 pounds) of rocks, minerals, and soils brought to Earth from the six United States missions that landed on the Moon. These have proved to be a source of immense usefulness to Earthbound astronomers and geologists. Not only have they revealed the geologic history of the Moon, but they have also helped to identify meteorites and to explain the early history of Earth itself.

Meteorites: Messengers from the Moon and Beyond

When the age of science started, scientists did not believe that meteorites existed at all. The idea that stones could fall from the sky seemed impossible, and they did not connect meteorites, which land on the Earth, with meteors, which merely flash through the skies. Eventually, however, evidence for meteorites became strong enough to eliminate skepticism and to provoke scientific explanations. In the early 1980s, however, scientific skepticism was tested again as the explanation of the origin of meteorites had to be expanded in several directions to accommodate new facts.

Most meteorites are believed to have been formed from matter left over after our solar system was formed billions of years ago. But in 1983 scientists reported that an unusual meteorite, found in Antarctica in December 1982, probably came from the Moon. Other meteorite fragments, found in Antarctica in 1979 and the subject of much controversy since then, may have come from Mars.

The ice of Antarctica has proved to be a bonanza for astronomers specializing in meteorite studies. Meteorites that fall to Earth in ordinary climates are soon lost among the regular Earth rocks. Not only is the presence of a rock that has nothing within meters of it but ice highly obvious, but also the processes of the ice flow in Antarctica have tended to concentrate meteorites in fields. As a result, the concentrations of identifiable meteorites on that continent is higher than anywhere else on Earth.

The meteor that came from the Moon was found to be a composite rock, called a breccia, identical with those of the lunar highlands. Astronomers theorize that a large meteor struck the Moon with such force that surface material was hurled out into space. Once outside the Moon's relatively weak gravitational field, this one chunk of rock, at least, was gradually drawn toward Earth and finally landed as a meteor in Antarctica. The lunar origin of the meteorite is relatively certain because we have the "Moon rocks" brought back by the Apollo missions.

The case for the supposed Mars meteor is more difficult. It is one of a small number of meteorites that are very young. Called SNC meteorites (shergottites, nakhlites, and chassignites), they are composed of rock that was formed from molten lava only about 1,300,000,000 years ago. This is too recent for them to have been formed on the Moon and far younger still than material formed at the birth of the solar system (4,500,000,000 years ago). It does coincide with the period when the outer crust of Mars was solidifying from molten rock, and this gave rise to speculation that the meteorite came from Mars.

The problem is that Mars has a much stronger gravitational field than the Moon. Some scientists maintain that the rock fragments could still have been launched into space by a meteor collision with Mars, while others hold that the force required to accelerate the rock to escape velocity would have vaporized it. There is a tantalizing shred of evidence in favor of the meteor impact theory though. In some of the young meteorites, the mineral plagioclase has been changed into glass-like maskelynite as a result of a sudden, tremendous impact. Scientists estimate the impact occurred some 180,000,000 years ago.

Further evidence for a Mars origin was reported in 1983 following analysis of the suspected Mars meteor fragments found in Antarctica in 1979. Classed as shergottites (they show signs of a tremendous impact), the fragments were found to have small quantities of gas trapped inside them. Using a method called stepped temperature extraction, scientists heated meteor samples to release and measure relative amounts of gases present. What they found were high ratios of isotopes of argon (argon 40 to argon 36) and xenon (xenon 129 to xenon 132) that closely resembled ratios discovered in the Martian atmosphere by the Viking spaceprobes. Other tests showed ratios of two nitrogen isotopes were also close to what had been found in Martian atmosphere.

Another type of meteorite is believed to contain particles of interstellar matter formed before our solar system. Called carbonaceous chondrites, these are the most primitive of all meteors found on Earth. The supposed interstellar particles, composed of matter synthesized in old stars and ejected into space, are believed to have mixed in with hot gases that were condensing to form the solar system. These particles remained in the relatively cool outer fringes of the gas cloud and in that way did not become vaporized. The first hint that some of the particles in meteorites might be not from the solar system came in 1969. Since then, fifteen years of work on the problem by many scientists has established that material from outside the solar system actually exists in this form on Earth.

Carbonaceous chondrites are grainy, dark gray rocks. Only about 40 of them have ever been found. They are divided into three classes (C1, C2, C3) on the basis of their mineralogy and consist of coarse grains of material embedded within fine-grained rock. The fine-grained material is made of silicates and carbonaceous substances. The coarse-grained matter consists of various minerals, including olivine, pyroxene, troilite, and nickel-iron.

Astronomers found it difficult to identify what material in the meteorites was actually presolar. Here investigation of the isotopic composition of fine and coarse grains of rock provided the means. Using stepped heating techniques, scientists looked for certain types of unusual isotopes and ratios between them (such as neon-20, neon-21, and neon-22; xenon isotopes; and even deuterium). Since isotopes are formed by nuclear processes, almost all within a star, and since stars produce isotopes at different levels depending on their age, it was possible to use the isotopes as tracers. Not only could the identification be made that the material was from outside the solar system, but in fact, its origin could be pinned down to the type of star in which the isotopes were formed. This clinched the argument, since the stellar origins had to be in stars that do not fit the description of our sun. The actual experiment performed was stepped heating. A chondrite grain that released isotopes in the appropriate ratios during stepped heating was in fact extrasolar in origin.

Using these methods, Roy S. Lewis and Edward Anders in 1983 have identified minute quantities of four new types of exotic carbon as extrasolar. They are carbon alpha, 5 parts per million, perhaps formed during a nova; carbon beta, 5 parts per million, formed by a red giant; carbon delta, 200 parts per million, formed by a supernova; and organic carbon, up to 3,000 parts per million, formed in an interstellar molecular cloud. It is clear that none of these carbon types could have formed in the solar system, since none of the parent bodies have ever been a part of the solar system.

WHY WETHERSFIELD?

Meteorites fall randomly on the surface of the Earth, but most of that surface is ocean and much of the land area is uninhabited. Not very many of the objects that enter the Earth's atmosphere survive to reach the ground in any case. It is not surprising, then, that fewer than a dozen houses are known to have been hit by a meteorite. In 1954 a meteorite passed though the roof of a house in Sylacauga, Alabama, hitting Mrs. E. Hulitt Hodges, the only person ever known to be actually struck by a meteorite. She lived to tell about it, since her only injury was a bruise on her thigh.

Lightning is falsely reputed never to strike twice in the same place. Meteorites, which are even rarer than lightning, should be even more unlikely coincidences. It is astonishing, then, that two of the houses hit by meteorites are in Wethersfield, Connecticut, a small suburb of Hartford. One struck Paul Cassarino's house on April 8, 1971, and another struck Bob and Wanda Donahue's house on November 8, 1982. The two houses are about a mile apart. Both times, the meteorite crashed through the roof and lodged in the house, from which it was recovered.

The only possible explanation that has been put forward for the occurrence of such an improbable event is that Wethersfield is the home of a great many actuaries, those people who calculate the odds of various disasters for insurance companies. The gods have always dealt severely with hubris.

Total Solar Eclipse

The longest total eclipse of the sun for this decade occurred on June 11, 1983. The path of totality, or line of places on the Earth's surface where the total eclipse was visible, began in the Indian Ocean and ran through the Indonesian islands to the South Pacific.

Total eclipses happen about once every eighteen months when the Moon passes in front of the sun, but they are visible only in isolated locales at any one time. The Moon, which is much smaller than the sun, is able to block out the sun almost completely because of its position relative to Earth. From Earth's surface, the Moon (386,000 kilometers; 240,000 miles distant) appears to be exactly the same size as the sun (150,000,000 kilometers; 93,000,000 miles distant). When the Moon is farther from the Earth, however, the sun is not completely blocked out. A ring of sunlight completely surrounds the totally blackened moon. The result is called an *annular* eclipse.

This temporary blotting out of the sun's inner mass allows astronomers to study the sun's corona, the thin outer layer of hot, electrified gas normally lost in the brilliant glow of sunlight. The corona is extremely hot—more than 1,500,000°C (2,700,000°F)—and one of the important questions scientists hope to answer by studying eclipses is why it is so much hotter than it should be.

The sun's inner mass, heated by fusion of hydrogen, is no more than 6,000°C (11,000°F), which is far less than the corona it is supposed to be heating. This appears to defy an important law of thermodynamics (heat cannot flow from a cooler to a warmer region), though scientists now believe that loops of the sun's magnetic field—the same explanation offered for sunspots—powered by the rotation of the sun provide the corona with its added heat energy.

Japanese astronomers, meanwhile, were concerned about producing an image of the ring of dust particles that surrounds the sun. The dust is thought to have been drawn toward the sun from the outer reaches of the solar system over a period of millions of years. It now forms a shell around the sun that emits infrared radiation, but in the past only parts of the dust ring had been photographed.

The Japanese scientists used a specially equipped high-altitude balloon to record infrared radiation from the dust ring during the total eclipse. Then, with the aid of computer enhanced photography, they produced an image of the entire dust ring, showing it to be somewhat egg-shaped, and thus confirmed its existence for the first time.

Solar Eclipses

Solar eclipses, when the Moon comes between Earth and the Sun, can occur from two to five times a year, but only three can be total in any one year. None will be visible in the United States in 1985 or 1986. The average time between a solar eclipse being visible from the same spot on Earth is 300 years.

Date	Type	Visibility
1985 May 19	Partial	Northeastern Asia, Japan, northern North America, Greenland, Iceland, extreme northwestern Europe, Arctic regions, North Pacific Ocean.
1985 Nov. 12	Total	Total in parts of South Pacific Ocean. Partial in most of Antarctica, eastern South Pacific Ocean, southern South America, southwestern part of South Atlantic Ocean.
1986 Apr. 9	Partial	Part of Indonesia, Australia, New Guinea, South Island of New Zealand, part of Antarctica, part of Indian Ocean.
1986 Oct. 3	Annular/Total	Annular/total in Iceland. Partial in North America except extreme southwest, extreme northeastern part of Asia, most of North Atlantic Ocean, Iceland, Greenland, northern South America and Arctic regions.
1987 Mar. 29	Annular/Total	Annular/total in southern South America, Africa near Equator, to Indian Ocean. Partial in southeastern South America, part of Antarctica, South Atlantic, Africa except extreme northwestern part, extreme southeastern Europe, extreme southwestern Asia, western part of Indian Ocean.
1987 Sept. 23	Annular	Annular in west central Asia, across China and Okinawa, between Micronesia and Melanesia, ending near Samoan Islands. Partial in Asia except northeastern and southwestern, northeastern half of Australia, western Pacific Ocean, and Oceania.

Source: U.S. Naval Observatory

Lunar Eclipses

While there are, on average, as many lunar eclipses, when the Moon is in Earth's shadow, as solar eclipses the lunar ones are more visible. A total lunar eclipse can be seen over half the Earth if nighttime skies are clear, while the path of a solar eclipse is always less than 269 kilometers (167 miles).

Date	Type	Duration (Min.)	Visible
1985 May 4	Total	70	Africa, Indian Ocean, eastern India, Australia
1985 Oct. 28	Total	42	Asia, E. Africa, W. Australia
1986 Apr. 24	Total	N.A.	N.A.
1986 Oct. 17	Total	N.A.	N.A.

Source: U.S. Naval Observatory

HOW HIGH THE MOON?

Although we know down to a few parts of a centimeter how far away from the Earth the Moon is today—thanks to mirrors left by American astronauts that reflect laser beams back to Earth—the question of where the Moon was in the past has been the subject of hot debate for a hundred years or so. George Darwin, the great evolutionist's son, thought the Moon had once been a part of Earth, but was ejected in some way, leaving behind the ocean basins. While the specifics of his theory have not held up to modern plate-tectonic views, the general idea that the Moon may once have been part of Earth has continued to be advocated by some. Until 1982.

That year, Kurt Hansen calculated from the laws of gravity that the Moon could never have been closer to Earth than 225,000 kilometers (140,000 miles). This result also put a serious crimp in another popular theory of the Moon's origin—that the Moon was a wandering small planet that drifted too close to Earth and was captured. Coming as close as 225,000 kilometers from the Earth with any momentum of its own, according to Hansen's figures, would not permit capture. At that distance, the Moon might have bowed to Earth, but it would have moved on.

The conclusion is that the Moon must have been formed—as most scientists suspected anyway—at the same time as Earth, revolving about the Earth as both condensed from a primordial cloud.

In the meantime, the Moon keeps moving away from its old friend Earth, even as a polite courtier, while facing toward the monarch, might slowly drift to a far corner of the room. However high the Moon was in the great days of bebop, it has moved about 4 centimeters (1.6 inches) farther away each year since.

Pioneer Spaceprobe Leaves the Solar System

The United States spaceprobe Pioneer 10 passed outside of our solar system at 5 A.M. Pacific Daylight Time on June 13, 1983, becoming the first manmade object to fly beyond the planets. Pioneer was launched on March 2, 1972, and was the first spaceprobe to reach the asteroid belt, and then Jupiter. It will continue to send back data on the extent of the sun's heliosphere (magnetic influence) and may also tell scientists if, as there are reasons to suspect, there is a tenth planet.

The Pioneer spaceprobe completed an important part of its mission during the Jupiter flyby in 1973. Its sensors indicated that metallic hydrogen is the major component of Jupiter's interior, provided maps of the planet's magnetic fields, and measured the size of Jupiter's moons.

The long journey out toward deep space began after the Jupiter flyby and Pioneer used Jupiter's strong gravitational field as a "slingshot" to propel it on its way. Thereafter it has sent back data on solar wind and has shown that the sun's heliosphere may extend more than 8,000,000,000 kilometers (5,000,000,000 miles) from the sun.

Pioneer was considered to have passed outside the solar system when it crossed Neptune's orbit on June 13, almost 4,800,000,000 kilometers (3,000,000,000 miles) from the sun. At that point Neptune was in fact farther from the sun than Pluto, which has a more elliptical orbit. A few enthusiasts celebrated when Pioneer passed the orbit of Pluto at about 2 P.M. PDT on April 25, 1983. Some purists insist that Pioneer would at least have to pass beyond the farthest point of the orbit of Pluto (which will occur on June 19, 1990), leave the heliosphere, or perhaps pass the Oort cloud, where comets come from, to be truly beyond the solar system. In the last two interpretations, Pioneer has still some unknown number of years of travel within the system to go. The probe is expected to keep transmitting data on its travels for another ten to thirteen years, so keep tuned in.

Three other spaceprobes, Pioneer 11, Voyager 1, and Voyager 2, are all due to leave the solar system.

Pioneer (and Voyager 2 as well) includes along with its scientific instrumentation a "message in a bottle." The plaque carried by Pioneer was designed by enthusiasts for the likelihood of extraterrestrial life. It is intended to be read by a finder who is intelligent, but knows no Earth language or facts about the Earth. The plaque indicates such information as the third planet from the sun and includes pictures of a man and a woman. Voyager 2, perhaps doubting the ability of extraterrestrials to read, includes a gold-plated phonograph record instead.

Titan's Atmosphere Like That of Primitive Earth

The discovery of carbon monoxide in the atmosphere of Titan, Saturn's biggest satellite, was reported in June 1983. Astronomers at Kitt Peak National Observatory in Arizona made the discovery during a spectrum analysis of light from Titan, and thereby added further evidence that Titan's atmosphere is similar to that of Earth before life evolved.

Titan's atmosphere is about 82 percent nitrogen and data from the Voyager spaceprobe indicated carbon dioxide (CO_2) is also present, along with methane (CH_4). The carbon monoxide content of the atmosphere is estimated to be about six to 40 molecules per 100,000 of nitrogen.

Given the presence of these essential chemical materials, scientists speculate that it is possible that some amino acids—the precursors of life—could have formed on Titan by chemical evolution, in much the same way they did on Earth. But because Titan is so cold, water remains frozen there, so it seems unlikely that the process of evolving life forms ever got beyond these tenuous beginnings. Nevertheless, scientists continue to be excited by finding such a close analog to Earth.

European Space Agency Back in Competition with Space Shuttle

The European Space Agency (ESA) got back into the race for satellite-launching business on June 16, 1983, just two days before the United States space shuttle was due to take two commercial satellites aloft.

ESA's Ariane 6, a conventional rocket that, unlike the American shuttle, is not reusable, was launched from a site in French Guiana and also put two commercial satellites into orbit. The success of this sixth Ariane rocket breathed new life into the faltering European space consortium. Prospects had looked good after three out of the first four test launches proved successful. But in September 1982, the first officially designated operational flight malfunctioned, sending both the rocket and its multimillion dollar satellite payload to the ocean bottom.

Needless to say, performance of this kind overshadowed the main advantage of the ESA rocket: namely, that right now it can deliver satellites into orbit for much less money than delivery by the United States space shuttle costs. Up until the success of Ariane 6, critics of the ESA program even went so far as to call the Ariane "Europe's first rocket-powered submarine."

Galaxies: More Mass There Than We Can See

Recently reported results of a six-year study of the rotational velocities of 60 spiral galaxies indicate that most of the mass in galaxies is not located at the luminous center, as had once been thought. In fact, the scientists concluded that as the amount of luminous mass, which consists of stars and hot gases, declines outward from the galaxy center, there is a corresponding increase in nonluminous, dark matter. This dark matter may be small dim stars, large planets, dust clouds, black holes, or even subatomic particles.

Because luminous matter alone is visible to optical telescopes, scientists until recently thought most of the mass in galaxies was concentrated at the galactic center. For this to be the case, however, the rotational velocity of stars and other objects in the galaxy would decrease steadily as the distance from the galactic center increased. This is the behavior expected from Kepler's Law for Planetary Motion.

To find out if that decrease in rotational velocity actually occurred, astronomers studied three types of spiral galaxies: Sa, large central bulge and tightly wound spiral arms; Sb, moderate-sized central bulge and moderately open arms; and Sc, small central bulge and widespread arms. Twenty galaxies of each type were chosen to provide a full range of sizes and luminosities within each category. Only those galaxies whose planes of rotation tilted sharply toward Earth were studied, to make it easier to measure rotational velocities at various points along the rotational plane.

What the scientists found was the opposite of what had been expected: rotational velocities of luminous bodies in galaxies either remain constant or increase out to the outer visible fringes of the galaxies. This result proved that mass could not be concentrated at the center of the galaxy; otherwise, the laws of gravity would not hold. Therefore, luminosity is not a reliable guide to the distribution of mass in a galaxy.

Scientists can only speculate about what exists beyond the visible fringe of galaxies, but it seems likely that galaxies are surrounded by spherical halos of dark matter that extend far beyond the visible galactic limits. The mass of unseen matter contained in the halo could then account for observed high rotational velocities of luminous objects.

The mass of dark matter that apparently exists in each galaxy is enormous and therefore is an important component in any attempt to calculate the universe's total mass. The mass of the universe is of interest since the amount of mass determines whether or not the universe will continue to expand, eventually become stable, or collapse.

X-ray Data on SS433

Since 1979, it has been known that one of the oddest—and therefore most interesting—objects in the sky is the one cataloged as SS433. SS433 started to arouse considerable interest in 1979 when it was discovered that lines in SS433's visible spectrum moved back and forth in cycles, indicating some sort of complex orbital motion. The analysis of the motion was complicated because there were two cycles of change superimposed. Their periods were identified eventually as thirteen days and 164 days. The radio and X-ray spectrums of SS433 showed a central source of radiation flanked by two tear-shaped lobes. The whole mass was surrounded by a thin cloud of the type left by a supernova.

An eighteen-month study of X-ray emissions by the Einstein X-ray Observatory satellite established the basic data that could be used to explain SS433's strange pattern. The Einstein data confirmed earlier theories on SS433.

The generally accepted view of SS433 was that it is a binary star system with a dense and compact neutron star or perhaps a black hole orbiting around an ordinary star. The neutron star or black hole emits a jet of high energy material from each pole, just as pulsars do. These emissions are detected on Earth as radio and X-ray sources. The jets are believed to be fed by material drawn from the ordinary star.

The analysis of the Einstein data conducted by Jonathan E. Grindlay and colleagues explains the changes in SS433's visible spectrum as resulting from the following combination of factors. The binary orbit of the neutron star or black hole is thirteen days. The precession period of the neutron star or black hole is 164 days. Because the axis of the neutron star or black hole is tilted, just as the axis of the Earth is, the amount of flow in the jets varies "seasonally"; the jets get a boost when one of the axes of rotation points more nearly toward the ordinary star. Since the light whose spectrum varies comes from the jets, it also varies with a combined period of thirteen days and 164 days, and furthermore, exhibits two flares during the thirteen-day period.

Grindlay's team has also concluded that the mysterious dark companion is probably a black hole with a mass about ten times that of the sun. If so, it would be the third black hole to be identified by calculating its mass. Calculations of the mass of a star or a black hole is possible only in a binary system, where the orbit can be used along with Newton's laws to deduce mass. A mass as great as SS433's combined with a small volume implies a black hole instead of a neutron star.

Jupiter and Saturn Satellites Get Official Names

The International Astronomical Union (IAU) in October 1983 released the official names for four satellites of Jupiter and Saturn that were discovered by Voyager 1 in 1979–80.

Three of the satellites belong to Jupiter. All three are small (40 kilometers or 25 miles across for the largest), and all complete their orbits in less than a day. The three are Jupiter XIV Thebe, Jupiter XV Adrastea, and Jupiter XVI Metis. All were first listed in 1979.

The satellite belonging to Saturn was first listed in 1980 and is now officially named Saturn XV Atlas.

The Langley Medal

The Langley Medal is awarded periodically by the Smithsonian Institution for achievement in aeronautics and astronautics.

1909	Wilbur Wright	1960	Robert H. Goddard
	Orville Wright	1962	Hugh Latimer Dryden
1913	Glenn H. Curtiss	1964	Alan B. Shepard, Jr.
	Gustave Eiffel	1967	Wernher von Braun
1927	Charles A. Lindbergh	1971	Samuel Phillips
1929	Charles Matthews Manly	1976	James Webb
	Richard E. Byrd		Grover Loening
1935	Joseph S. Ames	1981	R.T. Jones
1955	Jerome C. Hunsaker		Charles Stark Draper

IS THE SUN SHRINKING?

In 1979 two astronomers reported that the sun was shrinking, and fairly rapidly—its horizontal diameter was decreasing at a rate of about 0.1 percent per century. The astronomers used data from the daily observations of the sun's transit, taken since 1750 in Britain and since 1846 in the United States, to calculate the change in the sun's diameter. The observations could not always be assumed to be accurate because of variations in Earth's atmosphere and such imponderables as observer error. So along came three other astronomers with another approach. They used data derived from observations on the durations of solar eclipses and on transits of the sun by the planet Mercury.

Happily, the second group of astronomers was able to determine that, if the sun is shrinking, it is doing so at a much slower rate—perhaps only 0.008 percent (give or take 0.007 percent) per century. While the data are not conclusive, it seems likely that the sun's diameter both shrinks and expands, varying over a long period of time. So maybe the sun isn't shrinking after all. It's just palpitating!

1984

Update 1984: Astronomy and Space

GRAVITATIONAL LENSES FOUND

The year opened with a report in the January 6 issue of *Science* on the discovery of the fourth gravitational lens. Such lenses had been predicted by Arthur Eddington in 1920 and their mathematics worked out by Albert Einstein in 1936, but they had not been observed until 1979. Before the first one was discovered at radio wavelengths, most astronomers discounted the likelihood of the lenses' existence. Since then, however, there has been an average of one new lens discovered each year, and the fifth known lens was announced later in 1984.

A gravitational lens is caused by the effect that gravity has on light as predicted by relativity theory. An object with sufficiently massive gravity can refract light just the way that a glass lens does. In practice, all the five known gravitational lenses refract the light from distant quasars.

The lens announced first is identified as $2016 + 112$ and was found in a systematic survey conducted by radio telescopy by a team from Massachusetts Institute of Technology, California Institute of Technology, and Princeton University. The lens was then found to operate at visible-light wavelengths as well. Two images are the refracted images of a single quasar, while the other image is the lens itself.

The other lens, $1635 + 267$, was established by S. George Djorgovski and Hyron Spinrad of the University of California at Berkeley. They re-examined what had previously been thought to be two different quasars and found them to be images of a single object. Slight differences in the two images suggest, however, that the quasar is being seen at two different times in its existence. This is possible because gravity not only bends light, but it also can slow it down. If the light from one of the images is slowed by the lens, it would account for the differences observed.

BROWN DWARF FOUND?

Twenty-eight light-years from the sun there is an object that may be the first known "brown dwarf" star. If it were much farther, it might not be noticed at all, since the star is the coolest and least luminous object ever found outside the solar system. The star, known as LHS 2924, which was spotted by Ronald G. Probst of the Kitt Peak National Observatory and James Liebert of the University of Arizona, qualifies as a brown dwarf if it is so small that gravitational forces are not great enough to cause thermonuclear fusion. The possible existence of such dwarfs has been suggested as one way to account for the missing mass in the universe (see page 119).

FILAMENTS AT THE CORE

The Very Large Array (VLA) of radiotelescopes in New Mexico (see page 64) took a look at the center of our Milky Way galaxy in 1984 and revealed a surprising group of parallel filaments more than 130 light years long streaming out of it. The filaments are apparently formed by particles trapped in a magnetic field, the same effect that is believed to cause sunspots. No one knows exactly what is going on at the core of the Milky Way, since light cannot penetrate the clouds of dust and gas that surround the core. Radio waves, however, find the clouds no obstacle, and radio telescopes have revealed in the past that the galactic core is producing a tremendous amount of energy. Two possible sources for the energy are a giant black hole and a tight cluster of stars with a total mass of about a million suns.

Nothing exactly like the filaments has ever been seen before on this scale. It suggests that the galaxy has a magnetic field that is similar to that of the Earth.

The observations that revealed the filaments were made by Mark Morris of the University of California and Farhad Yusef-Zadeh and Don Chance of Columbia University and announced in July 1984.

THE AGE OF THE SHUTTLE: YEAR FOUR

Although NASA's schedule of shuttle flights planned for 1984 slipped from the ten originally planned to four, a major advance was supposed to come with the much postponed launch of the third STS (Space Transportation System) vehicle, the *Discovery*, in June 1984. Along with the usual scientific experiments in space technology—this time a study of the unfolding of a solar sail by Judith A. Resnik—and the launch of a communications satellite to be leased by the Navy, the *Discovery* was to carry the shuttle's first paying passenger. McDonnell Douglas and Johnson & Johnson paid for Charles D. Walker to spend 80 hours operating a system to make an unusually pure hormone. The exact nature of the hormone, which was to be produced by separation using low-gravity electrophoresis, was kept a commercial secret, however.

All bets were off, however, after countdowns were stopped two days in a row, the second just as the engines were starting to ignite. NASA scientists determined that the cause of the difficulty on June 26 was that some of the insulation had come loose in one of the main engines. This allowed liquid nitrogen to freeze the valves. When computers detected that the engine was not firing properly, they shut down the entire operation within four seconds of lift-off. The launch was once again postponed, then combined with the second *Discovery* mission.

Our New View of the Planets

Although six of the nine planets were known to the ancients, and the remaining three were found by early in the twentieth century along with many of the satellites, no amount of observation from afar prepared us for what we learned when we began exploring the solar system with spaceprobes in 1962. Furthermore, satellite observation has told us much that we did not know about our own planet, Earth. While this age of exploration is far from over, here is an interim report.

THE TERRESTRIAL PLANETS

The first four planets of the solar system—Mercury, Venus, Earth, and Mars—are called the terrestrial planets. The terrestrial planets all have a comparatively high density, a concentration of metallic elements, and hard, rocky surfaces. The Earth is the largest of the terrestrial planets, although it is dwarfed by the enormous size of the "Gas Giants" (Jupiter, Saturn, Uranus, and Neptune). All the terrestrial planets have magnetic fields. Earth and Venus have thick atmospheres, Mars has a thin atmosphere, and Mercury's atmosphere is almost nonexistent.

MERCURY

Mercury is the planet closest to the sun and in keeping with its namesake—Mercury, the winged messenger—it is the fastest. Mercury orbits the sun at a mean velocity of just under 48 kilometers (29.7 miles) per second and completes one revolution every 88 days. Its period of rotation is 59 days. (The spinning of a planet on its axis is its *rotation*; the trip of a planet or satellite in its orbit about another body is its *revolution*.) Mercury is usually obscured from view from Earth by the sun's glare, though it is visible on Earth's horizon just after sunset or just before dawn. About fourteen times every 100 years, Mercury can also be seen crossing directly in front of the sun's disk. The next such crossing, or *transit*, will take place in 1986.

The first detailed pictures of Mercury's surface were provided by the United

States Mariner 10 spaceprobe during flybys in 1974 and 1975. Mariner 10 mapped about 35 percent of the planet's heavily cratered, moonlike surface.

Mercury is a waterless, airless world that alternately bakes and freezes as it orbits the sun. The tenuous atmosphere that has been discovered there is one trillionth the density of Earth's atmosphere and is largely composed of argon, neon, and helium. On Mercury's sunlit side temperatures reach 510°C (950°F) and plummet to −210°C (−346°F) on the dark side. These extremes are largely due to Mercury's slow rate of rotation; because it takes 59 Earth days to complete just one rotation, there is plenty of time for the surface to heat up or cool off. Another way to look at this is to say that one Mercury day (or *sol*, as astronomers call it) is two-thirds of one Mercury year.

Previously, it was thought that one Mercury sol was exactly equal to a Mercury year, so that one face of the planet was perpetually toward the sun, while the other was always away from it.

Mercury's surface is scarred with hundreds of thousands of meteor craters, many of which were probably formed during the asteroid showers that are believed to have occurred in the early period in the formation of the solar system. Many areas have been smoothed over by ancient lava flows, however, indicating extensive volcanic activity on Mercury during and after the time of the asteroid showers. The surface is also crisscrossed by huge cliffs—called *scarps*—that were probably formed as Mercury's surface cooled and shrank. Some of these scarps are up to 2 kilometers (1.2 miles) high and 1,500 kilometers (932 miles) long.

Mercury's outer crust is made up of light silicate rock like that of Earth. Because Mercury has a magnetic field, scientists believe the planet has a hot core of molten metal, perhaps iron with some nickel. The molten core is thought to make up about half the planet's volume.

VENUS

Venus, as seen in the night sky from Earth, is second only to the Moon in brightness. Venus undergoes phases like the Moon (full phase at superior conjunction, when it is farthest from Earth) and is the planet that passes closest to Earth (42,000,000 kilometers, or 26,000,000 miles, at inferior conjunction). The discovery of the phases of Venus by Galileo (perhaps suggested to him by a former student) eventually led to his public support for the Copernican system. While a Venus circling the sun would show phases similar to those of the Moon, as a result of partial illumination of the sphere of Venus by the sun, the Ptolemaic theory had Venus traveling in a small circle between Earth and the sun; thus, only the crescent phases could be seen from Earth. When Galileo observed "full Venus," he knew that Ptolemy's theory had to be wrong.

Because of its proximity to Earth, Venus became in 1962 the first planet beyond Earth to be scanned by a spaceprobe (Mariner 2) and has since been visited by numerous United States and Soviet spacecraft. Soviet spaceprobes Venera 13 and Venera 14 recently sent back pictures from the Venusian surface (see Science Update 1982).

Terrestrial Planets

	Mercury	Venus	Earth	Mars
Position among planets	first	second	third	fourth
Mean distance from the sun	57,900,000 km 35,960,000 mi. 0.387099 AU	108,200,000 km 67,200,000 mi. 0.723332 AU	149,600,000 km 92,900,000 mi. 1 AU	227,900,000 km 141,500,000 mi. 1.523691 AU
Rotation period	59 Earth days	−243.01 Earth days (retrograde)	23 hr., 56 min., 4 sec.	24 hr., 37 min., 23 sec.
Period of revolution	88 Earth days	224.7 Earth days	365.26 Earth days	687 Earth days
Mean orbital velocity	47.89 km/sec. 29.7 mi./sec.	35.03 km/sec. 21.75 mi./sec.	29.79 km/sec. 18.46 mi./sec.	24.13 km/sec. 14.98 mi./sec.
Inclination of axis	2°	3°	23°27′	25°12′
Inclination of orbit	7°	3.39°	0°	1.9°
Eccentricity of orbit	0.206	0.007	0.017	0.093
Equatorial diameter	4,880 km 3,030 mi.	12,104 km 7,517 mi.	12,756 km 7,921 mi.	6,787 km 4,215 mi.
Diameter relative to Earth	0.382	0.949		0.532
Mass	3.303×10^{23} kg 7.283×10^{23} lb.	4.87×10^{24} kg 10.7×10^{24} lb.	5.98×10^{24} kg 13.2×10^{24} lb.	6.42×10^{23} kg 14.2×10^{23} lb.

	Mercury	Venus	Earth	Mars
Mass relative to Earth	0.0558	0.815		0.1074
Mean density	5.42 g/cm^3	5.24 g/cm^3	5.52 g/cm^3	3.93 g/cm^3
Gravity (at equator surface)	3.78 m/sec.^{2} 12.4 ft./sec.^{2}	8.60 m/sec.^{2} 28.2 ft./sec.^{2}	9.78 m/sec.^{2} 32.1 ft./sec.^{2}	3.72 m/sec.^{2} 12.2 ft./sec.^{2}
Gravity (relative to Earth)	0.38	0.88	1.0	0.38
Escape velocity at equator	4.3 km/sec. 2.7 mi./sec.	10.3 km/sec. 6.40 mi./sec.	11.2 km/sec. 6.96 mi./sec.	5 km/sec. 3.1 mi./sec.
Average surface temperature	440°K	730°K	288°K	218°K
Atmospheric pressure at surface	10^{-15} bars	90 bars	1 bar	0.009 bar
Atmosphere (main components)	Virtually none (of that, .98 helium, .02 hydrogen)	Carbon dioxide 96%; nitrogen 3.5%	nitrogen 77%; oxygen 21%; water 1%; argon .93%	Carbon dioxide 95%; nitrogen 2.7%; argon 1.6%
Other atmospheric gases (ppm)		Water 100; sulfur dioxide 150; argon 70; carbon monoxide 40; neon 5; hydrogen chloride .4; hydrogen fluoride .01	Carbon dioxide 330; neon 18; helium 70; krypton 1.1; hydrogen .5; methane 1.5; nitrous oxide .3; carbon monoxide .12	Oxygen 1,300; carbon monoxide 700; water 300; neon 2.5; krypton .3; ozone .1
Planetary satellites	None	None	1 moon	2 moons

The Venusian atmosphere, thick with clouds that have shrouded the planet's surface from view and added to the mystery of the planet, has been studied extensively by a series of United States spaceprobes. The atmosphere is very dense, with atmospheric pressure at the surface level 90 times that of Earth, and largely made up of carbon dioxide (96 percent) and nitrogen, with small amounts of other substances. The Venusian clouds range from about 45 to 60 kilometers (28 to 37 miles) above the planet's surface and are differentiated into three layers. Droplets of sulfuric acid and water have been identified in the clouds.

The clouds and high level of CO_2 in the atmosphere have combined to trap heat in the lower atmosphere of Venus as a result of the greenhouse effect. The clouds and CO_2 are thereby responsible for the extreme temperatures in the lower atmosphere, 482°C (900°F)—hot enough to melt lead. In fact, radiation of heat from the lower atmosphere is so inefficient that there is little variation of temperature between night and day.

One feature of the Venusian upper atmosphere is markedly different from that of Earth. The atmosphere superrotates on Venus—that is, the atmosphere above the clouds moves 60 times faster than the planet rotates—while the Earth and its atmosphere rotate at the same speed. So, high winds are a dominant part of Venusian weather. Even at only 50 kilometers (31 miles) above the surface, which is just under the cloud layer, the winds blow steadily at 175 kilometers (109 miles) per hour.

Though Soviet spaceprobes that softlanded on Venus provided a precious few photographs of the planet's surface (1975: Venera 9, 10; 1982: Venera 13, 14), the clouds have obscured most of the landscape from view. But radar maps of 93 percent of the Venusian surface, completed by the United States Pioneer Venus spacecraft (from 1978), now provide a detailed picture of the planet's surface. About 10 percent is highland terrain, 70 percent is rolling uplands, and 20 percent is lowland plains. There are two major highland areas; one is about half the size of Africa and located in the equatorial region, and the other, about the size of Australia, is located to the north. The highest mountain on Venus—Maxwell Montes—is located in the northern highlands and is higher than Earth's Mt. Everest. Two major areas of suspected volcanic activity—one larger than Earth's Hawaii-Midway volcanic chain—have been found on Venus and make Venus only the third body in the solar system thought to be volcanically active.

The outer crust of Venus is thought to be thicker than that of the Earth, which averages 20 kilometers (12.5 miles), and below that is a thick layer of basaltic rock. The hot core of Venus is believed to be composed of molten nickel-iron.

EARTH

Earth is the third planet from the sun and the only one in the solar system known to harbor life. From out in space, our planet appears as a bright, blue-and-white sphere. In fact, some 70 percent of Earth's surface is covered by water. At any given time, wispy, white clouds hide about half the planet's surface from view.

Earth is by far the most studied of all the planets. The coming of the space age and the orbiting space satellites provided scientists with powerful tools for gathering data on Earth's atmosphere, surface geology, magnetic field, and so forth. This research continues with such programs as Magsat (see "Magsat Maps Earth's Magnetism," p. 66), while other types of space satellites today monitor weather and other natural processes and provide highly sophisticated communications networks.

Earth's atmosphere is composed of about 78 percent nitrogen, 21 percent oxygen, and traces of other gases. The tenuous outer layer of the atmosphere—the thermosphere—begins about 500 kilometers (310 miles) above Earth's surface. Between about 100 kilometers down to about 50 kilometers (62 to 31 miles) is the mesosphere; below this is the stratosphere (down to about 13 kilometers, or 8 miles); finally, there is the troposphere, the bottom layer. The atmosphere, along with Earth's magnetic field, shields us from nearly all harmful radiation coming from the sun and from outer space.

Earth's interior is believed to be made up of three basic layers. The outer crust, largely made up of granite rock, varies from 40 kilometers (25 miles) deep under the continents to 5 kilometers (3 miles) deep under the oceans. The second layer, the mantle, extends down to about 2,900 kilometers (1,800 miles) below the surface and is composed of molten silicate rock rich in iron. Beneath this lies the Earth's core, composed of iron and nickel that is compacted under great pressure. Scientists believe the temperature at the center of the core could be 4,000°C (7,200°F).

THE MOON

The Moon is Earth's only satellite. Just over one-quarter the size of Earth, it is the brightest object in Earth's nighttime sky and undergoes phases as it revolves about Earth: *new moon*, when the Moon is between Earth and the sun; *full moon*, when Earth is between the Moon and the sun. The Moon rotate once during each revolution, so the same side of the satellite always faces Earth. In fact, the Moon has been found to be slightly egg-shaped, with its Earth-facing side being the elongated small end.

Over a decade of exploration of the Moon by United States and Soviet spaceprobes was capped by the landing of two United States astronauts on the Moon on July 20, 1969. A total of six two-man crews of American astronauts eventually landed on the Moon between 1969 and 1972, and they brought back over 360 kilograms (790 pounds) of samples.

The world these astronauts found was airless, waterless, and devoid of life. Temperatures on the Moon range from up to 134°C (273°F) on the bright side to −170°C (−274°F) on the dark side.

The lunar surface is pockmarked with craters up to more than 90 kilometers (56 miles) across and is broken by huge mountain ranges. The Earth side also has large seas (called *maria*) of solidified lava.

Scientists believe that the interior of the Moon is made up of three layers. The outer layer, the crust, is about 60 kilometers (37 miles) thick and composed of igneous rock that is rich in calcium and aluminum. A thick mantle lies underneath the crust, perhaps to 800 kilometers (497 miles) below the surface and is thought to be made up of a denser rock, perhaps olivine. The composition of the Moon's core is unknown, though scientists believe that it may still be at least partly molten.

The Moon has no magnetic field, but very old lunar rock samples bear traces of magnetism. Some scientists think the Moon may have had a magnetic field when the rock was formed from molten lava.

MARS

Mars is the outermost of the four terrestrial planets and has a distinctive reddish coloring. Martian soil is rich in iron oxide, which gives it this rust color. The Romans named the planet after their god of war, and the two irregularly shaped satellites of Mars have been named after the mythical horses—Deimos (terror) and Phobos (fear)—that pulled the war god's chariot. Mars is visible in Earth's nighttime sky and is lined up with Earth between it and the sun once every 780 days, though its closest approach to Earth (56,000,000 kilometers or 35,000,000 miles) comes at fifteen- or seventeen-year intervals.

The so-called Martian "canals"—later found to be optical illusions—were first observed by nineteenth century astronomers and led to the widespread belief that there was life on Mars. (For example, in 1900 the French Academy offered a prize to the first person to find life on any planet *except Mars*, presumably because everyone knew that there was life on that planet.) The planet thus became the target of numerous spaceprobes, both United States and Soviet.

The first successful flyby of Mars was achieved by the United States spacecraft Mariner 4 in 1965. The Soviets became the first to successfully land a probe on the surface of Mars in 1971, but the probe malfunctioned and stopped transmitting after only twenty seconds. It was not until 1976, when the United States Viking 1 and Viking 2 landers touched down on Mars, that extensive study of the planet from its surface became possible. In fact, the Viking 1 lander continued to function long after its mission had been completed and sent back information on Martian weather until 1982, when communications with Earth at last failed.

The big question of whether there is (or was) life on Mars has yet to be answered with certainty. The Viking landers conducted three experiments on Martian soil to check for biological processes. Some of the tests yielded positive results, but these could also be explained by the planet's soil chemistry. The lack of other evidence of organic molecules adds to the case against life on Mars.

The intense interest in Mars has made it the most studied of all planets except Earth. Orbiting satellites have mapped the entire planet down to a resolution of 150-300 meters (492-984 feet). The planet has ice caps at both poles (frozen carbon dioxide with some water) and the ice caps advance and recede with changes in the Martian seasons.

The planet's surface is heavily cratered, and there is extensive evidence of once-active volcanoes. There are also such spectacular features as Olympus Mons (an extinct volcano three times as high as Earth's Mt. Everest); mammoth canyons, one of which is four times deeper than the Grand Canyon; and a gigantic basin (larger than Alaska) in the Southern Hemisphere that was probably created by a single, huge asteroid.

But the most intriguing aspect of the Martian surface is that it appears that water once flowed there in great quantities. Parts of the terrain apparently have sedimentary origins, and there are many long channels, complete with smaller tributary channels and islands, that extend for hundreds of kilometers. Scientists speculate that Mars once had a much thicker atmosphere, made of gases vented during volcanic eruptions, and this made it possible for water to flow on the planet. Martian atmospheric pressure is now so low, however, that liquid water immediately vaporizes. Apparently the water in the past flowed through the channels to lowland areas and then sank into the Martian regolith, or upper soil layer, since there is no geologic evidence that standing bodies of water ever existed. Astronomers believe that at least some of the primordial water may still be trapped as ice.

The Martian atmosphere, tinted pink by wind-blown dust particles, apparently thinned out after volcanic activity ceased because Mars's gravity is so weak. Atmospheric pressure is now approximately one one-hundredth of that on Earth at sea level, and the predominant gas is carbon dioxide, which is a relatively heavy gas. There is a small amount of water vapor in the atmosphere, enough to form some clouds, small patches of fog in some valleys, and occasionally even patches of frost. Surface temperatures vary from about 25°C (39°F) at the equator to about −123°C (−189°F) at the poles.

By far the most pronounced feature of Martian weather (apart from the bone-chilling cold) are the dust storms. Whipped by hurricane-force winds, these dust storms sometimes engulf the entire planet for weeks on end.

THE OUTER PLANETS

Beyond Mars lie the asteroid belt and the five known outer planets of our solar system. Four of these planets—Jupiter, Saturn, Uranus, and Neptune—are called the Gas Giants. Many times larger than the terrestrial planets, these planets are huge, dense balls of hydrogen and other gases. Beyond them (most of the time) lies the solar system's outermost known planet, Pluto, which may be nothing more than a tiny ball of frozen gases.

The asteroid belt—first crossed by the United States spaceprobe Pioneer 10 on its way to Jupiter in 1973—lies between Mars and Jupiter and has the heaviest concentration of asteroids and small bits of debris (though many other such zones lie within the solar system). Astronomers once thought the asteroid belt had been formed by the breakup of a planet between Mars and Jupiter, but they now think it is debris left over from the formation of the solar system. For some reason, the asteroids did not accrete into a planet.

Gas Giant Planets

	Jupiter	Saturn	Uranus	Neptune	Pluto
Position among planets	Fifth	Sixth	Seventh	Eighth	Ninth
Mean distance from the sun	778,300,000 km 483,300,000 mi. 5.202803 AU	1,472,000,000 km 914,000,000 mi. 9.53884 AU	2,870,000,000 km 1,782,000,000 mi. 19.189 AU	4,497,000,000 km 2,793,000,000 mi. 30.0578 AU	5,900,000,000 km 3,664,000,000 mi. 39.44 AU
Rotation period	9 hr., 55 min., 30 sec.	10 hr., 39 min., 20 sec.	−23.9 hr. (retrograde)	22 hr. (or less)	6 days, 9 hr., 18 min. (retrograde)
Period of revolution	4,332.6 Earth days (11.86 yrs.)	10,759.2 Earth days (29.46 yrs.)	30,685.4 Earth days (84.01 yrs.)	60,189 Earth days (164.1 yrs.)	90,465 Earth days (247.7 yrs.)
Mean orbital velocity	13.06 km/sec. 8.1 mi./sec.	9.64 km/sec. 5.99 mi./sec.	6.8 km/sec. 4.2 mi./sec.	5.4 km/sec. 3.35 mi./sec.	4.7 km/sec. 2.9 mi./sec.
Inclination of axis	3°5'	26°44'	97°55'	28°41'	60°(?)
Inclination of orbit to the ecliptic	1.3°	2.5°	0.8°	1.8°	17.2°
Eccentricity of orbit	0.048	0.056	0.047	0.009	0.25

	Jupiter	Saturn	Uranus	Neptune	Pluto
Equatorial diameter	143,000 km 88,803 mi.	120,000 km 74,520 mi.	51,800 km 32,168 mi.	49,500 km 30,739 mi.	3,500 km (?) 2,173 mi.
Diameter relative to Earth	11.21 times	9.41 times	4.1 times	3.88 times	≈0.27 (?)
Mass	1.899×10^{27} kg 4.187×10^{27} lb.	5.686×10^{26} kg 12.538×10^{26} lb.	8.66×10^{25} kg 19.09×10^{25} lb.	1.030×10^{26} kg 2.271×10^{26} lb.	≈6.6×10^{21} kg ≈14.5×10^{21} lb.
Mass relative to Earth	317.9 times	95.2 times	14.6 times	17.23 times	≈0.0017 times
Mass of sun relative to planet mass (with atmosphere and satellites)	1,047	3,498	22,759	19,332	3,000,000 (?)
Mean density	1.314 g/cm³	0.69 g/cm³	≈1.2 g/cm³	1.8 g/cm³	≈0.7 g/cm³
Gravity (at equator surface)	22.88 m/sec.² 75.06 ft./sec.²	9.05 m/sec.² 29.69 ft./sec.²	7.77 m/sec.² 25.5 ft./sec.²	11 m/sec.² 36 ft./sec.²	≈4.3 m/sec.² ≈14.1 ft./sec.²
Gravity (relative to Earth)	2.34	0.92	0.79	1.12	0.43
Escape velocity at equator	59.5 km/sec. 36.9 mi./sec.	35.6 km/sec. 22.1 mi./sec.	21.2 km/sec. 13.2 mi./sec.	23.6 km/sec. 14.66 mi./sec.	5.3 km/sec. 3.29 mi./sec.

	Jupiter	Saturn	Uranus	Neptune	Pluto
Average temperature	(1-bar level) 165°K	(1-bar level) 140°K	(cloud tops) 58°K	(cloud tops) 43°K	≈0°K
Atmospheric pressure at surface	1 bar (apx.)	1 bar (apx.)	N.A.	N.A.	0.1 millibar (?)
Atmosphere (main components)	(near cloud tops) hydrogen 90%; helium ≈10%	hydrogen 94%; helium ≈6%	hydrogen, helium, methane	hydrogen, helium, methane	tenuous; methane & possibly neon
Other atmospheric gases	methane, water, ammonia, ethane, acetylene, phosphine, hydrogen cyanide, carbon monoxide	methane, ammonia			

JUPITER

Jupiter is the fifth and largest planet in the solar system. It has two and one-half times more mass than all the other planets of the solar system together, and is eleven times greater than Earth in diameter.

In fact, Jupiter is so large that scientists believe it almost became a sun. As the gases and dust contracted to form the planet, gravitational forces created tremendous pressure and temperature inside the core—perhaps as high as tens of thousands of degrees. But there was not enough mass available to create the temperatures needed to start a fusion reaction like that of the sun, which is above 10,000,000°C (18,000,000°F) at its core, and so Jupiter has been slowly cooling down ever since. Today, however, Jupiter still radiates about two and one-half times as much heat as it receives from the sun.

The first object to reach Jupiter from Earth was Pioneer 10. It returned the first close-up pictures of the giant planet in 1973. Subsequently, the more sophisticated spaceprobes Voyager 1 and Voyager 2 passed by Jupiter in 1979 and sent back still more photographs and data on the planet. One of the most exciting discoveries by Voyager 1 was that Jupiter has a faint but extensive ring system that extends almost 300,000 kilometers (186,000 miles) out from the planet's surface. And in 1983 a study of Voyager 2 readings indicated that Jupiter's magnetic field, stretched by solar wind into a long "tail" on the side away from the sun, had actually reached Saturn when Jupiter and Saturn were aligned with the sun (about once every twenty years).

Jupiter's thick atmosphere, which may extend downward by as much as 1,000 kilometers (620 miles), is made up primarily of hydrogen, some helium, and traces of methane, water, and ammonia. Because the planet spins so fast (one rotation in just under ten Earth hours), Jupiter's clouds tend to form bands. They give the planet its red, brownish, and white striped appearance. Bands of clouds at higher altitudes are carried eastward by jet streams, while those at lower levels are blown westward.

There are numerous eddies and swirls in Jupiter's atmosphere, but none can compare with the Great Red Spot, apparently a massive hurricane (rotating counterclockwise) located in the Southern Hemisphere near the equator. The Great Red Spot was first observed some 300 years ago, and this storm continues unabated today. Mathematical analysis based on theories of chaos shows that it should be stable for a long time. Since 1938 three smaller white ovals have been observed to the south of the Great Red Spot.

Jupiter's cloud tops are extremely cold (about –130°C or –202°F), but temperatures increase deeper inside the atmosphere. Pressure also increases and, at about 1,000 kilometers (620 miles) below the outermost atmospheric layers, scientists speculate there are great oceans of liquid hydrogen that form Jupiter's surface. These may be some 20,000 kilometers (12,000 miles) deep. Beneath them the hydrogen is so densely compacted it is in an essentially metallic state. Within this is the planet's core, thought to be an iron and silicate rock ball about the size of Earth.

Jupiter is now known to have sixteen moons, the four largest being the Galilean moons, so called because they were first observed by Galileo. These four moons are named Ganymede, Callisto, Europa, and Io—after the Roman god Jupiter's cupbearer (Ganymede) and three of Jupiter's girlfriends. Ganymede is the largest moon in the solar system and is larger even than the planet Mercury. It is a huge, cratered ball of ice and may have a core of solid silicate rock. Callisto, with an orbit outside that of Ganymede, is also covered with ice and is riddled with thousands of craters. Europa, which orbits inside Ganymede, is about the size of our Moon and has a smooth surface marked by networks of cracks.

The most interesting of Jupiter's moons is Io, which of all four Galilean moon orbits is closest to Jupiter. Voyager 1 actually photographed volcanoes on Io in the process of erupting. Orange-red patches on Io's mottled surface are apparently molten sulfur beds, but most other parts of Io's surface are apparently very cold (about −145°C or −229°F). The volcanic activity is believed the result of the opposing gravitational pulls from Jupiter, Europa, and Ganymede.

SATURN

Saturn is the sixth planet of the solar system and the second largest, after Jupiter. The outermost of the planets that can be identified easily in Earth's nighttime sky with the unaided eye, Saturn has a pale yellowish color and is not nearly so bright as Mars. Saturn's spectacular ring system, which makes it one of the most interesting planets of the solar system, is only visible through a telescope. Saturn's rings are more extensive than those of any other planet.

Like Jupiter, Saturn is composed of densely compacted hydrogen, helium, and other gases. Liquid or metallic hydrogen probably exists underneath the planet's thick atmosphere, and scientists believe there is a solid rocky core (about two times the size of Earth) at its center. Saturn's high rotational speed (once every ten hours, 40 minutes) makes it the most oblate (flattened) of all the planets; it is almost 11,000 kilometers (6,800 miles) wider at the equator than on a line through the poles.

Though exploration of Saturn began in 1979 with the first flyby (Pioneer 11), the Voyager 1 and Voyager 2 missions in 1980 and 1981, respectively, provided the first detailed look at the planet. Scientists will probably spend years sifting through the data and, though there were important new findings, many questions about Saturn remain unanswered.

Voyager found a huge storm thousands of miles across on Saturn, along with a wide band of extremely high winds—up to 1,600 kilometers (994 miles) per hour—at the equator. Winds in this band all travel in the direction of the planet's rotation (unlike bands of wind on Jupiter). The Voyagers also discovered a vast hydrogen cloud circling the planet above the equator.

Though Voyager data did discount one theory as to why Saturn radiates nearly three times as much energy as it absorbs from the sun, it did not provide any new clues as to the planet's heat source.

Planetary Rings

	Distance from Planet Surface at Equator		
	Inside Edge	Mean	Outside Edge
Jupiter			
Secondary Ring 1	0 km		51,800 km
			32,200 mi.
Primary Ring	51,800 km		58,200 km
	32,200 mi.		36,100 mi.
Secondary Ring 2	179,800 km		287,100 km
	111,700 mi.		178,300 mi.
Saturn			
D ring			12,600 km
			7,800 mi.
Guerin division	12,600 km		13,800 km
	7,800 mi.		8,600 mi.
C ring	13,800 km		31,800 km
	8,600 mi.		19,700 mi.
B ring	31,800 km		55,800 km
	19,700 mi.		34,700 mi.
Cassini	55,800 km		60,600 km
	34,700 mi.		37,600 mi.
A ring	60,600 km		76,200 km
	37,600 mi.		47,300 mi.
Encke division		72,600 km	
		45,100 mi.	
F ring		81,000 km	
		50,000 mi.	
G ring		90,000 km	
		55,900 mi.	
E ring	180,000 km		420,000 km
	111,800 mi.		260,800 mi.
Uranus			
Ring 6		15,755 km	
		9,784 mi.	
Ring 5		16,155 km	
		10,032 mi.	
Ring 4		16,455 km	
		10,218 mi.	
Ring alpha		18,655 km	
		11,585 mi.	
Ring beta		19,555 km	
		12,144 mi.	
Ring zeta		21,055 km	
		13,075 mi.	
Ring gamma		21,555 km	
		13,386 mi.	
Ring delta		22,155 km	
		13,758 mi.	
Ring epsilon		25,055 km	
		15,559 mi.	

Voyager's most exciting discoveries concern the planetary rings. Previously, about six different rings had been identified within the ring system, but Voyager 1 pictures show as many as a thousand separate rings. Narrow rings can even be seen within the Cassini Division, once thought to be an empty gap between the two major parts of the ring system. Even more fascinating is the fact that some rings are not circular, and that at least two rings are intertwined, or "braided." The sheer complexity of this ring system means that something more than just the gravitational effects of Saturn's moons must be responsible for its creation.

A strange new phenomenon was also discovered in the rings. Voyager 1 pictures clearly show dark, radial fingers—"spokes"—moving inside the rings in the direction of rotation. Scientists speculate that they are made of ice crystals.

Twelve moons of Saturn were known before the arrival of Voyager, and instruments aboard the spaceprobe helped locate three new ones. Most of Saturn's moons are relatively small and composed of rock and ice. All but one of the small moons are pockmarked by meteor craters and, in some cases, the moons appear to have been cracked by collisions with especially large meteors.

But Voyager pictures show that one moon, Enceladus, is smooth in large regions that are apparently unmarked by collisions with meteors. This, scientists believe, is due to the fact that Enceladus is being pulled and stretched by the combined gravities of a nearby moon and Saturn itself. The tidal forces have consequently heated the core of Enceladus and made its surface soft enough to smooth over any craters formed by meteor impacts.

Titan, Saturn's largest moon (5,150 kilometers, or 3,190 miles, in diameter) is currently of greatest interest to scientists. That is because it is the only moon in the solar system known to have an atmosphere of any substance. Furthermore, scientists suspect that at least some precursors of life may have formed there. As a result, Voyager 1 was guided to within about 4,000 kilometers (2,500 miles) of the moon during the Saturn flyby.

Though the surface of Titan was obscured by dense clouds, Voyager's sensors nevertheless returned a considerable amount of information about the moon and its atmosphere. Titan's atmosphere is composed mostly of nitrogen, like that of Earth, with only a small percentage of methane. (Carbon monoxide has since been discovered there as well. See 1983 Science Update, Titan's atmosphere). Atmospheric pressure is at least 1.5 times that on Earth and temperatures range around −181°C (−294°F). Titan, in fact, appears to be a frozen version of Earth before life evolved.

The possibility of oceans of liquid methane (or of nitrogen or methane rain) on Titan was a matter of considerable controversy for some time after Voyager 1 investigated the moon. But an analysis of Voyager data published in 1983 seems to indicate that Titan is "dry," at least in the regions investigated. Pools of liquid methane might still exist in other low-lying regions, but the study concludes that it is unlikely that either methane or nitrogen condenses to liquid form on Titan.

Though mechanical problems marred the Voyager 2 flyby in 1981, this mission still made a number of important discoveries. For example, Voyager 2 pictures show that Saturn has far more than a thousand rings—perhaps as many as a hundred thousand or more. One of the brightest rings is shown to be under 152 meters (500 feet) thick. Also the pictures do not show signs of moonlets that some scientists had thought were the cause of gaps between individual rings.

In addition to making further observations of Saturn's moons, Voyager 2 also found seasonal differences between the planet's two hemispheres and photographed a storm some 6,500 kilometers (4,000 miles) wide.

URANUS

Uranus is the seventh planet in the solar system and is the third of the so-called Gas Giants. The planet is barely visible in Earth's nighttime sky (it looks like a faint star) and for that reason it went undiscovered until 1781. Our knowledge of the planet is still limited, though scientists expect to learn a great deal when Voyager 2 passes by Uranus in 1986.

Nearly the same size as Neptune and only about 5 percent of Jupiter's mass, Uranus is a faintly greenish color, perhaps because its atmosphere contains methane. The planet's axis of rotation is tipped over on its side, and in 1977 a system of nine faint rings was discovered. There are five known moons of Uranus and they range between about 300 kilometers (186 miles) to 1,000 kilometers (621 miles) in diameter.

There have been hints in the past that Uranus may have a magnetic field. In 1983 a study of hydrogen Lyman alpha emissions from the planet indicated the possibility of an aurora in Uranus's atmosphere. This in turn means that Uranus may well have a magnetic field, since an aurora is associated with a planet's magnetosphere as on Earth.

The atmosphere of Uranus is composed of hydrogen, helium, and methane and is very clear and cold (−215°C or −355°F). No clouds and little haze have been observed in the atmosphere.

Scientists speculate that, as on Jupiter and Saturn, temperatures and pressures increase dramatically down through the outer layer of atmosphere. At some point the hydrogen and helium is sufficiently compressed to form a liquid or slushy surface "crust." Underneath this crust they believe is a mantle of solidified methane, ammonia, and water. Inside this mantle is a rocky core of silicon and iron that is about fifteen times as massive as Earth. The core is hot, probably about 7,000°Kelvin (12,000°F).

NEPTUNE

Neptune, the last of the Gas Giants, is the eighth planet in the solar system. It was discovered in 1846 after mathematical calculations based on irregularities in the orbit of Uranus provided astronomers with the correct location of the as yet undiscovered planet. Neptune, like Uranus, is surrounded by considerable uncertainty because of its enormous distance from the Earth. Neptune has never been visited by spacecraft from Earth, though Voyager 2 is scheduled to reach the planet in 1989.

Neptune is a pale bluish color and its atmosphere is composed of hydrogen and helium. Unlike Uranus, Neptune's atmosphere is often hazy—perhaps because of ice or other particles formed from an unknown substance—and it is very cold at the cloud tops (about −230°C or −328°F).

Scientists believe that Neptune has a three-layered structure similar to that of Uranus: a crust of solidified or liquid hydrogen and helium that gradually thins outward into an atmosphere; a mantle of solidified gases and water; and a hot, rocky core (about 7,000°Kelvin or 12,000°F) some fifteen times as massive as Earth. But one aspect of Neptune remains a mystery. Despite similarities with Uranus, Neptune has been found to radiate more heat than it receives from the sun (at a rate of 0.03 microwatts per ton of mass). Uranus, on the other hand, does not emit excess heat.

Neptune has two known moons (Triton and Nereid) and may also have a faint ring system. Triton is the largest moon and may have an atmosphere. Its orbit around Neptune is decaying, however, and sometime between 10,000,000 and 100,000,000 years from now it will come close enough to the planet to be torn apart by gravitational forces.

PLUTO

The ninth and outermost (most of the time) known planet of the solar system is Pluto, a ball of frozen gases that is probably only about the size of Earth's Moon. Because of its relatively small size and highly elliptical orbit (which at times crosses inside Neptune's orbit), some scientists think that Pluto is not really a planet at all. Instead, they theorize that Pluto is only a former moon of Neptune that has been pulled out of orbit by some other celestial body. This is one of the pieces of evidence that suggest that there may be an unknown tenth planet.

Pluto was discovered in 1930 as a result of an extensive search by Clyde Tombaugh. Although his search was based upon mathematical calculations derived from deviations in the orbit of Neptune, most astronomers now think that Tombaugh found Pluto as a result of a lucky accident. Pluto cannot, in fact, account for the deviations in Neptune's orbit by itself. This is another reason to suspect that there may be a tenth planet.

Frozen methane and a thin atmosphere of methane and some other gas have been detected on Pluto. Pluto has one known moon, Charon, which was discovered in 1978. Charon is about one third the size of Pluto. Pluto and Charon rotate and revolve synchronously like a double planet system.

NEMESIS

Scientists are also looking for a star that orbits the sun. The star, if it exists, has been named *Nemesis* since it is believed to cause periodic mass extinctions on Earth (see p. 272).

Sun

Position in solar system:	Center	Diameter relative to Earth:	109 times
Mean distance from Earth:	149,600,000 km (92,960,000 mi.)	Mass:	2×10^{27} metric tons
Distance from center of Milky Way Galaxy:	8,500 parsecs (27,710 light-years)	Mass relative to Earth:	333,000 times
		Mean density:	1.4 g/cm³
Estimated velocity of revolution:	282 km/sec. (175 mi./sec.)	Gravity relative to Earth:	27.8 times
		Temperature at core:	20,000,000°K
Period of rotation:	27 days (?)	Temperature at surface:	5,700°K
Inclination (relative to Earth orbit):	7°	Main components:	Hydrogen and helium
Equatorial diameter:	1.392×10^6 km (865,000 mi.)	Expected life of hydrogen fuel supply:	5,000,000,000 more years

Principal Planetary Satellites

Satellite	Diameter	Mean Distance from Planet	Mass	Sidereal Period (days)	Orbital Eccentricity
Earth					
Moon	3,476 km 2,160 mi.	384,400 km 238,860 mi.	7.38×10^{22} kg 1.6×10^{22} lb.	27.3	0.0555
Mars					
Phobos	23 km 13.7 mi.	9,380 km 5,825 mi.	9.6×10^{15} kg 21.2×10^{15} lb.	0.319	0.018
Deimos	12 km 7.4 mi.	23,500 km 14,593 mi.	2×10^{15} kg 4.4×10^{15} lb.	1.262	0.002

Satellite	Diameter	Mean Distance from Planet	Mass	Sidereal Period (days)	Orbital Eccentricity
Jupiter					
Amalthea	240 km	181,300 km	N.A.	0.489	0.003
	149 mi.	112,587 mi.			
Io	3,632 km	412,600 km	8.92×10^{22} kg	1.769	0.0
	2,255 mi.	256,225 mi.	19.67×10^{22} lb.		
Europa	3,126 km	670,900 km	4.87×10^{22} kg	3.551	0.0
	1,941 mi.	416,629 mi.	10.74×10^{22} lb.		
Ganymede	5,276 km	1,070,000 km	1.49×10^{23} kg	7.155	0.001
	3,276 mi.	664,470 mi.	3.28×10^{23} lb.		
Callisto	4,820 km	1,880,000 km	1.06×10^{23} kg	16.69	0.007
	2,993 mi.	1,167,480 mi.	2.34×10^{23} lb.		
Leda	2–14 km	11,110,000 km	N.A.	240	0.147
	1.2–8.7 mi.	6,899,000 mi.			
Himalia	170 km	11,470,000 km	N.A.	250.6	0.158
	106 mi.	7,123,000 mi.			
Lysithea	6–32 km	11,710,000 km	N.A.	260	0.12
	3.8–20 mi.	7,272,000 mi.			
Elara	80 km	11,740,000 km	N.A.	260.1	0.207
	49.7 mi.	7,290,000 mi.			
Ananke	6–28 km	20,700,000 km	N.A.	617.	0.169
	3.7–17.4 mi.	12,855,000 mi.			
Carme	8–40 km	22,350,000 km	N.A.	692	0.207
	5.0–24.8 mi.	13,879,000 mi.			
Pasiphae	8–46 km	23,300,000 km	N.A.	735	0.40
	5.0–28.6 mi.	14,469,000 mi.			
Sinope	6–36 km	23,700,000 km	N.A.	758	0.275
	3.7–22.4 mi.	14,718,000 mi.			
Saturn					
Mimas	390 km	185,000 km	3.76×10^{19} kg	0.942	0.02
	242 mi.	115,000 mi.	8.29×10^{19} lb.		
Enceladus	500 km	238,000 km	7.4×10^{19} kg	1.37	0.004
	310 mi.	148,000 mi.	16.32×10^{19} lb.		
Tethys	1,050 km	295,000 km	6.26×10^{20} kg	1.89	0.0
	652 mi.	183,000 mi.	13.8×10^{20} lb.		
Dione	1,120 km	377,000 km	1.05×10^{21} kg	2.74	0.002
	695 mi.	234,000 mi.	2.3×10^{21} lb.		
Rhea	1,530 km	527,000 km	2.28×10^{21} kg	4.52	0.001
	950 mi.	327,000 mi.	5.03×10^{21} lb.		
Titan	5,150 km	1,222,000 km	1.36×10^{23} kg	15.94	0.029
	3,190 mi.	759,000 mi.	3.0×10^{23} lb.		

Satellite	Diameter	Mean Distance from Planet	Mass	Sidereal Period (days)	Orbital Eccentricity
Hyperion	290 km	1,481,000 km	1.10×10^{20} kg	21.28	0.104
	180 mi.	920,000 mi.	2.42×10^{20} lb.		
Iapetus	1,440 km	3,560,000 km	1.93×10^{21} kg	79.33	0.028
	894 mi.	2,211,000 mi.	4.26×10^{21} lb.		
Phoebe	140 km	12,930,000 km	N.A.	550.4	0.163
	87 mi.	8,030,000 mi.			

Uranus
Miranda	320 km	130,000 km	3.4×10^{19} kg	1.4	0.000
	199 mi.	81,000 mi.	7.5×10^{19} lb.		
Ariel	860 km	192,000 km	6.7×10^{20} kg	2.5	0.003
	534 mi.	119,000 mi.	14.78×10^{20} lb.		
Umbriel	900 km	267,000 km	7.6×10^{20} kg	4.52	0.001
	559 mi.	166,000 mi.	16.76×10^{20} lb.		
Titania	1,040 km	438,000 km	1.2×10^{21} kg	8.7	0.002
	646 mi.	272,000 mi.	2.65×10^{21} lb.		
Oberon	920 km	586,000 km	8.2×10^{21} kg	13.5	0.001
	571 mi.	364,000 mi.	18.08×10^{21} lb.		

Neptune
Triton	≈3,600 km	355,000 km	5.7×10^{22} kg	5.9	0.00
	≈2,236 mi.	220,000 mi.	1.74×10^{22} lb.		
Nereid	940 km	5,562,000 km	1.3×10^{15} kg	359.8	0.75
	584 mi.	3,454,000 mi.	2.87×10^{15} mi.		

Pluto
Charon	≈800 km(?)	17,000 km	N.A.	6.4	N.A.
	497 mi.	10,557 mi.			

THE FIRST INTERPLANETARY MISSION

Typical of the problems plaguing the space program in the early days was the fate of *Mariner 1*, the American probe that did not make it to Venus. When the computer aboard the Atlas-Agena was programmed originally, the programmer inadvertently missed putting in a bar above the letter R in one spot in the program. When the computer could not find the crucial command, the rocket went haywire. For want of the bar, *Mariner 1* now lies rusting at the bottom of the Atlantic Ocean.

BIOLOGY

Biology is, of course, the study of life. At its most fundamental level, the study of life becomes the study of organic molecules. If we can understand what is happening at the molecular level, then we can understand the behavior of the organism—or so many biologists believe. While not everyone agrees with this dictum, molecular biologists can point to physics as a case in point. Understanding of the behavior of solids, liquids, and gases did not truly come until physicists began to develop models based on the motions of and forces between particles. Thus, in today's biology, work is proceeding on the molecular basis of learning, behavior, growth, development, and all other life processes.

Molecular biology, however, is having its most immediate influence on medicine. Biology and medicine have, since the nineteenth century at least, often pursued similar goals and complemented each other in vital ways. The problems of human health have often directed the attention of biologists to areas of fundamental concern; the basic understandings developed by biologists have led to better vaccines, improved drugs, and new approaches to treatment.

Seldom has this relationship been closer than in recent research on genetics and cancer. Cancer has turned out to be intimately involved with the basic biochemical reactions that govern growth and development in all living organisms. The revelation of how certain genes, believed to be responsible for normal growth processes in the cell, are changed to become cancer genes has been the big story in both biology and medicine of the early 1980s. Here, the story is told from the biologist's point of view, since we have so far gained much understanding but are still far from practical applications. One can only hope that the next edition of *The Science Almanac* will have good reason to report on the continuation of the story in the Medicine section.

Of course, there is more to biology than the investigation of genes. More traditional research produced evidence of ecosystems under the ground that were previously unknown, of new forms of predation, and of communication between trees. Two particularly fruitful areas of research concerned the process of learning and the biological differences between masculine and feminine behavior.

Names in the biological news included pandas, bowerbirds, cheetahs, alligators, saddleback wrasse, and skunk cabbage, although horsefly larvae and female fireflies had the stories that produced the most general interest. Another organism that was studied extensively—as always—was the human being. Studies of humans tended to revolve around hormones and related chemical messengers.

Almost every week, however, brought forth some new gene-connected item. First, the promise of the recombinant DNA technique, popularly known as gene splicing, began to come true. In fact, 1980 began with the announcement that genes producing a human protein (interferon) had successfully been transferred to a bacterium where they were functioning. This success was duplicated many times over, with commercial production of another human protein (insulin) started in the early 1980s and many other "gene-spliced" products on the way.

Not everyone approved of what was happening. Author Jeremy Rifkin led an attack on gene splicing, obtaining signatures from many prominent individuals on petitions to limit the practice and also leading a legal drive that stopped a planned field experiment with genetically transformed bacteria. Despite these concerns, most biologists felt that recombinant DNA techniques were not dangerous and were likely to be immensely helpful in the future.

Work with genes proceeded so rapidly that it became possible to identify genes more easily than the proteins formed by the genes. When the first protein's structure was determined (insulin by Frederick Sanger in 1954), it had taken ten years of patient labor by a large team. Now, starting from the gene for a protein, the sequence of a protein can often be found in weeks.

One other important advance will undoubtedly affect the biological and medical fields immensely in the future: the development of monoclonal antibodies produced by hybridomas. Hybridomas are cell lines that combine a cancer cell with a cell that normally produces antibodies. The result of the combination is a line of cells that produces an unlimited supply of antibodies of a specific type, called monoclonal antibodies. For basic biological research, monoclonal antibodies offer a means to identifying single proteins wherever they appear in the body, an enormously useful tool. Medicine has already begun to apply monoclonal antibodies to improved diagnostic tests and also to treatments for cancer.

1980

Gene Splicing Used to Produce Interferon

Interferons, a group of proteins produced by several kinds of cells, fascinate scientists because they appear to kill, or at least debilitate, viruses. They may also help cancer patients fight their disease. But obtaining interferon of any type has been costly, in part because interferons are species specific; mouse interferons do not work in humans and vice versa. Also, the supply of human interferons, from white blood cells, has been extremely limited, with annual production being only enough to treat about 600 cancer patients. In addition, the natural supply is a crude mixture of different types of interferon of varying effectiveness.

Charles Weissmann of the University of Geneva announced a breakthrough in 1980. His group was producing pure amounts of one form of interferon. Through complex and tedious procedures, genetic material carrying the instructions for a human interferon had been isolated and inserted into laboratory bacteria. These bacteria and their offspring now produce interferon that can kill viruses and protect cultured human cells. (For a further discussion, relating to applications mainly, see Medicine, page 384.)

Interferon is only one of several amino acid and human protein products that were first made in 1980 by bacteria as a result of gene-splicing techniques. The other proteins include thymosin alpha-1, a hormone that stimulates the human immune system; insulin, a hormone that regulates blood sugar level; proline, an amino acid; urokinase, a kidney enzyme that can dissolve blood clots in the lungs; human growth hormone; and beta-endorphin, a brain protein.

SAVE THOSE GENES!

Scientists warn that species extinction and the increasing reduction of the variability of existing species may pose a problem for future generations. Genetic variability provides us with the possibility of breeding organisms for our future needs for fiber, food, medicine, and energy. Small, easily grown and stored organisms such as bacteria, fungi, and algae, as well as some plants, are being saved in laboratories or seed banks. But large organisms, especially animals, are poorly represented in programs aimed at maintaining genetic diversity. Numerous breeds of chickens, four breeds of sheep, two breeds each of cows and draft horses, and a breed of hog are already near extinction.

Genes from One Mouse Function in Another

A dream of science is to be able to cure genetic diseases, such as sickle-cell anemia, hemophilia, and Tay-Sachs disease, by replacing nonfunctioning genes with functioning genes that operate normally within the body of the affected person. The first successful step toward that end was taken in 1980 at the University of California at Los Angeles, when Martin Cline and his co-workers successfully transferred a gene from one mouse to another.

Genetic material for resistance to methotrexate, a drug used in cancer therapy, was taken from bone marrow cells of a resistant strain of mice. The genetic material was then mixed with cells from the bone marrow of nonresistant mice and the cells were incubated. Under these conditions, some of the cells can be expected to take up the added genetic material. The incubated cells were then replaced into the bone marrow of the nonresistant mice and allowed to multiply. Finally, the mice were exposed to high, prolonged doses of methotrexate. They survived with few ill effects.

For a gene transplant to be successful, the genes must get into the cells in which they normally function and actually operate. So far the focus of research has been on gene transplants into types of cells that divide throughout a person's lifetime, such as blood tissues, because these tissues are easy to remove from a person's body and replace. But even with the limitation on types of cells used, the potential is tremendous. Gene replacement might be used with cancer patients undergoing chemotherapy to increase their drug tolerance and with people suffering from a variety of inherited blood disorders. The same research group is already at work on sickle-cell anemia; they are trying to transfer the human beta globin gene, which is malfunctional in sickle-cell, into mice. (See the other genetic approach to treating genetic blood diseases in Medicine, page 442).)

Later in the year Dr. Cline tried a related, but essentially different, method of gene insert in human beings in an attempt to cure inherited blood diseases. Because he had not obtained proper permission for the treatment, the treatment was broken off while still incomplete. It did not seem to be working in any case.

In experiments related to the mouse gene transfer, scientists, using hair-fine glass needles, inserted genes from the *Herpes simplex* virus into the nuclei of fertilized mouse eggs. Two of the 78 newborns were found to have incorporated the gene. Although these mice were sacrificed to do the DNA analysis, further tests were planned to see if incorporated foreign genes will actually function in new individuals. (See page 157 for further progress.)

Human Hybridomas Produced

Antibodies protect the body against invasion by "foreign" proteins, but they have other important uses as well in both research and treatment of disease. A supply of antibodies that is specific to a given protein can be used to seek out that particular protein wherever it is. If the antibody has been "tagged" with radiation or fluorescent compounds, the location of the protein can be determined. Similarly, the antibody can be combined with a poison that will destroy the protein. But to do this, you need a supply of pure antibodies specific to the protein.

One way to produce a supply of pure antibodies is to take a cell that produces the antibodies and fuse it with a cancerous cell. Most cells will not reproduce by dividing more than a few times, if at all, when they are grown in a culture medium. Cancer cells, on the other hand, because something has gone wrong with their regulatory system, continue reproducing forever, in what are called "immortal" cell lines. The idea behind fusing the two cells is to retain both the ability to produce antibodies and the ability to form immortal cell lines. In fact, the method works. The fused cells, which are called *hybridomas*, will continue dividing, and it is possible to collect the antibodies produced. This feat was first accomplished in 1978 with mouse cells.

In 1980 Henry S. Kaplan and Lennart Olsson of Stanford University reported having produced a hybridoma of human cells. They fused spleen cells that could produce antibodies against 2-dinitrochlorobenzene (DNCB) and cancerous bone marrow cells. Antibodies to DNCB were produced by some of the hybridomas formed and by the descendants of these cells. The hybridomas are separated into clones, with the cloned cell line producing just one kind of antibody. Hence, the resulting antibodies have been called *monoclonal*.

Using human hybridomas to produce monoclonal antibodies was still in its infancy in 1980. One problem was that one of the kinds of fused cells must have been exposed while in a living organism to the agent that stimulates the production of the desired antibody. To eliminate this problem, scientists were working on ways to expose cells *in vitro* (literally, "in glass"; that is, outside of living organisms) to foreign substances in the hope that some cells would begin producing the desired antibodies and could be used in making the hybridomas.

Because the technique was so promising, many researchers flocked to it in the early 1980s. By 1984, human hybridomas were no longer at the fringes of research in biology, but close to its center.

Cell Structure and Function

Because most cells are very tiny—although a hen's egg is just one large cell and some human nerve cells are about a meter (3 feet) long—they usually cannot be observed at all without a microscope. As microscopes have been improved, so has our knowledge of cell structure grown.

CELL MEMBRANE
Forms boundary of cell; monitors the flow of materials into and out of the cell; protein receptor sites on its surface combine with hormones and other regulatory molecules that affect cell function.

CYTOPLASM
Watery fluid filling cell's interior in which chemical reactions occur and many materials, both raw and waste, are dissolved; cell organelles "float" in this medium.

NUCLEUS (EXCEPT IN MONERANS)
Dense spherical structure that contains chromosomes (DNA plus protein); because the instructions for directing all cell activities are encoded in the DNA, the nucleus is often called the cell's control center; in monerans, DNA floats in the cytoplasm.

NUCLEOLUS
Spherical structure within the nucleus that is involved in the production of ribosomes.

MITOCHONDRIA
Powerhouses of the cell; nutrients—primarily glucose—are broken down here and their stored energy is released; released energy is distributed throughout the cell by an energy-carrier molecule, ATP, to where it is needed; wastes—carbon dioxide and water—are released into the cytoplasm.

ENDOPLASMIC RETICULUM (ER)
Ribbon-like series of canals that begins at the membrane surrounding the nucleus and runs throughout the cell; some ER, called rough ER, has ribosomes on its outer surface and seem to be involved with the manufacture of proteins that are exported from the cell; smooth ER (that without ribosomes) seems to be involved with the manufacture of lipids and with the detoxification of potentially harmful substances.

GOLGI APPARATUS OR GOLGI BODY

A series of flattened sac-like structures that package proteins that are exported from the cell; proteins are concentrated and surrounded by a portion of the membrane of the Golgi apparatus; this pouch pinches off, moves to the cell membrane, fuses with it, and releases its contents to the outside.

RIBOSOMES

Granular bodies involved in the production of proteins; may be free-floating in the cytoplasm or attached to the endoplasmic reticulum.

LYSOSOMES

Pouches of strong digestive enzymes thought to be formed by the Golgi apparatus; they can break down all cell structures and, in white blood cells, seem to play a role in destroying invading organisms such as bacteria.

MICROTUBULES

Protein arranged in a helical fashion that plays a role in maintaining the shape of a cell.

MICROFILAMENTS

Fine protein threads found in most cells that play a role in maintaining cell shape and, in the case of cells that move by crawling, in producing motion.

CENTRIOLES (ANIMAL CELLS)

Formed from microtubules; they play a role in cell division in animals and may be necessary for the formation of spindle fibers, strands to which chromosomes attach when cells divide.

CILIA

Structures for motion formed by microtubules; their movement is described as being a rowing motion.

FLAGELLA

Structures for motion formed by microtubules; their motion is described as whip-like; they tend to be longer than cilia.

CHLOROPLASTS (GREEN PLANT CELLS)

Structures containing the green pigment chlorophyll; this pigment can trap energy from sunlight; the trapped energy is used to convert carbon dioxide and water into simple sugars.

CELL WALL (PLANT, FUNGI, AND MONERAN CELLS)

Material laid down by the cell outside the cell membrane; it gives rigidity and protection to the cell.

1981

Automated Production of Genes

The laboratory production of specific synthetic nucleic acid chains or genes has been a precise and painstaking process. And the end product has demanded a high price: upward of $10,000 per microgram ($\frac{1}{100000000}$ gram or $\frac{1}{35000000000}$ ounce). In 1981, this process saw a big change. Bio Logicals of Toronto announced a machine that automatically assembles DNA or RNA sequences to order. The heart of the machine consists of a computer that determines the sequence of nucleotides in the synthetic nucleic acid, a column of solid particles through which the chemicals used flow, and bottles containing the raw materials, which consist of the different nucleotides and chemical reagents.

Although this machine was not the first gene synthesizer, it was faster and less expensive (under $20,000) than the other existing machine. It was also advertised as so simple to use that nonchemists with only a short training period would be able to supervise its operation. However, the gene splicer from Bio Logicals turned out to have a high failure rate, and the machine ultimately failed to live up to its promises. In the meantime, other companies introduced in the early 1980s several improved machines to do the same thing. By 1984 the field of gene-splicing machines was a thriving one.

The gene synthesizers produced immediate benefits by reducing the price of laboratory-produced genetic material. These fragments of genetic material are used to locate a specific gene in a cell and as such provide much information for scientists.

Yeast-Produced Interferon

A year after an interferon gene was successfully transplanted into laboratory bacteria, a new step was taken. In 1981 an interferon gene was successfully transferred to a yeast. Although both yeast and bacteria produce about 200,000 molecules of interferon per cell, there are some advantages to harvesting interferon from yeast. Because of its role in brewing, the large-scale culture of yeast is well understood and techniques can be applied to commercial drug production. Also, the yield of 200,000 molecules per yeast cell was seen early in the research. Because this production level was reached in bacteria after only a year of research, more study of yeast may yield even higher levels of production.

Fish and Mice Cloned

The word *clone* gained a certain amount of currency in the late 1970s and early 1980s, partly as the result of a notorious book that claimed the existence of a human clone. But the concept of a clone is not well understood by the general public. A clone of an individual living thing is another living thing with exactly the same genes. For all traits governed by the expression of genes, the original individual and the clone would be identical. Simple one-celled creatures may form populations of clones by cell division, and cloning plants is a regular part of agriculture (for example, all varieties of apples, such as the McIntosh, are clones). Cloning higher animals, such as fish, amphibians, or mammals, is more difficult.

Chinese scientists in 1980 reported that they had cloned a golden carp by implanting nuclei from carp embryos into unfertilized carp eggs with their nuclei removed. Although 189 transplants were produced, only two developed into juvenile fish. This was the first recorded instance of cloning a fish. In 1981 Karl Illmensee of the University of Geneva also successfully produced clones by nuclear transplant. He transferred two nuclei from one mouse embryo into two eggs with their own nuclei destroyed. From another embryo, he successfully transplanted three nuclei into three egg cells. The identical twins and identical triplets were the first instance of cloned mammals being produced. Previous attempts with this technique in mice, called nuclear transplant, resulted in only a single live offspring from one embryo, and a single individual by itself cannot be called a clone.

Unfortunately, Illmensee's work was called into question by some of his co-workers. Although investigators found no fraud, they revealed such poor laboratory practices that they called for other experimenters to repeat Illmensee's work with greater care before his results could be accepted.

An undisputed cloning advance was achieved later in 1981 by an American biologist from the University of Oregon, who succeeded in cloning a familiar aquarium fish by methods that appear to be more productive in the long run, especially for research.

Fruit flies and bacteria have previously been the favorite organisms of most genetic researchers because these organisms have short lifespans and are also so small that millions can be raised in a limited space. Now, for certain kinds of genetic research, cloned zebra fish can be added to the list of useful research animals. Although somewhat larger than fruit flies and much, much larger than bacteria, it is easy to keep a tankful of hundreds of zebra fish on a laboratory shelf.

This new breed of fish consists of fish that are not only clones, but also homozygous; that is, both genes in each pair are identical. The result is that, once isolated and duplicated, recessive mutations are expressed and can be studied. In addition, variation of expression of the same genes in different individuals can be studied more easily.

The zebra fish clones were created by manipulating fertilization and egg cell division. Eggs are activated by infertile sperm; then the egg cells are prevented from dividing during the first duplication of their single set of genetic instructions. After duplication of genetic material, the eggs are allowed to divide and develop normally, but these eggs now contain two identical sets of the female's chromosomes so both genes for every trait are the same. When the female develops from these homozygous eggs, her eggs can then be manipulated in the same way to form a clone of up to 200 fish—all of which are genetically identical. Although sex determination in zebra fish is not well understood, males can be produced by treating some of the eggs with male hormones beginning the day after fertilization and continuing for two weeks. In addition, some clones, for reasons not understood, naturally develop into males. If some of the 200 eggs from a homozygous female are hormonally treated and others are left untreated, a breeding clone is produced that will spawn offspring that are genetically identical.

While the potential of cloned fish and mice is great for genetic research, other areas may benefit as well. Cloning could also improve the quality and uniformity of domestically grown food fish. Cloning of mice could lead to the development of cloned domestic animals, such as cattle, with superior genetic endowments.

However, there is a serious potential problem in reliance on cloned sources of food. Even today, with noncloned but genetically uniform food crops, a pest or a disease can sweep through fields attacking every plant or animal. When there is genetic diversity, some plants or animals have resistance that spares them.

THE MYSTERY OF MORELS

The morel is a wild mushroom fruit, related somewhat to the truffle although it grows above ground, that is prized by cooks and gourmets in both Europe and North America. For hundreds of years mushroom fanciers have been trying to produce cultivated morels with no success. In 1981, Ronald Ower discovered the secret. He is keeping it a secret, however, except for telling a small group of scientists who confirmed his feat. All he will admit about his method is something that morel fanciers already knew: The secret is "something that happens only in spring."

Effects of Fetal Exposure to Male Hormones

Is differing behavior in men and women environmental or is there some physical origin? Evidence collected by observing children whose mothers had taken hormones during pregnancy, reported in 1981, suggested that some differences were caused by the effect of those hormones on the fetus. Although observed behavioral differences between boys and girls are often subjective, another 1981 report of laboratory research on a nonhuman primate, which relied on measurements that were objective, tended to substantiate the work on humans.

Robert Goy, working at the Wisconsin Regional Primate Research Center, injected pregnant female rhesus monkeys with one of two male hormones: testosterone propionate (T) or dihydrotestosterone propionate (dhT). Of the offspring, seventeen were males and 36 females; sixteen of the males and seven of the females were neutered within three months of birth. By neutering some of the subjects, Goy had a group in which the effects of current hormones, as opposed to fetal exposure, would be minimized. Goy placed the infants and mothers in social groups of four to six monkeys that included untreated control infants that had been born within the same time period as the treated infants.

When the infants were three months old, observations began and continued until the offspring were about a year old. The researchers focused on mounting behavior, which is rare among female rhesus monkeys, and rough play. During the observation period, the researchers found that untreated females mounted peers an average of five times, female offspring of T-treated mothers mounted an average of 40 times, and female offspring of dhT-treated mothers mounted peers an average of 50 times. Untreated males were observed mounting peers an average of 110 times.

Another behavior observed was rhesus young mounting their mothers. Again, this is rare in female young, but the females who had been exposed to male hormones engaged in it about twenty times as much as the untreated females and nearly twice as much as the untreated males.

The results for frequency of rough play were T/dhT and neutered males, 95 times; control males, 90 times; T females, 61 times; dhT females, 55 times; control females, 25; neutered females, 14.

These observations not only provide evidence for the role of testosterone propionate and dihydrotestosterone propionate in accentuating mounting behavior and rough play in rhesus monkeys, but also suggest that exposure to hormones in the womb is important in determining personality later in life.

The Main Hormones

Hormones are proteins that are released, largely by the glands, into the body at large. They cause changes in metabolism, mood, growth, and other body processes. An important part of the current picture in biology is the recognition of new hormones and the increased understanding of new roles for hormones that have long been known. Here is a current list of recognized hormones found in vertebrates. Although hormones are slightly different from species to species, the main hormones in vertebrates, and especially in mammals, are recognizably similar in structure and function.

Principal Vertebrate Hormones

Hormone	Associated gland(s)	Function
Growth hormone (somatropin or somatrophic hormone or STP)	Anterior pituitary	Stimulates bone and muscle growth
Adrenocorticotropic hormone (ACTH)	Anterior pituitary	Stimulates adrenal cortex
Thyroid-stimulating hormone (TSH)	Anterior pituitary	Stimulates thyroid gland to produce and secrete thyroxin
Prolactin (lactogenic hormone or LTH)	Anterior pituitary	Stimulates milk production (lactation)
Follicle-stimulating hormone (FSH)	Anterior pituitary	Stimulates follicle development in females and sperm production in males
Luteinizing hormone (LH)	Anterior pituitary	Stimulates ovulation and formation of corpus luteum in females and testosterone production in males
Oxytocin	Produced in hypothalamus, stored in posterior pituitary	Stimulates uterine contractions during childbirth and milk release
Vasopressin	Produced in hypothalamus, stored in posterior pituitary	Controls reabsorption of water by the kidneys; increases blood pressure
Thyroxin (Thyroxine)	Thyroid gland	Controls rate of metabolism and growth

Hormone	Associated gland(s)	Function
Calcitonin (Thyrocalcitonin)	Thyroid gland	Lowers level of calcium in blood by inhibiting calcium release from bones
Parathormone (Parathyroid hormone or PTH)	Parathyroid glands	Increases level of calcium in blood by increasing calcium release from bones; decreases blood phosphate level
Insulin	Pancreas (Islet cells)	Lowers blood sugar level and increases the storage of glycogen
Glucagon	Pancreas (Islet cells)	Increases blood sugar level by stimulating breakdown of glycogen
Cortisol and related hormones	Adrenal cortex	Affects metabolism of proteins, carbohydrates, and lipids; reduces inflammation
Aldosterone	Adrenal cortex	Controls reabsorption of sodium and potassium by the kidneys
Adrenalin (epinephrine)	Adrenal medulla	Increases blood sugar, pulse, and blood pressure
Norepinephrine	Adrenal medulla	Increases metabolic rate and constricts blood vessels
Secretin	Glands in the small intestine	Stimulates secretion of pancreatic digestive juices
Gastrin	Glands in stomach	Stimulates secretion of gastric juice
Cholecystokinin (CCK)	Glands in the small intestine	Stimulates pancreatic secretions and contraction of the gall bladder
Testosterone (androgens)	Testes	Stimulates development of male sex organs and secondary sexual characteristics
Estrogens	Ovaries and placenta	Stimulate development of secondary sexual characteristics in females; help regulate the ovaries and uterus
Progesterone	Ovaries and placenta	Helps regulate the uterus during the menstrual cycle and pregnancy
Chorionic gonadotropin	Placenta	Stimulates the ovaries to continue producing estrogens and progesterone during the early stages of pregnancy; hormone detected in a pregnancy test

Bowerbird Spotted—Thought to Be Extinct

The last previous record of a yellow-fronted gardener bowerbird was in 1895, when Lord Walter Rothschild, a British zoologist, bought three skins of this type of bowerbird. However, in 1981, nearly a century later, Jared Diamond, an ornithologist from the University of California at Los Angeles, reported observing the "extinct" yellow-fronted gardener bowerbird in a rain forest in New Guinea. Not only did he catch a glimpse of this elusive bird, but he also watched the courtship ritual.

Bowerbirds fascinate scientists because of their showy plumage and elaborate behavior to attract a mate. Males are colorful, with feathers of yellow and rust. They build elaborate structures (bowers) of twigs and moss that, for some species, may be 2.6 meters (8 feet) tall. The male yellow-fronted bowerbird builds a more conservative bower, only about 1.3 meter (4 feet) high, but he places piles of colored fruit around it, then struts, screeches, and whistles when a female is nearby. Even with this elaborate display, scientists have found that male bowerbirds are successful in attracting a mate only once in 200 tries. As for the male that Diamond observed, he was unsuccessful.

Rabbit Gene Functions in Two Generations of Mice

One of the long-term goals of human geneticists is to be able to cure genetic disease by replacing faulty genes with repaired ones, while having these new genes accepted and functioning properly in the recipient. A step toward this goal was taken by Thomas Wagner of Ohio University and his associates. They fertilized mouse eggs with copies of a gene for beta-globin, a protein that forms part of hemoglobin, the oxygen-transport molecule of blood. What is most significant is that this gene was from a rabbit, and the gene still functioned.

After introduction of the rabbit gene, the eggs were incubated, then transplanted into foster mothers. Three hundred and twelve eggs were treated and 46 mouse pups were born. Of these pups, five were found to have red blood cells with both rabbit and mouse beta-globin. When two of these five pups were mated, more than half of their offspring were discovered to produce rabbit beta-globin.

The success of this experiment does *not* mean that science is close to gene-replacement in humans. But it does provide what may be a faster way to produce animals with specific human diseases, thus providing animal models for the study of the diseases.

1982

Relocation of Cancer Genes on Human Chromosomes

It has been known for a long time that cancer is caused by some mechanism that changes the genetic expression of cancer cells. For example, cancer cells that are cultured outside the body do not grow or behave like normal cells. Although some animal cancers are known to be caused by viruses, and although it is clear that radiation or certain chemicals could cause cancer, the exact mechanism of cancer has been unknown.

Under most conditions genes function normally, but under changed conditions—exposure to viruses, chemical carcinogens, and the like—some specific genes are transformed into cancer-causing genes, called *oncogenes*. Not surprisingly, genes capable of undergoing such a transformation are called *proto-oncogenes*. New discoveries about oncogenes were made rapidly during the early 1980s.

William Hayward and his associates at the Memorial Sloan-Kettering Cancer Center in New York reported in 1982 that they had located the *mos* gene on human chromosome No. 8. This gene was first known from a virus that causes cancer in animals. The *mos* gene joins another, the *myc* gene, also located on human chromosome No. 8, that has been implicated as an oncogene in humans, but which was discovered in a virus causing cancer in animals.

In Burkitt's lymphoma, a type of cancer, the *myc* gene is transferred to another chromosome. Usually the new site of the *myc* gene is chromosome No. 14, but it may also be No. 2 or No. 22. Cells with the *myc* gene on another chromosome appear to become dominant in this type of cancer. It is thought that the transfer of a normal gene to a new location may influence its function in a way that causes the gene to make the cell cancerous.

In related findings, Sen Pathak and colleagues of the M.D. Anderson Hospital and Tumor Institute in Houston studied cells from a patient with kidney cancer whose family had suffered from kidney cancer over three generations. The Houston team compared detailed pictures of cell chromosomes of 30 cancer cells from the patient. In 22 cells they found that one chromosome No. 3 and one chromosome No. 11 had exchanged pieces, producing abnormal chromosomes: a very short No. 3 and a very long No. 11. The other No. 3 and No. 11 chromosomes were normal. The researchers suggest that this cancer might be caused by the exchange of genetic material, but also that the deletion of genes, occurring during the transfer, might also be involved.

Mirror-Image DNA Occurs in Nature

Was it a laboratory anomaly or a natural phenomenon? That was the question raised in 1980 when Alexander Rich of the Massachusetts Institute of Technology examined X-ray pictures of crystallized DNA and found molecules that twisted counterclockwise instead of the usual clockwise. In other words, the DNA was a mirror image of normal DNA. While mirror-image organic molecules are commonplace, in general only one form is biologically active. For example, while one form of sugar is digestible, its mirror image is not metabolized at all.

Rich dubbed this molecular form of DNA as Z-DNA and began a search for the previously undescribed molecule in living things.

By 1982 he and colleagues at M.I.T. and Tufts had at least a partial answer. They identified Z-DNA in fruit flies and thought that it may exist in mice as well. In addition they discovered that normal DNA can be transformed to Z-DNA by specific chemicals. While the significance of Z-DNA has not been determined, the researchers think it might shed light on how genes go from an active to an inactive state, with Z-DNA representing the inactive state.

Long Search Leads to Discovery of GRF

Roger Guillemin and colleagues at the Salk Institute for Biological Studies began their search for the elusive chemical in 1968. Growth-hormone releasing factor (GRF) is produced by the hypothalamus. It signals the pituitary to release growth hormone, but it is present in such minute quantities that the quest to find it seemed almost impossible. Then, a few years ago, French scientists provided tissue from a tumor of the pancreas that was producing large amounts of GRF. From this tissue the research team was able to isolate GRF and determine its structure. GRF is a peptide chain of 44 amino acids, large compared with other peptides produced by the hypothalamus.

As important as this discovery is, one question remains. Is the structure of GRF from a pancreatic tumor identical to that produced normally by the hypothalamus? The answer to this question is crucial because of the wide range of applications normal GRF might have. Growth hormone is involved with pituitary dwarfism, wound healing, metabolism, and growth of livestock. All of these might be cured or enhanced by the use of GRF. GRF should have fewer unwanted side effects than growth hormone itself.

Controlling the Sex of Unhatched Alligators

Human beings are used to the idea that sex is determined at the moment of conception. For many other species, life is not so simple. Sex may be determined in a number of ways and at various different times in the creature's life. For further evidence on this point, see "Fish: There'll Be Some Changes Made" on page 171. In the meantime, consider sexual determination in the American alligator.

The sex of American alligators, like that of seven other kinds of reptiles, is determined by the temperature at which their eggs are incubated. The incubation period for American alligators is 65 days but days 14 to 28 are most crucial for sex determination. Average temperatures of 30°C (80°F) or below during this period produce all female hatchlings; temperatures at or above 34°C (93°F) produce only males.

These findings by Ted Joanen of the Louisiana Wildlife and Fisheries Commission and Mark Ferguson of Queen's University of Belfast can help in programs designed to rear American alligators in captivity. But their findings may have other implications as well. How and why dinosaurs died out some 65,000,000 years ago continues to puzzle scientists, but there is evidence that there was a change in climate at that time. If the sex of dinosaurs, like that of alligators and some other reptiles, were determined by temperature, it is possible that a sustained temperature change favoring the development of one sex over the other could have dramatically reduced the reproductive potential of the dinosaurs and led to their extinction. On the other hand, this theory does not explain why so many other species became extinct at the same time (see page 272).

ARE BACTERIA RESPONSIBLE FOR A GOOD NIGHT'S SLEEP?

The search for chemicals that play a role in inducing human sleep has been going on for decades. A variety of substances have been found, including the amino acid tryptophan, found in high-protein foods such as milk. This discovery confirms the old theory that a warm glass of milk before bedtime is helpful in getting to sleep.

Now a new chemical has been added to the list of sleep promoters: factor S. This peptide (a short chain of amino acids) seems to enhance deep sleep of the sort that follows long periods of wakefulness. But there is also a mystery about factor S—its chemical structure is more like that of peptides produced by bacteria than those produced by humans. This has led scientists to suggest that factor S, like certain vitamins, is produced by bacteria that normally inhabit the human intestine. The human body then uses the bacterial products for its own physiological purpose.

Fiber Optics in Plants

Nature always seems to be ahead of human invention. New fiber-optic networks in which information is carried by light from city to city instead of by electric signals are the latest thing in communication between humans. Plants seem to have discovered this basic principle first, however.

Fiber optics depends on the ability of certain materials to reflect totally all light that strikes it at less than a certain angle, the angle of total reflection. A fiber of such a material that has a beam of light directed down it will transmit essentially all of the light, since the reflections at the boundary of fiber will all be less than the angle of total reflection.

Dina Mandoli and Winslow Brigs of Stanford University have investigated how light reaches specific areas in plants that control the plants' growth. It is clear that many plants use light in this way. Gardeners, for example, are familiar with changes in spinach and onions that depend on day length. In many plants, however, the areas that control growth are not where one would expect them to be aware of long or short days.

Mandoli and Brigs investigated this phenomenon by shining a fine light beam at one end of stems of corn, mung beans, and oats. The light did not diffuse outward, but appeared out the other end of the stems with the same shape and size as the original beam. In fact, the light could even travel around bends in the stem, just as in fiber optic systems based on glass or plastic. These results suggest that the plant stems have a natural fiber-optic system that collects light and distributes it to the areas of the plant that are light sensitive.

WHY DO LEAVES CHANGE COLOR IN THE FALL?

No one knows for sure, but scientists are getting close to the answer. The actual color change comes when the protein that binds to chlorophyll is removed and broken into amino acids that will be stored in the stems and roots for reuse the following year. (The nitrogen in amino acids, along with phosphorus, is one of the elements that is in short supply for most plants.) Without the protein attachment, chlorophyll also breaks up. Without the green of chlorophyll, the underlying brownish yellow color of the leaves shows through. Leaves that then turn red do so because cool weather thickens the sugar that is also being transported out for storage and energy over the winter. If the sugar cannot get out of the leaves, sunlight turns it into a red compound, anthocyanin. When the weather is too warm, the leaves do not turn red.

All this answers the how and part of the why, but it still does not explain what tells the tree that autumn and winter are on the way.

Enzyme-Type Activity by RNA

Enzymes are proteins that act as catalysts. Life depends on enzymes completely, since catalysis is needed to make unlikely chemical transitions become likely at reasonable temperatures. One problem in determining the origin of life has been that DNA, the common genetic material for almost all life on Earth (except for a few bacteria and viruses), cannot make proteins without enzymes to act as catalysts. But where would the enzymes come from in the first place, since DNA generally initiates the formation of proteins?

It has been argued, without regard to this problem, that the genetic material in the most ancient organisms was RNA rather than DNA. The solution to the puzzle of the first enzymes could be solved if RNA, in addition to encoding information in those early cells, also served as an enzyme-like catalyst. Until 1982, however, all catalysts observed in living things were proteins (inorganic catalysts are a different story completely). Therefore, moving the puzzle from DNA to RNA did not seem to help solve it.

In 1982 it was found that one type of RNA could catalyze itself. Thomas R. Cech and coworkers at the University of Colorado found a form of RNA that could both break and form its own bonds when no proteins were present. Cech and his colleagues discovered the unusual form of RNA functioning in a living environment, *Tetrahymena thermophila*, a one-celled pond animal, but the RNA from the organism also functioned as an enzyme when left to itself in a test tube.

Research by Sidney Altman of Yale University and coworkers in 1983 turned up another exception to the rule of protein catalysis. They investigated the action of a compound called ribonuclease P. Ribonuclease P is a small molecule made up of a combination of protein and RNA. Under proper conditions, ribonuclease P can break the chemical bonds of nucleic acids, thus acting as a catalyst. But Altman and his group found that it was RNA in ribonuclease P that was the catalyst, not the protein, as was expected.

These two results tend to support the idea that in the most primitive chemical "soup" that preceded life, RNA, acting to some degree as its own enzyme, rearranged itself into various forms. Some of these forms were better able to survive the existing environment than others, and so were preferentially maintained. The ability to form the most useful arrangements would have, therefore, improved over time, eventually leading to life.

Want to Grow a Giant Mouse?

Building on techniques developed in cloning zebra fish, dramatic new results in the genetic manipulation of mice were announced in 1982 (see "Fish and Mice Cloned", page 152, for the earlier results). Among the results was the first hint of the kind of monster creature long predicted in science-fiction movies, although still far away from such mythical creatures as "The Rabbit that Devoured Cleveland." In fact, the "giants" involved a doubling in size only.

A group of scientists from four different U.S. laboratories working together have, however, created something quite new. The team engineered a gene from another species and inserted it into fertilized mouse eggs. They took a very active gene for growth hormone from a rat and removed its control region. They then replaced it with the control region of a mouse gene that functions in the liver. Of 170 eggs that received multiple copies of this composite gene, 21 mice were born. Of these, six grew heartily, reaching almost twice their normal size. The giant mice had the rat gene functioning, although in a different place than normal—the liver. From this site, it caused the mouse bodies to produce rat growth hormone.

The extra growth in the six mice was not a result of the fact that rats are larger than mice; instead, the extra growth was caused because the gene in its new location was not turned off when enough growth hormone was produced. The gene just kept on making growth hormone. When normal females were mated to the large males, approximately half of the offspring were also large, providing evidence that the gene had been passed from one generation to the next.

Later the experiment was repeated with the gene for human growth hormone instead of that for rat growth hormone. The human gene also functioned in mice, representing the first time that a human gene had been transplanted into another mammal as a living organism (as opposed to human genes in mammal cells, which had been achieved earlier).

The success of this experiment opens the way for transferring other mammalian genes, either in a normal or an altered state. The transfer of the rat and human growth hormone by genetic means may also provide a model in which human disorders can be studied. The largest mice had a level of growth hormone in their blood that was about 800 times that of normal. A similar pattern is also seen in human gigantism. While there may be no direct connection in terms of the cause of human gigantism, experiments of this sort can provide an animal model for some of the effects of uncontrolled production of growth hormone.

1983

Isolated, Thrashing Tail Protects Lizards

Sometimes scientific studies are needed to prove what to most people would seem obvious. The very good reason for this is that the "obvious" truth often turns out to be false when carefully examined. On the other hand, the "obvious" may turn out to be true after all.

Most people know that many lizards can lose their tails to escape predators. The tail is a likely place to catch a fleeing lizard, but when the tail drops off, the lizard can continue to flee. Assuming that the lizard makes good its escape, a new tail will grow to replace the one lost.

In 1983 scientists in Texas discovered more about this phenomenon. When some types of lizards lose their tails, the disembodied tails continue to thrash on the ground. The lost tails of other types of lizards simply lie still. What the researchers observed in a series of controlled experiments is that lizards with disembodied, thrashing tails survive predators' attacks better. Apparently the tail movement diverts the attention of the would-be predator, giving the lizard longer to escape.

Sinuses Make People Magnetic

Pigeons, fish, dolphins, and even bees have magnetic material in their bodies. Various researchers have proposed that these magnetic bodies are aids in navigation, orienting the creatures that have them to Earth's magnetic field. In 1983 a potential location for such magnetic material was found in humans as well.

R. Robin Baker at the University of Manchester in England studied recently deceased corpses. By using stains for iron, he found a layer of iron compounds in the human sinuses of four of his five subjects. The fifth subject was anemic, a condition in which the body has insufficient iron. The iron deposits were about 5 microns (2 ten-thousandths of an inch) below the bone surface in the sinuses. Iron, of course, is one of the elements that exhibit strong magnetism.

Baker, who has published work on humans' ability to find direction in response to a magnetic field, feels that this iron-compound layer may be related to his earlier findings. Other scientists, however, are skeptical, suggesting that the iron may simply be stored in the sinus bones or may play a role in bone growth and repair.

Teaching Slugs and Snails

Finding out how living things learn had its start in the nineteenth century with Pavlov's classic work on the conditioning of dogs. Pavlov demonstrated that dogs could be trained to salivate when they heard a bell ring. Today, that form of learning is called *associative learning*, and it is probably the simplest type of learning. Nevertheless, the complexity of even a dog's brain is far too great to understand what is happening at the chemical, neuronal, or electronic level when associative learning take place.

The nervous system of some invertebrates is much simpler than that of humans or dogs. If simple invertebrates can be made to show learning, then perhaps the chemical basis can be found. Snails and slugs, for example, have very simple nervous systems with unusually large nerve cells. Recent studies on the lowly garden slug, sea slug, and marine snail are helping unravel the biochemical basis of associative learning.

The first problem is to teach these simple creatures anything. Until it was tried, most scientists doubted that conditioned responses, or associative learning, could be demonstrated in creatures this simple. However, Alan Gelperin of Princeton University and his coworkers showed that garden slugs (*Limax* species) could learn to avoid one of their favorite foods, instant mashed potatoes, if the food was doped with bitter chemicals. The next step is to find out exactly what changes in the nerve cells, or neurons, corresponded to learning the association between instant mashed potatoes and a bitter taste. Work is still proceeding on this matter.

Nevertheless, this project showed that slugs and snails, with their simple systems and large nerves, could be taught *something*. As a result, learning in these species has undergone a close scrutiny by various researchers.

For example, other researchers have been making progress in finding the neuronal and chemical bases of associative learning in sea slugs. The initial problem is to determine what sea slugs particularly like or dislike. Daniel Alkon of the Marine Biological Laboratory at Woods Hole, Massachusetts, found one thing. He developed a method of conditioning the sea slug (*Hermissenda*), which is usually attracted to light, to avoid it. He placed a sea slug at one end of a glass tube and then set off a flash of light at the other end. After the light flash, the glass tube with its inhabitant was spun around on a record-player type turntable. The spinning was apparently viewed by the sea slug as punishment. Normally, slugs move toward light, but after conditioning, they remained stationary at the end of the tube closest to the center of the turntable even when the turntable was not spinning. The key advance in this experiment was not the conditioning itself, but finding the biological basis for the change in the sea slugs' behavior caused by the conditioning process.

In working out the chemical and cellular bases for this behavior, Alkon and his colleagues found a complex system. It appears that the type B cells within the eyes of the slug are the trigger organs for at least this particular example of learning. The type B cells become more light sensitive during conditioning due to blocking of potassium channels in the cell membranes. These light-sensitive cells then inhibit the muscles the slug would use to move toward the light.

The conditioning appears to occur in a step-wise sequence: an initial stimulus prepares the nerve cell for changes, then a second stimulus actually induces the change. In the sea slug, it is thought that the flash of light primes the B cells; then the ride on the turntable causes balance organs to stimulate the B cells a second time. The result is associative learning linking the flash of the light and the rotation of the turntable. In effect, the sea slug chemically "thinks" that light, which is nice, is the same as rotation, which is terrible. As in most associative learning, even in humans, this is carried out at the cellular level, not at the conscious level—if there is such a thing as a conscious level in sea slugs.

But the champion of all the learners, from a researcher's point of view, is *Aplysia*, a marine snail. Because of its very simple nervous system and very large neurons, *Aplysia* may become to biological learning theory what *Drosophila* (the fruit fly) has been to genetics. Not only has associative learning been studied in *Aplysia*, as it has in other slugs and snails, but also the chemical basis of nonlearned, instinctive behavior has been sorted out in detail for *Aplysia*.

Eric R. Kandel, Richard H. Scheller, and Richard Axel, among others, have sorted out the specific chemical signals that are involved in *Aplysia's* complex egg-laying behavior. This behavior involves certain movements of the head at specific times in the procedure by which the eggs are laid.

Furthermore, they have identified the genetic basis of that repertoire of behavior patterns. The genes they have identified produce specific peptides, each of which controls a particular part of the behavior. Although egg-laying is unlearned behavior, the researchers have also found how associative learning occurs in specific reflexes of *Aplysia*. The chemicals and mechanisms involved appear to be part of vertebrate behavior patterns also, suggesting that work with such simple creatures as slugs and snails will shed light on the more complex behavioral patterns of vertebrates.

Gene Transplant for Tobacco and Petunias

Genetic engineering with plant cells has proceeded more slowly than with animal cells, and the accounts of it in the popular press have been less glamorous. But nonetheless 1983 was a boon year for gene transplants.

In January, scientists from Monsanto's Molecular Biology Group announced their success in getting injected genes to function in cultured plant cells. Monsanto's group transplanted a bacterial gene that protects bacteria from the antibiotic methotrexate into carrots, petunias, tobacco, and sunflowers. The method they used was to attach the gene to a segment of DNA that the plant cells recognized. This "familiar" DNA was then incorporated into the bacteria that cause crown gall disease in plants. The bacterium was then used to transfer the gene into plant cells. The proof that plant cells incorporated the gene was their growth on culture medium containing the antibiotic in levels that normally kill plant cells. The plant cells lived. The important new technique involved was not developing resistance to an antibiotic; it was showing that the bacteria could transfer a gene into the plant cells.

In March, Andrew Binns of the University of Pennsylvania, Kenneth A. Barton of Cetus Corporation, and Mary-Dell Chilton of Washington University at St. Louis reported transferring a yeast gene into tobacco cells, then regenerating the cells into plants. This was a significant advance, since the Monsanto group had stopped with cells, not growing the cells into plants. The yeast gene did not function in the plants, probably because it lacked the necessary regulatory mechanism.

In April, however, the Monsanto group reported successfully regenerating mature petunia plants that expressed the bacterial gene against the antibiotic methotrexate. A gene expresses a trait when the trait actually appears in the organism, a distinction that is necessary because some genes (recessive ones) may never be expressed, while others, such as genes for sexual hormones, may be expressed at one time and not at another. In this case, the petunia plants looked normal, but leaf pieces from the plant grew on a culture medium containing high amounts of methotrexate, showing that the gene was not only present, but also was expressed.

The ramifications of these successes could be enormous. Researchers are now confident that virtually any plant gene can be transplanted. This could mean a whole new generation of superplants with better disease resistance, greater yield, and more tolerance to both frost and heat. Such advances would help both large-scale agricultural operations and backyard gardeners.

Common Plants That Are Poisonous

Name (scientific name)	Poisonous Parts
Amaryllis (*Hippeastrum*)	bulbs
Anemone	all
Autumn crocus (*Colchicum autumnale*)	all
Bittersweet (*Celastrus scandens*)	seeds
Bleeding heart (*Dicentra*)	all
Boxwood (*Buxus sempervirens*)	leaves
Buttercup (*Ranunculus*)	all
Buttonbush (*Cephalanthus occidentalis*)	leaves
Castor bean (*Ricinus communis*)	all
Columbine (*Aquilegia*)	all
Crown of thorns (*Euphorbia milii*)	milky sap
Daffodil (*Narcissus*)	bulbs
Dutchman's breeches (*Dicentra cucullaria*)	all
Ferns (most species)	mature fronds
Four-o-clock (*Mirabilis jalapa*)	roots, seeds
Foxglove (*Digitalis purpurea*)	all
Gloriosa lily (*Gloriosa*)	all
Goldenseal (*Hydrastis canadensis*)	leaves, rhizomes
Holly (*Ilex aquifolium*)	berries
Horse chestnut (*Aesculus hippocastanum*)	leaves, flowers, fruits
Hyacinth (*Hyacinthus*)	bulbs
Hydrangea	leaves, buds
Jack-in-the-pulpit (*Arisaema triphyllum*)	leaves, roots
Larkspur (*Delphinium*)	seeds, young plants
Lily-of-the-valley (*Convullaria majalis*)	bulbs
Lobelia	all
Lupine (*Lupinus*)	seeds and pods
Marsh marigold (*Catha palustris*)	all
Mayapple (*Podophyllum peltatum*)	all except ripe fruits
Mistletoe (*Viscum album*)	all
Monkshood (*Aconitum*)	all
Oleander (*Nerium oleander*)	all
Opium poppy (*Papaver somniferum*)	unripe fruits
Peach (*Prunus persica*)	fruit pit
Philodendron	stems and leaves
Plum (*Prunus*)	fruit pit
Poinsettia (*Euphorbia pulcherrima*)	milky sap
Pokeweed (*Phytolocca*)	roots, berries
Potato (*Solanum tuberosum*)	green parts, berries
Prickly poppy (*Argemone*)	leaves, seeds
Privet (*Ligustrum vulgare*)	fruits
Rhubarb (*Rheum raponitcum*)	leaves
Snowdrop (*Galanthus*)	bulbs
Snow-on-the- mountain (*Euphorbia marginata*)	milky sap
Star-of-Bethlehem (*Ornithogalum umbellatum*)	all
Sweet pea (*Lathyrus*)	seeds
Yew (*Taxus baccata*)	all

Trees "Talk" Over Problems

When a hungry horde of insects descends on a likely looking tree, the later arrivals may find the tree less inviting than the first insects did. Gordon H. Orians and David F. Rhoades of the University of Washington in Seattle report that when sugar maple trees and Sitka willows are attacked by pests, their chemistry changes, making them less tasty food.

This discovery was not unexpected, because it seemed likely that plants would respond to an attack by stepping up their defenses. But the next part of their study led to a big surprise. The trees that were under attack seem to communicate the danger to nearby trees. Orians and Rhoades found that nearby trees not yet attacked by the insects also undergo chemical changes making their foliage less desirable. It is thought that a chemical released by the trees that have been attacked signals the nearby trees to change their chemistry. Such chemical communicators are called *pheromones* and are common among animals and protists (such as amoebae) but had not been known to exist in plants. Using Sitka willows, Orians and Rhoades were able to duplicate this natural phenomenon in the laboratory.

When trees have problems, they "talk" them over.

Do Spiders Feel Pain?

Fishermen and people cooking live lobsters learn from their peers that putting worms or insects on hooks or dunking arthropods in boiling water does not cause pain. This common view of pain in invertebrates has now been challenged, at least with regard to spiders. Observations by Thomas Eisner and Scott Camazine of Cornell University make them think that some spiders may, in fact, feel pain.

Honeybee venom, wasp venom, and phymatid venom injected into the leg of an orb-weaving spider all cause the spider to detach the affected leg. Because the response is so swift—only a few seconds—the venom has little chance to reach the spider's body. Spiders that do not discard their legs when stung in the leg usually die. Thus, discarding the leg has definite survival value. Katydids and three other families of spiders also display similar leg-shedding behavior when stung. While this behavior in itself does not prove that some spiders feel pain, the components of the venom associated with leg detachment suggest that these spiders do feel pain. Melittin, histamine, phospholipase A_2, and serotonin, found in the venoms, are known to cause human pain.

Little Genetic Variability in Cheetahs

The cheetah is the epitome of feline grace and ability. The fastest mammal in the world, it can reach speeds of 112 kilometers (67 miles) per hour in short bursts. Despite this apparent ability, which can make it a formidable predator, scientists have been concerned about the cheetah for some time. It is endangered.

The total African population of cheetahs is estimated at between 1,500 and 25,000, which is not a large number for a species. Furthermore, the cheetah's habitat is being destroyed by farming. Also, the native cattle farmers consider the cheetah a threat to their herds and try to eliminate it. Some other animals, however, such as the American coyote, survive and even increase in the face of changing habitat and opposition by farmers. What makes the superb cheetah unable to compete as well as the ordinary coyote?

Stephen J. O'Brien and David E. Wildt of the National Cancer Center and their colleagues reported in 1983 that studies show one factor that may be contributing to the decline of the elegant cats. This group of biologists studied red and white blood cells of cheetahs from the Transvaal, from Southwest Africa, from hybrids of the previous populations, and from zoo populations. They were looking for diversity among proteins from known gene locations. Genetic diversity is a crucial element in enabling species to survive under changing environmental conditions, as in their habitats. What they found in the cheetahs was very little genetic diversity. Of the 47 gene sites studied, all appeared essentially the same, and of the 155 proteins studied, there was only a small amount of diversity.

The best explanation seems to be that a population bottleneck occurred in the species' past. That is, strong environmental pressures of some kind reduced the population size of cheetahs dramatically in the past. The individuals surviving this disaster were genetically similar. When these individuals interbred, the lack of genetic diversity was perpetuated.

Other findings that support the notion that cheetahs represent a highly inbred species come from cheetah sperm studies. Up to 70 percent of their sperm are abnormal in form, and their sperm counts are as much as ten times lower than those of other cats. These characteristics in sperm are more commonly seen in populations of inbred mice and livestock than in populations of wild animals.

Finally, the interpretation of this evidence is that the cheetah today, because of genetic similarity, is a less adaptable animal than another animal—say the coyote.

Fish: There'll Be Some Changes Made

Although the sex of alligators and seven other species of reptile is determined by the temperature at which the egg is incubated, at least—once the sex is determined—the alligator stays the same sex for most of its life. Some creatures, however, can change from female to male (and sometimes back to female again). Although such changes have long been known in invertebrates, it is still surprising to find vertebrates that switch sexes.

In the 1970s, fish that could change their sex were found along the Great Barrier Reef. The sex change occurred in females and seemed to be stimulated by the death or disappearance of the male in the group. Other fish were found that exhibited this same reproductive strategy. Fish, unlike mammals, do not have separate chromosomes for sex, and are much more likely to change sex depending on environmental factors. If the environment does not include a male, it seems to be a reasonable time for a change.

Two researchers from Hawaii, Milton Diamond of the University of Hawaii and Robert M. Ross of the Hawaii Institute of Marine Biology, reported in 1983 on a sex change in female saddleback wrasse that seems to have a different stimulus: size of companions of the same species. Whatever the grouping of these wrasse, the largest fish is always male.

In their study, Diamond and Ross set up a number of experimental situations involving saddleback wrasse. They used single females kept alone, a single female kept with one to three larger wrasse, and a single female kept with one to three smaller wrasse. In addition, they placed a single wrasse female with one to three smaller fish that were not wrasse. In each setup, all fish were sexually mature. Only when the female wrasse was with smaller wrasse did she undergo a sex change and begin producing sperm; this occurred even when the other fish were males.

One possible explanation for this change is that wrasse look very much alike. Perhaps the only way to find a male is to seek the largest fish in a group. In any case, the change is irreversible; females, once they become males, cannot change back.

HOME SELECTION BY HERMIT CRABS

Hermit crabs are very selective about the shells they choose for homes. Esthetic, size, or style considerations, however, are not important. These shell-less invertebrates choose a shell that gives off the most free calcium. This sensitivity to calcium may help the hermit crabs detect shells that are partially buried. Or, they may just be fond of calcium.

Bacterial Photosynthesis Uses Same Genes as Plants

Photosynthesis, the use of sunlight to change carbon dioxide and water into sugar, is the basic energy source for nearly all forms of life. While the biochemistry of photosynthesis is fairly well understood, the genetic basis for it is not. However, research at the University of California's Lawrence Berkeley Laboratory (LBL) has begun to shed light on the genetics.

In 1983 the LBL group identified five genes in the purple bacterium *Rhodopseudomonas capsulata* that produce proteins involved in the light reaction, the step in photosynthesis that traps and stores energy. The proteins are associated with pigments used in photosynthesis and seem to determine the particular role the pigments carry out in energy capture and storage. This is the first time that the bacterial genes involved in photosynthesis have been identified.

R. capsulata is not required to obtain its energy from photosynthesis. If free oxygen is handy, the bacterium is a decomposer that uses the oxygen to consume organic material at the bottom of lakes. When oxygen is in short supply, the bacterium switches on its photosynthetic apparatus to obtain energy. It is generally believed that life originally evolved in the absence of free oxygen, so perhaps *Rhodopseudomonas* is descended from photosynthesizers that found it was "cheaper" to decompose than to photosynthesize, but that still retained the capacity for photosynthesis.

One of the *Rhodopseudomonas* photosynthesis proteins has a segment that matches a segment of a protein in tobacco and spinach that is known to be affected adversely by herbicides. This finding has a dual significance: it may help researchers better understand how herbicides work and how plants develop resistance to them. It also shows an evolutionary link between the photosynthetic mechanisms of bacteria and higher plants. Some biologists believe that photosynthesis originated first in bacteria and that the chloroplasts, bodies in which photosynthesis takes place in plants, were originally symbiotic bacteria. The LBL findings tend to substantiate this view of the origin of photosynthesis.

CLEANING UP TRASH WITH BACTERIA?

The gene for the enzyme cellulase has been successfully implanted into the lab bacterium *E. coli*. With this enzyme functioning, the *E. coli* can digest cellulose, a major component of paper products, sawdust and plant debris, into sugar. Not only could this help with trash problems, but the sugar produced could be used in fuels, plastics, and drugs.

Experiment with Engineered Bacteria Stopped

The formation of ice crystals in plants, not the temperature itself, is what causes frost damage in plants. While this damage may cause disappointment for home gardeners, it means great financial loss—an average of $1,000,000,000 per year in the United States alone—to commercial growers. For ice crystals to form and cause damage, they must have a nucleus, or seed, around which to grow. This seed is now known to be bacteria of one of two species.

Researchers at the University of California (Berkeley) have engineered one of these bacteria so that it cannot participate in ice making. After identifying the segment of DNA responsible for ice formation, new DNA pieces were inserted that lacked the ice segment. Scientists think that by spraying crops with the genetically altered bacteria, the bacteria will be able to prevent frost formation.

Originally, these bacteria were due to be tested in a potato field in northern California during the early autumn of 1983. It would have been the first release of gene-spliced bacteria into the environment. Some people opposed the release, however, although it had been approved by the committee that supervises gene-splicing research in the United States, the Recombinant DNA Advisory Committee of the National Institutes of Health (NIH). In mid-September 1983, a public interest group sued the NIH to stop the experiment, scheduled to get under way within weeks or months.

The suit claimed that the Recombinant DNA Committee does not have the expertise to anticipate likely consequences of the experiment, and it suggested that such consequences could be far-reaching. Ice-nucleating bacteria are important not only in frost damage to plants but in the development of rain and other precipitation in the upper atmosphere. If the new non-nucleating strain were to replace the nucleating strain in the atmosphere, what effect would this have on weather? Scientists on the Recombinant DNA Committee believe that objections to the experiment are without foundation and that nothing in past experience suggests any likelihood that the experiment could be hazardous.

In May, 1984, the challengers gained a significant victory when Federal District Judge John J. Sirica not only granted a preliminary injunction against the experiment, but also indicated that the experimenters would have to file an environmental impact statement. The environmental impact statement, if needed in a final ruling, would delay the experiment by another year even if the experimenters succeed in getting permission for the field trial.

Sugar as a Drug

Recent research suggests that the intake of sugar probably causes the body to secrete endorphins, the brain chemicals that are also known as natural opiates. These substances reduce sensitivity to pain and induce a feeling of well-being.

The good news for dieters is that the same effect can apparently be achieved from artificial sweeteners such as saccharin. Israel Lieblich of the Hebrew Univerity of Jerusalem and his coworkers fed laboratory rats tremendous doses of saccharin and found that the rats developed a tolerance to morphine that "unsweetened" rats did not.

Endorphins are known to fill the same receptor sites on brain cells to which morphine can attach. Research has shown that endorphins are involved in how the body breaks down and uses sweets. Some of the rats used for the study had been bred for their ability to ingest large quantities of saccharin, and some of them were normal. For 28 days the rats were fed either a strong saccharin solution or water. Then their pain tolerance was tested by placing them on a hot plate and timing how quickly they responded to the heat. Next the rats were injected with morphine and the hot-plate test was repeated at fifteen-minute intervals.

Rats that drank saccharin showed little difference in their tolerance to pain before and after the morphine injection. The saccharin blocked the action of morphine. One explanation for this is that the saccharin stimulated the production of endorphins that bound with brain receptor sites. This would leave no sites open for morphine attachment and consequently, morphine could not act as a pain-blocker.

These findings may suggest the reason why many people find candy or other sweet things so—what shall we say?—addicting.

HEAT-PRODUCING PLANTS

Skunk cabbages often poke their leaves through the snow in late winter. How these plants actively grew under such unfavorable conditions was long a puzzle. But an answer has now been found. It seems that these plants are capable of producing their own heat and may maintain internal temperatures of 30°C (86°F) higher than their chilly surroundings. That is possible because skunk cabbages, unlike most plant species, store and "burn" fat-like substances. Not only do they enhance their own survival, but they become a winter oasis for other species. Insects and other invertebrates can live and reproduce on and around the skunk cabbages and the miniature pools among their leaves formed by melted snow.

That Silly Grin Changes How You Feel

It has long been suspected that there are complex interactions between the brain and the body. In the past fifty years or so, three different approaches to studying those interactions have been used. For many years the behaviorist psychologists dominated the field, and most studies involved conditioned responses to stimuli. Then cognitive psychology became important again, and the brain was viewed as a thinking-remembering machine with the biological interest largely in remembering (since it is easier to measure than thinking). Recently, a new school of psychology has been forming, one in which concern for cognitive processes has been replaced by study of the fundamental mechanisms involved in emotional responses.

While it is clear that emotional states can effect physical changes—after all, such changes were among the first problems studied by Sigmund Freud—and such physical changes as disease can affect emotions, it has not been clear to scientists until 1983 that ordinary physical changes, such as muscle positions, affected emotional states (although yoga practitioners have claimed this for centuries). The study was based on the positions of facial muscles.

If you want to know if the facial muscle combinations that convey an emotion affect the involuntary nervous system, who better to study than professional actors? And that is exactly what Paul Ekman and Wallace V. Friesen of the University of California School of Medicine (San Francisco) and Robert W. Levenson of Indiana University did. The researchers selected six emotions with universal expressions to study: anger, fear, sadness, happiness, surprise, and disgust. They directed the actors to assume these expressions by telling them which facial muscles to contract.

The involuntary nervous system not only responded, but did so in different ways. For instance, heart rate rose steeply for anger and fear, slightly less for sadness, showed only a slight rise for happiness and surprise, and actually decreased for disgust. Skin temperatures rose dramatically in anger, slightly for sadness and happiness, and declined for fear, surprise, and disgust.

In other words, the physical expression shown, no matter what its motive, changed the internal state of the body—somewhat the reverse of what might be expected. It is believed that somehow the muscles activate neurotransmitters that then make the change in state of the nervous system. In turn, the nervous system mediates the physical changes that were measured in the experiments.

New Approach Produces Quick Gains in Understanding

Knowledge of gene structure has progressed so fast since the 1950s that today it is easier to find out what the sequence of codons (the "letters" of the genetic code) are in DNA or RNA than it is to determine the structure of a protein. Protein structure can be inferred from the gene that encodes it. In 1983 this approach was used in an attempt to determine what proteins occur in a specific organ. No preconceived ideas about what proteins should be there or what the proteins do were involved. As a result, the information from the study may open up totally new ideas about how the organ, in this case the brain, functions.

The older way to do this kind of research—still used and still useful—is to observe an effect and then try to find out its cause. In the new method, the kinds of compounds that are known to cause effects are cataloged at random. Then, researchers can look for the effects caused by previously unknown compounds.

Because different genes are active in different types of cells, comparing DNA, RNA, or proteins from different cells enables scientists to discover which are specific to a given cell type. By comparing gene activity in cells from rat brains, livers, and kidneys, scientists have established that brain cells produce as many as 30,000 different types of messenger-RNA. In addition, more than half of these messenger-RNAs may be unique to brain cells. Each type of messenger-RNA encodes a single protein, so decoding the RNA will tell what proteins are active in the brain. Proteins are the basic chemicals that direct the functioning of the cells.

About 200 types of messenger-RNA had been studied by late 1983, and two new brain proteins were discovered as a result. The presence of one type of messenger-RNA throughout the brain suggests that the protein it encodes may be related to the synthesis of materials in the branches of nerves that carry messages to nerve cell bodies. The messenger RNA for the second brain protein is found throughout a system within the brain that signals the brain to start bodily movements.

Along the way the researchers, from the Salk Institute and the Scripps Clinic, discovered a special sequence in a part of the DNA that was thought to have no function. This sequence was found in all of the eight brain genes studied in sufficient detail to locate the sequence—thus, it may be common to all brain genes. If so, it might be a sequence that tells the gene that it should be expressed only in the brain.

The brain was chosen at random for this study. Similar studies using the same approach can identify the function of other organs at the protein level.

New Organisms—New Phylum

Just when biologists think they have all the existing groups of organisms accounted for, a totally new organism is discovered. The discoverer, Reinhardt M. Kristensen of the University of Copenhagen, had observed larval forms of one particular marine organism since the mid-1970s, but only recently did he find adults—and that was merely by chance. One day, while extracting specimens from sediment from off the Denmark coast, Kristensen was rushed, so he rinsed the sand with fresh water instead of going through the usual procedures. The shock of the fresh water apparently caused these tiny animals to release their tenacious hold on the grains of sand and reveal themselves.

So unique are these specimens that scientists have assigned these organisms to a phylum all its own. Loricifera, as these animals are called, are distributed widely and for their small size are structurally quite interesting. Both larvae and adults have spines on their heads; for mouths they have flexible tubes that can be retracted. Instead of shells or true external skeletons, they have plates of chiton forming a "corset" around their midsections.

Classification of Major Groups of Living Organisms

KINGDOM MONERA
> One-celled organisms with simple (procaryotic) cells that lack a membrane around the genetic material. There are 16 phyla of bacteria:
> APHRAGMABACTERIA, FERMENTING BACTERIA, SPIROCHAETAE, THIOPNEUTES, METHANEOCREATRICES, ANAEROBIC PHOTOSYNTHETIC BACTERIA, CYANOBACTERIA, CHLOROXYBACTERIA, NITROGEN-FIXING AEROBIC BACTERIA, PSEUDOMONADS, AEROENDOSPORA, MICROCOCCI, CHEMOAUTOTROPHIC BACTERIA, OMNIBACTERIA, ACTINOBACTERIA, and MYXOBACTERIA.

KINGDOM PROTOCTISTA
> One-celled or colonial; complex (eucaryotic) cells that have a membrane around their genetic material.
> Phylum: CARYOBLASTEA (a giant, primitive ameba)
> Phylum: DINOFLAGELLATA (a form of marine plankton; the cause of the well-known, poisonous red tides)
> Phylum: RHIZOPODA (amebas)
> Phylum: CHRYSOPHYTA (golden-yellow algae)
> Phylum: HAPTOPHYTA (marine golden algae)
> Phylum: EUGLENOPHYTA (Euglenas)
> Phylum: CRYPTOPHYTA (green algae)
> Phylum: ZOOMASTIGINA (flagellated protozoans)
> Phylum: XANTHOPHYTA

Phylum: EUSTIGMATOPHYTA
Phylum: BACILLARIOPHYTA (diatoms)
Phylum: PHAEOPHYTA (brown algae)
Phylum: RHODOPHYTA (red algae)
Phylum: GAMOPHYTA (conjugating green algae)
Phylum: CHLOROPHYTA (yellow-green algae)
Phylum: ACTINOPODA (radiolaria)
Phylum: FORAMINIFERA (shelled, marine plankton)
Phylum: CILIOPHORA (heterotrophic microbes with cilia)
Phylum: APICOMPLEXA (includes the malaria parasites)
Phylum: CNIDOSPORIDIA (parasites, mostly of invertebrates and fish)
Phylum: LABYRINTHULAMYCOTA (slime nets)
Phylum: ACRASIOMYCOTA (cellular slime molds)
Phylum: MYXOMYCOTA (plasmodial slime molds)
Phylum: PLASMODIOPHOROMYCOTA (plant parasites)
Phylum: HYPHOCHYTRIDIOMYCOTA
Phylum: OOMYCOTA (water molds, white rusts, and downy mildews)

KINGDOM FUNGI

One-celled or multicelled; cells eucaryotic; nuclei stream between cells giving the appearance that cells have many nuclei; do not produce their own food.
Phylum: ZYGOMYCOTA (black bread mold is an example)
Phylum: ASCOMYCOTA (includes Penicillium, truffles, yeasts)
Phylum: BASIDIOMYCOTA (includes mushrooms)
Phylum: DEUTEROMYCOTA (includes fungus that causes athlete's foot,
          ringworm, and molds used in cheese production)
Phylum: MYCOPHYCOPHYTA (Lichens)

KINGDOM PLANTAE

Multicellular land-living organisms with eucaryotic cells that carry out photosynthesis
Phylum: BRYOPHYTA (mosses and liverworts)
Phylum: LYCOPODOPHYTA (club mosses)
Phylum: SPHENOPHYTA (horsetails)
Phylum: FILICINOPHYTA (ferns)
Phylum: CONIFEROPHYTA (conifers)
Phylum: GNETOPHYTA (cone-bearing desert plants)
Phylum: CYCADOPHYTA (cycads)
Phylum: GINKGOPHYTA (ginkgos)
Phylum: ANGIOSPERMOPHYTA (flowering plants)
  Subphylum: DICOTYLEDONEAE (dicots)
  Subphylum: MONOCOTYLEDONEAE (monocots)

KINGDOM ANIMALIA

Multicellular organisms with eucaryotic cells that get their food by ingestion; most are motile.
Phylum: PLACOZOA
Phylum: PORIFERA (sponges)
Phylum: CNIDARIA (hydras, jellyfish, corals, and sea anemones)
Phylum: CTENOPHORA (comb jellies)
Phylum: MESOZOA
Phylum: PLATYHELMINTHES (flatworms)
Phylum: NEMERTINA (ribbon worms)
Phylum: GNATHOSTOMULIDA
Phylum: GASTROTRICHA

Phylum: KINORHYNCHA
Phylum: ACANTHOCEPHALA (spiny-headed worms)
Phylum: ENTOPROCTA
Phylum: NEMATOMORPHA (for example, the horsehair worm)
Phylum: NEMATODA (roundworms)
Phylum: ROTIFERA (microscopic wormlike or spherical animals)
Phylum: ECTOPROCTA
Phylum: PHORONIDA
Phylum: BRACHIOPODA (lamp shells)
Phylum: PRIAPULIDA
Phylum: SIPUNCULA (peanut worms)
Phylum: ECHIURA (spoon worms)
Phylum: ANNELIDA (segmented worms)
Phylum: TARDIGRADA
Phylum: PENTASTOMA (tongue worms)
Phylum: ONYCHOPHORA (velvet worms)
Phylum: LORICIFERA
Phylum: MOLLUSCA (soft-bodied animals with a mantle and foot)
    Class: MONOPLACOPHORA
    Class: APLACOPHORA (solenogasters)
    Class: POLYPLACOPHORA (chitons)
    Class: PELECYPODA (bivalves)
    Class: GASTROPODA (slugs, snails)
    Class: CEPHALOPODA (octopus, squid)
Phylum: ARTHROPODA (segmented animals with an external skeleton)
    Class: PYCNOGONIDA (sea spiders)
    Class: MEROSTOMATA (horseshoe crabs)
    Class: CRUSTACEA (lobsters, crabs)
    Class: ARACHNIDA (spiders, scorpions, mites, ticks)
    Class: INSECTA (insects)
    Class: CHILOPODA (centipedes)
    Class: DIPLOPODA (millipedes)
    Class: SYMPHYLA
    Class: PAUROPODA
Phylum: POGONOPHORA (tube worms)
Phylum: ECHINODERMATA (sea stars, brittle stars)
Phylum: CHAETOGNATHA (arrow worms)
Phylum: HEMICHORDATA (acorn worms)
Phylum: CHORDATA (chordates)
  Subphylum: TUNICATA (tunicates)
  Subphylum: CEPHALOCHORDATA (lancets)
  Subphylum: AGNATHA (lampreys, hagfish)
  Subphylum: GNATHOSTOMATA (fish and animals with four limbs)
    Class: CHONDRICHTHYES (sharks, rays)
    Class: OSTEICHTHYES (bony fish)
    Class: AMPHIBIA (amphibians)
    Class: REPTILIA (reptiles)
    Class: AVES (birds)
    Class: MAMMALIA (mammals)
      Subclass: PROTOTHERIA (monotremes)
      Subclass: THERIA
        Infraclass: METATHERIA (marsupials)
        Infraclass: EUTHERIA (placentals)

Cancer Genes

Mariano Barbacid of the National Cancer Institute reported in 1983 that a single base-pair, the building blocks of DNA, in the Ki-*ras* gene was different in the lung tumor of a patient from the Ki-*ras* gene in other tissues of the patient's body. Because the different base-pair was found only in the tumor, it was suspected that it arose as a mutation during the patient's lifetime rather than being inherited. Other research has suggested that the Ki-*ras* gene might be involved in cancer; single base-pair mutations have been shown to change a normal cell to a cancerous one.

Barbacid's report was one of the most definite clues in a trail of research into the genetic basis of cancer all through the early 1980s. The keys to this research were the discovery by Francis Peyton Rous in 1911 of a virus that causes a type of cancer called a sarcoma in chickens; the recognition in the early 1970s that viruses, such as the Rous sarcoma virus, work by copying RNA onto DNA, the reverse of the usual procedure; the discovery by several groups of investigators in the late 1970s that the key genes in such viruses are basically normal genes from the cells of multicellular organisms that have been "picked up" by the virus in the course of evolution; and, finally, the development in 1982 of important new methods for locating these genes in cells. Barbacid's work was built on these foundations.

src	5 Rous sarcoma virus (chicken)	*mos*	Moloney murine (mouse) sarcoma virus
abl	Abelson murine (mouse) leukemia virus	*sis*	Simian (woolly monkey) sarcoma virus
fes	ST feline (cat) sarcoma virus	Ha-*ras*	Harvey murine (rat) sarcoma virus
fps	Fujinami sarcoma virus (chicken; same gene as *fes*	Ki-*ras*	Kirsten murine (rat) sarcoma virus
fgr	Gardner-Rasheed feline (cat) sarcoma virus	*fos*	FBJ osteosarcoma virus (mouse)
ros	UR II avian (chicken) sarcoma virus	*myb*	Avian (chicken) myeloblastosis virus
yes	Y73 sarcoma virus (chicken)	*myc*	MC29 myelocytomatosis virus (chicken)
*erb*B	Avian (chicken) erythroblastosis virus	*erb*A	Avian (chicken) erythroblastosis virus
fms	McDonough feline (cat) sarcoma virus	*ets*	E26 virus (chicken)
raf	3611 murine (mouse) sarcoma virus	*rel*	Reticuloendotheliosis virus (turkey)
mil(mht)	MH2 virus (chicken; same gene as *raf*)	*ski*	Avian (chicken) SKV770 virus

Artificial Chromosome Placed in Yeast Cells

How do you find out what makes a chromosome work and be distributed during cell division? To answer such a question, Andrew W. Murray and Jack W. Szostak of the Dana-Faber Cancer Institute of Boston built their own yeast chromosomes. The scientists selected regions of natural yeast chromosomes that had already been identified as being important to yeast activity. They also selected regions known to be involved in cell division. They put these regions together, forming a "short" chromosome (about 50,000 base units compared with the average of about 150,000 units for natural yeast chromosomes). They then inserted their artificial chromosome into yeast cells.

When the yeast cells grew, the artificial chromosomes were sometimes duplicated in the new cells formed, just the way natural chromosomes are. However, although the artificial chromosomes replicated themselves during the early stages of cell division, they often made too many copies and were lost during division more frequently than natural chromosomes are.

Fly Bites Toad

"On the evening of 27 August 1982, by a small pond near Portal, Cochise County, Arizona" is the opening of a remarkable 1983 report by Rodger E. Jackman, Stephen Nowicki, Daniel J. Aneshansley, and Thomas Eisner. Tiny spadefoot toads seemed to be disappearing partially into the mud and dying awful deaths. Digging into the moist soil, the scientists found the answer: horsefly larvae just beneath the soil surface were trapping the toads, injecting them with venom, and sucking out their blood and body fluids.

What they observed on that summer evening was horseflies eating toads. The larvae hide in the mud, and emerge beneath an unsuspecting toad. Not only do they inject a paralyzing venom into the toads along with the initial bite, but they also hold the toad in place with a force as much as sixteen times the weight of the toad. Although toads commonly eat insects, this is the first record of any member of the horsefly's genus (*Tabanus*) eating a vertebrate.

Of course, toads that escape (by virtue of not hopping on a patch of mud containing a horsefly larva) will in the course of their lives eat the odd horsefly or two. How do the ecologists diagram the food chain for this one?

Firefly's Flash Brings Predators as Well as Mates

The image of airborne predators homing in on signal lights is fitting for a war movie, but now scientists have found that female fireflies of the genus *Photuris* exhibit precisely that behavior. These were the results of a series of investigations carried out in 1983 by James E. Lloyd and Steven R. Wing of the University of Florida at Gainesville. Using light-emitting diodes (LEDs) attached to poles, they counted the number of attacks by female fireflies under several conditions that simulated male firefly flight and flashing. In an experiment using three LEDs, one LED was unlit, one glowed constantly, and one flashed. Each LED was covered with an insect-catching paste. The glowing LED was attacked by 33 females; neither of the others was attacked. A steady glow is sometimes seen in nature by fireflies with defective light organs, and it gives the female *Photuris* a chance to locate the prey. In a second experiment, a flashing LED attached to a pole was moved in a trolling fashion. As long as the LED flashed in some manner, female fireflies attacked. When live males were attached to the LEDs, the females not only attacked, but they ate the males.

This work grew out of observations made while investigating another method of *Photuris* predation, imitation of the sexual signals of another species of firefly (see below).

This research is significant because it documents the first example of airborne, nighttime predation in insects. Firefly females are also alone in being the only predators known to use energy emission from their prey to sense its location.

GLOW, LITTLE GLOWWORM

When a male firefly of some species in the Americas sees a sexual invitation being blinked to him from a female firefly, he does not speed directly to the source of the invitation. Instead, he drops straight down to the ground. Eventually, of course, he approaches, but with great caution.

There is a good reason for this very unmasculine behavior. The female firefly may be a female of his own species, but she also may be a female cannibal. Several species of American fireflies are accomplished predators on other species of fireflies. These females are not only tricksters, mimicking the sexual signals of other species, but they are also voracious. In fact, if two of the female predators zoom in to attack the same male, they may end up attacking each other instead, with the winner consuming the loser, while the lucky male flies off to, one hopes, happier sexual encounters.

Understanding Masculine and Feminine Behavior

At one time scientists believed that in mammals the secretion of male sex hormones (androgens) produced one set of behaviors and the secretion of female sex hormones (estrogens) produced a different and separate set of behaviors. The effects were expected to be the same from species to species. Furthermore, it was assumed that behaviors produced by sex hormones were caused only by the actual presence of the hormone in the system, not by past influences of the hormone. Now these cherished beliefs are being modified as a result of recent experiments.

Take aggression, which has long been considered a typically male trait under the control of testosterone, a male hormone. Males were considered naturally aggressive, females nonaggressive. Work by Joseph F. DeBold of Tufts University casts doubt on this assumption. What he found when studying rodents was that male and female rodents are both aggressive, but that their patterns of aggression are different and the probable causes of aggressive behavior are different. Males are generally aggressive toward other males while females are aggressive toward either sex if intruded upon. Female aggression is most pronounced when the female is in her own territory.

Male aggression does appear to be determined by testosterone. That is, injection of testosterone makes male rodents more aggressive, while removal of the testes makes the males less aggressive.

Female aggression, on the other hand, is not affected by the injection of testosterone or by the removal of the ovaries, suggesting that female aggression is environmentally determined. However, if newborn female rodents receive testosterone injections, they develop the male rather than the female pattern of aggression.

Rodent research has been widely used as the model for what is known about the development of sexual differences. In rodents, androgens (the whole class of male hormones) are thought to have two separate effects. Not only do they stimulate development of masculine traits, but they also suppress the development of feminine characteristics. But do all organisms exhibit this pattern of response to androgens?

Research by Michael J. Baum of the Massachusetts Institute of Technology says no. Baum studied ferrets, which are not rodents, and found that while they are masculinized by androgens, there does not seem to be a suppression of feminine traits as in rodents. In fact, adult male ferrets injected with estrogens (female hormones) will exhibit behaviors considered female.

Since the adult males are producing androgens, this is evidence that the androgens do not suppress the estrogens.

Several lines of research lead to the conclusion that the hormones have a role in early development that causes certain traits to become fixed. These fixed traits are thereafter not modified greatly by the levels of hormones. Sometimes the effects of hormones on behavior can be shown to come from very subtle influences on development, as the research described in the next paragraph indicates.

Studying mice in utero, Frederick S. vom Saal, of the University of Missouri at Columbia, has made some interesting discoveries about the ratio of testosterone to estrogen, fetal position, and adult behavior. A fetus growing between two other male fetuses in the womb was found to have a higher testosterone to estrogen ratio than male fetuses in other fetal placements. In the adult males who had been in between two males as fetuses this placement resulted in more aggressive behavior, more paternal behavior, and enlargement of some areas of the reproductive system. In females this position resulted in more aggressive behavior, later onset of puberty, and less sexual receptivity. It is also worth noting that females placed under stress during pregnancy produce offspring that act as if they had been between two males in the womb. It was also found that males in a normal, unstressed pregnancy that are between two females show a greater sexual interest in females, but are also more aggressive toward the young.

All of this research supports the importance of sex hormones, but it also points up the variability of hormonal influence and the role of prenatal environmental factors in determining adult behavior. (See also related research on monkeys and people, page 189, and on the effects of testosterone on mathematical ability, page 369.)

MOTHER'S MILK

The average human mother produces about 1 kilogram (2 pounds) of milk each day while lactating, but the record is thought to be 6 kilograms (13 pounds) a day. This compares unfavorably with the record for a cow. The annual record yield for a cow was set by Beecher Arlina Ellen at 25,000 kilograms (56,000 pounds), or an average of almost 70 kilograms (150 pounds) a day. She reached 90 kilograms (200 pounds) on her best day. The bovine pace is still far short of the daily amount for a blue whale, estimated at 590 kilograms (1,300 pounds) a day. No one has actually milked a blue whale, of course.

Half of the whale's milk is fat, while human milk and cow's milk are only 3.8 percent fat. If you prefer low-fat milk, ass's milk may be for you—it averages only 1.4 percent fat.

Nerve Cell Sheath Gene is Found

Many electrical cables have a protective, insulating sheath around them. Some types of nerve cells have a similar covering called a myelin sheath. Degeneration of this sheath is one of the events that occurs in diseases such as multiple sclerosis and Guillain-Barré syndrome; it is considered the main cause of the symptoms.

Now biologists are on the track of a genetic basis for diseases of the myelin sheath. They have identified the specific gene that makes the sheath. The process they used to find the gene is typical of the way modern molecular geneticists work—often backwards, but with extreme effectiveness.

Scientists first synthesized sections of DNA coding for parts of myelin protein; this protein had been analyzed chemically. Then, using rat brains, which produce a myelin protein very similar to that of humans, biologists isolated messenger-RNA for myelin protein. Messenger-RNA carries the genetic blueprint for a protein from a gene to the cell's protein production area. Working backwards, the messenger-RNA was used to construct a strand of DNA, the gene for the sheath.

Having found the specific gene, scientists can now study brain cells of victims with myelin degeneration to see if messenger-RNA for myelin protein is produced. When such an experiment was conducted on mice suffering from a myelin deficiency disease, researchers found what they expected. No messenger-RNA for myelin protein was present and segments in the DNA that correspond to known sequences on the myelin protein gene were missing. Therefore, in the case of these mice at least, the disease has a genetic basis. Studies are now turning to human subjects in an effort to determine what, if any, genetic factors are involved in human diseases of the myelin nerve sheath.

CHEESE WITHOUT THE VEAL

The decline of veal consumption has been hurting the cheese industry. Why? Because the enzyme renin, used in cheese-making, is found in the stomach of unweaned calves slaughtered for veal. So the availability of renin is directly proportional to the popularity of veal. But the British have come to the rescue. J.S. Emtage and P.A. Lowe, with their associates at Celtech Limited in Slough, Berkshire, U.K., synthesized a gene for prorenin and inserted it into bacteria. The bacteria were cooperative; they produced 50,000 to 250,000 molecules of the enzyme. When treated with acid, the prorenin is converted to renin and actively converts the milk to cheese as effectively as natural renin.

Groundwater Ecology—Organisms Where None Were Thought to Be

Since 1916, ecological dogma held that the soil below the level of tree roots lacked any living things. If biologists found living things from greater depths, they kept the information to themselves out of fear of looking silly or misguided. However, recent work by John Wilson of the Environmental Protection Agency, William Ghiorse of Cornell University, and David Balkwill of Florida State University (Tallahassee) has proven the old idea to be wrong. The aquifers (natural underground water reservoirs) and water-laden subsoil they have studied are teeming with microbial life.

Their studies were initiated as a result of an earlier discovery: toxic pollutants also occur in aquifers. The significance of this finding is that, because of the depth—sometimes hundreds of meters (yards)—aquifers were assumed to contain pure, uncontaminated water. Now it is known that some aquifers are too polluted for human consumption and may remain so for thousands of years. Biologists are investigating the possibility of microbes' degrading some of these pollutants to nontoxic levels. Microbes at the surface of the Earth are essential in degrading pollutants.

Although in some aquifers pollution is found to be out of hand, in others the microorganisms seem to be degrading the pollutants fast enough to maintain water quality. Part of the reason for this difference seems to be the rate at which pollutants are entering and moving through the aquifer. Where pollutants are slow-moving, the microbes are better able to control pollution. Another factor seems to be the diversity of the microorganisms themselves, and whether the organisms present can, in fact, metabolize the specific pollutants.

Initial studies suggest that the food chains in deep underground water sources are quite simple. No protozoan, fungal, or invertebrate predators of bacteria were found. And the bacteria that were found appear to use quite different food sources from their surface and subsurface relatives. For the latter group, organic molecules such as cellulose are important food items, but these groundwater bacteria get their organic carbon from molecules bound to soil particles, a more difficult food source to exploit. In fact, they are consuming just those compounds that the surface microbes could not digest. What the surface microbes could digest was, in fact, digested long before reaching the aquifer.

Groundwater ecology is in its infancy, but its findings can provide a better understanding of how pollutants move deep underground and suggest ways to preserve these crucial freshwater sources.

The Catastrophic Evolution of Corn

The origin of corn is a mystery. Unlike other grains, which closely resemble wild ancestors, corn is not very much like anything else. Furthermore, corn arose suddenly in the archaeological record, while such grains as wheat and rice appeared slowly over thousands of years. Thus, an ancestor of corn has been postulated but never found. Did modern corn arise as a hybrid of the Central American grass teosinte and a now-extinct wild corn? Corn is clearly closely related to teosinte, and the two plants even now readily hybridize. Thus it is a widely held belief that the original corn was a similar hybrid with one ancestor now no longer known.

Hugh H. Iltis of the University of Wisconsin (Madison) has a different idea. Iltis has added his view of corn's origin to other hypotheses that have been debated by botanists and archaeologists for about a century in a confrontation dubbed the "corn wars."

Teosinte, like modern corn, has both male and female flowers on a single plant; the development of these flowers provides the basis of Iltis's argument. Teosinte is a branched grass with tiny ears containing two rows of about a dozen hard-shelled seeds, or kernels. These are produced from fertilized female flowers found along the sides of a stem. The male flower cluster of teosinte, called the tassel, is in the same position as the ears of modern corn, at the end of the stem, although the stems of modern corn are much shorter than those of teosinte. The ears of modern corn are fruit formed by the female flowers at the end of these short stems, while the male flower, or tassel, of corn is at the end of the topmost stem only.

As in mammals, the differentiation of sex in plants is accomplished by hormones. In the case of teosinte, the sex hormone is formed in the main plant and travels along the stems partway. The part of the stem affected by the hormone develops female flowers. The end of the stem, which is not reached by the hormone, develops the male flower. Thus, teosinte that developed very short stems, of the kind found in modern corn, would produce female flowers at the ends of the stems simply because the hormone could reach the ends of very short stems. The exception would be the topmost stem, which is farthest from the source of the feminizing hormone.

A change of this sort would explain another mystery surrounding the development of corn. Teosinte kernels are covered with a hard sheath, but corn kernels are not. The hard sheath is a part of the normal female flower, but it does not occur in male flowers that are sometimes feminized by accidents of growth.

It is known that stress, such as environmental harshness or disease, can cause flowers to take on opposite-sex characteristics. In fact, this process sometimes occurs in wild teosinte today. Modern corn may also develop partly formed ears at the base of the tassel if the sex hormone reaches that far. These observations form the core of Iltis's argument: Due to disease or environmental disaster, there was a sudden, drastic change in some teosinte plants with the tassel taking on female traits and a change in the kernels producing those typical of modern corn. All that would be required would be a change that shortened the stems. Noting these changes, all of which are desirable from the human's point of view, ancient farmers collected the seeds (kernels) for planting and further selected for the characteristics of modern corn. Such intervention was necessary for the corn to be perpetuated, since the altered form of teosinte would not be able to disperse its seeds properly. In fact, corn cannot be propagated without human help.

Iltis feels that his theory—the "catastrophic sexual transmutation theory"—will resolve the corn wars, but other scientists who have worked on the problem are not convinced that his theory is any better than others. In fact, some feel that the archaeological evidence is against Iltis. However, evolutionary biologists see in Iltis's theory a perfect example of how evolution can occasionally proceed by sharp jumps. This concept, known as "punctuated evolution," has been advocated by paleontologists who have previously been little interested in any events as recent as the development of corn. Now some of the paleontologists have joined the agricultural biologists in the debate over corn's origin.

GREENER PLANTS FOR BETTER YIELDS

Green plants grow by capturing solar energy and using it to make carbohydrates, fats, and other chemicals. The molecule that does the light-absorbing step is chlorophyll, which also makes green plants green.

It is commonly held by botanists that plants contain all the chlorophyll they can use. What places a limit on the rate of plant growth is the so-called "dark reaction," which follows the light absorption and is not dependent on the amount of chlorophyll.

A plant biologist at the University of California has found that this is not so. By increasing the chlorophyll content of plants by 62 percent through conventional breeding, he was able to increase the rate of photosynthesis by 36 percent. More photosynthesis means more carbohydrates, fats, and other desirable nutrients in the plants. He suggests plant breeding of common food crops for chlorophyll content may result in increased yields.

Learning in Humans, Monkeys, and Rabbits

Adult and young monkeys given a test of coordination do about equally well. But given a test of memory, the adults out-perform the young until the young are about six months old. At this age young monkeys do as well on memory tests as adults. But why? What does age have to do with brain-controlled skills?

Work with human amnesia victims may be providing an answer. Over the years, two amnesia victims, N.A. and H.M., have been studied extensively. Although they both experience amnesia, its expression is different in the two. H.M. cannot store new memories; he has not remembered anything new, with minor exceptions, since 1953. N.A. can store visual images but not verbal memories. Both, however, are able to learn new habits, including motor skills, such as hitting a tennis ball, and intellectual skills, such as puzzle-solving, that can be taught by trial and error.

H.M. had most of his hippocampus and amygdala, areas of the brain, removed to cure epilepsy. N.A. had a freak fencing accident that damaged the left side of his thalamus in the center of his brain. The differences in the areas of the brain damage may hold the key to the two types of amnesia and may suggest that humans—and other primates—have two pathways to memory. At least this is suggested by research by Mortimer Mishkin at the National Institute of Mental Health, who is studying memory in monkeys.

Mishkin believes that spatial memories, which give information a geography, are handled in this way: the higher brain stimulates the hippocampus, which triggers the thalamus to store the information. Emotional memories follow the second pathway: the higher brain stimulates the amygdala, which triggers the thalamus.

The view of memory as localized in specific parts of the brain was confirmed in 1983 when Richard F. Thomson of Stanford University was able to localize, for the first time, a specific memory trace in a mammal. He found exactly which tiny part of the brain contained a conditioned reflex in rabbits. Destroying that small part of the brain also removed the reflex, but had no other effect on the rabbit. When the memory trace was removed, however, the rabbit could not relearn the conditioned reflex.

But what about the age difference and learning intellectual versus motor skills in monkeys? Mishkin has an answer. The circuits for motor skills, habits, and trial-and-error learning may develop earlier than those for other types of memory. This is quite possible if different types of learning and different types of memory are in different parts of the brain.

1983: A Tough Year for Pandas

The population of pandas in their native habitat in China is declining, making breeding in zoos important to the survival of this species. Furthermore, 1983 saw the start of a major "die-off" of bamboo, the panda's main source of food. Bamboos of a given species are all genetically identical, so they tend to flower at the same age. After they flower and fruit, they die, giving rise to the next generation. The simultaneous flowering happens in the same year all over the world, no matter what the environmental conditions are. When such a dieback occurs in a species on which pandas depend for food, many of the pandas starve. Efforts by the Chinese government to preserve native pandas include supplemental feeding, but zoo preservation became more essential than ever.

In the United States, the only pandas are in The National Zoological Park in Washington, D.C. Since 1976, the National Zoo has been trying to mate Ling-Ling, the female, with Hsing-Hsing, the male, or with Chia-Chia, a male panda from the London Zoo. Until 1983 all such efforts had proved fruitless, although Hsing-Hsing seemed to be gradually improving in his ability to mate. In the spring breeding season in 1983, it appeared that Hsing-Hsing might have been successful for the first time, but the zoo took no chances. They also artificially inseminated Ling-Ling with sperm from Chia-Chia. Something worked, because the 250-pound Ling-Ling did become pregnant and gave birth on July 21. Ling-Ling did what panda mothers are supposed to do, but to no avail. The tiny cub had been born with a bacterial infection that killed it after only three hours.

In December 1983 Ling-Ling became ill with reduced kidney function, a condition that may have resulted from her pregnancy. Antibiotics helped her recover, but a 1984 mating led to a stillbirth in August.

The news on who fathered Ling-Ling's 1983 cub did not come in until late in 1983. It was Hsing-Hsing, her mate in Washington, D.C., not Chia-Chia, the male panda from the London Zoo.

WOULD YOU WANT INTERFERON FROM ARMYWORMS?

The latest in the repertoire of cells producing human chemicals are ovary cells of fall armyworms. The fall armyworm is a destructive moth caterpillar. Max D. Summers of Texas A&M University manipulated insect viruses to carry the instructions for producing human interferon, then used the viruses to get the gene into the ovary cells in laboratory culture. The result is abundant production of the gene product.

1984

Update 1984: Biology

THE DEVELOPMENT OF A WORM

Near the end of 1983, a major biological project reached a stage that had seemed unattainable at the outset. (The publications that relayed the news to nonspecialists did not appear until 1984.) Over a period of seven years the complete structure and development, cell by cell, for *Caenorhabditis elegans*, a tiny nematode (roundworm), was worked out. Although *C. elegans* has only 959 cells in its entire 1-millimeter (0.04-inch) length, fellow biologists recognized that working out the complete cell lineage was a monumental accomplishment.

The workers, mainly at the Medical Research Council's Laboratory of Molecular Biology in Cambridge, England, were aided by the fact that *C. elegans* is transparent and that its development time is only three and a half days. With special optical techniques, they could identify individual cells through a light microscope; if something was missed in one observation, there were plenty of other chances to observe the same development.

Perhaps the most surprising finding was that although *C. elegans* is bilaterally symmetrical, development is not symmetrical. Identical structures on each side of the mature roundworm begin from totally different cell locations in the embryo.

DEVELOPMENT OF SEGMENTS IN WORMS, INSECTS, AND MAMMALS

Major news continued to come from further investigations into the exact functioning of genes, especially their role in development. Development occurs as one gene after another is turned on and turned off. In a given cell (other than eggs or sperm), all of the genetic endowment of the creature is present, but only the part needed for that cell is activated. In development, a sequence of genes is activated, with each set of functioning genes directing an exact stage of that cell's growth. Uncovering the mechanism by which the genes are turned off and on has become a major goal of biology. Why, for example, do some cells become bone cells and others become muscle or nerve cells? In the beginning, all cells are alike. In some creatures, such as the lizard, cells can revert to an earlier stage, allowing the lizard to regenerate a lost tail. Other creatures (and other structures in the lizard) have cells that, once they have reached a certain stage of development, cannot go back.

The mysteries of development are still mysteries, but in 1984 there was considerable progress. Among the most amazing finds was the discovery by Allen Laughon and Matthew P. Scott in the United States (based on work of Thomas Kaufman) and a larger group of biologists in Basel, Switzerland, (headed by Walter Gehring) of the part of the genetic mechanism that apparently initiates the development of repeated segments.

Repeated segments have been identified as basic to the body plan of the "higher" animals. It is because vertebrates (mammals, birds, reptiles, amphibians, and fish) have body plans based on repeated segments—ribs and vertebrae, for example—that the earthworm has been considered a closer relative of humans than, say, the sponge, the octopus, or the roundworm.

What the researchers found was a section of DNA consisting of about 180 base pairs (see "Genes and Cancer," page 193, for background on terminology) that is almost exactly the same in vertebrates, annelid worms, and insects. Specifically, the section was located in fruit flies, frogs, earthworms, beetles, chickens, mice, and human beings—and also in some forms of yeast, the kind that reproduce sexually.

The evidence suggests that what this particular sequence does is to tell the cells that they should start making a new segment. For example, the sequence could be turned on to initiate the development of a segment. When that segment was complete, the conserved sequence would be turned on again to start another one. Presumably, one part of the sequence counts the segments, so that insects end up with three (quite different) segments as adults and human beings stop at twelve pairs of ribs. In fact, one lethal mutation in fruit flies occurs in a gene called *ftz*, an abbreviation for fushi tarazu (Japanese for "not enough segments") that is part of the conserved sequence.

RECOVERING EXTINCT GENES

Allan Wilson and Russell Higuchi of the University of California at Berkeley succeeded in cloning genes from the quagga, an African relative of the zebra that has been extinct for a hundred years. The genes were taken from a skin that has been preserved in salt. By inserting the genes into bacteria, Wilson and Higuchi were able to obtain multiple copies which then could be analyzed to determine the quagga's evolutionary history. Wilson was also able to obtain some parts of mammoth genes from the flesh of a mammoth that had been frozen for thousands of years.

Genes and Cancer

The early 1980s saw great strides in understanding the genetic basis of cancer. At one time, of course, no one knew what caused the devastation of cancer, a mysterious disease in which growths called tumors appeared in various organs of the body, leading almost invariably to death. Even after people had learned that if the first tumor was recognized early and surgically removed, cures could be effected, it was not clear what caused the tumor to develop in the first place.

It is still not certain at the most basic level what causes cancer, but experience in the twentieth century has gradually made the picture much clearer. After radioactive elements were discovered, it soon became tragically clear that exposure to radiation could cause cancer. Indeed, many of the early chemists working with radioactive substances died of cancer, including Marie Curie. Somewhat later, it became evident that certain substances, ranging from poison gases to tobacco, also caused cancer. Experimental work with animals detected viruses that caused cancer. Exposure to X-rays or to ultraviolet light, even sunlight, caused cancer. By the 1970s people were beginning to believe that everything caused cancer, especially as more and more common substances, ranging from foods such as peanut butter to fluorescent lights in offices were found to contribute to cancer. It was recognized, however, that none of these "causes" was actually the mechanism that started cancer growing in the body. Instead, the "causes" of cancer were simply a number of different mechanisms that must promote some single kind of change in a cell, making the cell change from a normal cell to a cancer cell.

When the cell became a cancer cell, certain changes occurred. The cell reproduced steadily, unlike normal cells, which reproduce only a few times. The cell developed the ability to float through the bloodstream and attach itself to another part of the body, where another tumor would start growing. The cell attracted the blood vessels to it, enabling the cancer to grow at the expense of surrounding tissues. The cause of any of these changes was not known. All that was known were some of the many environmental factors that could promote such changes in a cell.

In the meantime, cancer had become the second-leading killer in the United States. Massive efforts were launched to find the true cause of the disease. Finally, in the 1980s, the cause was definitely located in the hereditary mechanism of the cell.

To understand how this was accomplished and what it means, it is necessary to know how heredity works.

INHERITING TRAITS

When Gregor Mendel studied the inheritance of characteristics in pea plants in a monastery garden in the 1800s, he came to the conclusion that traits were transmitted from one generation to the next by physical particles in the eggs and sperm of the parents. In this century those particles were given the name "genes." As research on heredity progressed, genes were found to control virtually every function of a living organism, from its shape, color, and size to the composition of its body fluids, the way it metabolized food, and how it sensed the environment.

As our understanding of what genes control has expanded, our notion of exactly what a gene is has changed. More and more research and the development of new techniques have given us a clearer view of the chemical structure of a gene and of its intricate functioning. One major breakthrough was the now-familiar double-helix model of DNA, the chemical of which a gene is composed, proposed by James Watson and Francis Crick in 1953. Another was the proposition that one gene corresponds to one enzyme. Enzymes are the chemical regulators that control how life functions are expressed. However, as more was learned about genes and enzymes, this proposition was refined to "one gene, one polypeptide chain." The reason for this refinement was the realization that genes are related to proteins and that while enzymes are proteins, there are other protein classes as well, including structural proteins such as collagen and elastin, which have no enzyme activity. A peptide is any short string of amino acids, the substances that make up proteins. A polypeptide is a long peptide. Genes often make polypeptide chains rather than proteins, since many proteins are composed of several polypeptide chains, each of which is under the control of a separate gene.

WHAT IS A GENE?

Most cells have an organelle called the *nucleus*. The nucleus contains material called *chromatin*. When a cell is actively pursuing its function, the chromatin is seen as fine threads, but when the cell divides, these threads condense into rod-shaped bodies called *chromosomes*. In all organisms composed of complex cells, including humans, chromatin is made up of deoxyribonucleic acid (DNA) and associated proteins. (Bacteria and blue-green algae have simple cells and their chromosomes are almost exclusively DNA without associated proteins. Some viruses use RNA as their genetic material.) Each chromosome is a single molecule of DNA. Structurally, a gene is a segment of a DNA molecule. The 23 pairs of chromosomes in each human body cell carry an estimated 50,000 genes. Functionally, a gene is a template that has coded into its structure the directions for manufacturing a specific polypeptide chain used in the organism in which it resides.

THE STRUCTURE OF DNA

DNA has a double-helix structure. Essentially, this structure is what you get if you twist a ladder around its long axis. When the DNA double helix is unwound, it looks remarkably like a very long, very thin ladder of only a few repeating elements. The backbone, or sides of the ladder, are made of alternating units of phosphate and sugar (deoxyribose). The rungs are made of paired nitrogen-containing bases. Only four bases are present: adenine, thymine, guanine, and cytosine. The bases pair with each other in a specific way. Adenine always pairs with thymine, guanine with cytosine (Figure 1). The specificity of base-pairing provides the mechanism by which DNA can make exact copies of itself and transmit the instructions for protein manufacture from the nucleus to the cytoplasm of the cell, where proteins are formed.

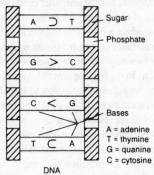

DNA

Sugar
Phosphate
Bases
A = adenine
T = thymine
G = quanine
C = cytosine

The DNA molecule, however, appeared to be too simple to carry enough coded information for all the many different inherited characteristics. Proteins and their polypeptide chains are composed of subunits called amino acids. Twenty different amino acids are the building blocks of living proteins. If each base coded for a different amino acid, only four amino acids could be accommodated. If pairs of bases coded for a single amino acid, the number of amino acids was sixteen. But if three bases coded for each amino acid, 64 different combinations were possible, far more than necessary for the twenty amino acids. As it turned out, this was how it worked. Three bases in a row, called a triplet or a codon, on the DNA strand code for a specific amino acid. In its simplest form, then, a gene is a sequence of DNA bases, taken three at a time, that codes for a polypeptide chain.

Genes can be thought of as lying side by side, like beads on a string along the DNA. Groups of genes that participate in the same reaction, say the genes for the enzymes necessary to convert substance A to substance B, and that also lie side by side, are called an *operon*. Preceding an operon on a chromosome are three other sequences of DNA responsible for controlling the function of the operon: a regulator gene, a promoter, and an operator. The promoter, operator, and operon are adjacent to one another; the regulator may or may not be adjacent to the promoter (Figure 2).

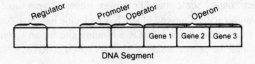

Regulator   Promoter Operator   Operon

| | | | | Gene 1 | Gene 2 | Gene 3 |

DNA Segment

How a Gene Works—Step 1: Transcription

The events that lead from the polypeptide code of a gene to the production of that polypeptide chain in the cytoplasm can be divided roughly into two parts, the first of which is called transcription. Simply stated, transcription is the process by which a mirror-image copy of a gene's instructions is made and sent to the polypeptide-manufacturing machinery in the cytoplasm of the cell. The production of copies of genetic instructions is controlled by the regulator, promoter, and operator. The operator is the on/off switch; the regulator produces a chemical, called a repressor, that controls the on/off switch; and the promoter binds with an enzyme that initiates the formation of copies of the genetic message when the switch is turned on.

The key idea is that when a repressor molecule is bound to the operator, the operon—that is, the gene that actually makes an enzyme—is repressed or turned off. There appear to be two ways a regulator gene can control the operator, which turns the operon off or on. In the first, to turn the operon off the repressor molecule attaches to the operator, switching the operon to the off position. However, if a second chemical, called an inducer, is present, the repressor binds with it instead; then the operator, free of its repressor, turns the operon on. In the second way, the repressor molecule alone cannot bind to the operator; this situation turns the operator on. If, on the other hand, a chemical called a corepressor is present, it bonds with the repressor and the two together bind to the operator, turning it off.

When an operator is turned on, DNA can begin to make a protein. The enzyme attached to the promoter combines with a unit of a second kind of nucleic acid—ribonucleic acid (RNA). Each RNA unit consists of a base attached to a sugar and a phosphate. The RNA joins with the exposed DNA base by base, forming a strand that is the mirror image of the sequence of bases on the DNA. This process continues until a stop sequence is encountered on the DNA strand. When the stop sequence is reached, the RNA strand is released and the DNA helix closes up. After the RNA strand is released it is pruned by enzymes in the cell's nucleus. Beginning and ending sequences are often removed, but even more surprising, stretches from the center of the strand are often snipped out. After being trimmed by enzymes to about half its transcribed length, the RNA strand is ready to leave the nucleus and participate in the formation of polypeptides.

How a Gene Works—Step 2: Translation

In the cytoplasm, the shortened RNA molecule, now called messenger-RNA (mRNA), is joined by two other forms of RNA. Ribosomal RNA (rRNA) and transfer RNA (tRNA) are assembled in the nucleus from DNA instructions. Ribosomal RNA is a major component of granular structures called ribosomes, which are the workbenches of polypeptide synthesis. Transfer RNA is present as small molecules shaped somewhat like clover leaves. These molecules act as ferries for the amino acids. One end of a tRNA molecule has three bases that code for a specific amino acid; at the other is a binding site for an enzyme. It is the enzyme that brings the proper amino acid to its specific tRNA.

Polypeptide synthesis begins when mRNA attaches to a ribosome. The ribosome reads the first three bases on the mRNA strand and bonds with the appropriate tRNA with its attached amino acid. The first tRNA is held in place while the ribosome reads the second triplet and accepts the second tRNA-amino acid complex. When the second tRNA is in place, a bond, called a peptide bond, forms between the first two amino acids. After the bond is formed, the first tRNA is released to pick up another amino acid. Still holding the second tRNA, the ribosome reads the next triplet, accepts the appropriate tRNA, the second peptide bond is formed, and the second tRNA is released. In this way, a polypeptide chain is assembled, amino acid by amino acid. The process continues until the ribosome reaches a sequence of mRNA bases that signals the termination of translation. At this time the polypeptide chain and the final tRNA are released into the cytoplasm.

Many copies of a polypeptide may be made off the same strand of mRNA. Ribosomes move along the mRNA strand one after another in follow-the-leader fashion, each producing a single copy of the coded polypeptide. When translation is completed, the ribosomes and RNA are broken down by the cell.

CANCER GENES

Among the plethora of processes encoded in a cell's DNA are the instructions for cell division. Normally cell division is an orderly process that simply replaces lost cells. These cells then mature and reliably carry out their specified functions. But sometimes something goes awry and cells divide out of control, a condition called cancer. These dividing cells do not mature; they remain in a juvenile state and continue dividing. In addition, the cells show another abnormal characteristic: they do not quit growing when they come into contact with other cells. These observations have led scientists to look to genes for an understanding of cancer. Cancer-causing genes, or oncogenes, are known to exist in certain viruses and it is clear that these viruses can transmit cancer to organisms other than humans. But the evidence definitively linking a virus and a human cancer has been elusive, although suggestive. Also, it is difficult to determine whether the oncogenes of viruses are naturally a part of the virus or a bit of DNA incorporated from a host. Currently the latter view seems more likely.

In the past few years twenty different genes that can cause cancer have been identified, and of these at least eleven are found in humans (see the table on page 180). The evidence for their cancer-causing ability comes from the laboratory. When DNA from these cells was placed in healthy mouse cells, for example, cancer developed. This research has resulted in the discovery of genes associated with lymph, breast, colon, lung, and bladder cancers in humans.

In their normal form, these genes, called proto-oncogenes, do not cause cancer. It is only when they are changed, or mutated, to oncogenes that they can contribute to cancer formation. The cause of the mutation might be any of the known carcinogens (cancer-producing agents): cigarette smoke, radiation, chemicals in food, pesticides, and so on.

The degree of mutation seems to be quite small, what biologists call a *point mutation* because the mutation affects only a single pair of DNA bases. The mutation also seems to be very specific, affecting only a given base pair in a gene.

For cancer actually to form, there appears to be at least a two-step process needed, a process affecting two different genes in a cell. The first step gives cells the ability to divide indefinitely, and the second step causes the cell to divide rapidly and alters the characteristics of the cell membrane. Genes thought to initiate the first step include *myc*, from a chicken leukemia virus; large T, from a rat polyoma virus; and *Ela*, from the human adenovirus. Step two seems to be associated with the *ras* gene of human bladder cancer, the *Elb* gene of the adenovirus, and the middle T gene from the polyoma virus. The other genes identified are expected to fall into these categories or to suggest new categories and new steps.

Researchers are excited about this line of research and expect to find a total of 30 to 50 proto-oncogenes. Once these have been identified, it may be possible to develop screening tests that will identify people at risk for different types of cancer. It is not unreasonable to expect that the identification of proto-oncogenes will also lead to an understanding of precisely how carcinogens interact with genes. Such information might enable us to neutralize known carcinogens by altering their chemical structure. This would certainly be desirable with carcinogens we tend to live with, such as pesticides and food additives.

Identifying proto-oncogenes and even understanding the mechanisms by which they become oncogenes does not mean an automatic end to cancer. The relationship between smoking and lung cancer has been well established for decades, but people still smoke. And the genetic basis for many inherited diseases is well documented, but these diseases can only be treated, not cured. However, understanding the genetic basis of cancer can help people think more clearly about the disease and make more informed decisions about their own exposure to carcinogens.

DETERRING PESTS WITH CITRUS SEEDS

A recent addition to the list of natural pesticides is citrus seeds. Actually, the deterring substances are bitter-tasting chemicals called limonoids found in the seeds, rind, and juice of citrus fruits. When placed on leaves, these substances were found to discourage fall armyworms and cotton bollworms from feeding on the leaves. Although sources of limonoids other than citrus seeds are available, they are inconvenient or expensive, while currently discarded grapefruit seeds alone might provide 300 metric tons of the chemicals.

CHEMISTRY

In terms of creating new chemicals, chemistry has probably never been so active as it is today. In fact, in March 1983 the American Chemical Society announced that it had recorded the six millionth chemical, while in May it supervised the tenth edition of *Chemical Abstracts* which contained short summaries of 25,000,000 reports, generally conceded to be a record. Chemistry continues to provide the material and techniques that make the modern technological society work.

At the same time, chemistry has come to occupy an unusual place among the sciences. Chemical reactions are explained by physicists. The applications of chemistry to biology and other sciences are so many that whole branches of these sciences, notably biochemistry, have become disciplines in themselves. Anthropologists, archaeologists, and geologists rely on chemistry for dating. Oceanographers use chemistry to trace ocean currents and to explain the formation of minerals. Astronomers are deeply involved in the chemistry of planets such as Venus, Jupiter, and Saturn, and of the planets' moons, such as Io and Titan. Other astronomers study the chemicals found in interstellar gas clouds. Chemists make the medicines and develop the tests that help us to remain or become healthy. About the only segment of the sciences that does not rely on chemistry on a regular basis is mathematics.

The result of this activity and pervasiveness is that chemistry as a discipline all its own is tending to disappear. The chemistry section of *The Science Almanac* is quite short, although that hardly reflects the amount of chemistry involved in other sections of the book.

In this section, we provide a broad picture of activity in the field of chemistry in the early 1980s, ranging from the upper atmosphere (an evaluation of the effects of fluorocarbons on the ozone layer) to the deep sea (the development of a chemical sponge for removing free oxygen from water). Probably the most notable single event during this time was one of the briefest: the creation for 5 milliseconds of another element.

1980

New Ideas on the Origin of Life

Scientists believe that some sort of chemical, nonliving evolution must have preceded the evolution of live organisms. Harold Urey and Stanley Miller in the early 1950s were among the first to suggest possible ways in which this proto-evolution may have come about. They proposed that primitive Earth had a strongly reducing atmosphere as the result of hydrogen left over from the gases and solid matter that coalesced to form the planet. Nitrogen, they felt, must have been in its most reduced form, which is ammonia, and atmospheric carbon would have been found in methane. The energy to initiate the evolution of life was assumed to have been supplied by lightning flashes and ultraviolet light from the sun. Miller filled a flask with these gases plus water and found that the passage of an electrical discharge through the vapor produced amino acids and the nucleotide bases required to form DNA. The gradual polymerization of these chemicals could have eventually led to living forms.

Some scientists today seriously question whether primitive Earth had the strongly reducing atmosphere used in Urey and Miller's experiment for any but a short time, if at all. For one thing, rocks 3,800,000,000 years old have been found in Greenland. These rocks contain carbonates that can only have come from an atmosphere containing appreciable amounts of carbon dioxide, an oxidized gas. The rocks must have solidified not long after the planet was formed, estimated at 4,600,000,000 years ago. Modern geochemists believe that the gases of the early atmosphere were spewed out of volcanoes and that they contained mostly free nitrogen, carbon dioxide, and water, with small amounts of carbon monoxide and hydrogen, but not methane and ammonia.

Laboratories in Washington, D.C., and Houston have generated amino acids and nucleotides from mixtures of nitrogen, carbon monoxide, and hydrogen. A computer model that assumes an early atmosphere of about the same composition as today, except for the absence of oxygen and the presence of small amounts of hydrogen and carbon monoxide, indicates that about 2,721,000,000 kilograms (3,000,000 tons) per year of formaldehyde would be produced under the influence of sunlight. In 10,000,000 years, this would accumulate in the oceans to a concentration high enough to begin to polymerize. The present oxygen-rich atmosphere had to await the invention of photosynthesis by green plants.

Contents of Atmospheric Air

Constituent	Percent by Volume
nitrogen	78.084
oxygen	20.946
carbon dioxide	0.033
argon	0.934
neon	0.00182
helium	0.00052
krypton	0.00011
xenon	0.0000087
hydrogen	0.00005
methane	0.0002
nitrous oxide	0.00005

There is also a variable amount of water vapor in the air.

Elements in Earth Surface

Constituent	Percent
oxygen	49
silicon	26
aluminum	7.5
iron	4.7
calcium	3.4
sodium	2.6
potassium	3.4
magnesium	1.9
hydrogen	0.88
titanium	0.58
chlorine	0.l9
carbon	0.09
all others	0.76

Elements Found in Normal Human Body

Element	Percent
oxygen	65.0
carbon	18.0
hydrogen	10.0
nitrogen	3.0
calcium	2.0
phosphorus	1.1
potassium	0.35
sulfur	0.25
sodium	0.15
chlorine	0.15
magnesium	0.05
iron	0.004
cobalt	0.00000016
copper	0.00015
manganese	0.00013
iodine	0.00004
cobalt	present
zinc	present
molybdenum	present

Major Dissolved Constituents of Sea Water

Constituent	Percent of Salt
chlorine	59.9
sodium	33.5
magnesium	4.0
potassium	1.17
silicon	1.3
calcium	1.2
zinc	0.0062
copper	0.000015
iron	0.0000062
manganese	0.00000062

Sources: **Air:** Reprinted with permission from the *Handbook of Chemistry and Physics.* Copyright CRC Press, Inc., Boca Raton, FL. **Surface:** Reprinted with permission of Richard Furnald Smith. **Body:** Reprinted from *Principles of Biochemistry* by White, Handler, Smith, and Stetten (1959) with permission of McGraw-Hill.

1981

Biological Electrodes

Probably the most frequently performed of all measurements in the average chemistry laboratory is for pH, which is commonly thought of as measuring the acidity or baseness of a compound, although it is actually a measure of hydrogen ion concentration (which determines acidity).

Measurements that are reliable, simple to use, and quick are essential both in laboratory work and in many industrial processes. This applies not only to pH, but also to the amount of various other elements or compounds in a chemical.

Before the invention of the glass electrode by Dole and MacInnes about 50 years ago, the measurement of pH was done by the use of pH-sensitive dyes, an inaccurate procedure (although familiar to most people through litmus paper), or by the use of a hydrogen electrode, which was accurate but slow and cumbersome. The glass electrode pH meter represented a significant improvement in simplicity and convenience. This device was also one of the first uses of a semipermeable membrane to measure chemical substances. The semipermeable membrane in this case is a thin film of glass fused onto the end of a glass tube. Inside the tube is an acid solution and a conducting wire. The glass electrode is placed in the solution whose pH is to be measured, along with a standard electrode, usually calomel (mercurous chloride) in potassium chloride. The glass membrane is permeable only to hydrogen ions and, to a slight extent, sodium ions. The different hydrogen ion concentrations on the two sides of the membrane produce an electrical potential which, when combined with the calomel electrode, creates an electric cell whose potential can be measured with an electrometer. Since the resistance across the membrane is high, the currents produced are small. Because of the small current, a sensitive electronic amplifier system is required to detect them.

The membrane is highly selective for hydrogen ions and will ignore all others, except sodium, for which a correction can be made.

The success of the glass electrode inspired chemists to search for other electrodes that would be specific for the substance they want to determine. Ideally, there would be a specific electrode for every substance that anyone would want to measure, but that goal is almost surely beyond our reach. However, much progress has been made in designing specific electrodes for various ions other than hydrogen. There are now electrodes for ions of potassium, calcium, silver, lead, and many other elements.

A major advance in measurement technology using electrodes was made in the 1970s with the development of electrodes sensitive to ammonia, carbon dioxide, hydrogen sulfide, and other gases. These devices are basically hydrogen-ion sensitive glass electrodes surrounded by a microporous synthetic membrane through which the gas molecules can diffuse. The gases change the pH of the solution inside the membrane, the ammonia toward the alkaline side, the others toward greater acidity. The pH changes are then measured by the glass electrode.

An ingenious use of electrodes of this type is seen in simple and specific methods for determining substances of biological importance. An example is the determination of glutamine, a derivative of the amino acid glutamic acid. A thin slice of pig kidney is attached to the end of an ammonia-sensing electrode by means of a piece of dialysis tubing. The electrode is then placed in the solution to be measured. An enzyme in the kidney slice, specific for glutamine, liberates ammonia from the glutamine. This travels through the dialysis membrane to the ammonia electrode, where it is detected. Many possible interfering substances were tested and found to give no response.

The life of the electrode is about 30 days, after which the kidney slice must be renewed. In a similar method, the tissue from the small intestine of a mouse was used to determine the compound adenine, important in energy production in cells.

Tissues of plant origin can also be used for specific organic electrodes. A method for determining the amino acid glutamic acid has been developed, for instance, in which a piece of tissue from the skin layer of the growing portion of yellow squash is used in conjunction with a carbon dioxide-sensing electrode. An enzyme in the skin specifically liberates carbon dioxide from the glutamic acid, which is then detected. Electrodes of this type are potentially so inexpensive that in the future they may be used once and discarded.

The most recent development is the replacement of animal and plant tissue with living bacterial cultures. For example, a histidine sensing electrode has been made by smearing a paste of the bacterial strain *Pseudomonas sp.* on the end of an ammonia electrode. It is not essential that pure strains be used. In fact, the amount of sugar in a substance has been measured using plaque from human teeth: the carbon dioxide liberated is detected by an appropriate sensor. When a bacterial electrode shows signs of exhaustion, it can be renewed simply by placing it in a suitable culture medium and letting the organisms grow. Eventually, however, it has to be replaced.

Glassy Metal Alloys Developed

It is easy to think of crystals only as gemlike minerals, but in fact the flat-sided cubes, tetrahedrons, and other shapes are the basic form that most objects we deal with in daily life take. Solids are formed from crystals, usually such tiny crystals that we cannot notice them without a microscope. Metals, for example, are all made of such tiny crystals.

But not every ordinary solid object is a solid in this sense. Glasses, for example, feel like solids, but in reality they are liquids that have been cooled rapidly. (Careful studies have shown, however, that ordinary glass at room temperature does *not* slowly flow, although you may read that it does.) Cooling increases the viscosity of liquids, and with many materials, such as the sodium calcium silicates that are the chief constituents of ordinary glass, the viscosity becomes so great that the atoms and molecules are frozen in place and are unable to form the crystal arrays that are characteristic of true solids. Metals crystallize very rapidly, so it is not easy to put a metal into the glassy state by rapid cooling. But recent technology has succeeded in doing so, producing metal glasses with unusual and potentially valuable properties.

Metallurgists at Allied Corporation have made metal glasses by cooling melts at rates of nearly 1,000,000°C (1,800,000°F) per second. The composition of these glasses is typically iron and nickel mixed with carbon, boron, or silicon. Some show promise for use in electric motors, because they are easily magnetized and thus do not suffer the energy losses of ordinary metals. Present powerline transformers, which undergo magnetization and demagnetization at the rate of 120 times per second, produce losses in the form of heat equivalent to 30,000,000,000 kilowatt-hours per year in the United States. Allied estimates that this loss can be reduced by two-thirds by using glassy metal cores. Other compositions show high-yield strength or high strength combined with lightness.

Another group at Battelle laboratories has prepared metal glasses by sputtering the atoms as a gas on the surface to be coated. The composition of the metal is about the same as that of low-chromium stainless steel. These sputtered alloys are highly resistant to corrosion because they lack the crystal grain boundaries at which corrosion usually begins. They retain corrosion resistance at temperatures up to 600°C (1,100°F), which makes them potentially useful in hot, highly corrosive environments, as in nuclear power plant reactors or piping for geothermal wells. The low-chromium content of these alloys is another advantage, since chromium is a material that the United States must import.

1982

Ozone: An Endangered Molecular Species

One of the threats to our environment of the early 1980s seemed to be to the ozone molecule. Ozone is a form of oxygen. It has the formula O_3, compared with O_2. O_2 is the form we breathe. Fortunately, in 1982 the expected danger from ozone was found to be considerably less than had been feared.

It may not be immediately clear why we need ozone, which when breathed is a highly toxic material, but the fact is that it is literally saving our skins. The radiation we get from the sun covers a wide range of wavelengths, from the short ultraviolet to the infrared. The wavelengths between 290 and 320 nanometers are called the "damaging ultraviolet," or DUV, because they can penetrate the skin and alter the DNA of the cells, possibly causing cancer. What saves us is a thin layer of ozone in the stratosphere that absorbs the DUV, converting it harmlessly to heat. The problem is that we are introducing a number of substances into the atmosphere that have the potential of destroying this protective shield of ozone.

Among the more recent substances introduced are the chloro-fluoromethanes (CFM). These compounds, known commercially as Freon, are widely used in air conditioners and refrigerators and to make plastic foams. When they escape into the atmosphere they react with ozone, converting it to O_2. The effect is a catalytic one; that is, the CFM molecule emerges unchanged from the reaction. As a result, one CFM molecule can destroy many molecules of ozone. A report of the National Academy of Sciences in 1979 warned of the potentially serious effect of CFM release. Such an increase in the amount of DUV reaching the ground can lead to an increased incidence of skin cancer, reduce crop yields, damage marine life, and contribute to the "greenhouse effect," which can make our planet warmer and warmer. (See the article on page 287 for more on the greenhouse effect.)

Such a conclusion warranted further study, which revealed that the danger was not so serious as it had seemed. Perhaps partly because the use of CFM as a propellant in sprays was stopped shortly after the threat to ozone was first observed, the panel reported in 1982 that the decrease in ozone would be only in the range of 5 to 9 percent. This estimate was further reduced two years later, in 1984, to a decrease of from 2 to 4 percent by late in the twenty-first century, not enough of a decrease to cause serious concern at present.

The Elements

Element	Symbol / Atomic No.	Type	Density	Boiling/ Melting Point(°C)	Parts per Million in Crust	Year Disc.	Derivation of Name
Actinium	Ac 89	R	10.07	B:3200 M:1050	Rare	1899	Greek *aktis*, a ray.
Aluminum	Al 13	M	2.70	B:1800 M:660	83,600	1827	Latin *alumen*, a substance having an astringent taste.
Americium	Am 95	TU	13.67	B:2607 M:994	Synthetic	1944	From *America*.
Antimony	Sb 51	M	6.7	B:1380 M:630.5	0.2	1450	Greek *antimonos*, opposed to solitude; symbol Sb is from Greek *stibi*.
Argon	A 18	G	1.784*	B:-185.7 M:-189.2	Rare	1894	Greek *argos*, neutral, inactive.
Arsenic	As 33	NM	grey:5.73 blk.:4.7 yel.:2.0	M:817	1.8	1649	Greek *arsenicos*, valiant or bold; from its action on other metals.
Astatine	At 85	R	N.A.	B:337 M:302	Rare	1940	Greek *astatos*, unstable.

Name	Symbol	Class	Density	B/M	Abundance	Date	Notes
Barium	Ba 56	M	3.5	B:1140 M:850	390	1774	Greek *baros*, heavy; because its compounds are rather dense.
Berkelium	Bk 97	R	N.A.	N.A.	Synthetic	1949	First made at U. of California at Berkeley.
Beryllium	Be 4	M	1.84	B:2970 M:1350	2	1797	Latin *beryllus*, Greek *beryllos*, gem.
Bismuth	Bi 83	M	9.8	B:1450 M:271	0.008	1450	German *weisse masse*, white mass; changed to *bismat* when Latinized.
Boron	B 5	NM	crys.:2.54 amor.:2.45	B:2550 M:2300	9	1808	Aryan *borak*, white.
Bromine	Br 35	G	3.12	B:58.8 M:-7.2	2.5	1825	Greek *bromos*, a stench; because of odor of vapor.
Cadmium	Cd 48	M	8.6	B:767 M:320.7	0.16	1817	Greek *cadmia*, earthy.
Calcium	Ca 20	M	1.54	B:1170 M:810	46,600	1808	Latin *calx*, *calcis*, lime.
Californium	Cf 98	R	N.A.	N.A.	Synthetic	1950	First made at U. of California.

Element	Symbol / Atomic No.	Type	Density	Boiling/ Melting Point(°C)	Parts per Million in Crust	Year Disc.	Derivation of Name
Carbon	C 6	NM	diamond:3.5 graphite:2.3 amor.:1.9	B:4200 M:>3500	180	PH	Latin *carbo*, coal.
Cerium	Ce 58	RE	6.8	B:3257 M:795	66.4	1804	From asteroid *Ceres*, discovered in 1801.
Cesium	Cs 55	M	1.87	B:670 M:28.5	2.6	1860	Latin *caesius*, bluish-gray.
Chlorine	Cl 17	G	3.214*	B:-34.6 M:-100.6	126	1774	Greek *chloros*, grass-green; from the color of the gas.
Chromium	Cr 24	M	7.1	B:2200 M:1615	122	1797	Greek *chroma*, color; because many of its compounds are colored.
Cobalt	Co 27	M	8.9	B:2900 M:1480	29	1735	Greek *kobolis*, a goblin.
Copper	Cu 29	M	8.92	B:2310 M:1083	68	PH	Latin *cuprum*; from which the symbol Cu is derived; for island of Cyprus.

Element	Symbol / No.		Density	B / M	Abundance	Year	Etymology
Curium	Cm 96	R	13.5	M:1340	Synthetic	1944	After Pierre and Marie Curie.
Dysprosium	Dy 66	RE	8.54	B:2330 M:1407	Rare	1886	Greek *dysprositos*, difficult of access.
Einsteinium	Es 99	R	N.A.	N.A.	Synthetic	1952	After Albert Einstein.
Erbium	Er 68	RE	4.77	B:2500 M:1522	3.46	1843	From *Ytterby*, region in Sweden.
Europium	Eu 63	RE	5.24	B:1489 M:826	2.1	1896	From *Europe*.
Fermium	Fm 100	R	N.A.	N.A.	Synthetic	1953	After Enrico Fermi, Italian physicist.
Fluorine	F 9	G	1.70*	B:-187 M:-223	544	1886	Latin *fluere*, to flow.
Francium	Fr 87	R	N.A.	B:677 M:27	Rare	1939	From *France*.
Gadolinium	Gd 64	RE	7.87	B:3000 M:1312	6.1	1886	After Johan Gadolin, Finnish chemist.

Element	Symbol / Atomic No.	Type	Density	Boiling/ Melting Point(°C)	Parts per Million in Crust	Year Disc.	Derivation of Name
Gallium	Ga 31	M	5.89	B:1700 M:29.75	19	1875	Latin *Gallia*, France; also Latin *gallus*, a cock, pun on first name of Lecoq de Boisbaudran, discoverer.
Germanium	Ge 32	M	5.46	B:2830 M:958.5	1.5	1886	From *Germany*.
Gold	Au 79	M	19.3	B:2600 M:1063	0.002	PH	Anglo-Saxon *gold*; Sanskrit *juel*, to shine; Symbol Au is from the Latin *aurum*, shining dawn.
Hafnium	Hf 72	M	13.1	B:>5400 M:1700	2.8	1923	From *Hafnia*, ancient name of Copenhagen.
Hahnium	Ha 105	R	N.A.	N.A.	Synthetic	1969	After Otto Hahn, German physicist.
Helium	He 2	G	0.1785*	B:-269.0 M:-272.2	Rare	1868	Greek *helios*, the sun; helium was first observed in the atmosphere of the sun.
Holmium	Ho 67	RE	8.8	B:2720 M:1470	1.26	1878	From *Holmia*, Latinized form of *Stockholm*.

Element	Symbol / No.	Group	Density	B.P. / M.P.	Abundance	Year	Etymology
Hydrogen	H 1	G	0.0899*	B:−252.7 M:−259.1	1,520	1766	Greek *hydor*, water, plus *gen*, forming.
Indium	In 49	M	7.28	B:1450 M:155	0.24	1863	Latin *indicum*, indigo.
Iodine	I 53	NM	4.94	B:184.4 M:113.5	0.46	1811	Greek *iodes*, violet; from the color of the vapor.
Iridium	Ir 77	M	22.42	B:4500 M:2350	0.001	1803	Greek *iris*, a rainbow; from changing color of salts.
Iron	Fe 26	M	7.86	B:3000 M:1535	62,200	PH	Anglo-Saxon *iren*; the symbol Fe is from Latin *ferrum*.
Krypton	Kr 36	G	3.708*	B:−152.9 M:−157	Rare	1898	Greek *kryptos*, hidden.
Lanthanum	La 57	RE	6.15	B:3454 M:826	34.6	1839	Greek *lanthanein*, to be concealed.
Lawrencium	Lr 103	R	N.A.	N.A.	Synthetic	1961	After Ernest Lawrence, American physicist.

Element	Symbol / Atomic No.	Type	Density	Boiling/ Melting Point(°C)	Parts per Million in Crust	Year Disc.	Derivation of Name
Lead	Pb 82	M	11.34	B:1620 M:327.5	13.0	PH	Anglo-Saxon *lead*; the symbol Pb is from the Latin *plumbum*.
Lithium	Li 3	M	0.534**	B:1200 M:186	18	1817	Greek *lithos*, stony.
Lutetium	Lu 71	RE	0.849	B:3327 M:1652	Rare	1906	Latin *Lutetia*, city in Gaul (now Paris).
Magnesium	Mg 12	M	1.74	B:1110 M:651	27,640	1755	Latin *Magnesia*, a district in Asia Minor.
Manganese	Mn 25	M	7.2	B:1900 M:1260	1,060	1774	Latin *magnes*, magnet; because of confusion with magnetic iron ores.
Mendelevium	Md 101	TU	N.A.	N.A.	Synthetic	1955	After Dmitri Mendeleev, Russian chemist.

		LM		B/M		PH	
Mercury	Hg 80		13.546**	B:356.9 M:-38.87	0.08		For Roman god *Mercurius*; symbol Hg is from Latin *hydrogyrum*.
Molybdenum	Mo 42	M	10.2	B:5560 M:2622	1.2	1778	Greek *molybdaina*, galena (lead ore).
Neodymium	Nd 60	RE	6.96	B:3030 M:840	39.6	1885	Greek *neo*, new, plus *didymos*, twin (with lanthanum).
Neon	Ne 10	G	0.9004*	B:-245.9 M:-248.7	Rare	1898	Greek *neo*, new.
Neptunium	Np 93	TU	N.A.	B:3902 M:640	Rare	1940	For planet *Neptune*.
Nickel	Ni 28	M	8.9	B:2900 M:1452	99	1751	German *Nickel*, Satan (Old Nick).
Niobium	Nb 41	M	8.4	B:4742 M:1950	20	1801	Latin *Niobe*, daughter of Tantalus; formerly named columbium.
Nitrogen	N 7	G	1.2505*	B:-195.8 M:-209.9	19	1772	Latin, forming niter, a compound containing nitrogen.
Nobelium	No 102	TU	N.A.	N.A.	Synthetic	1957	After Alfred Nobel; made at Nobel Institute in Stockholm.

Element	Symbol / Atomic No.	Type	Density	Boiling/ Melting Point(°C)	Parts per Million in Crust	Year Disc.	Derivation of Name
Osmium	Os 76	M	22.48	B:5500 M:2700	0.005	1803	Greek osme, smell; for its vile odor.
Oxygen	O 8	G	1.42904*	B:−183 M:−218.8	456,000	1774	Greek oxyx, sharp, as in acids, plus gen, forming; false idea that oxygen is the characteristic element of acids.
Palladium	Pd 46	M	12.16	B:2927 M:1553	0.015	1803	Greek goddess Pallas; asteroid Pallas.
Phosphorus yellow (alpha) Yellow (beta) violet black red	P 15	NM	1.82 2.36 2.70 2.20	B:280 M:44.1 M:600 M:590	1,120	1669	Greek phosphoros, light-bringer; glows in the dark because of rapid oxidation.
Platinum	Pt 78	M	21.37	B:3827 M:1773.5	0.01	1803	Spanish plata, silver; from the color of the metal.

Element		TU	N.A.	N.A.			
Plutonium	Pu 94				Synthetic	1940	For planet *Pluto*.
Polonium	Po 84	R	9.4	B:962 M:254	Rare	1898	Named by discoverer Marie Curie for her native Poland.
Potassium	K 19	M	0.87	B:760 M:62.3	18,400	1807	English *potash*, a compound of potassium. Symbol K is from Latin *kalium*.
Praseodymium	Pr 59	RE	6.48	B:3020 M:940	9.1	1885	Greek *prasios*, green, plus *didymos*, twin.
Promethium	Pm 61	RE	7.2	B:2460 M:1160	Rare	1941	For Greek god *Prometheus*, who stole fire from heaven.
Protactinium	Pa 91	R	15.37	M:<1600	Rare	1918	Latin *proto*, first, plus actinium.
Radium	Ra 88	R	5?	B:1140 M:960	Rare	1898	Latin *radius*, ray.
Radon	Rn 86	RG	9.72 *	B:-61.8 M:-71	Rare	1900	*Radium* plus *on*, as in *neon*.
Rhenium	Re 75	M	10	B:5630 M:3000	0.0007	1925	Latin *Rhenus*, Rhine.
Rhodium	Rh 45	M	12.44	B:4500 M:1985	Rare	1803	Greek *rhodios*, rose-like; for red of its salts.

Element	Symbol / Atomic No.	Type	Density	Boiling/ Melting Point(°C)	Parts per Million in Crust	Year Disc.	Derivation of Name
Rubidium	Rb 37	M	1.53	B:700 M:38.5	78	1861	Latin *rubidus*, red; from red lines in its spectrum.
Ruthenium	Ru 44	M	12.06	B:3900 M:2450	Rare	1844	From *Ruthenia*, where ore was first found in Urals.
Rutherfordium	Rf 104	R	N.A.	N.A.	Synthetic	1969	After Ernest Rutherford, British physicist.
Samarium	Sm 62	RE	7.7	B:1900 M:>1300	7.0	1879	For Scandinavian mineral samarskite.
Scandium	Sc 21	M	2.99	B:2727 M:1200	25	1879	For Scandinavia.
Selenium	Se 34	NM	red:4.50 gray:4.84	B:690 M:220	0.05	1818	Greek *selene*, the moon.
Silicon	Si 14	NM	2.4	B:2600 M:1420	273,000	1823	Latin *silex*, flint.
Silver	Ag 47	M	10.5	B:1950 M:960.5	0.08	PH	Assyrian *sarpu*; Anglo-Saxon *soelfor*; the symbol Ag is from Latin *argentum*.

Element	Symbol	Category	Density	B/M	Abundance	Year	Notes
Sodium	Na 11	M	0.97	B:880 M:97.5	22,700	1807	English *soda*, a compound of sodium; the symbol Na is from Latin *natrium*.
Strontium	Sr 38	M	2.6	B:1150 M:800	384	1808	For *Strontian*, a town in Scotland.
Sulfur	S 16	NM		B:444.6	340	PH	Sanskrit *solvere*.
alpha-rhombic			2.07	M:112.8			
beta-			1.96	M:119			
monoclinic							
Tantalum	Ta 73	M	16.6	B:5425 M:2850	1.7	1802	For Greek mythical king Tantalus, condemned to thirst; because of its insolubility.
Technetium	Tc 43	R	11.50	B:4877 M:2172	Synthetic	1937	Greek *technetos*, artificial; first artificial element.
Tellurium	Te 52	M	6.25	B:1390 M:452	Rare	1782	Latin *tellus*, the earth.
Terbium	Tb 65	RE	8.33	B:2800 M:1356	1.18	1843	For *Ytterby*, region in Sweden.
Thallium	Tl 81	M	11.86	B:1457 M:303.5	0.7	1861	Greek *thallos*, a young green shoot.
Thorium	Th 90	RM	11.2	B:3000 M:1845	8.1	1828	For Norse god Thor.

Element	Symbol / Atomic No.	Type	Density	Boiling/ Melting Point(°C)	Parts per Million in Crust	Year Disc.	Derivation of Name
Thulium	Tm 69	RE	9.32	B:1727 M:1550	0.5	1879	Greek *Thoule*, northernmost region of the world.
Tin	Sn 50	M	white: 7.31** gray: 5.75**	B:2260 M:231.85	2.1	PH	Anglo-Saxon *tin*; the symbol Sn is from the Latin *stannum*.
Titanium	Ti 22	M	4.5	B:3260 M:1800	6,320	1791	For Titans of Roman mythology.
Tungsten	W 74	M	19	B:5927 M:3370	1.2	1783	Danish *tung sten*, heavy stone; symbol W is from German *Wolfram*.
Uranium	U 92	R	18.7	B:3818 M:1132	2.3	1789	For planet *Uranus*.
Vanadium	V 23	M	5.69	B:3000 M:1710	136	1830	For Scandinavian goddess Vanadin.
Xenon	Xe 54	G	5.851*	B:-107.1 M:-112	Rare	18 98	Greek *xenos*, strange.

Name	Symbol						Origin / Notes
Ytterbium	Yb 70	RE	7.01	B:1427 M:824	3.1	1907	From *Ytterby*, region in Sweden.
Yttrium	Y 39	RE	3.80	B:2927 M:1490	31	1794	From *Ytterby*, region in Sweden.
Zinc	Zn 30	M	7.1	B:907 M:419.4	76	PH	German *zink*.
Zirconium	Zr 40	M	6.44	B:4377 M:1900	162	1789	Arabic *zargun*, gold color.
Element 106 (unnamed)	N.A.	N.A.	N.A.	N.A.	Synthetic	1974	Claimed by Soviet Union and U.S.; from decay of nobelium.
Element 107 (unnamed)	N.A.	N.A.	N.A.	N.A.	Synthetic	1981	Identified in West Germany following earlier, disputed claim by Dubna (Soviet Union).
Element 109 (unnamed)	N.A.	N.A.	N.A.	N.A.	Synthetic	1982	Created in West Germany by bombarding bismuth with iron ions.

Key: *Grams per liter, 0°C; ** grams per liter, 20°C; G gas; M metal; NM nonmetal; PH prehistoric; R radioactive; RE rare earth; TU transuranic

The Periodic Table

6 — atomic number
C — chemical symbol
12.01 — atomic mass
Carbon — name of element

	alkali metals IA				transition metals				
Period 1	1 **H** 1.01 Hydrogen	alkaline earth metals II A							
Period 2	3 **Li** 6.94 Lithium	4 **Be** 9.01 Beryllium							
Period 3	11 **Na** 23.00 Sodium	12 **Mg** 24.31 Magnesium	III B	IV B	V B	VI B	VII B		VIII
Period 4	19 **K** 39.10 Potassium	20 **Ca** 40.08 Calcium	21 **Sc** 44.96 Scandium	22 **Ti** 47.90 Titanium	23 **V** 50.94 Vanadium	24 **Cr** 52.00 Chromium	25 **Mn** 54.94 Manganese	26 **Fe** 55.85 Iron	27 **Co** 58.93 Cobalt
Period 5	37 **Rb** 85.47 Rubidium	38 **Sr** 87.62 Strontium	39 **Y** 88.91 Yttrium	40 **Zr** 91.22 Zirconium	41 **Nb** 92.91 Niobium	42 **Mo** 95.94 Molybdenum	43 **Tc** 98.91 Technetium	44 **Ru** 101.07 Ruthenium	45 **Rh** 102.91 Rhodium
Period 6	55 **Cs** 132.91 Cesium	56 **Ba** 137.34 Barium		72 **Hf** 178.49 Hafnium	73 **Ta** 180.95 Tantalum	74 **W** 183.85 Tungsten	75 **Re** 186.2 Rhenium	76 **Os** 190.2 Osmium	77 **Ir** 192.2 Iridium
Period 7	87 **Fr** (223) Francium	88 **Ra** (226) Radium		104 (261)	105 (262)	106 (263)	107 (267)	108 (265)	109 (266)

rare earth elements

Lanthanide series	57 **La** 138.9 Lanthanum	58 **Ce** 140.12 Cerium	59 **Pr** 140.91 Praseodymium	60 **Nd** 144.24 Neodymium	61 **Pm** (145) Promethium	62 **Sm** 150.4 Samarium	63 **Eu** 151.96 Europium
Actinide series	89 **Ac** (227) Actinium	90 **Th** 232.03 Thorium	91 **Pa** 231.04 Protactinium	92 **U** 238.03 Uranium	93 **Np** 237.05 Neptunium	94 **Pu** (244) Plutonium	95 **Am** (243) Americium

A figure in parentheses is the isotope of longest known half-life. No stable isotope is known.

Notes on Periodic Table of the Elements

1. The creation in the laboratory of element 107, atomic weight 262, was reported in 1981. In 1982 element 109, atomic weight 266, was created by a similar procedure. These elements have not yet been named.

2. Elements in the same vertical column have similar chemical properties. For example, in column 1a, Li, Na, K, Rb, Cs and Fr are similar. The first element, hydrogen, is an exception to this rule.

3. Certain of these groups of elements have been given special names. Group 1a, with the exception of hydrogen, is called the alkali metals. The 2a elements are the alkaline earths; those in 7a are the halogens; the 0 column at the end are the rare or noble gases; elements 57 (lanthanum) through 71 (lutetium) are the rare earths; elements 89 (actinium) and beyond are the actinide series and are all radioactive; elements 44, 45, 46, 76, 77, and 78 are the platinum group.

			IIIA	IVA	VA	VIA	VIIA	noble gases O
					nonmetals			2 **He** 4.00 Helium
			5 **B** 10.81 Boron	6 **C** 12.01 Carbon	7 **N** 14.01 Nitrogen	8 **O** 16.00 Oxygen	9 **F** 19.00 Fluorine	10 **Ne** 20.18 Neon
	IB	IIB	13 **Al** 26.98 Aluminum	14 **Si** 28.09 Silicon	15 **P** 30.97 Phosphorus	16 **S** 32.06 Sulfur	17 **Cl** 35.45 Chlorine	18 **Ar** 39.95 Argon
28 **Ni** 58.71 Nickel	29 **Cu** 63.55 Copper	30 **Zn** 65.37 Zinc	31 **Ga** 69.72 Gallium	32 **Ge** 72.59 Germanium	33 **As** 74.92 Arsenic	34 **Se** 78.96 Selenium	35 **Br** 79.90 Bromine	36 **Kr** 83.80 Krypton
46 **Pd** 106.4 Palladium	47 **Ag** 107.87 Silver	48 **Cd** 112.40 Cadmium	49 **In** 114.82 Indium	50 **Sn** 118.69 Tin	51 **Sb** 121.75 Antimony	52 **Te** 127.60 Tellurium	53 **I** 126.90 Iodine	54 **Xe** 131.30 Xenon
78 **Pt** 195.09 Platinum	79 **Au** 196.97 Gold	80 **Hg** 200.59 Mercury	81 **Tl** 204.37 Thallium	82 **Pb** 207.2 Lead	83 **Bi** 208.98 Bismuth	84 **Po** (209) Polonium	85 **At** (210) Astatine	86 **Rn** (222) Radon

other metals

64 **Gd** 157.25 Gadolinium	65 **Tb** 158.93 Terbium	66 **Dy** 162.50 Dysprosium	67 **Ho** 164.93 Holmium	68 **Er** 167.26 Erbium	69 **Tm** 168.93 Thulium	70 **Yb** 173.04 Ytterbium	71 **Lu** 174.97 Lutetium
96 **Cm** (247) Curium	97 **Bk** (247) Berkelium	98 **Cf** (251) Californium	99 **Es** (254) Einsteinium	100 **Fm** (257) Fermium	101 **Md** (258) Mendelevium	102 **No** (255) Nobelium	103 **Lw** (256) Lawrencium

4. Elements on the left-hand side of the table are generally metals (for example, magnesium, calcium, strontium) and those on the right-hand side are generally nonmetals (for example, oxygen, sulfur, selenium). Elements in the middle (for example, phosphorus, arsenic, antimony) show a combination of metallic and nonmetallic properties. As before, hydrogen is an exception. With increasing atomic weight, the elements become more metallic in behavior.

5. The term *oxidation state* refers to the number of electrons an element gains or loses when it reacts with other elements. The metals, like sodium, tend to lose electrons and become positive; the nonmetals, like chlorine, tend to gain electrons and become negative. The term *electron configuration* refers to the way the electrons are arranged around the nucleus of the atom. For example, sodium has two in the first (K) shell, eight in the second (L) shell, and one in the third (M) shell. The total, eleven, is equal to the atomic number.

New Element Created

The extraordinary sophistication of many modern scientific techniques is well illustrated by the announcement of the creation on August 29, 1982, of element 109, the heaviest element so far, by a West German team. The yield was not great, only a single atom. Nevertheless, the experts think there is about a 99 percent chance that the atom of the new element actually was created.

The technique employed, called cold fusion, involved the bombardment of a target of bismuth-209, (bismuth having an atomic weight of 209) with nuclei of iron-58 in a heavy-ion accelerator. The idea behind the method is that if the energy is great enough, the repulsive forces between the nuclei will be overcome and the nuclei will fuse, producing a new atom. This is an event of low probability, about one in 100,000,000,000,000. It required a week of bombardment to produce the single fusion.

Results of four independent methods all showed that element 109 had indeed been made. The velocity with which it recoiled from the target was measured and found to agree with the assumed mass and the energy of the bombarding nucleus. Another part of the experiment was to use a velocity filter to separate 109 from other nuclei and to measure the time of flight to the detector and the energy with which it struck. Both measurements agreed with theoretical calculations. Finally, the decay products of the atom were determined and found to be as expected. Decay occurred after 5 milliseconds according to the following pattern (where the symbol for the unnamed new element is X; X is also used for the as yet unnamed element 107). First a neutron was emitted from the two particles as they fused.

$$_{109}X^{266} \text{ (alpha emission)} \blacktriangleright \, _{107}X^{262} \text{ (alpha emission)} \blacktriangleright$$
$$_{105}Ha^{258} \text{ (electron capture)} \blacktriangleright \, _{104}Rf^{258} \blacktriangleright$$
$$\text{(fission to smaller nuclei)}$$

The subscripts indicate atomic numbers, the superscripts, atomic weights. Thus, $_{109}X^{266}$ means the new element has atomic number 109 and atomic weight 266. The atomic number is the number of protons in the nucleus; it establishes the identity of the atom. Since element 109 was formed by the fusion of $_{26}Fe^{58}$ (iron) and $_{83}Bi^{209}$ (bismuth), a little arithmetic shows that a neutron must have been ejected to account for a loss of one in the atomic weight at the beginning of the reaction. An alpha particle is a helium nucleus, $_2He^4$. Electron capture is the capture of an electron by a neutron in the nucleus, forming an additional proton.

Trees as Important Sources of Energy and Chemicals

One of the many ways in which we are changing the surface of the earth is by removing trees and shrubs and replacing them with annual crops such as wheat and corn. One reason for the replacement of perennials by annuals is that the cultivation of the annuals has been seen as more profitable. This idea may be changing as research on trees progresses. The yield of biomass, that is, total weight of organic matter from trees, has recently undergone considerable improvement. An experimental forest in South Carolina that produced an annual yield of 3 dry tons per hectare (2.47 acres) before 1960 is now yielding 11 tons, and 30 tons is considered attainable. In Brazil, a eucalyptus species that formerly yielded 23 tons was increased gradually to 40, and 100 seems possible.

An important job for the chemist is to upgrade the value of wood so that the cultivation of trees becomes more profitable. The organic matter of wood is 50 percent cellulose, which can be broken down to give the sugar glucose; 20 percent hemicellulose, which can be broken down to sugars of other types; and lignin, which is a polymer of a substance called coniferyl aldehyde. The lignin can be extracted out with solvents and used as an adhesive.

Top 25 Chemicals Produced in the United States

Figures below are in kilograms × 1,000,000,000 and refer to 1982.

Rank	Name	Output	Rank	Name	Output
1	Sulfuric Acid	29.4	14	Propylene	5.6
2	Nitrogen	15.9	15	Urea	5.4
3	Ammonia	14.1	16	Ethylene dichloride	4.5
4	Oxygen	13.3	17	Benzene	3.5
5	Lime	12.9	18	Carbon dioxide	3.4
6	Ethylene	11.2	19	Methanol	3.3
7	Sodium hydroxide	8.4	20	Ethylbenzene	3.0
8	Chlorine	8.3	21	Vinyl chloride	3.0
9	Phosphoric acid	7.9	22	Styrene	2.7
10	Sodium Carbonate	7.2	23	Xylene	2.4
11	Toluene	6.9	24	Terephthalic acid	2.3
12	Nitric acid	6.9	25	Hydrochloric acid	2.3
13	Ammonium nitrate	6.7			

Reprinted in part with permission from *Chemical and Engineering News*, May 2, 1983. Copyright 1983 American Chemical Society

1983

A New Stone Age Dawning?

Synthetic materials made of plastics have given us many useful products, but they suffer from certain drawbacks. One is that most of them require petroleum or natural gas as a raw material, a nonrenewable resource that we are using up at a great rate. Another is that their manufacture requires considerable amounts of costly energy. It is estimated that it takes six times as much energy to make a block of polystyrene plastic as to make a block of Portland cement of the same size. Another problem is the flammability of organic materials and the highly toxic smoke produced when they burn. These concerns have led to renewed interest in inorganic materials as substitutes for plastics.

The problem with inorganic materials is that, except for some metals, they are stiff and lack toughness, or resistance to impact. Are these inherent properties of inorganics? The study of inorganics produced by living organisms suggests they are not. Abalone shells are 99 percent inorganic calcium carbonate, but the shells have a tensile strength and toughness comparable to Plexiglass. The abalone seems to have achieved this by producing calcium carbonate in uniform sheets separated by thin layers of protein, and stacking the sheets in neatly arranged layers. This implies that the weakness of the manmade inorganics may not be inherent but may lie in the way they are prepared.

Modern research on Portland cement, which has been used in construction on a vast scale for many years, indicates that this is so. Portland cement, made by heating limestone with clay, is mostly calcium silicate. On adding water, the calcium silicate hydrates and binds the particles into a rock-like mass. But as the material dries, some of the water evaporates, leaving pores that range in size from less than a millionth of a millimeter to a millimeter or more. The total volume occupied by the pores is 25 to 30 percent in ordinary cement. A recent study shows that it is the large pores that produce brittleness in cement.

Researchers in England have found that the maximum pore size can be drastically reduced, to a few thousandths of a millimeter, by proper preparation. This is achieved by starting with materials that are graded in size, adding a small amount of organic polymer to improve flow, then kneading thoroughly after adding water. The result is macro-defect-free (MDF) cement, a material that has bending strength comparable to aluminum and is so flexible it can be made into a spring! Its fracture toughness is about ten times as great as that of ordinary cement and is about equal to that of jade. This makes it possible to produce tubes of the material on a lathe. By reinforcing with organic fibers, MDF cement can be made very resistant to impact and bendable, like a metal. Attractive cups and saucers can be made of it.

The new cement, however, is not a refractory, which is a material that retains its desirable properties on heating, the way that porcelain does. Inorganic research has turned up ways to make refractories at low temperatures, saving energy. Some natural inorganic materials, of which vermiculite is an example, consist of sheets of silicon and oxygen atoms connected in chains or rings. The stacks of sheets are separated by layers of *cations*, which are positively charged atoms or groups of atoms. A research team in Australia has found that if the cations are removed, the sheets can be readily separated by stirring in water. The thick slurry that results can then be dried to make strong, flexible, translucent, and refractory films. An inorganic foam, much like foams made of polystyrene, can be prepared by whipping air into the suspension and allowing it to dry. The foam retains the refractory property.

Another low temperature way to make refractories is to react aluminum chloride with phosphoric acid in ethyl alcohol at temperatures below the freezing point of water. A precipitate is obtained consisting of a cubic array of aluminum, oxygen, and phosphorus atoms, which are prevented from linking up into a three-dimensional polymer by the alcohol present. Gentle heating drives off the alcohol and permits linkage, forming a glassy refractory that is stable at temperatures as high as 1600°C (3,000°F). By adding aluminum oxide to the alcohol precipitate, a ceramic can be formed at a temperature as low as boiling water. Using similar techniques, low-temperature manufactured glasses of outstanding clarity have been formed.

Such developments, which utilize an inexhaustible supply of raw materials and only modest amounts of energy, have encouraged some to believe that a new Stone Age may be in our future.

SYNTHETICS AND ENERGY

Although it takes 25 percent more energy to make a cotton-polyester shirt than one of pure cotton, the energy cost of the cotton shirt is actually 90 percent higher if the longer lifetime and lower maintenance cost of the cotton-polyester is taken into account. The production of synthetics actually takes only a small fraction of the annual oil and gas production, about 3 percent.

---

pH (Acidity or Alkalinity) Values of Various Materials

The pH of completely pure water is considered to be neutral and has the value 7.0. Anything below this value is acid, anything above is alkaline. A change of one pH unit corresponds to a tenfold change in acidity or alkalinity. For example, a solution of pH 5 is ten times as acid as one of pH 6.

Acids	pH
sulfuric	1.2
citric	2.2
lactic	2.4
acetic	2.9
carbonic	3.8
prussic	5.1
boric	5.2

Bases	pH
sodium hydroxide	13.0
lime	12.4
sodium carbonate	11.6
ammonium hydroxide	11.1
borax	9.2
sodium bicarbonate	8.4

These values are all for 0.1N solutions, that is, solutions containing 0.1 gram equivalents per liter. For sulfuric acid, for example, this would correspond to 4.9 grams per liter. The carbonic acid and lime values are for saturated solutions.

Biological	pH
blood plasma	7.3-7.5
duodenal contents	4.8-8.2
feces	4.6-8.4
gastric contents	l.0-3.0
milk, human	6.6-7.6
saliva	6.5-7.5
urine	4.8-8.4

Foods	pH
apples	2.9-3.3
asparagus	5.4-5.8
bananas	4.5-4.7
beans	5.0-6.0
beer	4.0-5.0
cabbage	5.2-5.4
cheese	4.8-6.4
cherries	3.2-4.0
cider	2.9-3.3
eggs	7.6-8.0
grapefruit	3.0-3.3
grapes	3.5-4.5
lemons	2.2-2.4
limes	1.8-2.0
milk, cow's	6.3-6.6
oranges	3.0-4.0
peaches	3.4-3.6
pears	3.6-4.0
pickles	3.0-3.4
plums	2.8-3.0
potatoes	5.6-6.0
raspberries	3.2-3.6
rhubarb	3.1-3.2
soft drinks	2.0-4.0
spinach	5.1-5.7
strawberries	3.0-3.5
tomatoes	4.0-4.4
vinegar	2.4-3.4
water, drinking	6.5-8.0
wine	2.8-3.8

Reprinted with permission from *Handbook of Chemistry and Physics*, copyright CRC Press, Inc., Boca Raton, FL.

Chemical Processes and Nuclear Magnetic Resonance

Nuclear magnetic resonance (NMR) is a good example of how seemingly theoretical advances in physics make possible very practical advances in chemistry and other sciences. (See also the article on direct medical applications of nuclear magnetic resonance in the Medicine section, page 437.)

NMR depends on the fact that in many processes in nature, odd and even numbers produce quite different results. For example, atoms with odd numbers of nucleons (protons and neutrons, when considered together in atoms, are called *nucleons*) act like tiny magnets and are therefore affected by magnetic fields. For the scientist studying biological processes, the important atoms are hydrogen H-1, carbon C-13, and phosphorus P-31. The numbers indicate the number of nucleons in the atom. H-1 is ordinary hydrogen, with a single proton; C-13 is a form of carbon (an *isotope*) that is naturally present as 1.1 percent of carbon samples, which consist mostly of C-12; and P-31 is ordinary phosphorus.

In the NMR apparatus, the material or living thing being studied is subjected to a powerful magnetic field, which causes the atoms of H-1, C-13, and P-31 (as well as many other atoms with odd numbers of nucleons) to line up parallel to the field. Then a beam of radio waves is sent through the sample perpendicular to the magnetic field. If you gradually change the frequency (or wavelength) of the radio waves, you will observe a certain frequency where resonance takes place. At this specific frequency, the magnetic atoms will absorb the radio energy and change their orientation relative to the magnetic field.

What makes this of great interest to the chemist is that the precise frequency at which resonance occurs depends on the chemical environment of the resonating atom. For example, the H atoms in ethyl alcohol, $CH_3 \cdot CH_2 \cdot OH$, produce three different peaks in an NMR spectrum, one for the H atoms in the CH_3 group, another for those in the CH_2 group, and another for the H in OH. The height of each peak is proportional to the number of atoms of a given kind. As a result, the NMR spectrum can often tell the chemist what sorts of groups are present in the molecule and in what amount.

The great advantage of NMR for the biochemist or physiologist who wants to study the chemical reactions that occur in living organisms is that it is noninvasive and as far as we know does no harm to the cells or tissues. Thus, if a culture of an organism such as *E. coli*, a common bacterium, is placed in an NMR instrument, three separate P-31 peaks can be observed for the three phosphorus atoms in adenosine triphosphate, ATP, a chemical that is vital for energy production in the cell. Another peak is observed for inorganic phosphate. Two peaks can be observed for nicotinamide adenine dinucleotide, a molecule with two phosphorus atoms that is important in intracellular oxidation.

NMR has been used to supply a simple answer to the question, What is the pH (hydrogen ion concentration) inside the living cell? This is done by observing the position of the peak for P-31 in inorganic phosphate. Inorganic phosphate ions in solution exist either as $H_2PO_4^-$ or $HPO_4^=$, the dashes indicating the negative charge on the ions. The pH depends on the ratio of the two forms. The position of the inorganic P peak in the NMR spectrum also depends on this ratio and is an accurate measure of the cell pH. Using this technique, it was found that the pH in the *E. coli* cell remains constant at 7.4 as long as an energy source such as glucose is present. The peak due to P in ATP is seen to increase as the cell stores energy.

The chemical transformation of glucose in the animal body can be followed by feeding the subject glucose that has been labeled with C-13, the form of carbon that is visible in NMR. Labeling is done by synthesizing glucose in the laboratory using an intermediate that is high in C-13. A rat was fed this labeled glucose, and the progress of the compound was followed by NMR. It was possible to watch the glucose disappear from the stomach and appear in the liver, where it became stored as glycogen. (The glycogen produced this way was also labeled, since the glycogen obtained its carbon from the carbon in the glucose.) Then the C-13 in the liver started to disappear as the animal was exercised, which requires breakdown of glycogen.

More recently, NMR has been used to study the metabolism in muscles of human patients. In one case, a patient suspected of having a genetic disease in which phosphorylase does not function was studied with NMR. Phosphorylase is an enzyme needed to convert glycogen, the form in which glucose is stored in the liver, to glucose. Without glucose, the muscle tissues do not have an adequate source of energy. The patient's arm was placed in an NMR instrument, and the P-31 spectrum was followed as he exercised his arm. In normal muscle, the pH drops during exercise from 7.0 down to 6.4, then returns on resting. The subject showed a pH of 7.1, which went up to 7.2 instead of going down. The explanation is that normal muscle tissue obtains energy by converting glucose to lactic acid; this accounts for the drop in pH. In the patient, the absence of phosphorylase prevented this process from taking place.

Carbon Chains Cut Using Special Catalysts

Petroleum is mostly a mixture of many different hydrocarbons, compounds containing only carbon and hydrogen. Most of these hydrocarbons are of the saturated type, called *alkanes*. In alkanes the carbon atoms are united to one another by single bonds only, in contrast to the alkenes, which contain at least one pair of carbons that are united by a double bond, and the alkynes, which contain at least one pair of triple-bonded carbons. A typical alkane is ethane. The corresponding alkene is ethylene, and the alkyne of this series is acetylene.

The alkanes make excellent automobile fuels, but for another important use of petroleum, as a chemical raw material, they present serious problems. Specifically, the alkanes resist chemical attack by all but powerful reagents at high temperatures. These reagents usually attack the carbon chains indiscriminately, producing mixtures that may be low in the desired compounds and hard to separate. It is this relative lack of reactivity that gives the alkanes their other name, paraffins, meaning "little affinity."

Research at Yale, Berkeley, the University of Michigan, and other places has recently shown that by using special catalysts consisting of platinum, iridium, or rhenium bonded to chelating agents such as triphenylphosphine, the carbon chains can be attacked under much milder conditions and with greater specificity. At present, most of these catalyzed reactions break only the carbon-hydrogen bonds, permitting the introduction of other groups in place of the hydrogen. But others go further and break carbon-carbon bonds, thus making possible a greater variety of reactions. These developments can open up many new uses for petroleum as a chemical raw material.

Ultrasonic Waves Speed Up Chemical Reactions

Chemists at North Dakota University found in 1983 that many chemical reactions, especially those in which metal catalysts are used, proceed faster and give better yields if the reaction mixture is subjected to the action of ultrasonic waves. Reactions in which improvements have been noted include the addition of hydrogen to alkenes and alkynes using palladium catalysts, as well as the production of anilines and organic silicon compounds. In some cases, ultrasonics permitted lower temperatures and pressures to be used than ordinarily. It may be that the ultrasonic waves bring the components of the reaction into more intimate contact.

An Oxygen Sponge for Use Under Water

Animals live by oxidizing sugars and other substances with oxygen from the air. To capture the necessary oxygen, animals have developed a protein, hemoglobin, which is able to take up oxygen from the lungs and release it to the tissues where it is needed. Fish have gills instead of lungs and use the oxygen that is dissolved in water. Chemists at Duke University in 1983 produced a kind of artificial gill that can trap the oxygen from water, suggesting various interesting possibilities.

They have succeeded in immobilizing hemoglobin in a polyurethane sponge. When seawater is flowed past the sponge, oxygen is extracted by the hemoglobin. Experiments are under way aimed at using this "hemosponge" to provide oxygen for submarines and divers. The chemists believe that if water were passed through a canister 0.9 meters (3 feet) wide by 3 meters (ten feet) long, filled with hemosponge, enough oxygen for 150 people could be extracted. To free the oxygen from its combination with the hemoglobin, either an electrochemical process or a mechanical vacuum may be used. A diver with a small canister of this type would be able to move freely under water without air hoses connecting him with the surface or an oxygen tank. One of the first uses, however, may be in conventional submarines. These vessels operate on batteries when submerged, since they cannot operate combustion engines without using valuable oxygen. The hemosponge would permit such submarines to recharge batteries without surfacing.

Another possible application of great promise is under development by an engineering firm in Hawaii. This envisions the use of oxygen extracted from seawater to operate fuel-burning underwater power plants. Such engines could be as much as 300 times more efficient than battery-powered sources and could result in a revolution in underwater vehicles.

THINGS TO THINK ABOUT WHILE EATING ICE CREAM

Ice cream is less than 50 percent air by law, and most producers run as close to the 50 percent limit as they can. Most ice cream contains 10 percent milk fat, the most expensive ingredient. The milk fat in a quart of ice cream is produced when a half-ton cow eats 30 pounds of feed and turns part of it into fat. Still, most of ice cream consists of millions of tiny air bubbles and tiny fat particles held together by a small amount of sugar water. The sugar keeps the water from freezing completely. In a quart of ice cream there are less than 9 tablespoons of sugar water.

It still tastes great.

1984

Update 1984: Chemistry

GAP IN PERIODIC TABLE IS BRIEF

The same West German team that created a single atom of Element 109 in 1982 obtained triple the yield of Element 108 two years later, closing the gap in the periodic table that they had opened by skipping 108 earlier. The new element has an atomic weight of 265, and was created by bombarding lead-208 with iron-58, essentially the same method used to create element 109, although lead was substituted for bismuth as the target (see "New Element Created," page 222).

Not only did they obtain three atoms of element 108, but each one lasted three times longer than had been predicted by theory. Although the lifetime was still only milliseconds for each atom, the longer life than expected was encouraging for those who hope to create still heavier atoms. A longer lifetime means that atoms may continue to last long enough to be detected.

In the meantime, some doubt has arisen about element 109. The single atom that was produced in 1982 has been the only one created. Even the West German team has failed to reproduce the result.

BOND...H. BOND

Any student of chemistry since the 1950s has been aware of the importance of the different kinds of chemical bonds. Bonds are, among other things, what holds matter together. What is surprising is that chemists keep finding new types of bonds. One kind of bond, for example, causes two hydrogen atoms to link with an oxygen atom, forming water. This is the familiar covalent bonding. Another kind of bond occurs when the hydrogen in one atom of water attracts the oxygen from another atom. This is the less familiar bonding known as hydrogen bonding, but it is what gives liquid water some of its important properties, such as its boiling point, which would be much lower without the attraction of the hydrogen bonds, and the fact that ice floats in water.

Hydrogen bonds occur in compounds with oxygen, fluorine, and other reactive chemicals. The new discovery—actually made in 1983, but not fully explored until 1984—is that hydrogen bonds can form with unreactive carbon. Ferdos Al-Mashta and coworkers at the University of Anglia (United Kingdom) discovered the new bonding effect while experimenting with the polymerization of ethylene.

The Nobel Prize for Chemistry

Date	Name (Nationality)	Achievement
1901	Jacobus van't Hoff (Netherlands)	Laws of chemical dynamics of weak solutions.
1902	Emil Fischer (Germany)	Sugar and purine synthesis.
1903	Svante Arrhenius (Sweden)	Theory of ionic dissociation.
1904	Sir William Ramsay (England)	Discovery of inert gas elements and placement in periodic table.
1905	Adolf von Baeyer (Germany)	Work on organic dyes.
1906	Henri Moissan (France)	Isolation of fluorine.
1907	Eduard Buchner (Germany)	Discovery of noncellular fermentation.
1908	Sir Ernest Rutherford (England)	Studies of radioactivity, alpha particles, and the atom.
1909	Wilhelm Ostwald (Germany)	Work on catalysis.
1910	Otto Wallach (Germany)	Work with terpenes.
1911	Marie Curie (France)	Discovery of radium and polonium.
1912	Victor Grignard (France) Paul Sabatier (France)	Grignard's discovery of Grignard reagents and Sabatier's catalytic hydrogenations compounds.
1913	Alfred Werner (Switzerland)	Study of the bonding of atoms.
1914	Theodore Richards (U.S.)	Determination of atomic weights.
1915	Richard Willstätter (Germany)	Research on chlorophyll in plants.
1916	No award	_____
1917	No award	_____
1918	Fritz Haber (Germany)	Synthesis of ammonia from atmospheric nitrogen.
1919	No award	_____
1920	Walther Nernst (Germany)	Study of thermodynamics.
1921	Frederick Soddy (England)	Discovery of isotopes of radioactive elements.
1922	Francis W. Aston (England)	Discovery of isotopes with mass spectrograph and related work with atomic weights.
1923	Fritz Pregl (Austria)	Microanalysis of organic substances.
1924	No award	_____

Date	Name (Nationality)	Achievement
1925	Richard Zsigmondy (Germany)	Study of colloid solutions.
1926	Theodor Svedberg (Sweden)	Development of and work with the ultracentrifuge.
1927	Heinrich O. Wieland (Germany)	Development of steroid chemistry.
1928	Adolf Windaus (Germany)	Study of cholesterol.
1929	Sir Arthur Harden (England) Hans von Euler-Chelpin (Sweden)	Research in sugar fermentation and the role of enzymes in it.
1930	Hans Fischer (Germany)	Analysis of blood heme.
1931	Karl Bosch (Germany) Friedrich Bergius (Germany)	Invention of high-pressure methods of chemical production.
1932	Irving Langmuir (U.S.)	Study of monomolecular films.
1933	No award	_____
1934	Harold C. Urey (U.S.)	Discovery of heavy hydrogen.
1935	Frédéric Joliot-Curie (France) Irène Joliot-Curie (France)	Synthesis of new radioactive elements.
1936	Peter J.W. Debye (Netherlands)	Study of dipolar moments.
1937	Sir Walter N. Haworth (England) Paul Karrer (Switzerland)	Haworth's work on carbohydrates and vitamin C and Karrer's on carotenoids, flavins, and vitamins.
1938	Richard Kuhn (Germany)	Carotenoid and vitamin research (was not allowed to accept prize until after World War II).
1939	Adolf Butenandt (Germany) Leopold Ružička (Switzerland)	Butenandt's study of sexual hormones (was not allowed to accept prize until after World War II) and Ružička's work with atomic ring structures and terpenes.
1940	No award	_____
1941	No award	_____
1942	No award	_____
1943	György de Hevesy (Hungary)	Use of isotopes as tracers.
1944	Otto Hahn (Germany)	Discovery of atomic fission.
1945	Artturi Virtanen (Finland)	Invention of fodder preservation.
1946	James B. Sumner (U.S.) John H. Northrop (U.S.) Wendell Stanley (U.S.)	Sumner's first crystallization of an enzyme, Northrop's crystallization of enzymes, and Stanley's crystallization of a virus.

Date	Name (Nationality)	Achievement
1947	Sir Robert Robinson (England)	Study of plant alkaloids.
1948	Arne Tiselius (Sweden)	Research on serum proteins.
1949	William Francis Giauque (U.S.)	Study in low-temperature chemistry.
1950	Otto Diels (Germany) Kurt Alder (Germany)	Synthesis of organic compounds of the diene group.
1951	Edwin M. McMillan (U.S.) Glenn T. Seaborg (U.S.)	Discovery of plutonium and research on transuranium elements.
1952	Archer J.P. Martin (England) Richard L.M. Synge (England)	Separation of elements by paper chromatography.
1953	Hermann Staudinger (Germany)	Study of polymers.
1954	Linus C. Pauling (U.S.)	Work on chemical bonds.
1955	Vincent Du Vigneaud (U.S.)	Synthesis of polypeptide hormones.
1956	Sir Cyril Hinshelwood (England) Nikolai Semenov (U.S.S.R.)	Parallel work on kinetics of chemical chain reactions.
1957	Sir Alexander Todd (England)	Study of nucleic acids.
1958	Frederick Sanger (England)	Discovery of structure of insulin.
1959	Jaroslav Heyrovský (Czechoslovakia)	Polarography for electrochemical analysis.
1960	Willard F. Libby (U.S.)	Invention of radiocarbon dating.
1961	Melvin Calvin (U.S.)	Work on chemistry of photosynthesis.
1962	John C. Kendrew (England) Max F. Perutz (England)	X-ray study of the structure of hemoproteins.
1963	Giulio Natta (Italy) Karl Ziegler (Germany)	Synthesis of polymers for plastics.
1964	Dorothy M.C. Hodgkin (England)	Analysis of structure of vitamin B_{12}.
1965	Robert B. Woodward (U.S.)	Synthesis of organic compounds.
1966	Robert Mulliken (U.S.)	Study of atomic bonds in molecules.
1967	Manfred Eigen (Germany) Ronald G.W. Norrish (England) George Porter (England)	Study of high-speed chemical reactions.
1968	Lars Onsager (U.S.)	Theoretical basis of diffusion of isotopes.

Date	Name (Nationality)	Achievement
1969	Derek H.R. Barton (England) Odd Hassel (Norway)	Determination of three-dimensional shape of organic compounds.
1970	Luis F. Leloir (Argentina)	Discovery of sugar nucleotides and their biosynthesis of carbohydrates.
1971	Gerhard Herzberg (Canada)	Study of geometry of molecules in gases.
1972	Christian B. Anfinsen (U.S.) Stanford Moore (U.S.) William H. Stein (U.S.)	Pioneering research in enzyme chemistry.
1973	Ernst Otto Fischer (Germany) Geoffrey Wilkinson (England)	Chemistry of ferrocene.
1974	Paul J. Flory (U.S.)	Study of long-chain molecules.
1975	John W. Cornforth (Australia) Vladimir Prelog (Switzerland)	Cornforth for work on structure of enzyme-substate combinations and Prelog for study of asymmetric compounds.
1976	William N. Lipscomb (U.S.)	Study of bonding in boranes.
1977	Ilya Prigogine (Belgium)	Nonequilibrium theories in thermodynamics.
1978	Peter Mitchell (England)	Study of biological energy transfer.
1979	Herbert C. Brown (U.S.) George Wittig (Germany)	Brown for study of boron-containing organic compounds and Wittig for phosphorus-containing organic compounds.
1980	Paul Berg (U.S.) Walter Gilbert (U.S.) Frederick Sanger (England)	Berg for development of recombinant DNA and Gilbert and Sanger for development of methods to map the structure of DNA.
1981	Kenichi Fukui (Japan) Roald Hoffmann (U.S.)	Application of laws of quantum mechanics to chemical reactions.
1982	Aaron Klug (South Africa)	Developments in electron microscopy and study of acid-protein complexes.
1983	Henry Taube (U.S.)	New discoveries in basic mechanism of chemical reactions.

EARTH SCIENCE

The earth sciences include all aspects of the study of Earth, from its minerals and geologic processes to its oceans and atmosphere. Because the Earth's past can only be understood in terms of the development of life—for example, the 20 percent free oxygen in the atmosphere is believed to be entirely the product of living things—paleontology is also included among the earth sciences. Some earth sciences, such as meteorology, affect us every day, while others, such as major geologic processes that shape the Earth, are so slow that we usually do not notice them at all.

In the 1980s, however, the immediate effects of geology became very apparent to people living in the United States, for a volcano erupted in the state of Washington. The eruption of Mount St. Helens, coming as it did early in 1980 and continuing at least until the present (most geologists think that it will continue for another twenty years or so) focused attention on how slow geologic processes can suddenly become very fast ones. Significant earthquakes in California and Idaho, with predictions of more serious California quakes to come, as well as the strong possibility of a major volcanic eruption in California, continued to remind us of the potential for destruction in geology.

The range of the earth sciences is expanded to include biology, with the discovery of a new—to us—ecosystem. Under the seas, there are a number of places, deep and shallow, where sulfur-laden hot water vents into the ocean. Around these vents, bacteria live without photosynthesis at temperatures much higher than that of boiling water. Strange tube worms and giant clams live on the bacteria. In shallower waters, other, more familiar creatures also make the sulfur bacteria one of their main sources of food.

As far as more conventional paleontology is concerned, there was a lot of progress in understanding how the dinosaurs actually lived. While earlier excavations revealed a lot about individual dinosaurs and their species, a new trend focuses on determining what assemblages of dinosaur remains reveal about nesting habits and life styles in general.

In the more commonplace area of climate and weather, the early 1980s were, more than any time in the past, a period when everyone talked about the weather and wondered if anything could be done about it. The immediate crisis was a weather pattern, called El Niño, that caused far more destruction of life and general misery than all the volcanoes and earthquakes of the period—although some think that the strong weather pattern was caused by a volcano. The volcano El Chichón, which erupted in southern Mexico, has probably affected the weather in the whole world, an effect that may still be continuing; and El Chichón may have provided the spark that turned a weather disturbance in the Pacific into the giant El Niño.

While the short-term effects of El Chichón and El Niño were devastating, the long-term concern was about the "greenhouse effect." Carbon dioxide, released into the air by burning fossil fuels and by the reduction of forests, has been identified as the major one of several gases that trap the sun's heat in the atmosphere. Prestigious scientific committees studied the problems that the trapped heat might cause, and all came to the same essential conclusion: Earth is going to get warmer in the late twentieth and early twenty-first centuries.

Although the role of carbon dioxide in producing a greenhouse effect had been known for many years, there had been considerable debate as to how serious it would be. As of now, it appears to be very serious, and is the major earth-science story of the early 1980s, and perhaps of the next fifty to a hundred years.

On a longer time scale, there may be even a greater hazard in the future that was also recognized in the early 1980s. In 1980 Walter Alvarez and associates discovered evidence that caused them to conclude that the mass extinctions of 65,000,000 years ago, which included the extinction of the dinosaurs, were probably caused by an asteroid hitting Earth, causing a year of darkness that extinguished photosynthesis and reduced the number of species drastically. While not everyone agreed with the Alvarez team, attention was focused on bombardment as a cause of mass extinction, leading in 1984 to theories that periodic mass extinctions are caused by a rain of comets from a "death star," named Nemesis by some of the scientists who proposed the idea. The good news is that, if Nemesis exists, its next visit is not expected for 15,000,000 years.

The bad news is that two groups of scientists from many disciplines have shown that a nuclear war would have the same effect as Alvarez's meteorite or a visit from Nemesis. A "nuclear winter" would cause the same types of mass extinctions—and there is no guarantee that such an event is far in the future. The earth sciences have become very relevant in the 1980s.

1980

Mount St. Helens Erupts

In March, 1980, frequent earthquakes below Washington State's Mount St. Helens alerted scientists that a predicted eruption of the volcano was about to begin. (Dwight Crandell and Donal Mullineaux had published a 1978 U. S. Geological Survey report that said that the volcano would erupt by the end of the twentieth century.) On March 27, volcanic activity began with vents of smoke and ash. Scientists converged upon the volcano, anxious to take advantage of a chance to study the process from its beginning. The location, less than 100 kilometers (70 miles) from Portland, Oregon, meant that many scientists could be near the scene and make frequent visits from Portland. One of these scientists, David Johnston, lost his life in the pursuit of knowledge.

Human interest before the eruption focused on Harry Truman, 84 years old, who refused to leave his lodge at Spirit Lake on the north slope of the volcano. When the eruption came, Spirit Lake was nearly completely covered with either hot rocks or mud that formed a landslide down the mountain as a result of melting ice. Mr. Truman was buried in the flows.

The main eruption occurred at 8:32 A.M. Pacific Standard Time on Sunday, May 18. Dust and ash weighing 400,000,000 tons were ejected 20,000 meters (65,000 feet) into the air and down the north slope of the volcano. The mountain lost roughly 3 cubic kilometers (1 cubic mile) of rock, and was reduced in height from 2,950 meters to 2,550 meters (9,677 feet to 8,364 feet). A crater was formed that was 3.9 kilometers (2.4 miles) long and 1.9 kilometers (1.2 miles) wide. An area of about 96 square kilometers (23 square miles) was covered with material that flowed down the north slope, in what is known as a *pyroclastic flow*, a mixture of hot rocks, dust, and gases. The temperature of the pyroclastic flow was about 340°C (650°F) at its hottest. Sixty-one people were killed in the eruption, along with an estimated 5,200 elk, 600 black-tailed deer, 200 black bears, 11,000 hares, 15 mountain lions, 300 bobcats, 27,000 grouse, 1,400 coyotes, and 11,000,000 fish. Ash from the eruption covered 49 percent of Washington State as well as large areas of Oregon and other neighboring states. In places, the ash was 13 centimeters (5 inches) thick.

Smaller eruptions continue. The most serious occurred in 1980 on May 25, June 12, July 22, August 7, October 17–18, and December 27; and in 1981 on February 5–7, April 10, June 18–19, and September 6–7, and in 1982 on March 19, and again in April.

Active U.S. Volcanoes

Active volcanoes are found in several major chains around the world. (Volcanoes are classified as "active" only if they have erupted since written records were kept, a definition that is useful but not very conclusive about the state of a given volcano.) The largest number of active volcanoes is in Indonesia, where there are 76, with Japan following and the United States, with 53, a surprisingly strong third. While many of the active U.S. volcanoes are in Alaska and Hawaii, the Cascade Range along the coasts of California, Oregon, and Washington had a number of major eruptions in the nineteenth century, with two eruptions so far in the twentieth century, Mount Lassen in California and Mount St. Helens in Washington.

Volcano	Height (ft. above sea level)	Eruptions Since 1700	Types of Activity	Last Reported Eruption
California				
Cinder Cone	6,907	1	c.e.f?	1851
Lassen Peak	10,453	4	c,e,f,g,m	1914–21
Mt. Shasta	14,161	2?	e?	1855?
Oregon				
Mt. Hood	11,245	1	c,e	c.1801
Washington				
Mt. Baker	10,778	5	c,l,e	1870
Mt. Rainier	14,410	6	c,e	1882
Mt. St. Helens	9,671	21	c,e,f,d,g,m	1984
Hawaii				
Haleakala	10,025	1	l,f	c.1790
Hualalai	8,251	1	l,f	1801
Kilauea	4,090	51	c,l,f,p	1984
Mauna Loa	13,680	38	c,l,f	1984
Alaska & Aleutians				
Kiska	4,025	5	l,e,f	1969
Little Sitkin	3,945	2	e	1828
Cerberus	2,560	5	e?	1873
Gareloi	5,370	8	e,f	1930
Tanaga	7,015	4	e,f	1914
Kanaga	4,450	6	e,f	1933
Great Sitkin	5,775	5	e,f	1945
Keniuji	885	3	e,f?	1828

Volcano	Height (ft. above sea level)	Eruptions Since 1700	Types of Activity	Last Reported Eruption
Korovin	4,885	3	e	1844
Sarichef	2,015	1	e?	1812
Seguam	3,465	5	e	1927
Amukta	3,490	3	c,e,l,f	1963
Yunaska	1,980	4	e	1937
Carlisle	5,315	3	e	1838
Cleveland	5,710	6	e,f	1944
Kagamil	2,945	1	e?	1929
Vsevidof	6,965	5	e	1880
Okmok	3,540	10	e,f	1945
Bogoslof	c.150	11	e,d	1931
Makushin	6,720	14	e	1938
Akutan	4,265	22	e,f	1948
Pogromni	7,545	4	e,f	1830
Westdahl	5,055	1	l,e,f	1964
Fisher	3,545	1	e?	1826
Shishaldin	9,430	30	c,e,f	1979
Isanotski	8,185	4	e	1845
Pavlof	8,960	24	c,e,f	1963
Pavlof Sister	c.7,050	1	e?	1786
Veniaminof	c.8,450	8	e	1944
Aniakchak	4,450	1	e,f	1931
Chiginagak	7,985	2	e	1929
Peulik	5,030	2	e	1852
Mageik	7,295	4	e	1946
Trident	6,830	4	l,d,e,m	1974
Novarupta	N. A.	1	d,e,g	1912
Katmai	7,540	8	c,e,f	1974
Augustine	3,995	6	e,f,d	1976
Iliamna	10,140	6	e	1947
Redoubt	10,265	7	c,e	1966
Spurr	11,070	1	l,c,m	1953

Source: Gordon A. Macdonald, *Volcanoes,* ©1972. Adapted by permission of Prentice-Hall, Inc. Englewood Cliffs, N.J.

HOW THEY DIED

Autopsies performed on victims of the eruption at Mount St. Helens in 1980 are providing insights into the way volcanoes claim their victims. Most of the victims suffocated in a few minutes when they inhaled volcanic ash that mixed with mucus and blocked the larynx and trachea. Individuals near the blast received severe burns from scalding water or molten substances ejected by the volcanoes. Others died when they were struck by trees or rocks.

Undersea Geothermal Vents

One of the most fruitful areas of research in Earth science came about as a result of the discovery that hot springs, known as "geothermal vents," were an important feature of the ocean floor. Although the first undersea geothermal vents were spotted in the Red Sea in the 1960's and major explorations of the vents began in the East Pacific in the late 1970's, it was not until 1980 that scientists began to put together the data and produce complete descriptions of what we now recognize as a major influence on the composition of the sea, the deposition of minerals, and (perhaps) the origin of life.

These vents are found at centers of sea-floor spreading. Geologists have found that the Earth's crust is broken into large regions called "tectonic plates." These plates are in motion, some pushing into each other and others pulling away from each other. Tectonic plates on the East Pacific Rise, at the Galapagos Rift, and in the Red Sea are moving apart at rates from 6.5 centimeters to 18 centimeters (2.5 to 7 inches) each year. It is easy to think that such spreading *causes* oceans to become wider by pushing plates apart. In fact, it is the other way around. Undersea geothermal vents arise as a *result* of the plates moving apart. As cracks widen between the plates, the partially or completely melted rock from the layer of the Earth below the plates rises to fill the cracks.

Such spreading leads to magma rising close to the bottom of the ocean and also opens cracks for sea water to reach the superheated rock. Under pressures as high as 275 atmospheres at depths around 2600 meters (8500 feet), the water does not boil, but it soon reaches temperatures that exceed 340°C (650°F). At those temperatures, minerals from the crust dissolve in the water, forming new compounds, especially sulfur compounds. The compounds, such as zinc or iron sulfides are carried by the hot, rising water back into the sea, where the temperature of the water quickly drops, precipitating the sulfides in thick layers on the nearby ocean floor. This is the mechanism believed to produce large ore deposits such as the copper sulfide deposits of Cyprus (whose very name means *copper* because it was one of the principal mining centers of the ancient Mediterranean).

The sulfides form the basis of a food chain that does not depend on photosynthesis, impossible at those depths in any case because of the lack of sunlight. Bacteria metabolize the sulfur compounds, swaying tube worms 45-centimeters (18-inches) long live symbiotically with the bacteria, giant clams filter-feed on the bacteria, and crabs scavenge debris and occasionally attack the plumes of the tube worms. The tube worms are also known by the wonderful name of vestimentiferan pogonophorans, although their scientific name is *Riftia pachyptlia*. The bacteria, the base of the food chain, were shown in laboratory tests in 1983 to thrive in pressures of 265 atmospheres and temperatures of 250°C (482°F), and scientists have yet to find temperatures or pressures that kill the bacteria.

Other vents, found in the Guaymas Basin of the Gulf of California, were found in 1982 to be producing a type of petroleum as a result of heat activity on sediments formed from the skeletons of algae. These vents turned out to have a different ecology from those in the open ocean. White mats covering the vents turned out to be bacteria that may metabolize the petroleum. Black abalones use the bacterial mats as a major food source in northern California waters. In this way, the bacteria become part of the food chain for humans.

It has long been suspected that life originated in processes that did not depend on oxygen. There was probably no free oxygen in the atmosphere before the development of photosynthesis in bacteria, algae, protists, and green plants. All the free oxygen that now makes up about a fifth of the atmosphere is a result of billions of years of photosynthesis, which could not start until life developed. The undersea geothermal vents, however, have an ecology that does not depend on free oxygen. As a result, by 1983, the undersea geothermal vents were being touted as a possible locale for the origin of life on earth.

HOT SPRINGS OFF OREGON

Researchers have uncovered undersea hot springs about 422 kilometers (270 miles) off the coast of Oregon. The springs are the result of a ridge of volcanic activity where the Pacific and Juan de Fuca plates meet. Volcanologist David Clague explained that activity along the vent is probably fairly recent because rocks have been gathered there which are "less than 100 years old."

TUBE WORMS IN SULFIDE MINES

At least 95,000,000 years ago, tube worms were living in communities beneath the sea. This conclusion is based on recent discoveries of the fossilized remains of worm tubes in a sulfide mine in Oman in the southeastern part of the Arabian peninsula. These fossils are similar to animals currently found living in mineral-rich areas around seafloor vents.

Cause of Cretaceous Catastrophe Found?

In 1980 Walter Alvarez (with Luis Alvarez, Frank Asaro, and Helen Michel) discovered a thin layer of clay enriched in iridium that has been a source of controversy ever since. The iridium layer is found around the world—geologists have located it at about 50 different sites—at the boundary between the Cretaceous and Tertiary periods, a boundary that marks a time 65,000,000 years ago and also mass extinctions all over the earth. While the most famous of these extinctions is that of the dinosaurs (although they were seriously in decline before this time), many other land and marine species also were lost. One of the most dramatic extinctions was among a group of marine plankton at the height of their domination of the seas, the calcareous phytoplankton. This group lost 80 percent of its species at the end of the Cretaceous period, while 19 percent of the land species were lost (and only 3 percent of freshwater species).

Alvarez suggested that the iridium was the residue of an asteroid or comet that was, in fact, the cause of the extinctions. Dust from the asteroid, according to Alvarez, reduced photosynthesis to such an extent that the green plants died, destroying the base of the food chain.

Iridium is common in the universe, but on the earth's crust it is rare. Most of earth's iridium is believed to have sunk to the core. In the Cretaceous-Tertiary boundary layer, however, iridium levels (as well as those of other uncommon crustal elements) are 25 times or more higher than normal, matching levels found in meteorites.

While this theory of a cause for the Cretaceous-Tertiary mass extinctions was initially greeted with great skepticism, evidence since 1980 has tended to confirm Alvarez's theory, although skepticism still exists.

While many scientists agree that an extra-terrestrial body did indeed hit Earth about 65 million years ago, they have expressed severe reservations about whether it caused the mass extinction of the dinosaurs. First, some paleontologists claim that the last dinosaur fossils are found 3 to 10 meters (10 to 30 feet) below the iridium layer, indicating that the dinosaurs may have already died out before the asteroid struck. Secondly, other species of animals (species of reptiles, for example) seem to have survived the mass extinction, for their fossils are found in layers of sediment from both the Cretaceous and Tertiary periods. This seems to rule out in the minds of some critics a single catastrophic event.

Some critics of the Alvarez theory believe that instead of one catastrophic event bringing an abrupt end to the dinosaurs, they died out gradually. The reasons are still open to speculation, however. Some scientists believe that as the climate cooled and the vegetation changed, the dinosaurs simply could not adapt rapidly enough. In fact, the most popular alternative to the Alvarez theory is that such changes in global temperature can account precisely for which families become extinct and which survive. Scientists who subscribe to global temperature change as a cause of extinctions base their claim partly on the fact that families of sea creatures known to survive in cooler waters did not become extinct, while families found only in warm tropical waters were snuffed out.

The large number of extinctions at the Cretaceous boundary has caused other critics of the impact theory to propose yet another idea. The mass eruption of large volcanoes could have caused both the iridium layer and the extinctions. Evidence for this idea was discovered in 1983 when an eruption of Kilauea on Hawaii produced ejecta with a high iridium content. Evidence against the volcanic theory, also in 1983, included measurements of isotope ratios from the iridium layer that suggested an extraterrestrial source. Jean-Marc Luck and Karl Turekian of Yale University studied the ratios of osmium-187 to osmium-186 in material laid down 65 million years ago—the boundary between the Cretaceous and Tertiary periods. The ratios of the isotopes found in this material, which included marine sediments and crustal rocks, were 1.29 to 1 and 1.65 to 1. These are far more consistent with the ratios in extra-terrestrial bodies than the 10-1 ratios found in Earth's crust. Finally, in 1984, studies of the quartz in the iridium layer definitely showed that it had been stressed by an impact, ruling out a volcanic origin.

Yet another claim for a cause developed in 1984. Two groups of scientists proposed that the Earth is periodically bombarded by rains of comets, caused by a companion star to the sun. Since the exact composition of the cores of comets is unknown, it is not clear whether such a disaster could have caused the iridium layer, although otherwise the effect would be similar to an asteroid striking the Earth.

WHERE'S THE CRATER?

If a large asteroid did strike the Earth 65 million years ago leading to mass extinctions, as some scientists currently believe, where is the crater left by the impact? The odds are about 7 out of 10 that the crater still remains either beneath the sea or on land. One possibility is the Kara crater in the Soviet Union. Actually this is two craters that might have been caused by a pair or a group of asteroids. Created approximately 60,000,000 years ago, one of the craters is 60 kilometers (37 miles) wide while the other is 25 kilometers (15 miles) wide.

Ancient California Caldera Shows Magma Intrusion

Earthquake activity in the Long Valley caldera near Mammoth Lakes, California indicated in 1980 that a new intrusion of magma was taking place. This activity continued frequently during the next four years, reaching Richter-scale levels as high as 5.6 at times. Typically for magma intrusions, the earthquakes came in clusters, called *swarms*, sometimes as often as 10 quakes per second. One such swarm, in the spring of 1982, led the U. S. Geological Service to issue a warning of volcanic activity, since similar swarms preceded the eruption of Mount St. Helens. Although the eruptions did not take place, the earthquakes continued, with a particularly strong swarm in January of 1983.

If an eruption did occur in the area, it could destroy Mammoth Lakes and disrupt the water supply to Los Angeles.

Accompanying the earthquakes was another sign of a magma intrusion and possible eruption. The land surface over a distance of 4 kilometers (2.5 miles) was raised by about 4 centimeters (1.6 inches).

Studies of the nearby Mono and Inyo craters where eruptions have taken place over the past several thousand years also contributed to fears of an eruption at Long Valley. The eruptions at Mono and Inyo emanated from a series of vents over short time spans—perhaps twenty years or less. In the Mono craters, for example, the latest eruptions—which happened 250 years ago—came from a line of vents. At that time, movements in the earth's crust permitted magma to rise toward the surface. A similar situation now exists at Long Valley where an 8 kilometer (5 mile) dike of magma has pushed into openings beneath the caldera. After Mount St. Helens, the Mono-Inyo craters are the most frequently active volcanoes in the Pacific West.

SEEING AROUND CURVES

In 1597 a group of explorers were forced by bad weather to anchor off the Novaya Zemlya islands in the Arctic Ocean for the winter. There, during the long Arctic winter, they reported a seemingly incredible event: the sun began shining approximately 14 days before the dark winter was supposed to end.

Now modern scientists have an explanation for the "Novaya Zemlya effect," as it is called. Observers can see below the horizon when the atmosphere bends light. Such bending can occur when an acute temperature inversion takes place in the atmosphere. The difference in refraction of light between the warmer air and the cooler air it overlies is actually what bends the light rays. This apparently occurred in 1597, and similar events continue to be observed and recorded today.

Terrain or Terrane?

In recent years scientists have seriously questioned the theory of plate tectonics (a theory which itself has only been accepted for about two decades). According to the plate tectonics theory, the continents rest on huge crustal plates which have floated about the globe on liquid mantle for eons. Now some geologists believe that the continents, instead of consisting of relatively homogeneous crustal plates, are made up of crustal pieces that have come together at different times from various parts of the globe. Convincing evidence for many geologists in favor of the controversial crustal pieces theory was presented at the 1980 meeting of the American Geophysical Union in San Francisco, conveniently close to some of the regions said to be involved in buildup from crustal pieces.

Davy L. Jones of the U.S. Geological Survey and David B. Stone of the University of Alaska are two of the chief proponents of the concept that continental areas are built from separate pieces, with Stone's presentation of Alaska data one of the chief features of the meeting.

These geologists base their theory on evidence that adjacent sections of the continents do not match. Differences in fossils and minerals, for example, show the continental sections to be of different ages. One of the most important forms of evidence comes from the fact that many rocks can act as "frozen compasses." These rocks preserve their orientation to the Earth's magnetic field when they form. Studies of the magnetic orientation of rocks frozen when the rocks solidified, called paleomagnetism, indicate that the rocks formed in regions far removed from their present locations. Instead of a uniform terrain, these continental pieces make up a *terrane*, a word recently coined by scientists.

One section of the North American continent studied by geologists is the Salinian Terrane between San Francisco Bay and Bakersfield in California. After examining the magnetic field of this area, geologists believe it may have been formed somewhere in Central America and then drifted northward, eventually crashing into the California coast.

Other areas in North America formed in the same way are thought to include most of Alaska, the Wrangell Mountains in British Columbia, parts of Florida and Nevada, and New England. Much of the northern West Coast of North America seems to have come from tropical regions in the Pacific Ocean. Central America may have been constructed of four terranes that floated together, while Siberia's landscape also suggests that it is formed from terranes. Other sections of continents, such as the shield formation in eastern Canada, are probably not formed this way.

1981

Giant Thunderstorm Complexes

Prediction of thunderstorms is even less accurate than most weather forecasting, but in 1981 J. Michael Fritsch and Robert A. Maddox proposed a new approach to thunderstorms in the U.S. Midwest. They have observed that Midwestern thunderstorms are often organized into what they call Mesoscale Convective Complexes (MCCs). An MCC can be identified from satellite pictures and may consist of a dozen storms acting together.

Apparently, most of the thunderstorms in the Midwest in the summer form into these complexes. Individual thunderstorms can also form, however. Some of these are local phenomena that are much less predictable. Others, however, are individual storms but far too large to be called local.

Each spring, giant thunderstorms develop in an area roughly between the Rocky Mountains and the Ohio Valley. They begin as huge thunderclouds that stretch upward 16 kilometers (10 miles) into the atmosphere. Wind speeds in these clouds reach as high as 160 kph (100 mph). As warm moist air ascends through these clouds and the moisture condenses, huge amounts of heat are produced. This heats a broad area, eventually producing more thunderclouds which come together creating a storm that may stretch over three or more midwestern states.

Most thunderstorms develop during the hot midday hours, and they come and go in just a few hours. But these vast thunderstorms are different. While they may begin at the same time, they continue well past nightfall, often lasting 12 to 18 hours. During the daylight hours, the storms grow in intensity, then continue growing after dark, often reaching full force between midnight and 2:00 a.m., and finally ending about sunrise.

WHY NOBODY HEARD

Survivors of the eruption at Mount St. Helens report that everything around them seemed quiet while this terrible catastrophe was underway. Yet, we know that forests were falling with a thunderous thud and the noise from the eruption itself must have been deafening. Why then did all seem quiet? Scientists point out that sound, which usually travels 10 to 20 kilometers (6 to 12 miles) only traveled about 10 meters (33 feet) within the enormous dust cloud created when Mount St. Helens erupted. The large number of pine needles from the evergreen trees surrounding the volcano also acted to muffle the sound.

Beaufort Scale of Wind Strength

The Beaufort Scale is commonly used by sailors to describe winds. It is also a handy way for landlubbers to determine the approximate speed of winds from the observed effects of the winds. Commonly the word *force* is used to describe the winds; as in, "that looks like a force 5 breeze to me."

Beaufort Number	Description of Wind	Wind Speed	Description
0	Calm	<1.6 km/hr <1 mi./hr.	Still; smoke rises vertically.
1	Light Air	1–5 mi./hr. –5mi./hr.	Wind direction shown by smoke drift; weather vanes inactive.
2	Light Breeze	9–18 km/hr 6–11 mi./hr.	Wind felt on face; leaves rustle; weather vanes active.
3	Gentle Breeze	19–31 km/hr 12–19 mi./hr.	Leaves and small twigs move constantly; wind extends light flags.
4	Moderate Breeze	32–45 km/hr 20–28 mi./hr.	Raises dust and loose paper; moves twigs and thin branches.
5	Fresh Breeze	47–62 km/hr 29–38 mi./hr.	Small trees in leaf begin to sway.
6	Strong Breeze	63–80 km/hr 39–49 mi./hr.	Large branches move; telegraph wires whistle; umbrella difficult to control.
7	Moderate Gale	81–99 km/hr 50–61 mi./hr.	Whole trees sway; somewhat difficult to walk.
8	Fresh Gale	100–120 km/hr 62–74 mi./hr.	Twigs broken off trees; walking against wind very difficult.
9	Strong Gale	121–143 km/hr 75–88 mi./hr.	Slight damage to buildings, shingles blown off roof.
10	Whole Gale	144–165 km/hr 89–102 mi./hr.	Trees uprooted; considerable damage to buildings.
11	Storm	166–190 km/hr 103–117 mi./hr.	Widespread damage; rarely occurs inland.
12–17	Hurricane	>190 km/hr >117 mi./hr.	Extreme destruction.

The Two Pangaeas

Since the time of Alfred Wegener's "drifting continent" hypothesis of 1912, it has been believed that a supercontinent, called *Pangaea*, existed from about 300,000,000 years ago until it broke up around 150,000,000 years ago. Pangaea consisted of the plates containing all of the present continents.

Patrick Morel and Edward Irving used magnetism trapped in old rocks to propose a revision of this theory in 1977, although evidence for their new version was slight. By 1981, however, they had extended their data to make a convincing case that Pangaea existed in its classic form for only about 50,000,000 years, a relatively short time geologically speaking. For nearly 100,000,000 years before that time, the continents that formed Pangaea were either in a different formation or else on the move.

Specifically, from about 290,000,000 to 250,000,000 years ago, Pangaea B matches the Eastern shore of North America with Northwestern South America, while other rearrangements place Europe closer to North Africa, and the Arab peninsula below the Horn of Africa. Then for 40,000,000 or 50,000,000 years, the pieces moved around with respect to each other, finally stabilizing for 50,000,000 years as Pangaea A, the Pangaea that is familiar from textbooks.

Morel and Irving's theory solves several problems of geology, including providing good matches between older mountain chains in Europe and Africa, and between North and South America. Unlike younger chains of mountains, which match properly across present-day continental boundaries when Pangaea is invoked, these older chains had been seen as anomalies, since there was no reason to expect them to end abruptly at what were not boundaries when Pangaea existed.

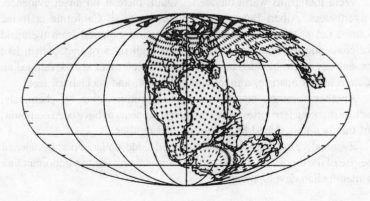

Therapsid Fans Confer

Sometimes it takes a bit of a show to focus attention on a particular set of problems in science. Such was the case with the therapsids when Nicholas Hotton III and others in 1981 organized the first major scientific conference to deal with the topic.

In the popular mind, all the large, scaly living forms that inhabited the land before the rise of the mammals are "dinosaurs," but to the paleontologist, there are many different kinds of early reptiles and amphibians. The dinosaurs were simply the last group of reptiles to occupy a dominant place in the global ecosystem. The assumption, conscious or unconscious, that each evolutionary development represents an improvement or progress— rejected by most biologists today—suggests that the dinosaurs were the "highest form" of reptilian life. This common notion may not be true, as the example of the therapsids shows.

Therapsids should be of interest, since they may be our distant ancestors. These mammal-like reptiles dominated our planet about 250,000,000 years ago, before the age of dinosaurs. Fossil remains show that the therapsids came in all sizes, ranging from about the length of a squirrel to the enormous bulk of a hippopotamus. They were herbivores as well as carnivores. Instead of moving with the sprawling gait of a lizard, therapsids walked in a more erect manner with their four legs directly under the body. Therapsids seemed to have a smooth skin rather than the scaly reptilian skin of their ancestors, based on fossil casts that have been found. The casts showed no evidence of hair, however. Therapsid fans also believe that therapsids laid eggs, and did not produce live offspring or nurse them as mammals do. Unlike the fairly abundant dinosaur eggs, however, no therapsid eggs have ever been found.

Were therapsids warm blooded? Again, there is no direct evidence. Nevertheless, Albert Bennett of the University of California at Irvine pointed out at the conference that information gathered from therapsid skeletons suggests that later species had high rates of metabolism like present-day warm blooded animals. Some species had also developed an acute sense of hearing, a more advanced jaw, and specialized teeth.

Another common mammalian trait, communication by chemicals, such as marking territories with scents, also appears to have been common, but this is a trait shared by mammals and reptiles.

Eventually, the therapsids were pushed aside, in large part because of the rise of the dinosaurs, but those that survived provided an important link in mammalian development.

"Here's Your Lousy Jaw!"

Early mammals are those that lived and died before the Tertiary Age began. They are not plentiful, since the dinosaurs did not become extinct until the end of the Cretaceous, the period before the Tertiary, and the mammals were both few and small so long as the dinosaurs ruled the earth. The end of 1980 and 1981 both significantly increased the number of known early mammals. In 1980 the first North American representative of one of the two (possibly three) families known from the Eastern Hemisphere was discovered, and in 1981 the number of known families of early mammals was increased to a definite three.

Both the new finds are from a site in Arizona, believed to be 180,000,000 years old, around the time of the first appearance of dinosaurs. In November of 1980 William R. Downs of the Museum of Northern Arizona in Flagstaff found some teeth from a creature called a Morganucodontid, thought to be an ancestor of the monotremes (such as the duck-billed platypus).

Fossil hunting is often carried out in the summer, since that is when school is out. At some sites, it can get very hot in the summer. The following July, Kathleen Smith, frustrated, hot, and tired from working the dig, finally found another mammalian fossil, handing it to Farish A. Jenkins of Harvard University, in charge of the dig, saying, "Here, Jenkins. Here's your lousy jaw!"

The lousy jaw turned out to belong to an apparent ancestor of the insectivores, called *Dinnetherium nezorum*, whose genus name means "wild beast of the Navajo," while the species name is for the Nez family of Gold Spring, Arizona, where the find was made. *Dinnetherium* was identified by Jenkins as a member of a new family of early mammals.

North America yielded additional fossil mammals, not quite so early as these, in 1982. Kenneth Carpenter and Robert Bakker found the fossils in rock that had been taken from hills near Como Bluff, Wyoming, in 1978. One, roughly 110,000,000 to 120,000,000 years old, is the earliest leaf-eating mammal found. It is not a direct ancestor, however, of any of the present leaf-eating mammals, but is, instead, a type of early mammal known as multituberculate. The multituberculates are believed to have been pushed out of their ecological niches by the arrival of the rodents, and became extinct about 25,000,000 years after the dinosaurs did (that is, 40,000,000 years ago). The proposed name for the newly found multituberculate is *Zofiabataar geographica*, while the other early mammal found, a tiny insectivore, was christened *Simpsonipauper fortis*.

The Newest Hawaiian Island (Coming Soon)

Scientists in 1981 conducted underwater examinations of Loihi, a volcanic seamount that will become the newest Hawaiian Island in about 2,000 to 20,000 years (depending on the level of volcanic activity). It is now 950 meters (0.6 mile) beneath the surface of the Pacific Ocean. A seamount is any undersea mountain.

All of the present Hawaiian Islands, as well as former Hawaiian Islands, were formed as similar underwater volcanoes and grew to their present size as the volcanoes continued to erupt after emerging from the sea. There is a chain of extinct volcanoes that stretches from Hawaii to the northwest reaching almost to the Kamchatka Peninsula, including the Emperor seamounts as well as the Hawaiian Islands themselves. Plate tectonic theory accounts for this long chain of seamounts and islands by saying that they are produced as the Pacific Plate moves over a hot spot. Magma from the hot spot continually rises. Where it breaks through the plate, a seamount forms, becoming an island if the magma reaches the surface of the ocean. After the volcano becomes extinct, the island, if formed, gradually erodes to seamount status again.

Unlike the ecosystems found along with volcanic activity produced where plates are moving apart, no specialized forms of life have been found that cluster around Loihi's eruptions.

Dredging and underwater examinations have led, however, to revisions in the accepted theories of how the undersea volcanoes erupt. Magma from Loihi is of a type previously believed not to form until near the end of the volcanoes eruption cycle. Apparently the magma from the hot spot comes from deep in the earth's mantle, and is fundamentally different from that found at plate boundaries.

WHEN PLATES COLLIDE

What happens to sediments when continental and oceanic plates clash together? Perhaps, as the oceanic plate sinks, sediment is skimmed off the top onto the continental plate. Or, perhaps the entire plate, complete with its sediment, plunges beneath the continental plate. Explorations in the Lesser Antilles, where the Atlantic Ocean plate plunges below the Caribbean plate, indicate that both things happen. A top layer of sediment is scraped off into the Caribbean plate. Lower layers stay with the Atlantic plate. What prevents the upper layer from taking the plunge is water. It keeps these sediments, which are much less compacted than the lower layers, afloat.

1982

Could Giant Meteorite Have Caused Continents?

John Klasner and Klaus Schulz announced early in 1982 that maps they had prepared revealed a circular gravity anomaly, or region of high gravity, 2,800 kilometers (1,736 miles) in diameter centered between the Great Lakes and Hudson Bay. The cause, they suggest, was bombarding of the Earth some 4,000,000,000 years ago by a meteorite, which may have occasioned the origin of continents on Earth.

The other planets in the solar system do not have features that correspond to continents. Continents in the sense we usually conceive—large bodies of land surrounded by water—could not exist because of the absence of liquid water on the surface of other planets. Even if the water were present, however, the difference between light continental rock and heavier oceanic crust is not found on other planets. Klasner and Schulz suggest that North America might be a result of geothermal activity caused by the impact. Other continents might be a result of similar impacts from the same meteorite bombardment.

Ages of rocks over the gravity anomaly vary in a way consistent with the hypothesis that the anomaly is the result of an impact that shattered the earth's crust. If the original crater were about 1,000 kilometers (600 miles) in diameter, geothermal activity in the crater could have continued for as much as 2,000,000,000 years, while the affected areas furthest from the crater would have cooled much more quickly. Since dating of the inner circle shows an age of around 1,800,000,000 years to 2,400,000,000 years, while the outer circle shows an age of 3,800,000,000 years, Klasner and Schulz consider the results as a good confirmation that the anomaly was caused by such an impact.

DON'T BLAME THE INDIANS

One thousand years ago a magnificent system of canals stretched across the desert along the Peruvian coast. It was built by the Chimú whose civilization thrived between 500 and 1500 A.D. These canals were necessary to bring water to the parched landscape so the Chimú could cultivate their land. But the canal system proved no match for the forces of nature. Crustal plate movements which caused a vertical movement of the Earth of 1.8 meters (5 feet 11 inches) 500 years ago caused the canals to collapse. Although the Chimú tried to rebuild them, heavy rains, probably caused by El Niños, overcame their efforts. Eventually the system had to be abandoned, marking the decline of the Chimú and their civilization.

Mammals in Antarctica

Paleontologists now have proof that marsupials once lived in Antarctica. Two days before the end of a 1981–82 expedition to Seymour Island, a particularly good site for fossils, Michael Woodburne from the University of California at Riverside was taking a break from the expedition's organized search for fossils. Since Woodburne is a vertebrate paleontologist, his idea of a "break" was picking up rocks and looking at them for fossils. He found the proof, which scientists had been seeking for years.

What Woodburne found were the fossilized jaw fragments of a small rodent-like marsupial which lived 40,000,000 to 45,000,000 years ago. The finding supports theories about the migration of animals and shape of global land masses billions of years ago.

Scientists have long suspected that Australia, South America and Antarctica were all part of a single large continent known as Gondwanaland. In addition, they have theorized that marsupials evolved in North America and then traveled to South America, where there were many marsupials in the past and a few remain today. Today, the marsupials are found mainly in Australia, but there is no evidence that they originated on that continent. From South America, the marsupials must have reached Australia in some way, since the major mammalian fauna of Australia consists of marsupials. The only likely route for the marsupials to reach Australia from South America is via Antarctica. Both hypotheses have been given support by the fossil discovery. The fossil is younger than those of similar marsupials found in South America, but older than those unearthed in Australia, indicating that Antarctica served as a land bridge between them, just as the hypothesis predicts.

The marsupial belongs to the genus *Polydolops*, an herbivore that lived along the coast of Antarctica. Remains of the genus have been found in South America, but not, so far, in Australia.

The expedition, led by William Zinsmeister of Ohio State University, brought back a wealth of other scientific data. Seventy million years ago the climate in that region was much warmer. Researchers found petrified wood on Antarctica from that period, evidence of the large forests that once existed there. In addition, fossil remains of large ocean reptiles have been discovered, evidence that the seas were much warmer as well as the land. By the time of *Polydolop*, however, the climate was changing, although there were still forests where *Polydolop* probably foraged for berries. Whales, sharks, and penguins, with some of the penguins as much as 2 meters (6 feet) tall, exploited the rich ocean life around the margin.

El Chichón Erupts

On March 28, 1982, an earthquake measuring 3.5 on the Richter scale occurred beneath El Chichón in Mexico. El Chichón (*the lump*) rises 1,260 meters (4,134 feet) in the Mexican state of Chiapas. Shortly afterward, an eruption followed that blasted gases, ash, and rock into the sky. Approximately a dozen people died locally as a result, but this was to prove just the beginning.

The following Saturday, April 3, another eruption occurred at 7:32 p.m. which soon subsided. This was the prelude to the major blast that shook the volcano at 5:20 a.m. the next day. Gases and ash rocketed into the sky traveling as far north as Texas and as far south as Guatemala. The death toll in the area around El Chichón was originally placed at 187 by Mexican officials, although estimates ran much higher. After scientists had investigated the eruption in more detail, the estimates were that 2,000 people had died in the eruption, many more than at Mount St. Helens, where official eruption predictions kept most people far away from the volcano.

The major eruption of El Chichón created a cloud of gases, dust, and ash that blackened the sky in the region for 44 hours. This eruption and the earlier ones belched an estimated 500,000,000 metric tons of dust, gas, and ash into the atmosphere. The explosions have been compared with the eruption of Mount St. Helens in 1980. In each instance, the amount of ashfall was about the same, approximately half a cubic kilometer (one tenth of a cubic mile). However, the total volume of material ejected was much greater at Mount St. Helens.

Nevertheless, the eruption of El Chichón had much greater consequences for our atmosphere than Mount St. Helens. The atmospheric conditions above El Chichón were just right at the time of the blast for materials to rocket upward all the way to the stratosphere and to be carried around the globe. In addition, the major force of El Chichón's blast was upward (not outward as it was at Mount St. Helens), making penetration more effective. As a result, a veil of ash from El Chichón began to circle the globe at a level of approximately 18 kilometers (60,000 feet) above the surface.

The ash veil of El Chichón, much larger than Mount St. Helens', not only contains volcanic ash, but also holds salt crystals, which have never been seen before from land volcanoes. These come from deposits of salt under the volcano. There was also an unusually high amount of sulfur produced, most of it as sulfur dioxide gas. The sulfur dioxide gas emitted by the eruption combined with moisture in the atmosphere to form aerosols, suspensions of very fine particles or droplets that can remain in the atmosphere for a long time. These aerosols and the volcanic ash found in the veil reduced the amount of sunlight penetrating to the earth—by how much and for how long was not really known. And what the total effect on our weather was could only be guessed at by scientists.

Some scientists believe that the veil could have had the effect of cooling the Earth by as much as 1°C (2°F). On the other hand, the veil could have warmed the Earth by blocking sunlight that is reradiated from the earth into the atmosphere. One problem with making the determination of whether or not the cooling or warming has actually taken place is that the normal annual variation in the Earth's temperature is well within this range.

In any event the effects of El Chichón did not match the incredible aftermath of the eruption of Tambora in Indonesia in 1815. The following year, which was marked by a June snowstorm in New England, was aptly named the "year without a summer." Nevertheless, some experts believe that the veil of ash created by El Chichón may have been implicated in the severe effects of the 1982–83 El Niño, which dramatically changed weather patterns around the globe (see "Strongest El Niño of the Century," page 259).

GRANDMOTHER WAS RIGHT, AFTER ALL

Older people usually remember that the weather was better when they were younger. Often they blame such innovations as aircraft or nuclear weapons as the cause of the decline in the weather.

Stanley Changnon, Richard Semonin, and Wayne Wendland of the Illinois State Water Survey have collected data that suggest that jet airplanes really do alter weather. The team compared a region over northern Illinois that gets considerable jet traffic with another region where traffic is far less. In the former area, the amount of cloud cover is 10 percent greater. Since cloud cover reduces temperature extremes, this could conceivably result in cooler days and warmer nights.

On days when there were no clouds, the trails of condensed vapor (contrails) left by the jets created effects that might be similar to those resulting from cloud cover. For example, contrails were shown to reduce the normal rate at which temperatures rise during the hours between sunrise and noon. In addition, they reduced the normal rate of decline in humidity during these same hours.

Contrails are presumably formed when water vapor from the jet engines condenses. At first, the contrails are narrow white bands, but gradually they spread out across the sky, although scientists do not yet understand the reasons.

Dinosaur Life Styles Revealed

John R. Horner of the Museum of the Rockies at Montana State University reported in the June 24, 1982, issue of *Nature* that at least one type of dinosaur nested in colonies. This was the hadrosaur, or duck-billed dinosaur (*Maiasaura*). Investigations in Montana have unearthed a group of eight nests located about 7 meters (24 feet) from each other in what appears to be a colony. Seven meters is about the length of one hadrosaur, so the nests are much closer together from the hadrosaur's point of view than from ours. The close grouping may indicate that the hadrosaurs relied on each other for protection from enemies while they were raising their young.

Skeletons of the young dinosaurs date from about 75,000,000 years ago and their size ranges from ⅓ to 1 meter (13 inches to 39 inches). Because of their size and the probable length of time they spent in the nest, Horner believes that they may have been warm-blooded. Assuming that they were cold blooded like present-day reptiles would imply that hadrosaurs grew slowly, the way reptiles do. In that case, the baby hadrosaurs would have been nestlings for about eighteen months, far longer than is reasonable, before they reached the size found. On the other hand, if the hadrosaurs grew at the rate of birds, then the nesting period would have been only a month and a half, which is not unlikely.

Another nesting site nearby was occupied by a similar type of dinosaur (perhaps a hypsilophodon). The nests, found at various levels in the sediment, indicate that these animals used the same site in successive years.

Footprints found in the mud of a prehistoric lake provide evidence that some carnivorous dinosaurs could swim. Approximately 43 prints from the claws of the dinosaurs were found. The larger prints were probably made by megalosaurus or teratosaurus, while the smaller scratchings could easily have been made by a dinosaur of the Coeluridae family. Popular belief has previously been that only herbivore dinosaurs swam as a means of avoiding their enemies.

In 1982 a flash flood exposed the fossilized tracks of a dinosaur that could have broken the Olympic record for the 100-meter dash (which is 9.9 seconds). Most likely the dinosaur was a carnivore called *Acrocanthosaurus*. A formula has been developed that measures a creature's speed from the distance between its footprints and the size of the prints. Paleontologist James Farlow used the formula to calculate that the dinosaur was loping along at nearly 40 kph (25 mph), more than three times as fast as the previous speed record for dinosaurs.

Earthquakes in the Eastern U.S.

Over the past decade, scientists have vastly increased their knowledge of earthquake activity east of the Rockies. They now predict that within the next 20 years there is a 25 to 50 percent chance of a major earthquake in this part of the U.S..

The earthquake zone in the East does not lie where two plates come together as it does on the Pacific coast. Instead the zone is found beneath the single North American plate along fault systems that were formed during the Precambrian period. In the East, these are usually strike-slip faults, or faults that move horizontally, probably under pressure from the spreading Atlantic sea floor. An area of earthquakes stretches from Arkansas to Canada and another is found from Alabama northward to New England.

It was in these areas that severe earthquakes occurred in the past. In 1811–12 a series of earthquakes that registered between 8.4 and 8.7 on the Richter Scale occurred near New Madrid, Missouri. These earthquakes were the largest in the United States to have occurred in historic times. They destroyed property in an area 240 kilometers (150 miles) by 80 kilometers (50 miles) in Arkansas, Missouri, and Kentucky and were felt for hundreds of miles. In 1886 another major earthquake struck Charleston, South Carolina, leaving 60 dead. The last major earthquake caused by the New Madrid fault occurred in 1895. More recently—June 1982, in fact—a series of 19,000 small quakes occurred in north-central Arkansas.

In 1984, the seven states likely to be affected by another large New Madrid earthquake formed the Central U. S. Earthquake Consortium. The consortium, heeding warnings that another major earthquake in the area could come by the end of the century, is developing a response plan.

Earthquakes have also been felt in an area around Moodus, Connecticut. In 1981, 500 small quakes were recorded near Moodus at depths of a mile or less. On June 17, 1982, a quake occurred that registered 2.1 on the Richter scale. Tape recordings made during the larger quake reveal rumbling noises, while the smaller quakes have been accompanied by popping sounds.

Earthquakes occur in the East every year. In 1983, for example, there were four earthquakes strong enough for people to feel in New York and in Maine, while there were two in New Jersey. Of course, these numbers are still small compared with California, where the 1983 count of felt earthquakes reached 119, over a third of the total of 342 earthquakes felt in the U.S. that year.

Strongest El Niño of the Century

El Niño (*the infant*) is a climatic phenomenon of the Pacific Ocean that has appeared many times in the past. It usually begins around Christmas and generally affects the western coast of South America. The benign name for this phenomenon comes from the nearness to Christmas and also because the current bears "gifts" in the form of exotic tropical species to the normally cold waters off the shore of South America. This benevolence is misleading, however, for El Niño also brings death and destruction. In 1982 the largest El Niño on record appeared, and its effects over much of the Earth were unusually devastating.

There are actually two related events that occur during one of these episodes. One is the Southern Oscillation, a change in the atmosphere. The other is El Niño, a change in ocean currents and temperature. Since the two events are intimately related, both meteorologists and oceanographers have come to refer to the combined phenomenon as an El Niño–Southern Oscillation event, or ENSO for short. For the rest of the world, however, the name *El Niño* has continued to win out over the somewhat more accurate but unromantic name *ENSO*.

Unlike previous El Niños, this one began earlier than Christmas, with noticeable changes in July rather than December, 1982, and what are now recognized as precursors in the spring of 1982. Changes in sea level and temperature in the eastern Pacific in July and August were preceded by a moderation of the trade winds, the global winds that normally blow from east to west and keep the warm Equatorial countercurrent at bay. When these winds slacken, the countercurrent begins to flow more rapidly and extends farther south. Evidence that this was happening appeared in July 1982 as sea levels in the mid-Pacific rose 15–25 centimeters (6–10 inches).

Although we think of "sea level" as definite measure, the sea is not "flat"—or spherical, if you think on a global basis. Currents cause the sea to be higher where the currents flow toward a land mass, lower where they flow away from land. Even gravitational differences in the Earth's crust affect the level of the oceans.

By the fall, the surge of warm water brought by the countercurrent had reached almost to the coast of South America. Water temperatures near the coast rose 6°C (11°F) by the end of 1982 and stayed that far above normal for months. This proved catastrophic for the vast storehouse of microscopic marine life that can only live in the cooler waters that usually flow along the coastline. They either floated away into cooler water or died, mostly the latter. The other animal life—fish, birds, and mammals—that directly or indirectly depends on these microscopic organisms died too. Birds that nest off Christmas Island, for example, declined to nest as they do in normal years; the number of great frigate birds on the island declined from 20,000 in June 1982 to fewer than 100 by November. All of the Galapagos fur seal pups born in 1982 died, and other seals in the eastern Pacific showed similar, if less drastic, losses. Another major disaster was the death of coral polyps across the Pacific. The coral depends on algae that live symbiotically with the polyps, but the high water temperature killed the algae, and the polyps soon followed. Since the coral reefs depend on living coral to maintain them, this change could lead to long-term consequences from El Niño. As far away as the coast of California, giant kelp also died and were replaced by smaller forms. Like the death of the corals, changes in the kelp population may affect parts of the ocean for many years. In addition, the warm water added heat and moisture to the atmosphere, creating torrential rains and flooding on the South American coast. In Ecuador, Peru, and Bolivia, the rainfall exceeded all previous records. Hundreds were killed in floods or mud slides. Across the Pacific, on the other hand, usually moist areas were wracked by devastating droughts. The worst drought in history struck Australia, and similarly droughts hit the Philippines, parts of India and even Africa. Many lost their lives while billions of dollars in crop damage resulted. In the United States, El Niño created storms that wracked the Pacific Coast, Louisiana, and Florida, taking lives and destroying homes and crops.

Some scientists believe that the present El Niño was connected to the eruption of El Chichón volcano in Mexico in April, 1982. They theorize that the ash and other materials sent aloft by the volcano absorbed sunlight and heated the area high over the equator more than usual. This had the effect of reducing the velocity of the trade winds. But Dr. Eugene Rasmusson of the National Weather Service's Climate Analysis Center in Maryland said that the first indication that El Niño was coming occurred before the eruption of El Chichón. This was a lowering of air pressure over Tahiti and Easter Island indicating a slackening of the trade winds. However, even if El Chichón did not start El Niño, the changes it caused masked the start of the phenomenon, confusing the picture until July.

El Niños usually flow for only a single year. Exceptions were 1877 to 1879 and 1940 to 1942 when the current flowed for a two-year period. Records from the period show that the winter of 1877–78 was warm in the U.S. and the summer was unusually hot much like the summer of 1983. Despite these precedents, the most devastating El Niño on record finally came to an end late in 1983.

Methane on the Rise

Two studies of polar ice show that methane levels have risen over the past few hundred years. One study of ice cores in Greenland and Antarctica examined methane in air bubbles over a period stretching back 3,000 years to as recently as 100 years ago. These core samples show that methane levels have doubled over the past two centuries. A second study of ice samples in Greenland offered confirming evidence. Here methane levels in gas bubbles started increasing 400 years ago.

Therefore, something seems to be adding a lot of methane to the atmosphere—but what?

In 1982 Pat Zimmerman of the National Center for Atmospheric Research, calculated that the 240,000,000,000,000,000 termites on Earth produce 224,000,000,000 cubic meters (8,000,000,000,000 cubic feet) of methane per year, or 150,000,000 tons each year, accounting for half the methane in the atmosphere. This result surprised everyone, and set a number of scientists to the task of measuring the amount of methane produced by termites and re-estimating the number of termites in the world.

One study, by Ralf Conrad of the Max Planck Institute for Chemistry in Mainz, West Germany and his associate Wolfgang Seiler, reported that the amount of methane contributed by termites is only about 10,000,000 tons. Conrad and Seiler based their estimates on measurements of methane emissions from termite mounds in Transvaal, South Africa. While this is still a lot of methane, it no longer makes termites the principal source of the gas.

Another study showed that swamps not only contribute methane to the atmosphere but also remove the gas. Research at the Great Dismal Swamp in North Carolina revealed that during those months when the swamp is dry, it acts as a huge receptacle, removing methane from the atmosphere. Presumably, the same process would occur in other swampy areas which experience a dry season. Scientists theorize that as marshes and other wet areas are drained for farmland they may serve to reduce the amount of methane in the atmosphere.

Such findings may have a significant impact. Scientists warn of a global warming trend, a contributing factor to which is the increasing amounts of methane in the atmosphere which trap heat radiated from Earth. However, if the soil can act as an important sink for methane absorption—contrary to what was previously thought—the amount of global warming may be reduced.

1983

Kilauea Erupts Again

Kilauea volcano, on the island of Hawaii, has erupted many times over the ages, the most recent eruption beginning in January 1983 and continuing on and off throughout 1983 and into 1984. This eruption is considered the most powerful Kilauea activity since 1977. During 1983 about 130,000,000 cubic meters (170,000,000 cubic yards) of lava poured out of the volcano.

Scientists can now predict an eruption at Kilauea about a day before it begins. The eruption is preceded by a series of small earthquakes caused by rising magma through the dikes inside the volcano. This time, the earthquakes occurred in an area called the East Rift Zone. Soon after the eruption begins, the huge caldera atop the volcano diminishes. In this case, it subsided 800 millimeters (31.5 inches) in five days. Researchers have been able to study eruptions of Kilauea because the eruptions are comparatively gentle.

Kilauea shares the island of Hawaii with another huge volcano—Mauna Loa. Both are gently sloping shield volcanoes, and when one is active, the other is usually not. The two volcanoes were thought to share the same reservoir of magma in the Earth's mantle, which would account for the two volcanoes' not erupting at the same time. However, late in 1983, with Kilauea still at it, USGS Hawaii Volcano Observatory scientists started preparing for another eruption of Mauna Loa. Mauna Loa last erupted in 1975 after a silence of 25 years, so it was due again. Both small earthquakes and a swelling in the summit of Mauna Loa indicated that magma was moving into the volcano.

An eruption of Mauna Loa could have been touched off by an earthquake of magnitude 6.7 that rumbled along the valley between Mauna Loa and Kilauea on November 16, 1983, and the scientists went on immediate alert. But no eruption occurred.

The volcanologists were right, however, about the nearness of Mauna Loa's next major eruption. By March 1984, Mauna Loa was erupting in full force in a spectacular show that could be seen live all over the island of Hawaii (and on television in the rest of the world). For a time Hilo, the largest city on the island, appeared to be threatened by lava, but the flows stopped before reaching the city. Since Kilauea continued to erupt, scientists now assume that the magma reservoirs for the two volcanoes are at least partly separated.

Obsidian Dating Introduced

The age of an archaeological site, when there are no written records found, is a key number that is often hard to determine. Until fairly recently, the absolute date could not be determined, although relative dates based on strata or types of pottery could be used. More recently, methods based on radioactive carbon-14 or on a chronology of tree rings have yielded absolute dates. Yet, suitable sources of tree rings are often absent, as are sources of organic carbon. Furthermore, these sophisticated methods must often be carried out in a laboratory months or even years after the archaeologist has done the digging.

Archaeologists now have a cheap, simple method of dating ancient objects made from obsidian, a volcanic glass used by early civilizations to make a variety of tools. The new method was developed in 1983 by Joseph W. Michels, Ignatius S. T. Tsong, and Charles M. Nelson.

For years, archaeologists have known that as obsidian ages, it forms a hydration rim. The rim, which is visible with an optical microscope, is a layer of obsidian where hydrogen ions from water replace other ions in the glass that is just below the surface. If they knew the rate of hydration and measured the rim, archaeologists would be able to determine the age of the artifact. But this method was not accurate enough to use, because the rate of hydration was different for each site and object. As a result archaeologists had to rely on carbon-14 dating, using wood or other organic samples instead of obsidian. In many cases, they could not date the site absolutely, but could only make comparative estimates.

The new method is based on accurately determining the rates of hydration for different types of obsidian and different sites. Michels, Tsong, and Nelson gathered artifacts of obsidian from different sites. In the laboratory they put the samples under high temperature and pressure to speed up the rates of hydration. They measured the rim thickness of the artifacts and determined the rate for a given type at a given site. To test the method, they compared dates from obsidian with carbon-14 dates from two of the sites. Both methods produced almost the same results.

The obsidian dating method is cheaper than the carbon-14 technique, since it does not require careful measurements by laboratories of radioactivity. A simple microscopic measurement of the obsidian's hydration layer is all that is needed, which can be done at the site. Obsidian dating is also potentially far more useful at many sites. The obsidian method has dated objects as old as 120,000 years, or even older, while carbon-14 can only be used with objects up to about 50,000 years old.

A Warmer Arctic

A haze that forms over the Arctic each spring could mean a warmer climate to come for that part of the world. The haze consists of minuscule particles of carbon, believed on the basis of trace elements found in the haze to come from the industrial complexes inside the Soviet Union. In the spring, when the sun shines in the Arctic, the haze absorbs solar radiation, producing temperatures that are warmer by an amount estimated between a few tenths and 1°C (2°F). Although particles are also present in winter, this is the period of the long Arctic night, so the effects are not the same. By the summer months, the particles have largely disappeared, and a new haze does not arrive until the following spring.

Scientists began a study in March 1983 to determine what the long-term effects of the Arctic haze might be. The haze absorbs sunlight; and when the particles fall on the snow, they reduce its ability to reflect sunlight. Both of these could have the effect of producing warmer Arctic climates. This, in turn, could contribute to the worldwide warming presently underway.

Researchers have also discovered that the effects of the eruption of El Chichón in 1982 have reached the Arctic. Samplings of the stratosphere have revealed concentrations of sulfur dust reaching as high as 2,000 ppm, where 35 ppm is normal. The effect on climate is as yet undetermined.

Harmonic Rolls

Data from the oceanographic satellite SEASAT has indicated "harmonic rolls" in the seafloor, an apparent reflection of undulations in Earth's gravitational force. These patterns, which are up to 1,000 kilometers (621 miles) in length, occur at regular intervals of about 200–250 kilometers (124–155 miles). They are found on the fastest moving crustal plates such as the one under the South Indian Ocean.

The rolls may be caused when a plate compresses, causing the lithosphere below to move up and down, or when the mantle and lithosphere interact as a plate moves rapidly over the mantle.

The SEASAT satellite did not observe the ocean floor directly, but instead observed changes in sea level. These changes reflect the gravitational changes in the crust beneath the sea. SEASAT is capable of measuring the distance between itself and the ocean with a precision of 10 centimeters (4 inches).

Oldest Animals, Plants, and Rocks

The year 1983 was a banner year for the discovery or announcement of a number of important finds from very long ago, ranging from the oldest land animals and seeds to the oldest solid chunks of the Earth's surface.

The discovery of a group of fossils of very early land animals, 380,000,000 years old, suggests that life came ashore long before scientists previously believed. The fossils, embedded in rocks found at a site in upstate New York, include the earliest known centipede, a mite, some arachnids, and perhaps the oldest-known insect. The fossils show that these ancient creatures already had primitive lungs and some also contained tracheal tubes. Since these animals were already well adapted to life on land, the beginnings of adaptation must have occurred much earlier.

The fossils were found in 1977 by Patricia M. Bonamo and James D. Grierson of the State University of New York in Binghamton, but they required six years of analysis before a definitive report could be given on them in May of 1983. In addition to the New York site, fossils of early land animals have also been found in Scotland and Germany. However, these latter two sites were probably swampy areas, because they contained both land and aquatic creatures. These three sites were all from beds that were located on the so-called Old Red Continent—a tropical land mass found at the equator 400,000,000 years ago that is now broken into parts of North America and Europe.

The fossils of the tropical animals found there look, in some cases, much like animals of today. For example, the centipede unearthed at the New York site bears a striking resemblance to its modern counterpart. One of the arachnids is similar to a daddy longlegs.

In southern France, scientists uncovered the oldest octopus fossil ever discovered. The creature—about 5½ inches long—roamed the seas about 155,000,000 years ago and goes under the name *Proteroctopus*.

A French-American group of researchers working in South America discovered the fossil remains of marsupials thought to be 70–75 million years old. This makes them the oldest land mammals to be found on that continent and indicates that marsupials did not arrive there until late in the Cretaceous period, when they invaded from North America after a land bridge formed between the two continents. Then the marsupials traveled farther along land bridges that existed then, but which are now long gone. The marsupials went on through Antarctica and into Australia.

In Pakistan, Phillip D. Gingerich and Neil A. Wells, working with colleagues from France and Pakistan, unearthed in 1978 the remains of the world's most primitive whale to date—*Pakicetus inachus*, which thrived about 50,000,000 years ago. Since the fossil was discovered in fresh water, scientists theorize in their April 22, 1983, report on the find that *Pakicetus* may have been a land carnivore that went into the water only to catch its prey. In addition its ears and its teeth indicate that the whale lay somewhere on the evolutionary timeline between a land carnivore and a present-day whale. Thus, this creature may be an evolutionary link between land mammals and ocean mammals.

Scientists have also found seeds dating back 360,000,000 years—the earliest so far to be unearthed—at a West Virginia site. The seeds were contained in the cupules of seed ferns—forerunners of many of today's seed-bearing plants. Finding seed plants in this early period is somewhat unusual since most plants reproduced by spores, a more primitive method of reproduction.

An Australian scientific team has discovered minerals 4,100,000,000 to 4,200,000,000 years old—the oldest ever unearthed by 200,000,000 to 300,000,000 years. Discovered in western Australia in 1983, the minerals are zircons. They were found by students of William Compston, head of the isotope group in Australia's National University's Research School of Earth Sciences in Canberra.

Scientists believe that the zircons were scraped from the ancient silica-granite crust, and deposited in sedimentary beds. Since most experts agree that the Earth is only 4,500,000,000 years old, the zircon discovery tells us something of the planet's early years. It also raises hopes that scientists may be able to find rocks of an even earlier age from which the zircons were scraped so they can determine such things as what the Earth's chemical composition was soon after the Earth was formed, and when the Earth's crust and mantle separated.

The zircons were dated by comparing the isotopes of uranium and lead that were found in them. Since uranium breaks down at a known rate into specific lead isotopes, these ratios can be used to determine how long the uranium has been fixed in the mineral.

A STRANGE GLOW

Many observers have reported seeing a strange glow when earthquakes occur. Researchers have now identified the cause of that glow. Friction that takes place at the fault-line heats up a strip of rock at the surface. This in turn causes moisture in the atmosphere to vaporize. The heated rock and the vapor barrier surrounding it produce an electric field which discharges electricity that is seen as light.

Changing Shape of the Earth

The Lageos satellite was launched in 1976 solely to provide a stable Earth-orbiting body that can be used as a reflecting surface in space to measure changes in the Earth. Its 9,700 kilometer (6,000 mile) high orbit is too high for the drag of any part of the atmosphere to affect it, and the satellite is small and heavy, giving it momentum and little surface to be affected by such forces as the solar wind. As a result, Lageos is supposed to remain in its orbit for hundreds of millions of years. It contains no instruments at all.

A 1983 study of the actual orbit of Lageos over five and a half years showed that the gravity field of the Earth, which is linked to the distribution of its mass, is changing. This led scientists to conclude that the planet's shape is slowly changing, too. The discovery would not have been possible had scientists not been able to observe the orbit of the Lageos satellite almost exactly with the help of lasers. The shape of the satellite's orbit is pegged to the shape of the Earth. When the orbit of Lageos changed unaccountably, scientists reasoned that fluctuations in Earth's gravity field and shape were the cause.

The change in shape is likely due to the so-called post-glacial rebound. According to this theory, the Earth is rebounding in the higher latitudes as the glaciers retreat, a process that began 15,000 years ago. The crust is somewhat elastic, so after the immense pressure of the glacial ice was removed, the crust began to rebound.

The effects of the ice ages, however, account for only about 10 percent of the planet's shape. The movement of molten rock in the planet's interior and its rotation are the primary reasons for its shape.

Antarctic Glacial Cycles

In 1983 Peter N. Webb and David Harwood discovered marine microfossils in rocks that had been taken from the Transantarctic Mountains in Antarctica in 1964. The discovery of the fossils hints at Antarctic glacial cycles. The rocks were scraped from the ocean floor and carried up the mountain by the receding Antarctic glacier. When the glacier began to move forward toward the sea again, the rocks were left behind. Webb and Harwood also point out that rocks were repeatedly transported from the sea to the mountains in a series of glacial cycles: 70,000,000 to 65,000,000 years ago; 50,000,000 to 40,000,000 years ago; 30,000,000 to 20,000,000 years ago; and 7,000,000 to 3,000,000 years ago.

An Important Boundary

Paleontologists have designated a site along the Aldan River in Siberia as the official boundary between the Cambrian and Precambrian geological periods. The boundary line—570,000,000 years old—marks the emergence of animals with skeletons or shells. The river bluff site is unusual in that it contains limestone outcroppings that clearly chronicle development from the Precambrian period (when animals were only soft-bodied) to the Cambrian period, the first with fossil remains of shelled animals.

Geologic Time Scale

Time	Era	Period	Epoch	Time Before Present (Millions of Years)
Precambrian	Archean			4,600
	Proterozoic			2,500
		Cambrian		570
		Ordovician		500
	Paleozoic	Silurian		425
		Devonian		395
		Carboniferous		350
		Permian		290
		Triassic		235
	Mesozoic	Jurassic		190
		Cretaceous		130
Phanerozoic			Paleocene	65
			Eocene	55
		Tertiary	Oligocene	38
	Cenozoic		Miocene	26
			Pliocene	6
		Quaternary	Pleistocene	1.8
			Holocene	.01(11,000 yrs)

Earthquakes in California

Although earthquakes occur in all parts of the United States, nowhere do they occur with greater frequency and destructiveness than in California. As a consequence, much of the earthquake research in the United States is carried out in that state.

The 6.5 magnitude earthquake that occurred at Coalinga, California on May 2, 1983, may have been caused by a fault lying beneath a hill in the area. After the quake took place, scientists looking for a fault along the surface found nothing. But they did notice that the Anticline Ridge—a hill in the area that stretches for 15 to 20 kilometers (9.3 to 12.4 miles)—rose by half a meter (a foot and a half). Meanwhile the adjacent valley where Coalinga is located had lowered 0.2 meter (9 inches). The main shock of the Coalinga earthquake is thought to have occurred along the hidden fault.

Geologists now speculate that some of the hills in the Coalinga area are folds created by movement in the earth's crust along hidden faults that may lie many kilometers below the surface. Some of these hills run at right angles to the San Andreas fault west of Coalinga. The fault apparently slipped about 2 meters (6.5 feet) at a depth of 4 to 12 kilometers (2.5 to 7.5 miles) below the surface with very little breakage of the ground at the surface. An earthquake of this type causes ground motion at the surface, however, that is two to three times higher than the more common kind of slippage along a surface fault. As a result, it can do more damage than a surface earthquake of similar energy.

Work on the Coalinga earthquake has been useful in understanding the seismic events at a number of other earthquakes of the same type, including the 1980 El Asnam, Algeria, quake; the 1964 Niigata, Japan, quake; and the 1952, Kern County, California, quake.

Researchers working along California's San Andreas fault have come up with a series of predictions for the next great earthquakes in that region of the country.

A study of the records by Steve Kilston and Leon Knopoff indicated in 1983 that the position of the moon might be implicated in earthquakes that occur in areas such as Southern California. There seem to be two reasons: the moon's gravitational pull has a major effect on the Earth, causing tides of 15 centimeters a day in the "solid" Earth itself (not to mention the tides in the oceans); and the northwest to southeast direction of the San Andreas and other faults in the area. If it were not for this second factor, the effects of the Moon's gravitational pull on different parts of the Earth would tend to cancel each other.

Every 18.6 years the moon reaches what astronomers call its maximum lunar declination—that is, its northernmost position. At this time, the moon's gravitational pull would be strongest in the same direction in which the western side of the Southern California faults are already moving. If the moon is implicated in earthquake activity, the activity in the California area would tend to occur at the times of the maximum lunar declination. Examination of the records has, indeed, revealed that earthquakes measuring 6.0 or more on the Richter Scale occurred in 1857, 1933, 1953, and 1971 when the moon was at its maximum declination. The next time this occurs is November 1987, so there could be another major earthquake about that time.

Researchers have also discovered that major earthquake activity in Southern California occurs during full and new moons. At these times, the sun and moon exert gravitational pull in the same direction along the faults. One side is probably pulled apart by the combination of the sun's and moon's gravity, making earthquake activity more likely.

In the fault zone around San Francisco—where the massive earthquake occurred in 1906—geologists believe a similar size quake is unlikely over the next three decades because there is too little stress. During the next 20 years, predictions are that there is a 30 to 60 percent chance of an earthquake of magnitude 6.0 to 7.0 in the San Jose to San Juan Bautista area. (In actual fact, however, a slightly smaller earthquake occurred in the vicinity of San Jose on April 24, 1984, causing no deaths and limited damage.) Geologists believe that in the fault zone stretching from Cajon Pass to the Salton Sea (the southern third of the San Andreas fault) chances of a great quake of 8.0 magnitude are 25 percent, making this area the most vulnerable to massive earthquake destruction.

In the area around Parkfield—which lies between Los Angeles and San Francisco—scientists rate the chances of a 5.5 to 6.0 magnitude quake in the next 20 years as high as 75 percent. Since 1922, Parkfield has experienced three moderate earthquakes. And Coalinga lies just north of Parkfield.

Scientists base these predictions on earthquake histories of the California sites. Pioneering work by Kerry Sieh of the California Institute of Technology into the history of the Pallett Creek site, has revealed that earthquakes follow recurring patterns. It is assumed that earthquakes follow such patterns because the movement of tectonic plates causes steadily increasing pressure along a fault. If the fault breaks when the pressure reaches a certain level, the steady increase of pressure combined with a stable breaking level will produce roughly periodic earthquakes.

A New Dinosaur

In May 1983, the British Museum of Natural History sent a team to continue the excavation of a project that had been started in January by Bill Walker, an amateur fossil hunter. He had found the claw of a large dinosaur in Surrey in Great Britain. The 30-centimeter (1-foot) claw—twice the size of that of Tyrannosaurus Rex—belonged to a carnivorous dinosaur.

Assuming that the rest of the dinosaur matches the claw, the creature was 3 to 4.5 meters (10 to 15 feet) in height when it lived, 125,000,000 years ago. This beast must have put fear into the heart of even the great king of dinosaurs.

Arctic Evolution Claimed and Criticized

In September, 1983, Leo J. Hickey, Robert M. West, Mary Dawson, and Duck K. Choi proposed the theory that many plants and animals, including early horses, turtles, and tortoises, evolved during warm spells in the Arctic between 65,000,000 and 45,000,000 years ago. At that time Europe and North America were joined in the arctic regions, and the organisms that evolved there then radiated south into the two continents. The dates for this proposed evolution range from 2,000,000 to 18,000,000 years earlier than in the competing theory, which has the plants and animals evolving in the south and radiating north during the warm spells.

The new theory is not based upon new fossil finds, since early plants and animals have previously been found in the far north, especially on Ellesmere island, the northern tip of Canada. Instead, it is based upon new dates for the fossils derived from a study of magnetic reversals in the rocks on the island.

Most of the criticisms of the new theory centered around the long polar night. Critics pointed out that plants and animals would be less likely to evolve where it was dark for six months of the year.

A HEAD TRANSPLANT

In 1879, paleontologists excavating a site in Wyoming found the headless remains of a brontosaurus. They created a skull for the animal, based on dinosaur jawbones found at a site nearby. This blunt-nosed broad skull has remained popularly associated with the animal; until recently, that is. Scientists have now unearthed a genuine brontosaurus skull. This one is longer and slimmer and significantly changes our perception of the huge animal's appearance.

Extinction: A Fact of Life?

Fossil studies indicate that mass extinctions may have happened throughout Earth's history. In fact, according to J. John Sepkoski, Jr., of the University of Chicago and David M. Raup, they seem to occur roughly every 26,000,000 years. This has led to speculation about some extraterrestrial event as the cause. Many scientists now believe that an asteroid did indeed collide with Earth about 65,000,000 years ago.

Discoveries in 1983 reveal that the earliest mass extinction took place among the algae 650,000,000 years ago. This precedes by 200,000,000 years what was previously thought to be the first mass extinction—that which occurred among various species of shell-covered marine animals. The findings came when studies of sedimentary rocks in a variety of areas revealed a large drop-off in the numbers of species in the same geologic period. The areas under study were islands in the Arctic Ocean; Scandinavia; the Baltic; and part of Greenland. Dr. Andrew H. Knoll, who reported the discoveries, said that there is no explanation for the extinction. However, Dr. Knoll said environmental changes brought on by the beginning of an ice age about the same time may hold the answer.

There is only one other known mass extinction of this extent in Earth's history. A mass extinction took place about 220 to 225,000,000 years ago and destroyed 50 percent of the animal families present at the time. Lesser extinctions, however have been shown by Sepkoski and Raup to have happened every 26,000,000 years, with the last one occurring 11,000,000 years ago.

In 1984, two groups of scientists proposed specific theories to account for the Sepkoski-Raup pattern, although they modified the period to 28,000,000 years. (This number is also close to a period of 31,000,000 years, found by two other groups of scientists, working independently, for periods of heavy meteorite bombardment of the Earth.) Both theories assume that disturbances in the Oort cloud of comets that surrounds the solar system cause comets to be displaced toward the sun. Inevitably, one or more of these comets hits Earth, causing the extinctions.

NASA accounts for the disturbances in the Oort cloud by saying that the solar system passes through the plane of the Milky Way at these times. A group including Richard A. Muller, Marc Davis, and Piet Hut has an even more dramatic idea. They believe that the sun has a companion star with a period of 28,000,000 years. The star, which they have named Nemesis, has an orbit that brings it though the Oort cloud when it is closest to the sun, creating the disturbance that sends the comets sunward. Efforts are being made to locate Nemesis.

Glomar *Challenger's* Last Leg

It would be hard to underestimate how much deep-sea drilling has added to our understanding of the Earth. The entire theory of plate tectonics rests on information originally derived from ocean cores. Such topics as the global climate in the past and the confirmation of the 65,000,000-year-old iridium layer could not have been investigated in sufficient detail without the cores. Additionally, the cores on the continental shelves have enriched our pockets as well as our minds by helping to locate petroleum deposits beneath the oceans. Foremost in these efforts has been the United States Deep Sea Drilling Project.

For the fifteen years from 1968 to 1983 the Deep Sea Drilling Project consisted almost entirely of the 96 voyages, or "legs," of the American drilling ship Glomar *Challenger*. The last leg began in October 1983 to takes cores from the Gulf of Mexico near the Mississippi Delta. In October 1984 the Deep Sea Drilling Project was scheduled to be replaced by the Ocean Drilling Project and a new ship. During its tenure, the *Challenger* added more to the knowledge of the sea and the past than any vessel since its predecessor, the H.M.S. *Challenger*, made the first systematic study of the oceans on a voyage that started in 1872 and lasted two and one half years.

On its 93rd leg, for example, Glomar *Challenger* revised the theory of how offshore oil deposits are formed. Here is what it found from its cores of the ocean floor:

Located on the continental rise at a depth of 1,212 meters (4,000 feet) below the ocean floor are mats of sand and sandstone replete with organic matter. If heat were present to break down the organic matter, this site, located 422 kilometers (270 miles) east of Cape Hatteras, North Carolina, would be an ideal source of petroleum. This near-miss has led scientists to urge further exploration of the continental rise as an oil source.

The sands at the recently explored site are interspersed with layers of black shale and covered by a 900 meter (3,000 feet) shale cap which would be sufficient to hold the hydrocarbons in the mat once they were produced. The entire deposit was layed down 115,000,000 years ago during the Cretaceous period. The sands could have been deposited in river deltas found along the continental shelf.

Before the *Challenger's* find, it was believed that offshore coral reefs prevented organic material from reaching the ocean this far from shore. Since the organic matter located by the *Challenger* undoubtedly exists, either the coral reefs did not exist, or else they did not stop the deposits of potential oil-bearing shale from forming where they are now located.

Positive Lightning

Lightning bolts that shoot between clouds and Earth usually carry a negative charge that may reach as high as 80,000 amperes. Most textbooks still record that all lightning bolts are negative. But National Oceanic and Atmosphere Administration (NOAA) scientists have also discovered that in the midst of violent storms, positive bolts may shoot earthward. A spokesman for NOAA explains that these bolts may be caused by intense updrafts that are "highly sheared," that is, their wind speed changes significantly with height. This may cause the usual positive-over-negative layers in the atmosphere to tilt, permitting bolts from the positive layer to strike the Earth. In the United States, positive lightning has been found to be common east of the Rocky Mountains, although many meteorologists had been skeptical about it until 1983.

Martin Unman, director of the Lightning Research laboratory at the University of Florida in Gainesville, reported in October, 1983, that he and his colleagues have proof positive of positive lightning in Florida thunderstorms. Positive lightning has also been reported by other observers, including the NOAA Severe Storms Laboratory at Norman, Oklahoma, but there was some doubt whether or not it was common or caused by special conditions. The Florida studies helped resolve the question in favor of positive lightning as a common component of severe summer thunderstorms.

Studies indicate that positive lightning has a maximum current of 100,000 to 250,000 amperes, a factor of 10 higher than the average current for negative lightning. David Rust of the NOAA center points out that the current for positive lightning is continuing; that is, when lightning makes its connection with the ground, the connection holds for about ¼ second. This is much longer than the millisecond connection of negative lightning and means that positive lightning is far more destructive than the ordinary negative variety.

LIGHTNING UP THE SKY

Lightning is often seen traveling within a cloud or downward from a cloud to the Earth. But now a rarer phenomenon has been spotted: lightning traveling upward from a cloud. Not surprisingly, the phenomenon was reported by airplane pilots flying in the stratosphere who saw the lightning bolts shoot an estimated 35,300 meters (120,000 feet) into the air, nearly reaching the ionosphere.

Earthquake in Idaho

An earthquake registering 6.9 on the Richter Scale occurred October 28, 1983 in central Idaho, sending out tremors felt in all the contiguous states and Canadian Provinces. Termed the strongest earthquake to hit the area in a century, it was also the most powerful in the lower 48 states since a California quake in 1952. The strength of an earthquake does not always relate directly to its destructiveness. The population of the area, the way the houses are built, the time of day, and the depth of the earthquake are all factors in the destruction equation.

The Idaho quake took the lives of two young children. Seismologists believe the earthquake's epicenter was located along the Big Lost River fault, an area of earthquakes as well as volcanic activity in the past.

The United States Geological Survey (USGS) calls an earthquake "significant" if it measures as much as or more than 6.3 on the Richter Scale. In 1983, there were four significant earthquakes in the United States. In addition to the Idaho quake, the May 2 earthquake in Coalinga, California (which injured 45 people) measured 6.5, while a November 16 earthquake on the big island of Hawaii (which injured six people) measured 6.6. The other earthquake of 6.5 was actually off the coast of Alaska in the Pacific, but counted as a U.S. quake by the USGS.

Worldwide, significant quakes have been on the rise since 1981, going from 50 that year to 56 in 1982 to 70 in 1983. However, this is not likely to represent any real trend, since there were 71 earthquakes in 1980.

It often seems, however, that two volcanoes erupt or two earthquakes happen quite close to each other. For example, in the same week as the Idaho earthquake, a 7.1 Richter scale earthquake also hit Turkey, killing at least 1,200 persons and probably more.

OTHER EARTHQUAKE LIGHTS

At lookout stations on the Yakima Indian Reservation in Washington State, observers have reported seeing white balls of light, about the size of luminous baseballs. These may be associated with a series of small earthquakes that have occurred in the region. In the past, scientists have observed lights along the fault lines where earthquakes have occurred. These "luminous baseballs" may be another manifestation of the fault-line phenomena. In the Yakima region, 82 of these balls of light were reported between 1972 and 1977. The monthly increase in these reports coincided with an increase in the number of quakes during the same period. Scientists speculate that the light may be caused by frictional heating when the quakes occur.

Measuring Earthquakes

The Richter Scale

Richter Magnitudes	Earthquake Effects
2.5	Generally not felt, but recorded.
4.5	Local damage.
6.0	Can be destructive in populous region.
7.0	Major earthquake. Inflicts serious damage. Roughly ten occur each year.
>8.0	Great earthquakes. Occur once every 5–10 years; produce total destruction to nearby communities.

Modified Mercalli Intensity Scale

I. Not felt except by a very few under specially favorable circumstances.

II. Felt only by a few persons at rest, especially on upper floors of buildings.

III. Felt quite noticeably indoors, especially on upper floors of buildings, but many people do not recognize as an earthquake.

IV. During the day felt indoors by many, outdoors by few. Sensation like heavy truck striking building.

V. Felt by nearly everyone, many awakened. Disturbances of trees, poles, and other tall objects sometimes noticed.

VI. Felt by all; many frightened and run outdoors. Some heavy furniture moved; few instances of fallen plaster or damaged chimneys. Damage slight.

VII. Everybody runs outdoors. Damage negligible in buildings of good design and construction; slight to moderate in well-built ordinary structures; considerable in poorly built or badly designed structures.

VIII. Damage slight in specially designed structures; considerable in ordinary substantial buildings with partial collapse; great in poorly built structures. (Fall of chimneys, factory stacks, columns, monuments, walls).

IX. Damage considerable in specially designed structures. Buildings shifted off foundations. Ground cracked conspicuously.

X. Some well-built wooden structures destroyed. Most masonry and frame structures destroyed with foundation. Ground badly cracked.

XI. Few, if any, (masonry) structures remain standing. Bridges destroyed. Broad fissures in ground.

XII. Damage total. Waves seen on ground surfaces. Objects thrown upward into air.

Major American Earthquakes

Date/Time	Mercalli (Richter) Rating	Center/Range	Effects
2/5/1663 17:30	X	St. Lawrence River/ All NE U.S.; 750,000 sq. mi.	No definite accounts from that time. Landslides along St. Lawrence left water muddy for a month. Houses on Massachusetts Bay were shaken, pewter fell from shelves.
11/18/1755 4:11 5:29	VIII	E. of Cape Ann, MA/ New York to 200 mi. E. of Cape Ann; Chesapeake Bay to Nova Scotia	Came with a roaring sound like thunder. Two minute shock cracked beams, swayed treetops and threw people to the ground. Ships at sea thought they had run aground.
12/16/1811 1/23/1812 2/7/1812	XII	New Madrid, MO/ E. to Boston, Canada to New Orleans; 2,000,000 sq. mi.	Small amount of damage because of sparse population. 12/16: people awakened by creaking of timbers in houses and crashing of chimneys. Repeated shocks through night and next day as ground rose and fell, toppling trees. Waves on Mississippi River capsized boats and caved in banks. Equally intense shocks on 1/23 and 2/7. Topographical changes over 30–50,000 sq. mi.
12/21/1812 11:00	X	Transverse Range, CA/ Total area unknown	Foreshock sent many fleeing buildings and saved lives in Santa Barbara, Ventura and northern Los Angeles counties. Destroyed or damaged missions and churches.
6/?/1838	X	San Francisco area	Shock comparable to great 1906 quake. Extensive displacement along San Andreas fault. Mission walls and churches extensively damaged.

Date/Time	Mercalli (Richter) Rating	Center/Range	Effects
1/9/1857 8:00	X–XI	Fort Tejon, CA	Violent shock. Ground crack 40 miles long between Los Angeles and San Francisco. Many large fissures. Changed flow of artesian wells.
11/18/1867 14:50	VIII	Virgin Islands/ Dom. Repub. to Guadeloupe	Much property damage in Puerto Rico. Sea withdrew 150 yards and then advanced an equal distance.
4/2/1868 16:00	X	S. coast of Hawaii	Shook down every wall in Hilo and stopped clocks in Honolulu. Trees bent over, tsunami over 60 feet tall, many drowned.
3/26/1872	X–XI	Owens Valley, CA/ Most of CA, NV, AZ; 125,000 sq. mi.	At Lone Pine, destroyed most adobe houses and killed 27. An area 200–300 ft. wide sank 20–30 feet, leaving vertical walls. Much disturbance of terrain.
8/31/1886 21:51 21:59	IX –X	15 mi. NW of Charleston, SC/ 800 mi. radius	Started gradually and built to a roar, then a rude, rapid quiver. All movable objects shook and rattled. Second shock 8 min. later. People gathered in Charleston public square to escape falling buildings. 60 killed. Cut off from aid by damaged railroads and telegraph lines. Many fled.
5/3/1887 14:13	VIII–IX	Sonora, Mexico/ 300 mi. area; NM & AZ to El Paso	Sparsely populated area. Water in tanks slopped over. Railroad cars set in motion on tracks, buildings cracked.

Date/Time	Intensity (Magnitude)	Location	Description
			Important topological changes to unpopulated area.
9/3/1899 9/10/1899	XI (8.3 & 8.6)	Yakataga, AK/ Yakutat Bay, AK; total area unknown	
4/18/1906 5:12	XI (8.3)	NW of San Francisco/ Destructive to 400 miles; felt in 375,000 sq. mi. area	Greatest earthquake known in California. Extent of displacement along San Andreas fault was about 180 miles. Motion of 10–15 feet not uncommon. Buildings in all parts of San Francisco damaged or destroyed. Shock and resulting fires caused at least 700 deaths.
10/2/1915 22:54	X (7.8)	Pleasant Valley, NV/ OR to San Diego; W to Salt Lake City; 500,000 sq. mi.	Felt over larger area than 1906 quake; it did little damage. People thrown from beds, adobe houses destroyed.
10/11/1918 10:15	VIII–IX (7.5)	NW Mona Passage	Most destructive Puerto Rican earthquake. Tsunami drowned many and destroyed native dwellings. 116 died.
9/29/1921 10/1/21	VIII	W. Elsinore, UT/	After several weeks of preliminary shocks, first shock threw down chimneys and fractured walls. Third shock on 10/1 was like the silent blow of a great hammer. Great clouds of dust arose from the dry desert.
2/28/1925 21:19	VIII	St. Lawrence River region/ felt S to VA, W to Mississippi River	Remarkable for moderate intensity over great area (2,000,000 sq. mi.); damage was comparatively small.

Date/Time	Mercalli (Richter) Rating	Center/Range	Effects
6/27/1925 18:21	VIII (6.8)	E of Helena, MT/ violent to 600 sq. mi.; felt to 310,000 sq. mi.	Large schoolhouse badly wrecked, though reinforced concrete buildings not damaged. Rockfalls disrupted train service.
8/16/1931 5:40	VIII 6.4	Mt. Livermore, TX/ 450,000 sq. mi.	Would have been severe in more populous region. Adobe buildings damaged, cracks in earth, tombstones rotated.
12/20/1932 22:10	X (7.3)	Cedar Mountain Dist., NV/ Rockies to Pacific; San Diego to San Francisco	Little property damage; extensive and complicated faulting occurred over wide area.
3/12/1934 8:06 11:20	VIII 6.6 6.0	N. end of Great Salt Lake/ 170,000 sq. mi.	Little damage due to sparse settlement. Large quantities of water emitted from fissures and craterlets.
10/12/1935 11/28/1935	VI– VIII	Helena, MT/ felt in WY, ID, WA	Six quakes between these dates almost directly beneath city and each shock further damaged weakened buildings. Many large buildings severely damaged, including new high school. Only 2 killed.
5/18/1940 20:37	X (7.1)	Imperial Valley, CA/ 60,000 sq. mi. in CA, AZ, NV	Epicenter at El Centro. Surface slipping over 40 mile distance. Damage in all towns in Imperial Valley. Canal damage disrupted irrigation. Nine died.

Date / Time	Intensity (Magnitude)	Location / Area	Description
7/21/1952 3:52	XI (7.7)	S. of Bakersfield, CA/ 160,000 sq. mi.	Largest quake since 1906. 12 killed. Reinforced concrete tunnels with walls 18" thick were cracked or caved in. Water splashed from swimming pools in Los Angeles.
12/16/1954	X (7.1)	Frenchmen's Station, NV/ 200,000 sq. mi.	Little property damage; spectacular surface ruptures. 5 to 15 ft. vertical movement in Dixie Valley fault zone.
7/9/1958 20:16	XI (7.9)	SE Alaska/ S. to Seattle; 400,000 sq. mi.	Massive rockslide in Lituya Bay caused waves which swept boats to sea. 3 killed.
8/17/1959 23:37	X (7.1)	Hebgen Lake, MT/ Seattle to ND; Canada to UT	Great avalanche cascaded from Madison River Canyon, blocking flow and creating 175 ft. deep lake. 28 killed. New geysers erupted at Yellowstone.
3/27/1964 17:36	IX–X (8.3)	S. Alaska/ 700,000 sq. mi.	One of most violent ever recorded; caused vertical displacement over 200,000 sq. mi. Downtown Anchorage severely damaged. Caused tsunami which devastated towns in Alaska and reached as far as Hawaii. 131 killed.
2/9/1971 14:01	XI (6.4)	N. of San Fernando, CA	Epicenter in sparsely populated area of San Gabriel Mts. 58 died, 49 of them in San Fernando V.A. Hospital. Extensive property damage. New earthquake-resistant hospital buildings were total loss. Created zone of discontinuous surface faulting.

Date/Time	Mercalli (Richter) Rating	Center/Range	Effects
4/26/1973 20:26	VIII (6.2)	NE Hawaii Island/ Felt on all Hawaiian Islands	Extensive property damage to Hilo. Downtown area closed. Landslides and ground cracking damaged roads and structures over wide area.
2/2/1975 8:44	IX (7.6)	Near Islands, AK	Caused severe damage to Shemya Island and injured 15. Large crevasses and landslides.
11/29/1975 14:48	VIII (7.2)	Hawaii/ Felt on all Hawaiian Islands	Largest in Hawaii since 1868. Extensive structural damage in Hilo. Ground cracking damaged roads in Volcano National Park. Generated tsunami 6 m high. Caused small eruption at Kilauea Volcano 45 min. later.
10/15/1979 23:17	VII–IX	Imperial Valley, CA/ 50,000 sq. mi. of CA, NV, AZ	Injured 91; destroyed 6-story reinforced concrete building in El Centro. Severe agricultural losses.
5/19/1981	N.A.	Mendocino, CA	Deepest earthquake ever reported in California; damage slight.
5/2/1983	(6.5)	Coalinga, CA	Heavy damage to town of Coalinga. May have caused .2 inch movement on San Andreas fault.
10/28/83	(6.9)	Central Idaho/ felt in 8 adjacent states	Two children killed by falling building in Challis, ID. Left cliff 15 ft. high and 25 miles long. Many changes in area hydrology.

Source: U.S. Dept. of Commerce and U.S. Dept. of Interior

Giant Landslides Common Volcanic Phenomenon

The magnificent sea cliff on the north coast of east Molokai, Hawaii—the world's highest—was created by a gigantic landslide that thundered in the ocean over 400,000 years ago. Robin Holcomb of the United States Geological Survey in Vancouver, Washington, advocated this idea late in 1983 to explain the sea cliff, which rises to an altitude of almost a kilometer (3,300 feet). Holcomb claims that an oval-shaped area set on top of the cliffs is actually half of the volcano's caldera. The other half fell into the ocean during the giant landslide that removed approximately 500 cubic kilometers (120 cubic miles) of the volcano. Deposits in the ocean floor near Molokai are similar to material on the volcanic cliffs.

Erosion, Holcomb believes, would have not been sufficient to remove so much of the volcano. In addition, he points out, landslides are not uncommon along the volcanoes that make up the Hawaiian Islands. Eight others are known to have taken place in the past. Most recently, at Kilauea, during the 1970s, landslides have been observed removing vast quantities of volcanic material into the sea.

Major landslides of this type occur where the volcano is over a "hot spot" in a plate, putting them on old crust that is thick with unstable sediments. Similar landslides do not occur, for instance, in the Galapagos, which are on younger crust. Since Mount St. Helens, however, geologists have also realized that major landslides can be a part of the eruption of land-based volcanoes as well.

Texas Quarry Yields Possible Dinosaur Ancestor

The finds of some bones of ancient animals were announced by Sankar Chatterjee in November 1983. The fossils were unearthed in a quarry in Texas. One of the animals, named *Postosuchus*, was a reptile that lived 200,000,000 years ago. About 4 meters (13 feet) in length and weighing approximately 600 pounds, *Postosuchus* resembled a small *Tyrannosaurus*, of which it may be the ancestor.

The jaw of another animal, thought to be a plant-eating dinosaur, was also unearthed at the quarry site. Researchers think this may be the earliest ornithischian, or bird-hipped dinosaur, which included armored and duck-billed species. Other animal bones found at the site were a jaw that resembles one belonging to a modern snake, as well as a mammalian-type animal known as an ictidosaur.

A Glacier Retreats

The Columbia glacier near Valdez, Alaska is retreating more and more rapidly, scientists of the U.S. Geological Survey report. In 1983 the glacier showed the greatest rate of retreat on record, and this rate will increase with the glacier setting loose between 20 and 27,000,000 tons of icebergs daily until 1985. This is an increase of 600 to 800 percent since 1978. Over the next three to five decades scientists expect the "drastic" retreat to be underway when the glacier will slim down to half its present size from a length of 64 kilometers (41 miles) to 32 kilometers (22 miles).

The retreat is speeding up because of the depth of the water underneath the glacier. The leading edge of the glacier has now reached a part of the fjord where deeper water encourages faster iceberg formation.

According to Mark Meier of the United States Geological Survey in Tacoma, Washington, the melting of the Earth's mountain glaciers, such as the Columbia, may account for a gradual rise in sea level that has been observed. Over the past century, sea level has risen about 6 inches on the globe. Changes in the polar ice caps do not satisfactorily account for the rise. Meier cites records of mountain glaciers over the past century which indicate that their mass has decreased sufficiently to account for the increase in sea level.

A Weakening Field?

Recent analyses of sediments at various depths in a Minnesota lake bottom indicate that the Earth's magnetic field has weakened over 50 percent during the past 4,000 years. This report was presented by Dr. Subir K. Banerjee, professor of geophysics, and Donald Sprowl, a graduate student, at a meeting of the American Geophysical Union in late 1983. A weakening of the magnetic field may indicate that the magnetic poles are reversing. In the past these reversals have occurred a number of times, taking thousands of years for each—a short time on the geologic time scale. Magnetic-field reversals have become one of the most useful ways to date events in the past four or five million years. During this period, at least ten reversals have occurred, so the times of events between reversals are known within about a half-million years.

If the field really is reversing, the result will be that your compass needle, which now points to magnetic north, will point to the new magnetic north, which will be to the south.

1984

Update 1984: Earth Science

THE EDIACARAN SOLUTION

It is not always necessary to find some new object or perform some new experiment to make a scientific discovery. Adolf Seilacher of the University of Tübingen (W. Germany), for example, made big scientific news early in 1984 with his reinterpretation of some fossils that had first been discovered in 1947.

The fossils in question date from about 670,000,000 years ago, before the start of the Cambrian Period, classically the start of the fossil record. While other forms of pre-Cambrian life are known today, nearly all of them are tiny one-celled creatures that formed mats; tiny tubes, coils, and caps (the Tommotian fauna); or creatures known only from tracks left in soft sediments. One pre-Cambrian group, known as the Ediacaran fauna, are represented by fossils up to a meter (a yard) long. It was Seilacher's reinterpretation of the Ediacaran fauna that made news.

Previously, the Ediacarans had been interpreted as either ancestors of jellyfish, alcyonarian corals ("sea pens"), or segmented worms. Seilacher demonstrated that these identifications were not plausible. Instead, he suggested that the Ediacarans were related to each other, but not to modern life forms. All of the Ediacarans had an unusual body plan—very long and very thin. There is little evidence of internal organs or even of mouths.

Seilacher thinks that the Ediacarans represent one solution to a basic problem of life—how to grow to a large size and still have all your body parts in close contact with needed oxygen, water, and nutrients. The post-Cambrian solution is to have tubes of various kinds extending into the body (for example, lungs or intestines). The Ediacaran solution may have been to change the shape of the body so that there was a lot of surface for little volume.

Apparently an episode of mass extinction (see page 272) ended the Ediacaran experiment, while some of the still tiny creatures with a different solution survived to populate the Cambrian and later periods.

WHEN OLD PLATES DESCEND

The theory of plate tectonics holds that new ocean crust is being created at midocean ridges and the old crust descends into deep ocean trenches. Domenico Giardini and John H. Woodhouse of Harvard University have now partially solved one of the mysteries that relate to that theory: What happens to old plates after they disappear into the trenches?

In a report in the February 9, 1984, issue of *Nature* they tell what was observed by studying 49 deep earthquakes near the Tonga trench, north of New Zealand. The plate behaves as if it were being pushed into a plastic, but resistant medium. As the plate descends, it thickens. It also shortens, as if the plate seems to telescope in upon itself. These processes cause fractures, resulting in the deep earthquakes Giardini and Woodhouse studied.

The two scientists could only follow the plate to a depth of 650 kilometers (400 miles), since there were no earthquakes below that level. This was close to the 670-kilometer (415-mile) depth at which all earthquakes disappear, a depth that marks the boundary between the upper and lower mantle.

THE DESTRUCTION OF THE MINOAN CIVILIZATION

The ancient island of Thera in the Aegean Sea was destroyed sometime between 1200 and 1700 B.C. by a volcanic eruption that left the former "round island" (an ancient name for the island) as a sort of "atoll" around a giant caldera. The largest of the remaining parts of the island is now known as Santorini.

By either name, the volcanic island has occasioned much controversy since 1939, when Spyridon Marinatos first suggested that the eruption had led to the extinction of the Minoan civilization that was centered on Crete, a nearby island. The Minoan civilization had been the dominant one in the Mediterranean, but it suddenly disappeared, to be remembered only in legend for centuries.

In the June 7, 1984, *Nature* W.S. Downey and D. H. Tarling from the University of Newcastle upon Tyne (United Kingdom) reported upon careful studies of the magnetism that had been trapped in the rocks at the time of the original eruption—or eruptions, as it turns out. They found that there were two events, 30 years apart. In one, 5 centimeters (2 inches) of ash fell, which did not affect the Minoan civilization. Thirty years later, however, the rocks were again superheated. Downey and Tarling think that this was the result of fires set off by earthquakes related to the second Thera event, when the giant caldera formed. Other researchers have suggested that tsunamis caused by the formation of the caldera were the cause of the final collapse. Whether by fire or water, the cause of the destruction will no doubt continue to be debated.

Global Weather Changes

It used to be that everybody talked about the weather, but increasingly people are talking about the climate. Climate consists of a long-term pattern of weather. Because climate normally does not change very fast, people tend to think that it is stable, but geologists know that the climate has often changed in the past. Today, there is good reason to believe that the climate will change rapidly in the near future. It is an open question as to whether or not the changes will be harmful or beneficial and as to whether or not people can do anything to modify the climate trends that are developing.

Here is a summary of some of the climate problems now facing the Earth, along with a brief account of how climate has changed in the past.

THE "GREENHOUSE EFFECT"

The polar icecaps would melt producing a 5-meter (16.5-foot) rise in the level of the ocean. Half of Florida and coastal lowlands in the southern U.S. would be covered by the sea, fertile river deltas around the world would be flooded, and farmland in such places as parts of the Netherlands and Bangladesh would be lost to the rising seas. Climatic changes would include a shift in rainfall patterns, reducing river flow and groundwater in a number of areas of great agricultural importance and increasing it substantially in others. For example, parts of India, Bangladesh, Thailand, Cambodia, Laos, and Vietnam would be subjected to frequent flooding from swollen rivers.

These and other changes were believed likely with just a 3-degree Celsius (5-degree Fahrenheit) rise in the average world temperature that scientists thought could be caused by the "greenhouse effect." The President's Council on Environmental Quality issued a warning in early 1981 against the possible dire consequences of allowing carbon dioxide, a potential cause of the greenhouse effect, to build up further in the atmosphere, although the warning did not cause widespread public concern.

The report noted that carbon dioxide levels were estimated to be 15 to 25 percent above levels in 1800, before the Industrial Revolution began. Increased burning of coal, oil, and other fuels as the result of the spread of industry to underdeveloped countries might substantially increase even that amount, and the council recommended that industrialized nations work toward limiting carbon dioxide levels to 50% above those of 1800.

The greenhouse effect is a more exaggerated instance of a normal phenomenon of the Earth's atmosphere. It is an effect of rising levels of carbon dioxide (or certain other gases) on global temperatures. About a century ago, the Swedish chemist Svante Arrhenius theorized that if the concentration of carbon dioxide were to double, global temperatures would increase by 9°C (16°F). Carbon dioxide acts differently on various wavelengths of radiation. Visible wavelengths—which account for most of the sun's radiation—easily pass through carbon dioxide. However, the carbon dioxide molecules trap the longer wavelengths that are reradiated from the Earth back into space. This creates the greenhouse effect. A few scientists earlier in this century believed that the gradual increase in global temperatures might be due to rising levels of carbon dioxide. The prevailing scientific opinion at that time, however, was that the oceans would absorb any excess carbon dioxide produced by burning fossil fuels. This opinion seemed to be confirmed when global temperatures began to fall during the middle of this century. In fact, average temperature variations since careful records have been kept have remained within the normal plus or minus 2 degrees Celsius (3.6 degrees Fahrenheit). Given the current rate of increase of carbon dioxide pollution, however, some scientists think that a distinct warming trend will begin in the 1990s.

Two reports in 1983—issued within days of each other—reaffirmed that carbon dioxide levels in the environment are rising, and predicted that this rise would cause a gradual increase in global temperatures. Unlike previous warnings, these reports made the issue come home to the general public.

The reports were issued by the Environmental Protection Agency and the National Academy of Sciences during a single week in October, 1983. The EPA study is more pessimistic, saying that the possible changes that might result by the end of the next century "could be catastrophic." The National Academy of Sciences takes a milder approach in its report, calling for further study over the next 20 years so that all the possible effects of rising carbon dioxide levels can be better understood. Yet both reports agree that the global warming trend is inevitable, and humans should concentrate on the best ways to adapt to the situation.

These studies focus on an important issue that has concerned scientists for many years. As part of the International Geophysical year in 1957–58, Charles David Keeling and his associates at the U.S. National Oceanic and Atmospheric Administration began to measure carbon dioxide levels. One of the sites selected for these measurements was located near the summit of Mauna Loa in Hawaii. This site was relatively free from contamination and would presumably yield accurate readings. Researchers discovered that carbon dioxide levels varied with the seasons. During the growing season, plants consumed a large quantity of carbon dioxide resulting in a net removal of the gas from the atmosphere. At other times, carbon dioxide was returned to the atmosphere due to the oxidation of plant tissues. But discounting these seasonal fluctuations, measurements taken at the Mauna Loa site showed a steady increase in carbon dioxide levels. These levels rose from 315.8 parts per million (ppm) in early 1959 to 340 ppm by the early 1980s. Other studies have shown that over a much longer period—from 1860 to 1977—carbon dioxide levels rose only 40 ppm. Researchers have also come up with accurate methods of measuring carbon dioxide levels in air bubbles contained in the Antarctic ice cap. These bubbles, which were formed as long ago as 100,000 years, are examined by scientists who are trying to determine long-term changes in Earth's atmosphere. Evaluation of the bubbles reveals that carbon dioxide in the atmosphere was 16 ppm about 15,000 years ago. Taken together, these three studies suggest that while carbon dioxide rose less than 300 ppm over most of the time since the last ice age, a rate of less than 0.02 per year, the level has risen nearly 50 ppm since the American Civil War, a rate of about 0.4 ppm per year, and has risen about 25 ppm since 1959, a rate of about 1 ppm per year.

The rapid rise over the past 25 years is believed to be partly caused by increased use of fossil fuels along with the effects of more rapid deforestation—two conditions that are likely to persist. As a result, the National Academy of Sciences predicts that carbon dioxide levels will probably double to about 600 ppm by 2065.

The report by the National Academy of Sciences predicts "with considerable confidence" that the Earth's surface temperature and the temperature of the lower atmosphere will increase 1.5 to 4.5 degrees Celsius (2.7 to 8.1 degrees Fahrenheit) by the year 2065. By contrast, global temperatures have changed only 2°C (4°F) in the past millennium, and approximately 6° to 7°C (11° to 13°F) in the last 1,000,000 years.

Scientists can only speculate on the long-term global effects of these rising temperatures. They think that the rise will probably be greatest at the poles where a 3° to 4°C (5° to 7°F) increase, and the consequent ice melting, could raise sea levels by 60 to 70 centimeters (25 to 30 inches).

Temperature increases of 10°C (18°F) could occur in northern states of the United States, creating a climate in New York City similar to that of Daytona, Florida. Rising temperatures might alter rainfall patterns—producing higher levels in some parts of the globe and lower levels in other locations. The National Academy of Sciences report stated that drier conditions were possible in America's southerly farm belt. This might have the effect of reducing crop yields, but such an effect could be offset by the increased levels of carbon dioxide, which would make plant photosynthesis more efficient. The report went on to say that growing seasons might be somewhat longer in the northern part of the U.S., while the West might experience drier climatic conditions.

However, the National Science Academy's report cautioned that predictions are really impossible because there are many unknowns. For example, scientists are uncertain how much carbon dioxide can be absorbed by the oceans. The amount of absorption depends on the rate of turnover by water at or near the surface and at much lower depths. Turnover is tied to the vast undersea circulation of waters between the poles and the equator—a process that goes on over many centuries.

Another imponderable is the rate of fossil fuel consumption. If the industrialized world reduces its use of fossil fuels and relies more heavily on alternate energy sources, carbon dioxide levels may be gradually reduced.

Another unknown quantity is the amount of deforestation that will occur. Researchers have discovered that when deforestation occurs, one half of the carbon dioxide in the upper level of the soil is poured into the atmosphere. Scientists previously assumed that the soil and land biota acted as a storehouse of carbon dioxide. Recent studies estimate that between 1860 and 1980, 135–228 gigatons (billion tons) of carbon was released into the atmosphere by soil and land biota; and 80 percent of it was due to deforestation. An estimated 1.8 to 4.7 gigatons of this were released in 1980, as the rate of deforestation in Third World countries accelerated.

Other gases—methane, nitric oxide, ozone—also contribute to the greenhouse effect. Scientists must be able to project the levels of these gases in order to predict the full impact of the greenhouse effect in the future. Methane has been studied especially carefully (see page 261).

As a result of these imponderables, the National Science Academy cautioned against any doomsday predictions and urged further study. "We feel we have 20 years to examine options before we have to make drastic plans," the report said.

CLIMATE IN PERSPECTIVE

Over its more than 4,000,000,000-year history, the Earth has experienced many fluctuations in climate. While a gradual warming in the next hundred years might cause great changes for civilization, past climates have varied for many reasons as well.

In the beginning, there was greenhouse effect already. Paleontologists speculate that in very early Precambrian times carbon dioxide levels were much higher than they are today, resulting in warmer global temperatures than we have today. As the Precambrian time progressed, however, it was punctuated by glacial advances. The first was the Huronian glaciation about 2,700,000,000 to 1,800,000,000 years ago. This was followed by a warming period, then another glacial advance about 950,000,000 years ago. Two more glacial retreats and advances occurred during the Precambrian, with each glaciation lasting about 100,000,000 years and spreading over wide areas of the Earth. Throughout its history, warmer periods have been followed by such periods of ice formation.

During the Paleozoic era, the climate was not unlike that of today, although possibly somewhat drier. In the late Paleozoic—the Carboniferous period—climatic conditions grew cooler and moister. Almost 100,000,000 years of glaciation followed, spanning much of the Carboniferous and the Permian periods. The glaciation that took place in the Carboniferous period was centered near the South Pole. At that time, there was a single super continent, Pangaea; and the glaciation could have begun at one end of it and gradually moved northward.

Climates of the Past

Climates in the past can be deduced from evidence that persists today, ranging from isotope ratios and types of fossilized plants and animals to effects of glaciation, but the geological record is not complete and different methods of calculation give different results. Nevertheless, there is broad consensus about changes in climate. Here are some geological snapshots of the climates of the past.

Period	Years Ago	Global Mean Temp. (C)	Global Mean Temp.(F)	Possible Cause
Precambrian Time	4,250,000,000	37	98.6	High carbon-dioxide levels.
	3,500,000,000	25	77	
Paleozoic Era	500,000,000	15	59	Unknown.
Mesozoic Era	200,000,000	17–25	63–77	Formation of Pangaea.
Eocene Epoch	55,000,000	13	55	Meteorite; ice sheet forms in Antarctic.
Pleistocene Epoch	2,000,000–15,000	10–12	50–54	Glaciation in north and south.
Holocene Epoch	12,000	13–14	55–57	
	11,000	10	50	
	8,000–5,000	16–17	61–63	
	2,400–1,200	13–14	55–57	
	1,200–800	25	77	
	800–200	10–12	50–54	"Little ice age;" reduced solar activity
	200–present	15	59	

Analysis of deep sea sediments indicates that much of the Mesozoic era had a warm climate. A moderate cooling trend occurred during the Jurassic period, but analysis of deep sea cores reveals that the climate was growing warmer during the early Cretaceous. This was followed by a cooling trend which became more severe as the period ended about 65,000,000 years ago. This cooling coincided with the mass extinction of the dinosaurs, and the two events may be connected.

By the beginning of the Eocene Epoch, the globe was experiencing a long-range climatic cooling that was most pronounced in the polar regions. During the Miocene Epoch, the Antarctic ice sheet had probably started forming. Cooling effects occurred more slowly in the northern polar regions, but sea ice could be found through much of the year in the Arctic seas by the Pliocene Epoch.

The Pleistocene Epoch that followed was a time of fluctuating climatic conditions when enormous glaciers advanced and retreated four times over a period lasting from 2,000,000 to about 15,000 years ago. At one time or another, these glaciers covered 10,000,000 square kilometers (4,000,000 square miles) in North America, 5,000,000 square kilometers (2,000,000 square miles) of Europe and 4,000,000 square kilometers (1,500,000 square miles) of Siberia. Following the last glacial retreat, the Earth entered into a period of post-glacial—or inter-glacial climate if you think these massive ice sheets will return.

Scientists know that over the eons of Earth's history the levels of carbon dioxide have changed. Carbon dioxide levels were lower during the great ice ages and rose as the glaciers retreated. Michael B. McElroy of Harvard University believes he has an explanation for this phenomenon. During the ice ages, sea level is lowered, exposing nitrogen-rich sediments along the coast which are returned to the seas by rivers and streams. These sediments feed ocean plants which take carbon dioxide from the atmosphere, reducing the levels of this gas. As the glaciers retreat, the reverse process takes place and carbon dioxide levels increase.

Evidence of climatic trends during the Holocene Epoch is provided primarily from pollen samples. The variations in temperature are much better known and more accurate than for earlier times. The chief fluctuations were the age of the Viking expansion, when the temperature was quite warm, and the "Little Ice Age" that followed.

Scientists have developed a variety of theories that attempt to explain the climatic fluctuations that have occurred in the past. For example, it has been suggested that the glacial periods that occurred in Precambrian times might have been due to mountain building taking place on the continents. These mountains provided places where snow could collect. The gradual accumulation of snow might have eventually led to lower global temperatures, more snow, and finally glacial advances. This idea, however, does not explain all periods of glaciation.

Earlier in this century, the Yugoslavian astronomer and geophysicist Milutin Milankovich proposed an astronomical theory of climatic changes. According to Milankovich's theory, climatic fluctuations were due to changes in the shape of the Earth's orbit (eccentricity); changes in the angle the Earth's axis makes with the plane of its orbit (obliquity); and wobbling of the Earth's axis (precession).

The shape of the Earth's orbit changes over a 90,000–100,000 year cycle, becoming more circular, then more elliptical. When the orbit is most elliptical, the Earth receives 20 to 30 percent more radiation when it is closest to the sun (perihelion) than when it is most distant (aphelion). (The usual difference is about 3 percent.) This phenomenon could easily produce warmer summers and colder winters.

Over a period of about 41,000 years, the tilt of the Earth varies from 22.1° to 24.5°. A smaller tilt would result in a smaller temperature difference between winter and summer. This might result in cooler summers with less ice melting.

Finally, the Earth wobbles on its axis over a 26,000 year cycle. This changes the time in Earth's orbit when winter and summer occur. At a certain point in this cycle, summer will take place in the northern hemisphere when the Earth is at aphelion; at another point, when the Earth is at perihelion. This phenomenon affects the heat during the summer months, and conceivably the amount of ice melting.

Recent studies of deep sea sediments tend to add some support to the astronomical theory. They show that over the past 450,000 years climatic fluctuations have roughly coincided with changes in obliquity, eccentricity, and precession, allowing for time lags.

Finally, the Earth may have alternated through periods of greenhouse and icehouse conditions strictly because of the carbon dioxide levels in the atmosphere. At various times in the past, the carbon dioxide levels were much higher and the world climate was warmer. Fossil evidence, for example, indicates that vegetation and animal life once flourished at the poles. One greenhouse period occurred between 50–150,000,000 years ago when the continent Pangaea was breaking apart. Scientists believe this was a period of heavy volcanic activity, which increases the levels of carbon dioxide.

As evidence for the theory, scientists cite variations in calcium carbonate deposits layed down in the oceans during past epochs. These were largely in calcite formed during periods of high carbon dioxide levels, the theory states; but they have included aragonite types when levels were lower. We are living through such a latter period now, the theory points out, although recent reports show that carbon dioxide levels are rising.

NUCLEAR WINTER

While several groups of scientists were predicting higher temperatures in the next hundred years, other groups were worrying about the effects of much lower temperatures arriving unexpectedly and suddenly as a result of a nuclear war.

The idea that nuclear war might result in a "nuclear winter" started with Walter Alvarez's 1980 discovery of the worldwide iridium layer at the Cretaceous-Tertiary boundary (see page 243). Alvarez proposed that vast clouds of soil particles raised by a giant meteorite had not only caused the iridium layer, but also had caused the mass extinctions that occurred at that time. The extinctions were, in his theory, caused by the lowering of temperatures and the cutting off of photosynthesis that would result from such a dust cloud if the dust stayed in the air for a year or so. Follow-ups on this idea resulted by 1984 in various theories of mass extinction as a result of similar bombardment and consequent dust clouds (see page 272).

In 1982, P. J. Crutzen and J. W. Birks proposed that a nuclear war could cause similar climate changes as a result of ignition of urban and forest fires. Sooty smoke would act the same way as a giant dust cloud, especially if there were enough force in the flames to propel the soot into the stratosphere.

An interdisciplinary group of scientists was inspired by these ideas to calculate the climate effects of a nuclear war in detail. The group consisted of Richard Turco, Brian Toon, Thomas Ackerman, James Pollack, and Carl Sagan, and came to be known as TTAPS—after the initials of their last names, but also a name that suggests strongly what these researchers found. In 1983 they reported on their findings, making front-page news around the world. In short, their conclusions were that, ignoring the effects of blast, fire, and radiation, severe long-term effects of a "small" nuclear exchange would include "significant surface darkening over many weeks, subfreezing land temperatures persisting for up to several months,... and dramatic changes in local weather." Even a very small nuclear exchange, if it involved sooty fires in urban centers, would also trigger severe climatic effects.

What would the effect on life be? Another group of twenty scientists, including Paul Ehrlich, Thomas Eisner, Stephen Jay Gould, and Ernst Mayr among the most familiar names, worked on this part of the problem, mainly following an April 1983 conference on the subject. Their conclusions, published in December 1983, were based on the work of TTAPS. Light would be reduced to 1 percent of what it normally is. Surface temperatures away from the moderating effects of oceans could fall by 40°C (72°F) no matter what season of the year it was. It would take a year for light and temperature to return to normal. Over a ten-year period, the ecosystem would gradually recover from the "nuclear winter," but not to what it was before. There would be mass extinctions of species, especially in the tropics. Agriculture would be almost totally wiped out in the first year, but would have the capability to recover by the tenth year. However, human society would have been reduced to at least prehistoric levels. Humans that survived the initial blast would face "extreme cold, water shortages, lack of food and fuel, heavy burdens of radiation and pollutants, disease, and severe psychological stress—all in twilight or darkness." Despite this, if the nuclear exchange were not extensive, the collective opinion of the twenty scientists is that humanity would not be among mass extinctions.

There was general agreement among the members of the scientific community that the TTAPS papers and the paper on "Long-Term Biological Consequences of Nuclear War" had taken the important factors into account. However, it was stressed, the "nuclear winter" was based on a number of specific assumptions, including simplified atmospheric models. Other scientists tried to determine what would happen to climate after a nuclear war using other, more sophisticated models. The results were not substantially different from the TTAPS findings.

ENVIRONMENT

The 1980s started with the environmental movement in retreat, but gradually a new awareness grew of threats to the whole population of Earth.

In the 1960s the environmental movement had consisted largely of people who liked birds, plants, or wilderness, although there were increasing stirrings toward the end of the period about air pollution and water pollution. *Silent Spring*, published in 1962, also alerted the nation to the dangers of pesticides, especially DDT.

In the 1970s the environmental movement got off the ground with mass celebrations of Earth Day and sweeping new legislation and regulations to slow air and water pollution—even reversing some of it. DDT and other long-lasting pesticides were banned. The surprising new development of the 1970s, the oil crisis, turned the public's attention to energy problems, but, as the energy problems came to at least a temporary halt at the end of the 1970s, many people lost interest in environmental problems.

In 1980 Ronald Reagan was elected President of the United States with what amounted to a strong anti-environmental plank in his platform. Thus, the 1980s began with a fragmented environmental movement, one that had seen many successes in the past twenty years, but which now seemed tired and often irrelevant to pressing economic problems. Unlike many politicians of the past, however, President Reagan treated his platform promises seriously. After three years of reduced activity and possible politicizing of the Environmental Protection Agency (EPA), however, the top management of the EPA was replaced. The new management promised increased attention to environmental problems.

Gradually, however, the concern of the populace at large about the environment turned from indifference to fear. Previous perceptions had been that air pollution was merely annoying to most people, water pollution killed fish, and the oil crisis hurt the pocketbook and was an insult from the Third World. Although for some people in some circumstances air and water pollution had been known to take lives, for most people there was little cause for concern. Toxic wastes, however, could be life-threatening. Love Canal's problems with toxic wastes, reaching the public consciousness at the end of the 1970s, signaled the alert; but few believed that Love Canal, New York, was more than an isolated horror story. Soon afterward, however, it became apparent that there were "Love Canals" in all parts of the United States. Widespread dumping of toxic wastes from the 1940s to the present touched thousands and perhaps millions of Americans. A new word entered the vocabulary—*dioxin*. By the end of 1983, the Environmental Protection Agency had declared toxic wastes its number one problem, and dioxin the most serious toxic waste.

At the same time, the concerns of previous years continued, often in new guises. Air and water pollution problems became combined in the international issue of acid rain. While people feared toxic wastes, it was apparent that acid rain had the potential for widespread damage as well. Early in 1984, general slowing of the growth of forests in the Eastern United States suggested that acid rain did more than just injure fish. By mid-1984, environmental groups were claiming also that acid rain in the United States was not simply a problem in the Northeast. Acid rain was claimed for the Southeast and acid fog for the Ohio River Valley.

Because of its widespread effect on animal life, and its unknown effect on human life, acid rain became to the early 1980s what DDT had been to the 1960s—a major rallying point for environmentalists.

The energy crisis had spawned many plans for dealing with the energy problems of the future, ranging from solar power to nuclear power. But complications surfaced with all of the energy alternatives, particularly the difficulty of competing with oil, the cheapest and most portable source of energy. Nuclear power had public-image problems—and perhaps real problems—that went beyond the economics of energy. The synfuels business finally got started in a limited way in the early 1980s, but the initial enthusiasm for coal and oil-shale alternatives to petroleum had died down, especially as there now seemed to be enough oil to go around for a few more years.

With DDT and similar pesticides banned, the birds that Rachel Carson had lamented were returning, but a whole host of other species, from swamp plants to whales, were still in danger of extinction.

Finally, in 1983 a worry that had been frequently dismissed as of little consequence, the greenhouse effect, received sudden credibility from two separate scientific commissions. The environmental story of the greenhouse effect and the other alternative, nuclear winter, is told in the EARTH SCIENCE section, beginning on page 236.

1980

Solar One—Bringing Solar Energy Down to Earth

Converting sunlight to a usable form of energy—still a far-off dream of cheap, unlimited power—came a step closer to reality in October 1980 with the start of construction on Solar One, the world's largest solar energy plant. Located in the Mojave Desert near Daggett, California, Solar One consists of 1,818 heliostats (computer-controlled mirrors) arranged in a 40.5-are (100-acre) circle around a central tower. The 2.14-square-meter (23-square-foot) heliostats, installation of which was completed in 1981, concentrate sunlight on a boiler mounted on top of the central tower. Superheated steam from the boiler at temperatures of 516°C (960°F) is used to drive a turbine generator. An energy storage unit consisting of a holding tank filled with oil and crushed rock provides enough heat to operate the generator for up to seven hours without sunlight.

By the end of 1982, Solar One was generating 10 megawatts of electricity when operating at full power, enough power to supply electricity to 6,000 homes. But the price of solar power in the early 1980s was far above the price of conventional sources. Solar power so far certainly does not fulfill the dream of cheap energy. For example, Solar One was built at a cost of $142,000,000, 80 percent of which was supplied by the Department of Energy. The plant initially produced electricity at five to ten times the cost of conventionally generated electricity. This is partly because electrical plants always figure in capital costs in determining unit cost.

Plants like Solar One could become competitive by the late 1980s, however, if savings from mass production of equipment and improvements in heat storage units live up to expectations. In fact, the Southern California Edison Company, which built Solar One, is hoping to build a full-scale version of Solar One by 1988. If all goes as planned, the new plant will produce electricity at competitive prices.

Large areas of Earth's surface, like the Mojave Desert, are climatically unsuited for human use at present. Solar power could change this, but not everyone believes that the way to do this is to build giant, expensive, steam-driven plants.

At the same time that Solar One was being built, many scientists were working on making less expensive and more usable photoelectric cells as an alternative form of solar power.

A photoelectric cell uses more of the energy of the sun directly; processes that produce steam just use the heat energy. At noon on a clear day, the sun produces a total energy density of about 100 watts per square foot. Most photoelectric cells have efficiencies of around 10 percent, so 1 square foot of cells could produce about 10 watts of energy. Scientists calculate that photoelectric cells could produce electricity at one-fifth the cost of conventional sources if the cells could be built more cheaply and last longer.

Two different approaches to photoelectric cells are being used. In one, electricity is produced directly from sunlight. In the other, sunlight is used to produce fuels. Fuel production can even involve using sunlight to split water into hydrogen and oxygen, although other fuels can also be produced in this way.

Because about 55 percent of the population of the world is not connected to an electric grid, successful photoelectric production of electricity could be an important factor in the development of the Third World.

When photovoltaic cells were first introduced, they cost about $600 a watt, as compared with less than 50 cents a watt for conventional power. In the twenty years since, the cost of electricity from conventional sources has risen by a factor of four and the cost of standard photovoltaic cells, based on silicon crystals (like the ones used in computer chips), has fallen to $8 a watt. New types of cells, however, based on silicon glasses instead of crystals, are expected to lower the cost to $2.50 a watt before the end of the 1980s. If conventional sources continue to rise in cost, the two different sources may be about equal by that time. (See also "Wet Photoelectric Cells—Dunking Your Semiconductor," p. 334.)

THE COLUMBIA—THAT DAMMED RIVER

The Columbia River runs more than 1,930 kilometers (1,200 miles) from the Canadian Rockies through Washington and Oregon to the Pacific Ocean. It is the largest river in North America that flows into the Pacific—or, more to the point, is *allowed* to flow into the Pacific.

There are a total of eleven hydroelectric dams blocking the waters of the Columbia, beginning with the famous Grand Coulee Dam in northern Washington and ending with the Bonneville Dam just above Portland, Oregon. Columbia River hydroelectric plants produce a whopping 80 percent of all electricity used in the Pacific Northwest. Every gallon of water in this river is accounted for and allocated to a specific purpose: hydroelectric power, irrigation, maintaining downstream flow, and so forth. Nearly all of the energy potential of the river is used. The Columbia drops about 396 meters (1,300 feet) as it courses through the United States into the Pacific. In all, only about 24.5 meters (80 feet) of this drop have not been utilized for generating electric power.

Eleven Ways to Solve the Energy "Crisis"

The oil glut of the early 1980s helped take the "crisis" out of the energy crisis, and some mammoth projects, such as the Colorado Shale Oil synfuels plant, have been abandoned. But efforts at developing new energy sources—and conserving energy—continue. Among these efforts are

SOLAR POWER

There are various forms of solar power, not all of them obvious. While the number of experimental solar generating plants is growing, there has been more emphasis on passive solar heating to help conserve fossil fuels. Passive solar heating consists of letting the sun warm the interiors of houses constructed so as to retain the warmth. Solar generating plants are more costly than passive designs. The plants use either heliostats to gather and focus sunlight on a steam boiler, or banks of solar cells to convert sunlight directly into electricity. Despite the development of new processes for manufacturing solar cells (thin films and liquid-crystal interfaces), solar cells are still more expensive than conventional sources. And, of course, there is the problem of storing power for use at night and on cloudy days. In any case, here are a few variations on the solar-power theme:

Solar power beamed from space   The idea is to build great orbiting solar-energy collectors in outer space. The energy would then be converted and beamed to Earth in the form of microwaves. A ground installation would convert the microwaves to electricity for distribution. Because an enormous facility in space would have to be built, this idea will have to wait until we have had some experience with a space station. There is no night and there are no clouds if you are far enough away from Earth, which makes the idea attractive. On the other hand, one has to be concerned about the effects of a powerful microwave beam on Earth.

Temperature differences in the ocean   Water heated by the sun provides a storage tank for solar power. Ideas have been around for a hundred years on how to use this form of solar power. While good possibilities exist for tapping this source of power in warm areas of the ocean, this approach has yet to pay off in a commercial application. Here is the way the solar power of the ocean can be tapped. A floating platform in the ocean uses pipes that reach deep below the surface to take advantage of differences between the temperature of surface and deep waters. Any temperature difference can be exploited to produce energy, and various schemes have been devised to use this one. One such scheme was shown to work for a model generator off Hawaii in 1979. Scaling the model up to commercial ranges, however, requires moving a volume of water about as large as the flow of the Nile. Although this approach is nonpolluting in the ordinary sense, the thermal pollution of the lower layers of the ocean could be serious. Finally, no one knows how much fouling of the system might occur because of the living organisms in the ocean; it could halt the whole operation. At present, the goal is to have commercial plants (probably in Hawaii and Puerto Rico, where the temperature differences are sufficient) by the end of the century.

Wind power A fanciful and free source of power, wind-powered generating plants on a small scale once dotted the Great Plains (although many of these only pumped water, and did not make electricity). The more recent versions, set up during the rush for energy alternatives, have included giant windmills and wind "farms," which are large groups of conventionally sized windmills. Some of these are producing surplus electricity that can be sold to the local power grid, but wind seems to be too unreliable to become a sole source of energy. Perhaps a great many windmills that sold surplus energy would make a difference. Since winds are generated by solar heating, wind power is another form of solar power.

Biofuels Biological sources use the energy of the sun to make chemicals. Among the chemicals produced this way are such fuels as alcohols and methane gas. Other biological sources (bacteria, yeasts, fungi) can complete the conversion. There are even plant species that directly produce burnable liquid substances, but we are a long way from large-scale "fuel-farming." A project to convert peat moss to methanol is getting under way in North Carolina, however; and in Brazil, where there is a lot more forest than there is petroleum, half the automobiles run on biologically produced alcohol. Methane generators based on animal wastes are becoming popular in the Third World. Like wind power, small scale biofuels seem to work, but it is difficult to see a large-scale future for any of them.

Water power While on the subject of small-scale operations, there is a movement afoot to convert the old water mills of New England and New York State into small power plants that would power a local factory and also provide electricity to the grid. The ultimate source of such power is the energy of the sun, which powers the water cycle. The conversion of mills from grain grinding or textile manufacture to generating electricity is not very complicated or expensive, and we certainly have the technology already, since this is how it was done nearly a hundred years ago.

FUSION POWER

Not every potential source of energy is solar, although the wood, coal, oil, and natural gas that we have used in the past all received their initial energy from the sun. One possible new source of energy bypasses the middle step, by going directly to the source of the sun's power—nuclear fusion. Fusion power is often hailed as the great technological hope of the future. After years of research, however, controlled nuclear fusion is only just this side of the purely theoretical. Fusion uses a thermonuclear reaction (the sun's energy source as well as the "kick" in the hydrogen bomb) in which deuterium (heavy hydrogen) is brought to temperatures of 100,000,000°C (180,000,000°F) to start the reaction. Huge amounts of energy could be produced, mostly in the form of neutrons, but the technological problems are awesome. Achieving and maintaining such high temperatures and controlling the reaction have only taken place for tiny fractions of a second so far. Progress is being made, however. By 1983, two of the three conditions needed to produce an excess of power over the amount spent in starting the reaction had been met. These achievements have been on a laboratory scale only, however. A full-scale power plant would require significant new engineering advances.

FUEL CELLS

Fuel cells generate electricity by using hydrogen and oxygen as reactants (without their burning explosively). The cells consist of two electrodes and a liquid electrolyte. This is an exceptionally "clean" way to produce power because water is the only byproduct. But a continuous supply of oxygen and hydrogen is needed. Furthermore, current fuel cells are neither efficient nor cost effective, except for applications in outer space. Nevertheless, both Tokyo and New York City got demonstration fuel-cell power plants in 1983. While the plant in Tokyo got off to a good start (largely because it was aware of mistakes made in New York and took corrective measures), the New York plant was slowly struggling into operation in 1984, behind schedule. These will be extensively tested with the idea of building full-scale plants in the future.

CATALYSIS OF WATER INTO HYDROGEN AND OXYGEN

Catalysis of water is exactly the opposite of a fuel-cell approach. In a fuel cell, oxygen and hydrogen obtained from somewhere produce electricity and water. In catalysis, electricity and water are used to produce hydrogen and oxygen. Hydrogen can then be burned with the oxygen in air as a fuel.

It has been said that someday we will have a hydrogen-based economy and someday there may be an efficient photolytic (utilizing sunlight) process for breaking down water into hydrogen and oxygen. So far, there are only a number of different reports of laboratory experiments in which sunlight has been used to catalyze water.

GEOTHERMAL POWER

This is a down-to-earth idea and there are operating geothermal electric generating plants in various parts of the world. It has long been known that the interior of Earth is very hot. Development of this energy source has been largely restricted to areas where heat from Earth's core rises naturally to the surface (in geothermal hot springs, for example). Iceland, which has the suitable heat sources, is a major user of geothermal power. However, some work is going on at more normal sites. One promising method is being tried in Cornwall, England, where two holes in an experimental site in Cornwall are 6,000 meters (20,000 feet) deep. Explosions and high water pressure crack the rock between the holes, permitting the water to flow from one hole to the other. In the second hole, it emerges as steam with a temperature of 260°C (500°F). At the moment, there seem to be no environmental or technological problems for the development of geothermal power.

ICE PONDS

While nearly everyone else is trying to produce energy from heat or along with heat, at least one researcher is working on producing cold. Air conditioning is a major user of energy. Ted Taylor, a theoretical physicist, has modernized the old New England approach to cooling; use ice from the preceding winter. Taylor uses a variant of the snow-making machines familiar to skiers to make large amounts of ice in the winter. He keeps the ice in insulated buildings, and then uses it to cool buildings in summer.

NUTS TO THE BIG POWER COMPANIES

California laws encourage the development of alternative sources of energy. The electric power produced is then sold to the state's large power companies, which have been discouraged from building large coal-fired or nuclear plants. As a result, power supplied to the large power companies from small alternative sources includes many small hydropower sources, 8 megawatts from the Magma Energy Company's geothermal sources, twice that much from the Oak Creek Energy Company's 340 wind machines, and 4.5 megawatts from the Diamond Walnut Growers. The growers produce their power by burning walnut shells.

Saving the Sea Turtles Proceeds at a Snail's Pace

The Convention on International Trade in Endangered Species of Wild Fauna and Flora includes several kinds of sea turtles on its list of endangered species, but there are numerous problems hampering efforts to save the turtles. One problem is the continued illegal killing of sea turtles, for soup and for their eggs, hide, and shells; tortoiseshell carvings are highly prized in Japan, for instance. Another is the destruction of sea turtle hatcheries on beaches, due to commercial and residential development along beachfronts. Two other important problems are lack of knowledge about sea turtles and, as some recent studies seem to indicate, the turtle itself.

Take the case of Kemp's ridley, the most seriously threatened of the seven sea turtle species. In the summer of 1980, only about 700 female turtles appeared at Rancho Nuevo, Mexico, the sole spawning ground for the species. Once a hundred thousand or more had buried their eggs on those beaches. As in other recent years, most of the eggs were relocated to a protected area of the beach and placed under a special guard of Mexican marines to prevent their being stolen. Another 2,000 eggs were packed and transported to Padre Island, Texas, where in recent years eggs have been hatched in the hope of creating a second spawning ground for Kemp's ridley.

These efforts, however, were carried out with precious little understanding of the turtles. For example, it was only in 1982 that the Endangered Species Office of the United States Fish and Wildlife Service funded a study to determine the mechanism of sex differentiation in sea turtle eggs. Preliminary findings of the study, reported in 1983, indicated that the temperature of eggs during incubation determines the turtle's sex. Only males are produced at temperatures below an as yet undetermined point, while only females are born at temperatures above it. This information is crucial to the effort to restore sea turtle populations, because one male sea turtle usually mates with several females.

Another big step was taken in 1984, when one of the outstanding mysteries of science was partly solved: Where do baby sea turtles go after they hatch? For decades scientists, led by Archie Carr, have been trying to find out what happens to sea turtles between the time they hatch as tiny silver-dollar-sized creatures and when they reappear near shore the size of dinner plates. Finally, Carr and others definitely located the tiny turtles that hatch on the Atlantic side of North America. They were where the scientists had long suspected they might be, riding on patches of floating sargasso weed, often hundreds of miles from shore. The turtles live on the patches of weed, preying on other weed hitchhikers and small sea creatures, until they are large enough to travel to the regular turtle feeding grounds.

However, pollution is threatening this way of life. The same currents that form the sargasso weed into patches, some tens of meters across, also bring floating pollutants to the weed patches, where they further endanger the already endangered turtles. Plastic, raw materials for plastic, and lumps of tar caused by oil spills are the main culprits.

Another sea-turtle problem lies within the biology of the turtles, however. While ongoing studies have yet to provide a definitive answer, indications are that newly hatched sea turtles take anywhere from 10 to 50 years to reach sexual maturity. If the reproductive cycle of sea turtles is indeed this slow, populations of an endangered species such as Kemp's ridley will recover extremely slowly. If the workers on Padre Island have to wait 50 years for the first turtles to return to breed, everyone may have long since forgotten what the experiment was about.

ENVIRONMENTAL BRAINTEASER

Q — Suppose you have the only five living dusky seaside sparrows left in the entire world and they are all male. How would you go about saving the species?

A — By mating the males with hybrid sparrows (half dusky and half Scott's seaside sparrow). Offspring are then mated with the full-blooded males and by this means each succeeding generation will have a higher proportion of dusky parentage.

This is precisely what researchers have been trying to do at the Santa Fe Community College Teaching Zoo at Gainesville, Florida. When efforts to find a female of the species in the wild failed, the dusky sparrow was declared functionally extinct, and it became time for a creative approach.

Enter the half dusky/half Scott's hybrids. These birds were created during experiments some years earlier and were being kept in a Gainesville, Florida, museum. Seven hybrids were sent over to the zoo in hopes of getting the first phase of the dusky rescue operation underway. But two of the females refused to have anything to do with their mates and one of the hybrids was actually a male. Of the other four, only one built a nest. And what a time of it she had. She went through six nests and laid several eggs before finally hatching the first of the experimental offspring in August 1983. If the dusky sparrow is to be saved, more than one of the hybrid females will have to get down to business—and soon. The full-blooded males are already quite old for sparrows, and, in fact, one of them has died since the 1983 efforts. In 1984, however, the attempt was to be repeated.

Endangered Mammals

The sea turtle species are all endangered, but there are also many more-familiar species on the list. The total list in 1983, on the tenth anniversary of the Endangered Species Act numbered nearly 800 species. Here, for example, is the 1983 list of endangered mammals from the United States Environmental Protection Agency.

It should also be noted that in 1983 Paul Ehrlich of Stanford University and Daniel Simberloff of Florida State University presented reports suggesting that the world was entering a period of major extinction of species. Simberloff predicted that by some point in the 21st century, the loss of species among flowering plants would be 66 percent of all species and among birds would be 69 percent of all species. On the other hand, the Endangered Species Act is credited with major gains in preventing extinction.

Common Name	Range	Common Name	Range
Anoa, lowland	Indonesia	Bat, Ozark big-eared	U.S.A. (MO, OK, AR)
Anoa, mountain	Indonesia		
Antelope, giant sable	Angola	Bat, Virginia big-eared	U.S.A. (KY, VA, WV)
Argali	China		
Armadillo, giant	Venezuela, Guyana	Bear, brown	China
		Bear, brown	Palearctic
Armadillo, pink fairy	Argentina	Bear, grizzly	Canada, W U.S.A.
Ass, African wild	Somalia, Sudan, Ethiopia	Bear, Mexican grizzly	Mexico
		Beaver	Mongolia
Ass, Asian wild	SE & central Asia	Bison, wood	Canada, NW U.S.A.
Avahi	Madagascar	Bobcat	Central Mexico
Aye-Aye	Madagascar		
Babirusa	Indonesia	Bontebok (antelope)	S Africa
Bandicoot, barred	Australia	Camel, Bactrian	Mongolia, China
Bandicoot, desert	Australia		
Bandicoot, lesser rabbit	Australia	Caribou, woodland	Canada, W U.S.A.
Bandicoot, pig-footed	Australia	Cat, Andean	Chile, Peru, Bolivia, Argentina
Bandicoot, rabbit	Australia		
Banteng	SE Asia		
Bat, gray	Central & SE U.S.A.	Cat, black-footed	S Africa
		Cat, flat-headed	Malaysia, Indonesia
Bat, Hawaiian hoary	Hawaii		
Bat, Indiana	E & MW U.S.A.	Cat, Inomote	Japan
		Cat, leopard	India, SE Asia

Common Name	Range	Common Name	Range
Cat, marbled	Nepal, SE Asia, Indonesia	Deer, swamp	India, Nepal
		Deer, Yarkand	China
Cat, Temminck's	Nepal, China, SE Asia, Indonesia	Dhole (Asiatic wild dog)	U.S.S.R., Korea, China, India
Cat, tiger	Costa Rica	Dibbler	Australia
Chamois, Apennine	Italy	Drill	Eq. W Africa
Cheetah	Africa to India	Dugong	E Africa to S Japan
Chinchilla	Bolivia	Duiker, Jentink's	Africa
Civet, Malabar large-spotted	India	Eland, Western giant	Senegal to Ivory Coast
Cougar, eastern	Eastern N America	Elephant, Asian	SC & SE Asia
Deer, Bactrian	U.S.S.R., Afghanist-an	Ferret, black-footed	W U.S.A., W Canada
Deer, Bawean	Indonesia	Fox, Northern swift	U.S.A., Canada
Deer, Barbary	Morocco, Tunisia, Algeria	Fox, San Joaquin kit	U.S.A. (CA)
		Fox, Simien	Ethiopia
Deer, Cedros Island mule	Mexico	Gazelle, Arabian	Arabian Pen., Palestine
Deer, Columbian white-tailed	U.S.A. (WA, OR)	Gazelle, Clark's	Somalia, Ethiopia
Deer, Corsican red	Corsica, Sardinia	Gazelle, Cuvier's	Morocco, Algeria, Tunisia
Deer, Eld's brow-antlered	India to SE Asia	Gazelle, Mhorr	Morocco
Deer, Formosan sika	Taiwan	Gazelle, Moroccan	Morocco
Deer, hog	Thailand, Viet Nam	Gazelle, Rio de Oro Dama	W Sahara
Deer, key	U.S.A. (S FL)	Gazelle, Pelzeln's	Somalia
Deer, marsh	S America	Gazelle, sand	Jordan, Arabian Pen.
Deer, McNeill's	China		
Deer, musk	C & E Asia	Gazelle, slender-horned	Sudan, Egypt, Algeria, Libya
Deer, North China sika	China		
Deer, pampas	S America		
Deer, Persian fallow	Iran, Iraq	Gibbon	China, India, SE Asia
Deer, Philippine	Philippines		
Deer, Ryukyu sika	Japan	Goat, wild	SW Asia
Deer, Shansi sika	China	Goral	E Asia
Deer, South China sika	S China		

Common Name	Range	Common Name	Range
Gorilla	C & W Africa	Langur, golden	India, Bhutan
Hare, hispid	India, Nepal, Bhutan	Langur, Pagi Island	Indonesia
		Lemur	Madagascar
Hartebeest, Swayne's	Ethiopia, Somalia	Leopard	Africa, Asia
		Leopard, clouded	SE & SC Asia, Taiwan
Hartebeest, Tora	Ethiopia, Sudan, Egypt		
		Leopard, snow	C Asia
Hog, pygmy	India, Nepal, Bhutan, Sikkim	Linsang, spotted	SE Asia
		Lion, Asiatic	Turkey to India
Horse, Przewatski's	Mongolia, China	Lynx, Spanish	Spain, Portugal
Huemul, North Andean	S America	Macaque, lion-tailed	India
		Manatee, Amazonian	S America
Huemul, South Andean	Chile, Argentina	Manatee, West Indian	SE U.S.A., S America
Hyena, Barbary	Morocco, Tunisia, Algeria	Mandrill	Eq. W Africa
		Mangabey, Tana River	Kenya
Hyena, brown	S Africa	Mangabey, white-collared	Africa
Ibex, Pyrenean	Spain		
Ibex, Walia	Ethiopia	Margay	U.S.A. (TX), C & S America
Impala, black-faced	Namibia, Angola		
Indris	Madagascar	Markhor, Kabal	Afghanistan, Pakistan
Jaguar	U.S.A. (TX, NM, AZ), C & S America	Markhor, straight-horned	Afghanistan, Pakistan
		Marmoset, cotton-top	Costa Rica to Colombia
Jaguarondi	U.S.A. (TX, AZ), Mexico, C America		
		Marmoset, Goeldi's	S America
		Marsupial, eastern jerboa	Australia
Kangaroo, Tasmanian forester	Australia		
		Marsupial-mouse, large desert	Australia
Kouprey	SE Asia		
Langur, capped	Bangladesh, India	Marsupial-mouse, long-tailed	Australia
Langur, entellus	China, India, SE Asia	Marten, Formosan yellow-throated	Taiwan
		Monkey, black colobus	W C Africa
Langur, Douc	Cambodia, Laos, Vietnam		
		Monkey, Diana	Coastal W Africa
Langur, François	China, Indochina	Monkey, howler	Mexico to S America

Common Name	Range	Common Name	Range
Monkey, L'hoest's	Africa	Panther, Florida	S U.S.A.
Monkey, proboscis	Borneo	Planigale, little	Australia
Monkey, red-backed squirrel	Costa Rica, Panama	Planigale, southern	Australia
		Porcupine,	Brazil
Monkey, red-bellied	W Nigeria	Possum, mountain pygmy	Australia
Monkey, red-eared	Nigeria, Cameroon	Possum, scaly-tailed	Australia
Monkey, spider	Costa Rica, Nicaragua, Panama	Prairie dog, Mexican	Mexico
		Prairie dog, Utah	U.S.A. (UT)
		Pronghorn, peninsular	Mexico
Monkey, Tana River red colobus	Kenya	Pronghorn, Sonoran	U.S.A. (AZ), Mexico
Monkey, woolly spider	Brazil	Pudu	S America
Monkey, yellow-tailed woolly	N Peru	Puma, Costa Rican	Nicaragua, Panama, Costa Rica
Monkey, Zanzibar red colobus	Tanzania	Quokka	Australia
Mouse, Australian native	Australia	Rabbit, Ryukyu	Japan
		Rabbit, volcano	Mexico
Mouse, Field's	Australia	Rat, false water	Australia
Mouse, Gould's	Australia	Rat, Morro Bay kangaroo	U.S.A. (CA)
Mouse, New Holland	Australia		
Mouse, salt marsh	U.S.A. (CA)	Rat, stick-nest	Australia
		Rat-kangaroo, brush-tailed	Australia
Mouse, Shark Bay	Australia		
Mouse, Shortridge's	Australia	Rat-kangaroo, Gaimard's	Australia
Mouse, Smoky	Australia		
Mouse, western	Australia	Rat-kangaroo, Lesuer's	Australia
Muntjac, Fea's	N Thailand, Burma		
		Rat-kangaroo, plain	Australia
Native-cat, eastern	Australia	Rat-kangaroo, Queensland	Australia
Numbat	Australia		
Ocelot	U.S.A. (TX, AZ), C & S America	Rhinoceros, black	Sub-Saharan Africa
		Rhinoceros, great Indian	India, Nepal
Orangutan	Borneo, Sumatra	Rhinoceros, Javan	SE Asia
Oryx, Arabian	Arabian Pen.	Rhinoceros, northern white	Africa
Otter, Cameroon clawless	Cameroon, Nigeria	Rhinoceros, Sumatran	Bangladesh, Vietnam
Otter, giant	S America		
Otter, long-tailed	S America	Saiga, Mongolian	Mongolia
Otter, marine	Peru	Saki, white-nosed	Brazil
Otter, southern river	Chile, Argentina	Seal, Caribbean monk	Carib. Sea, Gulf of Mexico
Pangolin	Africa		

Common Name	Range	Common Name	Range
Seal, Hawaiian monk	Hawaii	Tiger, Tasmanian	Australia
Seal, Mediterranean monk	Medit., African coast, Black Sea	Usksan	S America
		Urial	Cyprus
		Vicuna	S America
Seledang	SE Asia	Wallaby, banded hare	Australia
Serow, Sumatran	Sumatra		
Serval, Barbary	Algeria	Wallaby, brindled nail-tailed	Australia
Shapo	Kashmir		
Shou	Tibet, Bhutan	Wallaby, crescent nail-tailed	Australia
Siamang	Malaysia, Indonesia		
		Wallaby, Parma	Australia
Sifakas	Madagascar	Wallaby, Western hare	Australia
Sloth, Brazilian 3-toed	Brazil		
Solenodon, Cuban	Cuba	Wallaby, yellow-footed rock	Australia
Solenodon, Haitian	Haiti, Dom. Rep.		
		Whale, blue	Oceanic
Squirrel, Delmarva peninsula fox	U.S.A. (mid-Atl.)	Whale, bowhead	Oceanic (north)
Stag, Barbary	Tunisia, Algeria	Whale, finback	Oceanic
		Whale, gray	N Pacific Ocean, Bering Sea
Stag, Kashmir	Kashmir		
Suni, Zanzibar	Zanzibar		
Tahr, Arabian	Oman	Whale, humpback	Oceanic
Tamarou	Philippines	Whale, right	Oceanic
Tamarin, golden-rumped	N Brazil	Whale, sei	Oceanic
		Whale, sperm	Oceanic
Tamarin, pied	N Brazil	Wolf, gray	Holarctic
Tapir, Asian	SE Asia	Wolf, maned	S America
Tapir, Brazilian	S America	Wolf, red	U.S.A. (SE to TX)
Tapir, Central American	S Mexico to C America		
		Wombat, hairy-nosed	Australia
Tapir, mountain	S America	Yak, wild	China
Tiger	Temperate & tropical Asia	Zebra, mountain	S Africa

Source: U.S. Dept. of the Interior, Fish and Wildlife Service (Effective July 27, 1983)

PLEASE DO EAT THE CAMELLIAS

Consider the serow, a rare Japanese mammal that has a head like a goat and a body like an antelope. Although the two serows in the San Diego Wild Animal Park will eat the enriched alfalfa pellets and fresh vegetables that are their normal zoo food, they have been picky eaters. The remedy—camellias, which serows are known to eat in Japan. Five hundred *Camellia japonica* "Debutante" were rapidly devoured by the serows, right down to the ground. Not only that, but other park animals, including the sika deer and the markhor, also relished the camellias. The zoo has turned to local growers for a steady supply of the serow's favorite snack.

Eight Ways to Save a Dying Species

Much of the focus on endangered species in the United States has been on legislation that would limit environmental impact. In many cases, however, much of the damage to a species' habitat has already been done or is simply caused by the spread of human habitation. Methods to prevent endangered species from extinction often have to be more imaginative than simply keeping a dam from being built or setting aside ordinary nature reserves. Here are a few of the ways that are currently being tried.

BREED IN CAPTIVITY

This method has been used as a last ditch measure for dying species, as well as for restocking programs (such as fish and birds). Zoos and other institutions use captured adults to breed offspring in a setting that is free of predators and harmful pollutants. Once the young have reached the proper age, they are released into the wild to help restore the species population. For birds, they may use captured eggs, which are then raised to adulthood and then bred to produce birds that will be released.

RAISE AN ENDANGERED SPECIES AS LIVESTOCK

The farming approach has been used in some places and is really a more aggressive form of breeding in captivity. The principal differences between farming the species and breeding in captivity are in the numbers raised and in the fact that species raised as livestock are not intended to be released to the wild. Famous successes of the livestock approach include Pere David's deer and Przewalski's horse, both in China originally, but also farmed in game parks in England.

BACKBREED

Mating adults of a dying breed with a related species can be used to develop hybrids similar to the endangered species. Also, a domesticated species can be carefully bred to recapture the general characteristics of the species' wild ancestor, as the aurochs, ancestor of today's cattle, has been "recreated." This method has also been tried in a desperate attempt to save a species of bird in which only males were left alive. Pure males were mated with half-breed females. The offspring were then to be mated with pure males. Ideally, progressively purer females of the endangered species would then result.

PASS PROTECTIVE LEGISLATION

This has become a widely used device to protect dwindling populations of various species. The Endangered Species Act in the United States provides formal legal protection from direct and indirect threats by man (hunters, developers who destroy habitats, etc.). Legislation may be in the form of an outright ban or a limit on commercial killing, such as limits on whale catches. The idea is to allow the population time to increase in number. It may fail if the habitat of the species is also endangered.

ESTABLISH SPECIAL WILDLIFE PRESERVES

While parks and game preserves benefit the entire wildlife population, it has been recognized that the only way to save some species is to protect their habitat. A case in point is the small fish known as the snail darter. It made headlines some years ago when a proposed dam threatened to destroy its habitat. More recently, after the snail-darter debate was over, another population of the still endangered snail darter was located.

IDENTIFY THE CAUSE FOR THE DECLINE OF A SPECIES

A species may become endangered through indirect causes, such as pollution in the environment. One of the reasons for the ban on DDT was that it was found to be killing off various species of wildlife. For example, it caused the eggshells of fish-eating birds, such as eagles and pelicans, to become thin and break in the nest. When DDT was banned, the populations of these birds, which had been in a serious decline, started to rise again.

RELOCATE WILDLIFE POPULATIONS

When a species' habitat cannot be saved as a special preserve, an alternative is to move the entire population to a new area. In Sri Lanka, for example, elephants have been relocated hundreds of miles away from their original home.

ESTABLISH NEW SPAWNING GROUNDS

Some species, such as salmon and sea turtles, instinctively return to the place where they were born to spawn or lay their eggs. Some species of birds lay their eggs on only one beach in the entire world. When spawning grounds are threatened, it may help to create additional spawning grounds where the threat is less serious.

1981

Process to Neutralize PCBs Approved by EPA

Polychlorinated biphenyls, PCBs, had been manufactured since the late 1920s for use in a variety of products, including electric transformer oil, lubricants, inks, paints, and adhesives. But by 1977 researchers had linked PCBs to liver disorders, cancer, and birth defects and in that year further production of the chemical was banned.

But the ever-present problem of what to do with existing stocks of the toxic chemical remained. There were but a few approved incinerators and designated chemical landfill operations for disposing of PCB-contaminated oil and other materials, and by 1981 the Environmental Protection Agency (EPA) estimated there were still some 347,000,000 kilograms (765,000,000 pounds) of PCBs yet to be disposed of. What was worse was that the spread of PCBs, which break down very slowly in the environment, posed a very serious pollution problem. The Hudson River, for example, was found to be polluted by PCBs, and in May 1981 the EPA approved a dredging operation in the upper Hudson to clean up a number of PCB "hot spots" (bureaucratic wrangling held up the project until 1983).

Later in May of 1981, however, the EPA granted approval to a new process which promised to solve a large part of the PCB problem. The process, developed by Sunohio Company, a transformer oil recycling company, was designed to neutralize the PCBs in transformer oil. It involves adding a chemical to PCB-laden oil to strip chlorine atoms from the PCB molecule, reducing PCBs to a paste of harmless salts and an inert polymer. And the process is commercially profitable because it leaves the expensive transformer oil unharmed.

Unfortunately, Sunohio's system offers no solution to the problem of PCB contamination in the environment. But it was a good start, especially when you consider that there were then an estimated 10,000,000 PCB-contaminated transformers still in use.

BUCKSHOT: FOUL FOWL FOOD

It has been estimated that some 2 to 3,000,000 ducks and geese die annually as a result of lead poisoning. The source of the poisoning is shot that has missed the birds the first time. Apparently ducks and geese gobble up spent pellets (traditionally made of lead) along with their normal food in the mud at the bottom of lakes and streams. To combat the problem, many states now require use of steel pellets.

Medfly Mayhem

The Mediterranean fruit fly—*Ceratitus capitata* or Medfly as it is popularly known—has long ravaged crops in areas around the Mediterranean, as well as in parts of Southern Africa, South America, and Mexico. It is a nasty pest that attacks over 200 varieties of fruits and vegetables and destroys any of the foodstuffs it infests.

Because of this destructive bent, the Medfly aroused real concern among California's farmers when it turned up in Los Angeles and Santa Clara counties in 1980. This was all too close to California's huge commercial farming district and thus threatened the state's $14,000,000,000 agricultural industry. On the other hand, the most effective way of combatting the Medfly, aerial spraying, was bitterly opposed by environmentalists. After all, spraying was to be conducted over populated areas, and there were fears (though far from any conclusive proof) that the pesticide to be used was hazardous. What resulted was a confrontation between agricultural interests and environmentalists, with protests, hysterical outbursts, a cameo spot by California's governor, Jerry Brown, and finally the spraying.

The Medfly, the cause of the mayhem, is somewhat smaller than a common housefly. It might go unnoticed except that it uses fruit as a host medium for part of its reproductive cycle. (Some of the fruits that we think of as vegetables, such as the tomato and the bell pepper, may also be used.) The adult female inserts anywhere from five to fifteen of her eggs into a single fruit or vegetable by means of an egg-laying tube, or ovipositor. Such an insertion is repeated over and over again during an adult female's four-week lifespan. Each time a new piece of fruit is infested. The eggs soon become larvae, or maggots, which for eight days feed on the host fruit. Furthermore, the larvae excrete bacteria that cause the fruit to rot. In fact, the bacteria are important symbiotes for fruit flies, since the bacterial rotting is necessary for the larva to obtain proper nutrition from the fruit. Because of the rotting, instead of having a fruit with just a few "wormholes" in it, the whole fruit is rendered completely worthless. Like all insects, the medfly has a complex development pattern, albeit somewhat on the short side. At the end of this stage of its life, each larva leaves the host and burrows into the ground for a three- to four-day pupation period. Then the insects emerge as adult flies, and, after one or two days, begin to mate, which they do at an alarming rate. Furthermore, a female can produce up to 800 offspring. The short life cycle combined with the fecundity accounts for the seriousness with which California farmers viewed the infestation.

But there was another fly in the California ointment. In the 1950s and 1960s, other Medfly outbreaks had been checked by means of aerial spraying with malathion, an effective pesticide commonly used to fight the insect in Mediterranean regions. California environmentalists pointed out that malathion belongs to the family of organophosphates, used for chemical warfare gases, and that in very large doses it could kill humans. But in low concentrations, as in spraying operations or in pet flea collars, malathion appeared to be quite safe.

National Cancer Institute studies had failed to link malathion with cancer, though other studies, using heavy doses of the pesticide, had indicated possible chromosomal damage in human tissues, as well as a higher rate of mutation during cell division.

These doubts were sufficient to set the stage for the panicky campaign to prevent spraying with malathion. And the campaign worked. California officials at first tried to get rid of the pest without aerial spraying. After discovering the Medfly in June 1980, California first tried stripping trees of fruit and spraying the ground with malathion. The Medfly continued to thrive in Santa Clara county, however, and in 1981 officials tried an Integrated Pest Management Program. This included stripping fruit trees, ground spraying, releasing millions of sterilized female Medflies, and quarantining fruits and vegetables in the area around Santa Clara county. But by mid-1981 it was clear that the program had failed; new Medfly infestations were found outside the quarantined area and pressure mounted to begin the aerial spraying.

The Integrated Pest Management Program was further damaged by persistent reports that some of the "sterilized" Medflies released were still fertile after they had been supposedly sterilized by radiation. Plans were made to resume spraying with malathion from the air, as had been done effectively in earlier Medfly infestations.

Protesters, including pregnant women, took to the streets, decrying the planned spraying as an attempt to rain poison down on millions of innocent people. There was talk of blowing up helicopters to be used in the spraying and Governor Brown, dubbed "Lord of the Flies" by a political cartoonist, staunchly refused to permit the spraying to begin. Finally, mounting concern about the spread of the Medfly to other parts of the country brought President Reagan into the act. The President threatened to quarantine all California fruits and vegetables. Furthermore, Japan banned the importation of California fruit altogether. On July 10, 1981, Governor Brown relented and ordered the aerial spraying, which lasted for 33 weeks and finally eradicated the pesky Medfly (at a cost of $100,000,000).

Some scientists pointed out that the Medfly's habits might cause it to be less of a pest than expected in any case. Medflies have difficulty penetrating the skin of citrus fruits, for example, much preferring peaches. As semitropical insects, the Medflies might not survive even a California winter. The soil in California's main agricultural regions was too heavy for the larvae to enter or for the adult flies to emerge. The scientists' views were not tested.

Just to show how safe malathion was, one man went so far as to drink a glass of the chemical diluted to a concentration equal to the spray. He was quoted as saying, "I had a terrible taste in my mouth for days afterwards, but other than that I felt fine." This was probably true of most everyone in the aftermath of California's Medfly mayhem.

Two years later, however, it looked as if history might repeat itself. This time, the villain was the Mexican fruit fly, a far more dangerous pest than the Medfly. The Mexican fruit fly is larger than a housefly, can fly farther, and lives as long as sixteen months, laying eggs in fruit along the way with much the same effects on fruit as the Medfly.

With a new governor, George Deukmejian, action was swiftly taken. The first Mexican fruit fly ever found in California was discovered in a grapefruit in late October 1983. Aerial spraying began almost immediately, although the region where the Mexican flies were found was downtown Los Angeles. Perhaps because the source of the infestation was not near suburban communities, environmental voices were hardly heard. By the end of November 1983, it appeared that the infestation had been contained, although spraying and some reports of infestation continued into 1984.

THE RAPE OF THE MEDFLY

When the season is young, female Medflies are receptive to males, which they find by following a sex attractant (pheromone) released by the males. In fact, several males will get together, so that their combined scents will attract more of the willing females. But once a Medfly has mated, the females are no longer interested. They have a "headache" that lasts for about ten days. After a while, almost none of the females are interested.

The male changes his tactics. Instead of counting upon his perfume, he goes on the attack. He finds a female in the process of laying eggs on a fruit and rapes her.

As evolutionary strategy, this is a good thing. For reasons that are not clearly understood by scientists, the eggs (which are actually fertilized as they are laid) tend to be fertilized by the last male with whom the female mates.

1982

Nuclear Reactors—Emerging Technology or the New Dinosaurs

Just under three years after the notorious Three Mile Island accident, an emergency at the Robert E. Ginna nuclear reactor near Rochester, New York, threatened for a time to become America's second nuclear disaster. Slightly radioactive steam was released into the atmosphere (posing no health hazard) and over 41,000 liters (11,000 gallons) of radioactive water spilled inside the reactor containment building. A truly serious accident was averted, but not without equipment failures that turned a malfunction into a cliffhanger. There was little consolation for the nuclear energy industry, already plagued by technical problems, faulty equipment, increasing costs, public protests, and a widespread fear of what might someday happen.

The Ginna plant had a pressurized water reactor, similar in design to that of the Three Mile Island reactor. On the morning of January 25, 1982, a corroded pipe in the reactor's primary cooling system burst and released radioactive water under high pressure into the low-pressure secondary cooling system. Pressure relief valves automatically vented steam into the atmosphere to lower pressure in the secondary cooling system. But, as was the case at Three Mile Island, a valve malfunctioned and caused a steam bubble to form inside the reactor. At Ginna backup systems functioned to eliminate the steam bubble and keep the reactor core covered with water. Thus the core was undamaged, but another malfunctioning valve, again reminiscent of Three Mile Island, got stuck in the open position and poured the 41,000 liters of radioactive water onto the containment building floor.

The problems are not limited to the United States. In Canada, where the CANDU reactor, which is cooled with deuterium-rich heavy water, was supposed to be immune to failure, a giant 6-meter (20-foot) pipe suddenly developed a split nearly 2 meters (76 inches) long and 2 centimeters (¾ inch) wide. This accident at the Pickering plant, just 32 kilometers (20 miles) from Toronto, was followed in quick succession by two other accidents, one of which resulted in the release of a small amount of highly radioactive tritium into Lake Ontario.

The list of problems with nuclear power plants is growing and the question arises as to whether nuclear power will succumb to technical problems beyond our control. For example, consider the eight problems that are outlined on the following page.

Corrosion in pressurized-water reactor pipes Corrosion in primary cooling system pipes, which carry radioactive water at high temperatures and pressures, has been a problem since the 1970s. Each attempted solution has led to another problem, and because corrosion shortens the working life of the system, it has raised operating costs considerably.

Junk in the system Despite efforts during construction and maintenance to prevent it, bits of scrap metal get left behind in water vessels and pipes. Carried by fast-flowing water during operation of the reactor, scraps can crash into—and weaken or even puncture—pipe walls.

Cracks in pipes Boiling water-type reactors are especially prone to this problem; of course, cracking also raises operating costs.

Cracking of vessel walls in pressurized-water reactors Neutron bombardment appears to have made 20.3-centimeter (8-inch) thick walls brittle much sooner than expected, posing the possibility of cracking under some circumstances. A crack would allow primary cooling system water to leak out and thereby expose the reactor core. Again, a shorter working life here raises operating costs.

Faulty equipment Materials used in constructing critical parts of reactor systems must be able to withstand tremendous stresses and meet strict standards. In 1983 the Nuclear Regulatory Commission was trying to track down substandard pipes and fittings produced by a New Jersey company. The equipment was supposed to withstand pressures of up to 1,360 kilograms (3,000 pounds), but because of faulty manufacture they may not be able to withstand a mere 68 kilograms (150 pounds) of pressure.

Waste disposal There are some 30,000 spent fuel rods being stored in temporary holding tanks at nuclear power plants around the country. Costly commercial reprocessing plants (banned in the United States from 1977 to 1981) could solve this problem, but highly radioactive liquid wastes remain.

Decommissioning old reactors Disassembly and decontamination could require as much as $100,000,000 to complete for a single plant and must be figured into operating costs.

Fire damage A fire at a nuclear power plant can incapacitate electronic and other operating and emergency systems and thus cause a major disaster. New fire safety systems, ordered after a fire at the Brown's Ferry reactor in 1975, are very costly. The deadline for compliance has been moved up to 1984.

1983

Sick Building Syndrome—Indoor Pollution

A New Jersey science teacher was exposed to high concentrations of formaldehyde gas a few years ago and became very sick as a result of it. The incident did not take place following an accident in one of his science classes, as one might expect. Instead it happened in his home, which had recently been insulated with urea formaldehyde foam insulation (UFFI). His illness was a classic example of health problems caused by indoor pollutants—the "sick building syndrome."

UFFI, installed in over 500,000 homes in recent years, has been linked with many complaints of flu-like symptoms (1,200 cases in 1981). These illnesses were believed to be the result of formaldehyde gas given off by the UFFI insulation over a long period of time. As a result, UFFI manufacturers were hit with some well-publicized lawsuits. Early in 1983 the Consumer Product Safety Commission banned further sale of UFFI.

The health problems associated with UFFI aroused concern about indoor pollution and especially about formaldehyde, which is used in a wide range of products. Some 3,600,000 kilograms (8,000,000 pounds) of formaldehyde are produced each year and the compound is used, often in trace amounts, in building materials, such as particle board and paneling, plastics, permanent press clothing, some cleaners, and some personal care products, such as cosmetics and toothpaste. But the compound also occurs in the home as a result of cooking with natural gas and as a component of cigarette smoke. Despite this, serious health hazards associated with formaldehyde pollution in homes are rare. In addition, the extent of minor reactions is difficult to assess, because symptoms in these cases are so easily confused with influenza-like diseases. Late in 1983, however, as a result of a lawsuit, the Environmental Protection Agency started a complete review of formaldehyde that could result in banning at least some of the chemical's many uses.

Formaldehyde is not the only type of indoor pollution, and awareness of health problems arising from the so-called sick building syndrome has been increasing in recent years. Irritation of mucous membranes, headaches, dizziness, nausea, diarrhea, rashes, and abdominal and chest pains are all symptoms that can be associated with sick building syndrome. In recent years both the Consumer Product Safety Commission and the National Institute for Occupational Safety and Health have investigated cases of this type in offices, schools, and factories.

The recent trend toward energy-efficient homes is one reason for the problem, not only because of increased use of synthetic building materials (which may give off toxic fumes), but also because of the effort to seal up houses to prevent heat loss. This drastically reduces normal ventilation. As a house "breathes" more slowly—that is, as the air inside it turns over more slowly—it takes longer to disperse indoor pollutants, such as those from cooking and space heating appliances, tobacco, personal care products, and such natural contaminants as radon and microorganisms.

For example, in a house with slow air turnover and two or more heavy smokers, breathable particles will exceed the National Ambient Air Quality Standard for a 24-hour period. (Cigarette smoking, of course, produces a host of other contaminants, including NO_2, CO, nicotine, acrolein, and formaldehyde.)

Even cooking dinner on a conventional gas stove causes a sharp rise in such contaminants as NO_2, CO, and CO_2 for a short period of time. All byproducts of combustion, these contaminants may also be traced to leaky furnaces and unvented kerosene and gas heaters. Open fireplaces, attractive though they may be, are perhaps the worst polluters of all. Not only do fires produce dangerous gases, but the soot contains tiny POMs, polycyclic organic materials. There are more than a thousand different POMs given off by wood or coal fires, many of which are known carcinogens, or cancer-causing chemicals. Tobacco smoke also contains dangerous POMs.

Then there is radon, the colorless, odorless, but radioactive gas given off by radium. Radium is a trace element in many rocks and consequently in building materials made from rock. As radium decays, one of the elements that results is radon. Radon is typically found in highest concentrations in basements, especially when a basement is made of rock or masonry. Energy-efficient homes tend to have the highest levels of radon, especially if they are built in areas where there is a lot of rock near the surface of the earth. Since radon is radioactive, there is every reason to believe that exposure to it increases the risk of cancer to people living in such houses, although a direct connection has not been shown.

As if all this were not enough, there's more! Of course one would expect that paints and varnishes, cleaning fluids, and even personal care products add to the contamination of a house (they give off such substances as acetone, ammonia, benzene, chlorinated compounds, and toulene). But people pollute as well! Human beings give off such "bioeffluents" as butyric acid, ethyl and methyl alcohol, and acetone—not to mention CO_2 and water vapor.

Sources of Indoor Pollutants

Pollutant	Major emission sources
Origin: Predominantly Outdoors	
Sulfur oxides (gases, particles)	Fuel combustion, smelters
Ozone	Photochemical reactions
Pollens	Trees, grass, weeds, plants
Lead, manganese	Automobiles
Calcium, chlorine, silicon, cadmium	Suspension of soils, industrial emissions
Organic substances	Petrochemical solvents, natural sources, vaporization of unburned fuels
Origin: Indoors or Outdoors	
Nitric oxide, nitrogen dioxide	Fuel burning
Carbon monoxide	Fuel burning
Carbon dioxide	Metabolic activity, combustion
Particles	Resuspension, condensation of vapors, combustion products
Water vapor	Biological activity, combustion evaporation
Organic substances	Volatilization, combustion, paint, metabolic action, pesticides
Spores	Fungi, molds
Origin: Predominantly Indoors	
Radon	Building construction materials (concrete, stone), water
Formaldehyde	Particleboard, insulation, furnishings, tobacco smoke
Asbestos, mineral, and synthetic fibers	Fire retardant materials, insulation
Organic substances	Adhesives, solvents, cooking, cosmetics
Ammonia	Metabolic activity, cleaning products
Polycyclic hydrocarbons, arsenic, nicotine, acrolein, etc.	Tobacco smoke
Mercury	Fungicides, paints, spills in dental-care facilities or labs, thermometer breakage
Aerosols	Consumer products
Microorganisms	People, animals, plants
Allergens	House dust, animal dander, insect parts

Source:"Indoor Air Pollution: A Public Health Perspective," J.D. Spengler, *Science* Vol. ??1, pp. 9-17, 1 July 1983. Copyright 1983 AAAS.

Groundwater Contamination

Almost half the population of this country depends on wells or springs for drinking water and 75 percent of our major cities get most of their fresh water from wells. Serious pollution reported in Atlantic City, New Jersey, was only among the most recent in a growing number of cases of groundwater contamination.

Ninety percent of Atlantic City's municipal water supply once came from wells, but by 1982 toxic contamination had forced the closing of some of the city's wells and had reduced the wellwater use to 50 percent of the total. The contamination, a toxic soup of benzene, arsenic, and other carcinogenic chemicals, was said to be seepage from Price's Pit, listed by the Environmental Protection Agency (EPA) as one of the ten worst toxic dumps. Located 10 kilometers (6 miles) from Atlantic City, the dump had received an estimated 34,000,000 liters (9,000,000 gallons) of chemical waste before it was closed down. The pollutants had seeped into the ground and were moving slowly with the groundwater—about 18 centimeters (7 inches) a day—toward the remainder of Atlantic City's municipal wells. In 1981 contamination of groundwater in Nassau and Suffolk counties on Long Island, New York, resulted in the closing of 36 municipal wells that supplied over 2,000,000 persons with their water. In the town of Babylon, Long Island, the problem was particularly acute and a 1.6-kilometer-long (mile-long) plume of pollutants forced hundreds of homeowners to shut down their wells and switch to municipal water.

It might be viewed as somewhat ironic that one of the serious sources of groundwater pollution on Long Island and elsewhere is cleaning fluid. The toxic chemical dry cleaners use to keep our clothes spotless have an uncanny ability to leak into the wells and make them dirty.

In 1983 the Environmental Protection Agency added gasoline to the list of major groundwater contaminants. The EPA declared that 41,600,000 kiloliters (11,000,000 gallons) of gasoline annually were seeping from nearly 100,000 underground storage tanks. Four liters (1 gallon) a day is sufficient to pollute the water supply for 50,000 people who are served by a well-based water system. Gasoline contains the carcinogens benzene and ethylene dibromide. Also, much gasoline still contains lead, a potent poison that causes brain damage. Despite this, gasoline is not listed as a hazardous substance and at the end of 1983 was not regulated.

In cases like these, once groundwater becomes contaminated, well water may be undrinkable for many years. That is because underground water lies in aquifers—layers of porous rock, gravel, or sand, up to hundreds of meters thick—and consequently flows at a very slow rate. Aquifers vary in size according to the geology of an area (ranging from a single locale to several states). Factors such as depth below the surface and type of soil and rock formation above the aquifer determine vulnerability to contamination. Because groundwater moves so slowly, contamination from a single source tends to form a *plume*, an ever-widening area of contamination on the downstream side. Thus, depending on the type of flow and direction, nearby wells can sometimes escape being polluted. (For one of the possible hopes for control of groundwater pollution, however, see "Groundwater Ecology—Organisms Where None Were Thought to Be" on page 186.)

Toxic waste and other industrial contamination is only the most recently publicized form of pollution to seep into our groundwater supplies—and our consciousness. In fact, the problem is far from new and is as diverse as the regions in which pollution occurs. There are naturally occurring contaminants such as radioactive uranium and arsenic (in some Western states), pollutants from mining and agricultural operations (in the Midwest), pollutants from individual and municipal sewage disposal, and contamination by saltwater intrusion (in coastal areas) caused by heavy demands on freshwater aquifers. Available statistics on *reported* cases of contamination that caused illnesses between 1945 and 1980 give some idea of the scope: viruses, bacteria, and parasites caused 158 contamination incidents in which 31,000 persons suffered illness; inorganic chemicals caused 20 incidents of pollution in which 300 persons became ill; and organic chemicals were responsible for 57 persons becoming ill.

It is difficult to estimate the extent of pollution in this country's 125,000,000,000,000,000,000 to 379,000,000,000,000,000,000 liters (33,000,000,000,000,000 to 100,000,000,000,000,000 gallons) of groundwater, but it is clear that implementing measures to prevent pollution before it occurs is the only reasonable course of action. Methods such as capping and damming aquifers, digging up polluted material in aquifers, or pumping out polluted waters are either ineffective or enormously expensive.

For example, take the Rocky Mountain Arsenal in Colorado: when contaminants from nerve gas and pesticide production polluted wells as far as 13 kilometers (8 miles) away, the Army was forced to spend $6,000,000 on an elaborate system to halt further spread of the toxic materials. This included a 1.6-kilometer-long (1-mile-long) dam of clay, dug 6 to 18 meters (20 to 60 feet) below the ground surface, and a groundwater treatment plant that pumps 3,785,000 liters (1,000,000 gallons) a day.

Worldwide Ban on Whaling

If the International Whaling Commission can enforce its recently adopted worldwide ban on whaling (to take effect in 1986), it may finally put an end to the commercial whaling industry, which for years has threatened various species of whales with extinction. A ban on commercial whaling was first proposed in 1972. In subsequent years annual limits on catches were invoked to at least reduce the slaughter of whales. But this new total ban is bound to bring about a confrontation between its supporters and countries that still have active whaling industries.

Earlier attempts to save whales, by putting pressure on the commercial whaling industry, failed, and the sperm, fin, sei, and minke whales are now believed to be especially endangered. In past years quotas were introduced to sharply reduce catches. For instance, in 1960 about 67,000 whales were taken worldwide, but by the mid-1970s quotas were set at 46,000; in 1981 they were reduced to 14,500. In addition, the commission in 1979 put a total ban on whaling in the Indian Ocean, where a number of whale species spawn. In 1981 it banned hunting of the seriously endangered sperm whale in all areas except the seas near Japan.

While the movement to save the whales and the commission's steady reduction of whaling quotas has brought an end to commercial whaling by many countries, Japan, the Soviet Union, Norway, Iceland, Brazil, Peru, and South Korea still have a major interest in whaling. All are members of the International Whaling Commission and all voted against the recent total ban on whaling. Some of these countries threatened to defy the ban and continue whaling. In retaliation, the United States threatened to ban any fishing in its waters by nations that continued killing whales. Thus the stage is set for a power struggle. The commission itself has no power of enforcement, but the 27 nations that support the ban are free to impose economic sanctions as they see fit. This may help bring into line the countries that permit renegade whaling operations.

Beyond that there is the pressure of world opinion. The militant group Greenpeace seemed to be gearing up for battle in 1983 when it took on the Soviets in a highly publicized demonstration. Greenpeace protesters landed at a Soviet whaling station on July 18 and attempted to pass out antiwhaling leaflets to workers there. Six Americans and a Canadian were promptly seized by Soviet authorities. The Greenpeace trawler *Rainbow Warrior*, which had transported the protesters to the whaling station, managed to return to American waters after the incident, but the Soviets detained the seven protesters until July 22.

The Unraveling Dioxin Story—from Ignorance and from Bliss

The names dioxin and Times Beach, Missouri, became virtually synonymous in 1983, but the Environmental Protection Agency's (EPA) decision to buy out and evacuate 2,000 Times Beach residents was only one episode in the long history of this toxic chemical.

Dioxins, actually a whole family of compounds, are usually found as unwanted contaminants in some herbicides and in compounds called chlorophenols. However, only one of the dioxins, 2,3,7,8-TCDD (2,3,7,8-tetrachlorodibenzo-*p*-dioxin) is a serious public health hazard. That dioxin was found at Times Beach. The 2,3,7,8-TCDD compound is the most toxic of all dioxins; it has, in fact, been ranked as the fourth most toxic substance known to man.

One of the reasons it is so hazardous is that it does not break down quickly in the environment or in the human body (it collects in fatty tissues). Exposure to even minute quantities of the compound—one part per 1,000,000,000 is now considered a safe upper limit—can cause serious illness, including chloracne (skin lesions and rashes), liver disorders, very high blood levels of lipids and cholesterol, nerve damage, loss of weight, and psychological disorders. These symptoms may disappear in months or may linger for years. A cancer, soft-tissue sarcoma, has also been linked to exposure to 2,3,7,8-TCDD, and there are as yet unproved charges that it also causes birth defects.

The dioxin story—and investigation of the toxic effects of 2,3,7,8-TCDD—can be traced back to industrial accidents during the manufacture of chlorophenols in the late 1940s and 1950s. Though the existence of 2,3,7,8-TCDD was as yet unknown, workers who suffered overexposure to the chlorophenols developed severe cases of chloracne. Finally, in 1957, German scientists isolated 2,3,7,8-TCDD as a contaminant in chlorophenols and as the cause of chloracne. This led to new manufacturing processes that produced less of the unwanted 2,3,7,8-TCDD.

During the Vietnam War in the late 1960s and early 1970s, the United States used huge quantities of the notorious Agent Orange, containing the dioxin-contaminated herbicide 2,4,5-T, to defoliate an estimated 1,700,000 hectares (4,198,000 acres) in South Vietnam. This produced a twofold toxic chemical disaster. The spraying not only exposed tens of thousands of United States and Vietnamese soldiers to the toxic 2,3,7,8-TCDD, it also produced enormous stockpiles of contaminated chemical wastes at chemical companies that manufactured Agent Orange in the United States.

It was from one such plant at Verona, Missouri, that Russell Bliss—who was in the business of disposing of waste oil and chemicals—got hold of dioxin-contaminated wastes in 1971. He mixed them with waste oil and sprayed the toxic "soup" on unpaved roads in Times Beach and other parts of Missouri to help keep down dust. All in all, Bliss dumped 81,400 liters (18,500 gallons) of contaminated sludge around Missouri, and even used some on the roads around his own farm. When in 1971 birds, small animals, and horses suddenly began dying in Moscow Mills, a town near Times Beach, an investigation was launched. By 1974 2,3,7,8-TCDD was shown to be the cause, but at that time the compound was believed to break down quickly and nothing was done about the contamination.

Meanwhile, other indications of the dangers of dioxins were coming to light. In 1972 hexachlorophene—an antibacterial agent and member of the family of chlorophenols—was taken out of soaps and detergents when studies showed it caused brain damage in monkeys. Then in 1976 the explosion of a chemical plant at Seveso, Italy, resulted in widespread dioxin contamination. Though long-term studies are still being conducted, one early report showed an increased level of birth defects and miscarriages in the area of the explosion; there are no accurate records from earlier periods for comparison, however.

By 1981 the United States National Institute for Occupational Safety and Health had completed studies that showed 2,3,7,8-TCDD caused increased rates of soft-tissue sarcoma in workers exposed to the chemicals. But the real shock came in 1982, when EPA officials went back to Times Beach and found that the 2,3,7,8-TCDD had not broken down quickly, as expected. In fact, they found the chemical in the soil in concentrations as high as 100 to 300 times that which is considered safe.

Soon the news was out and the whole country had heard about Times Beach and the sinister menace posed by even barely detectable traces of 2,3,7,8-TCDD. Then reports of other toxic chemical waste dumps containing 2,3,7,8-TCDD made headlines, notably Midland, Michigan, and Newark, New Jersey, in addition to suspicions about contamination in many parts of Missouri.

Despite all the years of study, however, researchers are still unsure about just how much 2,3,7,8-TCDD is safe for humans. Animal studies have failed to produce results that can be translated to dosage levels for humans, and studies of levels in humans exposed to 2,3,7,8-TCDD are difficult (surgical removal of fatty tissues would be required). Even the mechanism by which the chemical poisons the body remains a mystery. Evidence indicates that symptoms such as chloracne appear when certain enzymes are produced in the body, and some scientists believe that 2,3,7,8-TCDD acts to trigger certain genes in cells, causing them to produce the destructive enzymes.

Late in 1983, the EPA declared that it had developed a dioxin policy. Likely sites of dioxin contamination had been identified. All would be tested, and if levels of dioxin reached one part per 1,000,000,000, measures would be taken. The measures would consist of a range of actions, from simply covering the site with a tarp to using chemical treatments to remove the dioxin from soil for subsequent incineration.

Meanwhile, a University of California scientist reported on what appears to be the best method for decontaminating soil that has been tainted by 2,3,7,8-TCDD—simply expose it to sunlight. What to do with all the contaminated soil at places such as Times Beach had been thought of as the biggest problem in a cleanup program. But it has been learned that a few hours' exposure to sunlight (or an ultraviolet light) breaks down 2,3,7,8-TCDD into a much less toxic, biodegradable substance. The soil still must be dug up and turned regularly for a period of time to complete detoxification, because only the dioxin directly exposed to the ultraviolet light breaks down.

SINKHOLES—NOT SO TERRA FIRMA

It is easy to take certain things for granted—the sunrise every morning, the solid earth beneath our feet—but not so in Winter Park outside Orlando, Florida. At least not since a huge sinkhole 104 meters (342 feet) wide and 30 meters (100 feet) deep opened up and swallowed a house, the surrounding lot, part of an auto repair shop, and a parking lot. Even the deep end of a municipal swimming pool slid into the newly formed pit.

Small comfort to those whose property suddenly disappeared, but sinkholes are very much a part of life in central Florida. That is because a layer of limestone—a brittle, porous rock—thousands of feet thick lies under all of Florida. Though covered by clay, sand, and topsoil, it is this layer of limestone that is literally at the bottom of the sinkhole problem.

That is because water that seeps down through the upper layers, after a rainstorm for example, becomes somewhat acidic. This tends to dissolve the limestone below, at first creating fissures in the rock. These fissures can become enlarged into a series of small openings and then great limestone caverns.

In Florida, under normal conditions, these honeycombs of fissures and caves are filled with water (the limestone layer forms Florida's precious freshwater aquifer). Because water cannot be compressed, it provides the support needed to prevent the honeycomb from collapsing. But in 1983 a severe drought, coupled with a great demand for water by an ever-expanding Florida population, lowered the water table enough to allow some of the caverns to drain—and collapse.

50 Worst Hazardous Waste Sites

(1983 National Priorities List)

ALABAMA
Limestone/Morgan
Triana Tennessee River

ARKANSAS
Jacksonville
Vertac, Inc.

CALIFORNIA
Glen Avon Heights
Stringfellow

COLORADO
Commerce City
Sand Creek

DELAWARE
New Castle County
Army Creek Landfill
Tybouts Corner Landfill

FLORIDA
Pensacola
American Creosote Works
Plant City
Schuylkill Metals Corp.
Tampa
Reeves SE Galvanizing
Corp.

INDIANA
Seymour
Seymour Recycling Corp.

IOWA
Charles City
LaBounty Site

KANSAS
Cherokee County
Cherokee County

MAINE
Gray
McKin Co.

MASSACHUSETTS
Acton
W.R. Grace Co. (Acton
Plant)
Ashland
Nyanza Chemical Waste
Dump
Holbrook
Baird & McGuire
Woburn
Industri-plex

MICHIGAN
Swartz Creek
Berlin & Farro
Utica
Liquid Disposal Inc.

MINNESOTA
Fridley
FMC Corp.
New Brighton
New Brighton/Arden Hills
St. Louis Park
Reilly Tar

MONTANA
Anaconda
Anaconda Smelter-Anaconda
Silver Bow Creek
Silver Bow/Deer Lodge

NEW HAMPSHIRE
Epping
KES-Epping

Nashua
    Sylvester
Somersworth
    Somersworth Sanitary
    Landfill

NEW JERSEY
Brick
    Brick Township Landfill
Bridgeport
    Bridgeport Rental & Oil
Fairfield
    Caldwell Trucking Co.
Freehold Township
    Lone Pine Landfill
Gloucester Township
    Gems Landfill
Mantua Township
    Helen Kramer Landfill
Marlboro Township
    Burnt Fly Bog
Old Bridge Township
    CPS/Madison Industries
Pitman
    Lipari Landfill
Pleasantville
    Price Landfill

NEW YORK
Oswego
    Pollution Abatement
    Services
Oyster Bay
    Old Bethpage Landfill

South Glens Falls
    G.E. Moreau

OHIO
Darke County
    Arcanum Iron & Metal

OKLAHOMA
Ottawa County
    Tar Creek

PENNSYLVANIA
Bruin Borough
    Bruin Lagoon
McAdoo Borough
    McAdoo Associates

SOUTH DAKOTA
Whitewood
    Whitewood Creek

TEXAS
Crosby
    French, Ltd.
    Sikes Disposal Pits
Houston
    Crystal Chemical Co.
Lamarque
    Motco

WASHINGTON
Kent
    Western Processing Co.,
    Inc.

Eight Ways to Get Rid of Toxic Wastes

Reduce wastes generated by the manufacturing process   Increased cost of toxic waste disposal has encouraged some companies to change manufacturing processes that produce toxic wastes, so that wastes are either recycled or drastically reduced. This is the best of all possible approaches.

Grow special-purpose microorganisms to eat toxic wastes   Though this method is still in the experimental stage, the idea is to develop a strain of microorganism that would eat a particular chemical, such as PCBs or dioxin, and use it to clean up a contaminated landfill. Once the contaminant had been entirely consumed, the microorganisms would supposedly die off from lack of food. But suppose they have a change of appetite?

Dissolve in supercritical water   Water that is subjected to high pressure and heat becomes "supercritical," and in this state dissolves toxic wastes, reducing them to chemically safe byproducts. This is an especially effective method of disposal.

Zap with plasma arc   In this method, air is superheated to its plasma state (25,000°C, 45,000°F) by means of an electric arc. Toxic wastes are then introduced and burned so completely that toxic residues cannot be detected.

Treat with molten salt bath   Sodium carbonate at 899°C (1,650°F) has proved effective for destroying toxic solvents and acids. It works at temperatures below those of ordinary incinerators.

Neutralize with chemicals   Some toxic chemicals can be reduced to safe chemical compounds by means of chemical treatment.

Incinerate   Possibly harmful pollutants may be released into the air during incineration, though presumably this is not the case at specially designated toxic waste incinerators. However, since the 1970s it has been legal to burn hazardous chemicals as boiler fuel, and there are no special controls over emissions. An estimated 21,000,000 tons of hazardous wastes are disposed of in this manner each year.

Secure toxic waste landfills   This is by far the cheapest method of disposing of toxic wastes, and is just a step up from the unregulated toxic waste pits of a few years ago. Landfills do not really dispose of wastes, they simply contain them. To that end they use layers of clay, plastic liners, and leachate (wastes that have seeped to the bottom of the pit) removal systems. Top layers of earth cap the landfill.

THE EPA'S LIST

Late in 1983, the Environmental Protection Agency announced a new list of priorities. In order:

1. Stabilize dangerous toxic wastes
2. Clean up toxic-waste sites
3. Enforce toxic-waste handling laws
4. Enforce toxic-waste disposal laws
5. Carry out a strategy on acid rain

There are 21 other items on the list, many dealing with toxic wastes.

The Sputtering Synfuels Industry

While the oil glut and stable oil prices of the past few years have provided much welcomed relief to most Americans, they have proved disastrous for what was hoped would be an important key to America's energy independence—the synfuels industry.

Synfuel is a loosely used term that refers to a wide range of alternative fuels, such as methanol and alcohol, that can be substituted for gasoline. It also refers to gasoline made artificially by chemical processes (such as those investigated during World War II). Even processes to convert energy sources from one form to another (coal gasification) and unusual extraction processes for naturally occurring oil (oil sand and shale oil projects) are lumped under the term *synfuel*.

In the late 1970s the government formed the United States Synthetic Fuels Corporation and funded it with billions of dollars to encourage the development of private synthetic fuel manufacturing operations. This along with other privately financed synfuels projects was expected to create a new American energy industry with an output of the equivalent of some 2,000,000 barrels of oil a day by 1992.

Nothing like that seems likely in the years to come. Conservation resulted in lower demand for oil, oil prices held substantially below the minimum $40 per barrel price required to make large-scale synfuel operations profitable, and high interest rates and soaring construction costs forced companies to abandon plans for big synfuel plants. Even the government seemed reluctant to venture into these troubled waters. The Synthetic Fuels Corporation did not make its first grant to private industry until 1983, when it released $120,000,000 in loan guarantees to the Cool Water Coal Gasification Project in Daggett, California and promised $800,000 for a methanol plant in Creswell, North Carolina.

Meanwhile, the massive synfuels projects envisaged earlier went the way of the dinosaurs. A privately funded project to extract oil from tar sands in Alberta, Canada, was abandoned in May 1982, and Exxon's huge Colony Shale Oil Project in Colorado was called off soon after. (Shale oil is not oil and does not come from shale; instead, it is a hydrocarbon called kerogen that is trapped in limestone, which is then called "oil shale." But kerogen burns—in fact, lightning sometimes sets rocks containing it on fire—and it can be used to make high-quality jet fuel.)

Most experts agree that even if the world oil market remains stable in the years ahead, it would be wise to have at least small commercial synfuels plants in operation as a hedge against any future energy crisis. These plants would be on the order of the Union Oil Company's Colorado shale oil project, which began manufacturing 10,000 barrels of oil a day in 1984, and which would provide a base for the rapid expansion of the synfuels industry, when and if it is needed. Union itself, with a guarantee from the Synthetic Fuels Corporation of up to $2,700,000,000, plans to expand production at the plant to 50,000 barrels a day by 1994.

While there are tremendous deposits of oil shale in Wyoming, Utah, and Colorado, as well as in other regions around the world, exploitation of the deposits in quantity might create more environmental problems than it would solve. All the known processes for liquifying the kerogen require vast amounts of water, already a scarce commodity in that part of the West. Furthermore, the processes expand the rock, leaving huge quantities of spent shale that cannot be returned to the holes from which the shale was originally mined. Current plans are to toss the spent shale into canyons, with unknown consequences for the wildlife of the area. However, as 1983 came to an end, the Synthetic Fuels Corporation funded an experimental project to extract shale oil from shale without removing the rocks from the ground.

THE DISAPPEARING LAKE TRICK

It really was an impressive show on a grand scale when the shallow 3.2-kilometer-long (2-mile-long) Lake Peigneur suddenly vanished from the Louisiana landscape on November 20, 1980. But no one wants to take credit for it. No wonder. It's not every day that a $50,000,000 salt mine, a $5,000,000 oil rig, eleven barges, one tugboat, and 65 acres of lakefront property all go down the same drain.

There was an extensive salt mine in a mile-wide salt dome situated under Lake Peigneur. Oil very often gets trapped in rock strata along the outer edges of salt domes, and so an exploratory well was started next to the salt dome. Whether the oil rig actually drilled into the mine shaft is a matter for litigation. But at 5 A.M., with the drill pipe down to 381 meters (1,250 feet), the drill suddenly became stuck, then mysteriously jerked up and down several times.

Events followed in quick succeson after that incident: the drill rig sank beneath the surface of the lake an hour or so later; miners reported serious flooding at the 396-meter level at around 8 A.M.; a whirlpool appeared in the lake by late morning and eleven barges were dragged into the swirling waters as the water level of the lake began dropping rapidly; the lake bed itself began sinking; and water in a small canal (connecting the lake with the Gulf of Mexico) started flowing backwards into the lake. By 2:30 in the afternoon the lake had disappeared down what had become a large hole in the lake bed. Fortunately, no one had been killed.

A few days later, when the salt mine had completely filled with water, the lake again reappeared, filled by backflowing water from the canal. As Lake Peigneur was being refilled, nine of the eleven missing barges bobbed back up to the surface.

New Approach to Coal Mine Fires—Controlled Burnout

Underground fires are one of the worst possible disasters that can befall a coal mining region. Once started, these fires burn at temperatures of up to about 556°C (1,000°F)—so hot that water cannot be pumped in fast enough to put them out. What is worse is that fires spread far beyond the immediate area of the shafts themselves, and drawing air through fissures in the ground, can follow coal seams over an ever-widening area.

There are over 500 fires burning in mines and waste heaps throughout the United States and until 1983 the only sure way to put them out was to dig up and extinguish all the burning material—a difficult and enormously expensive process. In mid-1983, however, the United States Bureau of Mines patented a new system called controlled burnout, which was developed by a bureau researcher, Robert Chaiken.

In this approach, a 1.2-meter- (4-foot-) wide vent pipe, which is lined with brick and has a water jacket, is driven down into the underground fire zone. Other smaller air vents are also drilled through the ground into the fire zone at specially selected points. Then a giant, mobile exhaust fan powered by a 350-horsepower motor is used to draw out gases and heat from the fire at a rate of 651 cubic meters (25,000 cubic feet) per minute. The overall effect is to create a current of air extending over an area of 12.1 hectares (30 acres) and running from the air sources (fissures and drilled air vents) to the vent pipe. The air current causes the fire to burn hotter—up to 1,000°C (1,800°F)—and much faster than it normally would. That may seem the opposite of what is wanted, but instead of taking perhaps a hundred years to slowly burn itself out, the fire travels quickly toward the air sources and, once the coal has been completely consumed, burns itself out.

That is only one advantage of the new system, however. The heat and exhaust gases from the burning coal are drawn up through the vent pipe and thus heat up water in the jacket surrounding the pipe. This produces steam to run a turbine-powered electric generating plant, so at least a portion of the energy stored in the coal, usually totally lost in these fires, is retrieved. Fumes from the fire, normally a serious hazard to those in the immediate area, are channeled through the single point of the vent pipe. In addition, the higher temperatures make the coal burn more completely, yielding mostly nonpolluting water vapor and carbon dioxide. Impurities are taken out before being exhausted from the vent pipe.

Yet another advantage of this system is that it can be used in abandoned coal mines that have played out. Mining operations do not retrieve more than 50 percent of the coal; the rest is left behind as rubble or as too unprofitable to mine. What the controlled burnout system would do in these cases is deliberately set the mines on fire and use the fire to burn up the remaining coal. Thus abandoned mines would become enormous furnaces to produce heat to drive steam-powered electric generators.

The controlled burnout system was tried out in 1982 on a small 1-acre mine fire at Calamity Hollow, Pennsylvania, and it proved highly successful. The power plant even provided twenty times the energy needed to run the system. But the big test will come if the system is used at Centralia, Pennsylvania, where a coal mine fire has been burning since May of 1962 and has spread over an 80-hectare (195-acre) area. After years of unsuccessful attempts at containing the fire, indecision, and bitter fighting, the townspeople of Centralia voted in August 1983 to abandon their homes in return for public funds for relocation. This may open the way for a large-scale controlled burnout operation if the people actually do abandon the town. However, even after Congress voted on November 19, 1983, to fund the move to the tune of $42,000,000, many residents were reported to be staying. In fact, the best estimate as of December was that about half would leave and half would stay.

A CURSE TURNED TO PROPHECY

Many of the residents of Centralia, Pennsylvania, believe that the ravages of the underground fire that has been burning beneath their homes are a tragic fulfillment of a curse on the town.

According to one version of the story, some local miners during the 1870s were using violent tactics to force mine owners to improve working conditions. The priest at Centralia's St. Ignatius Church denounced the tactics. In return, the miners gave him a severe beating. Furious, the priest called the townspeople to the church that night and placed the curse upon all those involved, as well as their descendants. Then he vowed that some day all of Centralia—except the church—would disappear from the face of the earth.

Today steam rises from fissures in the ground and carbon monoxide gas seeps into basements of homes over the fire. The underground inferno spreads inexorably under the town, and the fires have even broken through to the surface. On St. Valentine's Day in 1981, a twelve-year-old boy was almost swallowed up when a 30-meter- (100-foot-) wide pit suddenly opened up in a yard where he was playing. Now cracks have appeared on Locust Avenue, a main street in Centralia, and the people are leaving.

Oddly enough, the only building in Centralia that is in no danger at all is the church. Though the surrounding land is laced with mine shafts and coal seams, all now in flames, it so happens the church was built over an area of solid rock.

Wet Photoelectric Cells—Dunking Your Semiconductor?

Photoelectric cells are in many ways the best source of electric power—if only they were not so expensive—since they convert sunlight directly to electrical energy without any intermediate steps. So far the high cost of photoelectric cells has limited their use to specialized situations, such as recharging the batteries in artificial satellites.

Scientists have been working for some time on developing a less expensive alternative to the conventional solid-state photoelectric cell. The wet cell—a liquid electrolyte with a semiconductor submerged in it—has been around for a few years, but earlier models tended to suffer either from corrosion or from a low light-to-electricity conversion efficiency.

Late in 1983, a team of German and American scientists reported they had developed a wet cell with a conversion efficiency of 9.5 percent and minimal corrosion. The cell uses an electrolyte solution containing copper and iodide ions and molecular iodine. The semiconductor that is "dunked" in the electrolyte is a doped, n-type single-crystal semiconductor. The components of the new wet cell are relatively inexpensive.

Wet cells do not produce electricity by chemical processes, though an electrolyte must be present. Instead, the device takes advantage of a "liquid junction" that automatically forms between the submerged semiconductor's surface and the electrolyte. Light striking this liquid junction is converted into electricity without involving the semiconductor or electrolyte in a chemical reaction (as in a battery, for example).

Advocates of solar power continue to hold high hopes for this absolutely nonpolluting source of energy, even though many energy experts say that it will never make up a major portion of the nation's energy budget. Even if the experts are right, for many specialized uses, solar power exists already. The Japanese, for example, bypassed the effort to produce solar power on a large scale and instead have developed solar-powered watches and portable television sets.

BEAVERS FOR EROSION CONTROL

After spending millions trying to control creek erosion, the federal government tried a new approach. Eight beavers were released at two sites in Wyoming. The creeks involved had been carrying away the soil with swift water flow, partly because cattle had trampled the vegetation along their banks. In a few months, the beavers had built dams. The water flow was slowed and erosion was stopped. Today vegetation is returning to the banks of the creeks.

1984

Update 1984: Environment

THE GREAT EDB SCARE

Late in 1983 the Environmental Protection Agency (EPA) concluded that ethylene dibromide (EDB) was a cancer-causing chemical and that it should no longer be used to fumigate soil (an emergency ban) nor for fumigating grains or citrus fruits (a planned phase-out during 1984). This prompted some states, notably Florida, to test food already in distribution for EDB. They found enough in some products to pull them off shelves and to start what was to become the great EDB scare of 1984. By February the EPA had taken new emergency measures, although hampered by a 1956 law that exempted grains from any EDB limits. In 1956, the belief was that EDB quickly evaporated from stored grains after it did its job, so there was no need for the limits. This turned out to be wrong; EDB does not naturally evaporate in conditions under which grain is stored.

By Valentine's Day, the Food and Drug Administration was taking steps to permit radiation by gamma rays to replace EDB in many applications, causing new concerns among the public. A few days later, the EPA essentially limited the remaining exposures to EDB in the United States, including EDB found in imported fruit (where extraordinary high levels had been found).

The action in February seemed to end the public concern, although many of the possible replacements for EDB are believed to be equally dangerous.

INTERNATIONAL RADIATION ACCIDENT

On January 16, 1984, a truck loaded with steel rods turned into Los Alamos National Laboratory, and the alarm went off. A unique detection device had determined that the steel in the truck was radioactive.

Tracing the rods back to their source showed that they were made from scrap from a junkyard in Juarez, Mexico. Further investigation revealed that steel rods for reinforcing concrete and steel tablelegs made from the scrap—all radioactive—had been shipped all over the United States.

At the same time, the junkyard itself was so radioactive that two of the Mexican workers developed wounds and blisters on their hands and feet. The junkyard employed 60 workers, all of whom were exposed to some radiation.

What had caused this disaster? A machine for treating cancer with radiation, built before 1963, was decommissioned by a hospital in Lubbock, Texas, and sold. By 1977 it had been resold to Centro Medico in Juarez, although the clinic failed to use it. After six years in a warehouse, it was apparently stolen and on December 6, 1983, was resold as scrap. At the junkyard, the machine was cut apart, releasing 6010 pellets, each about a millimeter (0.04 inch) in diameter, each containing 70 microcuries of radioactive cobalt-60. The pellets bounced all over the junkyard, they lodged in various places around the truck—which was so radioactive on March 7 that it tripped geiger counters from 100 meters (300 yards away)—and they were carried around Juarez in the tires of vehicles that drove in and out of the yard.

MERCURY IN JAPAN

One of the first environmental crises caused by toxic wastes was Minamata disease in Japan. This mysterious disease of people living near Minamata Bay in southern Japan paralyzed, crippled, and killed hundreds of people, many of them children. The cause was traced to mercury discharged from a chemical plant into the bay. The mercury traveled up the food chain and underwent "biological magnification;" that is, the mercury was concentrated first in small fish, and then even more in the large fish that ate the small ones. The top of that food chain was the human population around the bay, whose traditional Japanese diet relied heavily on fish. To this date, the Minamata disaster remains the only fully-documented instance of toxic wastes directly causing human deaths.

Now Japan has another worry—and since it involves mercury once again, the Japanese people are especially sensitive to the issue. The problem—batteries. The consumer orientation of many Japanese products, ranging from battery-powered wristwatches to battery-powered television sets, has led to two kinds of problems. One is that infants tend to swallow the small, pill-shaped batteries used in watches and cameras. As many as 5,000 cases of battery ingestion occur each year. Although no deaths have resulted, a swallowed battery could lead to perforation of internal organs and also to mercury poisoning, since most batteries contain mercury.

Used batteries are thrown away with the garbage. Eventually, they wind up in landfills, which, because of the shortage of land in Japan, always get converted to other uses after they become full. The fear is that these landfills are becoming mercury time bombs.

Not all batteries use mercury, but switching to the other popular type of battery, nickel-cadmium, will not solve the problem. Cadmium is virtually as toxic as mercury.

Acid Rain: Clearing the Air

Nineteen eighty-three is likely to be remembered as the year in which the long acid rain controversy in the United States—"the acid rain game"—finally reached a turning point. It has been a game of sorts. In recent years, speculations, dire predictions, and new findings about acid rain have been volleyed back and forth between those who want federal regulations to control pollutants that form acid rain and those who oppose such controls.

Those in favor of regulations painted vivid and headline-grabbing pictures of the ecological damage caused by acid rain—dying lakes, melted buildings, and dead trees. Invariably it seemed we had only a day or so left to act before the whole environment dissolved around us. Meanwhile, opponents charged that it could cost up to $100,000,000,000 to install antipollution equipment. They also claimed that the acid rain phenomenon was far from being understood well enough to justify such an expense.

The Reagan administration, which was on record against new federal regulatory programs, sat on the fence by ordering more studies. There was some justification for this inaction. The scientific community, which had been the source of the controversy in the first place, was divided over the issue. That was largely because past studies still had not answered important questions about what nearly everyone agreed was a broad and complex problem.

But June 1983 turned out to be a great month for scientific certainty. Three separate government reports were issued that month, and while they addressed different aspects of the acid rain problem, they all favored concerted action to reduce levels of harmful pollutants that cause acid rain. A barrage of acid rain publicity of this magnitude from inside the government could hardly have been ignored, and by year's end it appeared that the Reagan administration had been forced to abandon its "wait-and-study" position. The Environmental Protection Agency, however, while promising imminent action, stalled until well into 1984 without announcing a program for change.

Early in 1984 new evidence came to light. While they could not absolutely point to acid rain as the culprit, scientists studying tree rings reported an unprecedented slowing of growth in various species of evergreens (and somewhat less of a slowdown in some hardwood species) in the United States. The only similar phenomenon known was in Europe, where it preceded a massive dieback in

European forests. The change in the U.S. trees' growth rates, as far south as Alabama, was occurring in a broader band than the dieback in lakes, which had held the early warning of problems. In fact, the area where the problem existed coincided nicely with a map produced by the Canadian government showing the extent of acid rain in North America. Later in the year a similar problem in the Ohio River Valley was attributed to the pollutants that also form acid rain.

While the consensus among government scientists that acid rain is a serious problem is new, the investigation of acid rain is not. Acid rain was discovered in England in the nineteenth century, and was first described (with remarkable precision) by Robert Angus Smith in 1872. He wrote, "Acidity is caused almost entirely by sulphuric acid." It was rediscovered in the 1950s by the Canadian ecologist Eville Gorham, who was studying peat bogs in Great Britain. His 1955 paper on the subject attracted little attention, although he pointed the finger of blame for acid rain at industry. Finally, in 1961, the Swedish scientist Svante Odén rediscovered acid rain, this time in Scandinavia. Odén did not publish his findings in a scientific journal. Instead, he took them to a Swedish newspaper, where the report attracted the attention that Gorham's paper had failed to engender. Since then, the concern about acid rain has been the subject of increasingly loud public debate.

Over the years scientists have broadened their research to include all forms of acid precipitation—rain, snow, fog, and "dry" deposition of acidic particles. Pollutants—especially sulfur and nitrogen oxides—from burning fossil fuels were definitely linked to the formation of acid rain in the earliest studies by Smith, Gorham, and Odén. But until recently drawing broad conclusions about acid rain was difficult, and assertions were easily challenged. This was partly because the earlier studies were so limited in scope, a problem that was almost inevitable considering the sheer number of processes involved in acid rain formation.

Unlike the localized clouds of smog that engulf cities such as Los Angeles, London, or New York, acid rain is regional in nature. Gaseous pollutants from one place, for instance, may rise hundreds or thousands of meters into the atmosphere and travel hundreds or thousands of kilometers before falling to the ground as acid rain. Acid rain also involves a whole series of physical mechanisms and chemical processes, including emission of pollutants, acid rain formation and precipitation, acidification of local ecologies, and wind and other weather phenomena that transport acid rain over wide areas. Even after years of research, the recently reported government studies notwithstanding, these processes are not yet fully understood.

Natural and Artificial Acid Rain

One of the principal problems confronting scientists studying acid rain is sorting out naturally produced acidic compounds from those that result from manmade pollutants. Unpolluted rainwater tends to be slightly acidic naturally. To get an idea of the magnitude of manmade acid rain, as well as to determine whether the problem is getting worse, considerable research has been done to find the natural levels of acids in rain, in lakes, and in soils.

The acidity of rainwater is determined by the amount of hydrogen ions (H^+) in the solution, and acidity (or alkalinity) is measured by the pH factor. The pH of a chemical is expressed in numbers ranging from 0 to 14; 7 is neutral, values above 7 are alkaline, and values below 7 are acidic. The pH scale is logarithmic: a decrease of 1 on the scale equals a change of 10 times the magnitude. Thus 6 is 10 times more acidic than 7, 5 is 100 times more acidic, and 4 is 1,000 times more acidic.

Rainwater begins with water vapor. The two major sources of water vapor in the atmosphere—evaporation from the Earth's surface and transpiration from plants—provide a continuous supply of moisture that is in effect distilled. The water vapor is pure H_2O and has a neutral pH. However, on entering the atmosphere, the water vapor condenses on particulate matter and mixes with other gases present. Here CO_2 plays an important role: when it is dissolved in water droplets, it forms carbonic acid (H_2CO_3). This in turn dissociates somewhat to form bicarbonate ions (HCO_3^-) and hydrogen ions (H^+).

Because of the interaction of water droplets, CO_2, and other naturally occurring substances, the pH of unpolluted rainwater is usually found to be 5.65, which is slightly acidic. Rainwater that has a pH value lower than this is taken to be acid rain, but natural sources of contaminants can be responsible for raising or lowering the pH and making the rain more acidic or alkaline than normal.

Soil particles, for example, are carried up into the air as wind-borne dust and can shift the pH one way or the other. Dust particles in the Western and Midwestern United States tend to be alkaline and are responsible for raising the pH above 6 in many locales. In the East, however, soil particles tend to be slightly acidic and help lower the pH. In addition, wind-borne seaspray is the source of a wide range of natural contaminants in the atmosphere near the coasts; it thus affects the pH of rainwater in these areas. Among the contaminants from seaspray are alkaline substances such as sodium, magnesium, calcium, and potassium; and acid-forming chlorides and sulfates.

Average pH at Remote Sites

Location	pH	Range	Years
Mauna Loa, Hawaii	5.30	3.84–6.69	1973–1976
Pago Pago, Samoa	5.72	4.74–7.44	1973–1976
Prince Christian Sound, Greenland	5.73	4.70–6.81	1974–1976
Valentia Island, Ireland	5.43	4.2–6.8	1973–1976
Glacier Park, Montana	5.78	2.60–7.10	1972–1976
Pendleton, Oregon	5.90	4.67–7.60	1972–1976
Ft. Simpson, NWT, Canada	6.27	5.22–7.14	1974–1976
Mauna Loa, Hawaii	5.84	4.76–6.60	1976–1978
Pago Pago, Samoa	6.00	5.55–6.51	1977–1978

Scientists have focused on sulfur and nitrogen oxides, the two major types of contaminants that contribute to acid rain, in their efforts to determine the "background" levels of natural pollutants. Sulfur dioxide (SO_2), hydrogen sulfide (H_2S), nitric oxide (NO), nitrogen dioxide (NO_2), and a host of other compounds of sulfur and nitrogen are variously released into the atmosphere by volcanic eruptions (and venting), seaspray, decomposition of organic matter by bacteria, reduction of sulfates in the soil, and even forest fires and lightning.

That is not to say that people do not make a sizable contribution to the problem. Burning of coal and oil for heat and electric power, automobile and truck exhausts, and heavy industry (smelting, petroleum refining, chemical manufacturing, and so forth) all are responsible for dumping massive amounts of sulfur and nitrogen oxides into the atmosphere each year.

Estimates of people-produced emissions in the United States for one year showed fuel combustion (particularly coal) by utilities produced over 60 percent of the total of sulfur oxides released into the atmosphere. Industrial processes gave off another 18 percent, industrial fuel burning produced almost 13 percent, and autos and trucks just under 3 percent. As for nitrogen oxides emitted during the same period, autos and trucks produced over 40 percent, utility fuel combustion 35 percent, industrial fuel combustion 12 percent, and industrial processes under 5 percent.

Estimates of the relative amounts of natural to industrial sulfur and nitrogen oxide emitted on a worldwide basis are rough approximations at best, but the industrial portion has been steadily revised upward over the years. Current estimates of total worldwide release of sulfur oxides are two-thirds people produced to one-third natural emissions; in the 1970s natural sources were assigned two-thirds the share. In the United States alone, about 30,000,000 tons of sulfur dioxide are released each year. A sharp rise in total emissions that began in the 1950s appeared to have leveled off by the 1970s.

The relationship of natural and industrial emissions of nitrogen oxides on a worldwide basis is far less certain, though it appears that natural sources contribute the majority of these contaminants to the atmosphere. Studies have variously estimated that natural sources account for anywhere from 50 percent to 95 percent of the annual nitrogen emissions worldwide. Nitrogen oxide emissions from industrial sources in the United States range around 25,000,000 tons annually, and the total has been rising steadily for years.

MAKING IT RAIN ACID

Release of natural or industrial sulfur and nitrogen oxides into the atmosphere is only the first stage in the creation of acid rain. These precursors, as the sulfur and nitrogen pollutants are called, undergo complex chemical processes while still in the air, until finally a part of the total volume in a cloud is transformed into sulfates and nitrates. Of these sulfuric and nitric acids are the most harmful.

The sulfates in acid rain are believed to be the end products of oxidation reactions of various types. One such reaction is the catalytic oxidation of sulfur oxides in water droplets by transition metals (vanadium pentoxide is an important one). Another reaction involves catalytic oxidation of sulfur dioxide on collision with carbon or other particulate matter. Yet another type of oxidation reaction takes place in heavily polluted air (such as that in a smokestack plume). This involves gas-phase oxidation of sulfur dioxide after collision with such strong oxidizing radicals as HO^*, $HO_2{}^*$, and $CH_3O_2^*$. These radicals are produced in the daytime by photo-oxidation of hydrocarbon-nitrogen oxide mixtures, and they can lead to oxidation of sulfur oxides both in the air and diluted in water droplets.

The transformation of nitrogen oxides is far less well understood. The most important reactions involve nitric oxide and nitrogen dioxide, which, after being oxidized in a complex chemical process, emerge as nitrates. Other oxides of nitrogen are not usually present in sufficient quantities to contribute significantly to the creation of nitrates.

It is possible, of course, for some of these oxidizing reactions to take place in relatively dry atmosphere. In this case dry sulfates and nitrate particles gradually fall to the ground to collect on surfaces or mix with the soil. This is called "dry deposition."

In the presence of water vapor, though, sulfate, nitrate, and other particles (such as dust) become involved in the atmospheric water condensation process. The particles provide the nucleus on which a water droplet (or snowflake) forms. During droplet growth, still other natural and industrial contaminants are absorbed and undergo chemical transformation.

The droplet in effect becomes a chemical factory, and its acidity or alkalinity is only a measure of the balance of chemical compounds present in the surrounding atmosphere. Where concentrations of sulfates and nitrates are low, alkaline substances—such as naturally occurring bicarbonates, mineral particles (as in the Western United States), or ammonia (from decomposition of organic matter)—may neutralize newly formed acids. But with high concentrations of acid-forming pollutants, these buffers do not have sufficient strength to alter the pH of the droplet.

Though the difficulties of unraveling the chemical processes in a moving cloud of pollution hundreds of feet above Earth's surface are enormous, scientists have identified some distinct patterns. For one, rates of transformation of sulfur oxides to sulfates in polluted atmospheres tend to be higher during the daytime (reflecting increased photo-oxidation processes), and in the summer season in general. Also, the longer sulfur and nitrogen oxides remain in the atmosphere, the more likely they are to be rained out as acid rain. In fact, the broad pattern of deposition of sulfates and nitrates tends to be in dry particulate form near a smokestack or other polluting source, and as acid rain at greater distances.

EFFECTS OF ACID RAIN

Once acid rain falls to the ground, it produces a wide variety of effects, depending on the strength of the acid content and the nature of the local ecology. Some of these effects are dramatic, such as the killing of fish in the high Adirondack lakes or the acid damage to stone buildings and statuary, and have been used to rally public opinion in the acid rain controversy. But while the more obvious types of ecological damage caused by acid rain are fairly well understood, the nature and extent of more subtle ecological changes are still under study.

An important factor in the extent of ecological damage in any given area is the nature of the soil itself. The fairly alkaline soils of the Western United States, for example, are able to neutralize much higher levels of acidity in rain than those of other areas.

On the other hand, the Northeastern United States is an area of high sensitivity to acidification. That is largely because of the geology of the region; areas with types of bedrock based on silica compounds, such as granite or quartzite, have soft water (low buffering capacity) because the rock resists weathering. In addition, areas with heavy forest cover tend to have a low capacity for neutralizing acids. In these areas, lakes and streams may quickly fall victim to acidification.

Northern sites are more vulnerable than southern ones in many cases because the glaciers of the last ice age have swept away much of the soil cover and such sedimentary rocks as limestone. The remaining thin soil and igneous or metamorphic rock cannot buffer the rain as well as in regions not affected by recent glaciation.

Extensive studies have shown conclusively that acidification has devastating effects on the ecologies of lakes and streams. In the acid-sensitive Adirondack region of New York, over 200 lakes have become ecologically dead because of acidification. Thousands of lakes in Scandinavia have suffered a similar fate and many others are threatened. In addition, fish have been reported disappearing in Pennsylvania streams, as well as in streams in parts of Canada.

Changes of acid levels in lakes and streams can occur in two ways. One is by acid shock, when acid-bearing ice and snow suddenly melt during the spring. The other is by slow increase as the acid content of the local watershed rises in response to continued precipitation of acidic contaminants.

In both cases, it is the surrounding soil and bedrock that determines the ability of a lake to neutralize acids and maintain a life-sustaining pH level. In ecologically sensitive areas such as the Northeast of the United States, the soil and bedrock do not usually contribute alkaline substances to rainwater as it filters through the soil and into a lake. Consequently, lakes contain soft water, and bicarbonate ions provide most of the limited buffering capacity of these lakes. Then, as the precipitation becomes more acidic, neither the soil nor the lake itself is capable of neutralizing incoming water.

The effect that the composition of surrounding soils can have in preventing acidification of lakes is amply illustrated in certain areas of Maine. Despite the fact that the bedrock in these areas is granite (low-buffering capacity), and that there is continuing precipitation with an average pH of 4.3, the lakes have not become acidified. Scientists believe that this is because soils overlying the bedrock around the lakes contain limestone and marine clay. Both these substances have a high-buffering capacity.

Acid shock occurs in lakes and streams with low-buffering capacity when acid-bearing snows melt. The sudden influx of acidic substances overcomes the limited buffering capacity in acid-sensitive lakes and results in sharp swings in acid levels. Some buffering capacity is restored in the lake, however, as acids become diluted after the thaw. But when the pH of a lake is lowered gradually over a long period of time, bicarbonates and other natural buffering agents are permanently depleted and are replaced by sulfates. Thereafter the pH remains below 5 permanently, and aquatic life is seriously threatened.

Fish are very sensitive to change in pH, and their disappearance from acidified lakes and streams was observed in Scandinavia as early as the 1930s. As a result of extensive studies, it has been shown that changing acid levels affect fish in various ways. Some species are more resistant than others to continued concentrations of acids (rainbow trout disappear at pH 6 to 5, while pike can survive in pH 4 to 3.5). However, acid shock can kill off species of fish even though pH levels remain

Effects of pH Changes on Fish

pH	Effects
11.5–11.0	Lethal to all fish.
11.5–10.5	Lethal to salmonids; lethal to carp, tench, goldfish, and pike if prolonged.
10.5–10.0	Roach, salmonids survive short periods, but lethal if prolonged.
10.0–9.5	Slowly lethal to salmonids.
9.5–9.0	Harmful to salmonids, perch if persistent.
9.0–6.5	Harmless to most fish.
6.5–6.0	Significant reductions in egg hatchability and growth in brook trout under continued exposure.
6.0–5.0	Rainbow trout do not occur. Small populations of relatively few fish species found. Fathead minnow spawning reduced. Molluscs rare. Declines in a salmonid fishery can be expected. High aluminum concentrations may be present in certain waters causing fish toxicity.
5.0–4.5	Harmful to salmonid eggs and fry; harmful to common carp
4.5–4.0	Harmful to salmonids, tench, bream, roach, goldfish, common carp; resistance increases with age. Pike can breed, but perch, bream, and roach cannot.
4.0–3.5	Lethal to salmonids. Roach, tench, perch, pike survive.
3.5–3.0	Toxic to most fish; some plants and invertebrates survive.

Source: Department of Energy Report

above the minimum for their survival. When the pH is gradually lowered, adult fish may continue to live, but they cease to reproduce at pH levels well above minimums for survival, and fish populations quickly dwindle.

In addition, acid rain passing through soil has been found to leach out aluminum and other metals, such as mercury (reported in the United States, Canada, and Scandinavia). Aluminum, which is highly toxic to fish, thus finds its way into lakes and streams and contributes to the killing of fish populations.

Other aquatic life suffers in a manner similar to fish populations when the pH of the water is lowered. Frogs and other amphibians diminish in number, being unable to reproduce in pH lower than anywhere from 6 to about 4, depending on the species. Large numbers of species of green, golden brown, and blue-green algae quickly disappear as the pH drops. In addition to killing off these food producing

Effects of pH Changes on Aquatic Organisms

pH	Effect
8.0–6.0	Long-term changes of less than 0.5 pH units in the range 8.0 to 6.0 are very likely to alter the biotic composition of freshwaters to some degree. The significance of these slight changes is, however, not great.
8.0–6.0	A decrease of 0.5 to 1.0 pH units in the range 8.0 to 6.0 may cause detectable alterations in community composition. Productivity of competing organisms will vary. Some species will be eliminated.
6.0–5.5	Decreasing pH from 6.0 to 5.5 will cause a reduction in species numbers and, among remaining species, significant alterations in ability to withstand stress. Reproduction of some salamander species is impaired.
5.5–5.0	Below pH 5.5, numbers and diversity of species will be reduced. Many species will be eliminated. Crustacean zooplankton, phytoplankton, molluscs, amphipods, most mayfly species, and many stone fly species will begin to drop out. In contrast, several pH-tolerant invertebrates will become abundant, especially the air-breathing forms (e.g., Gyrinidae, Notonectidae, Corixidae), those with tough cuticles that prevent ion losses (e.g., Sialis lutaria), and some forms that live within the sediments (Oligochaeta, Chiromomidae, and Tubificidae). Overall, invertebrate biomass will be greatly reduced.
5.0–4.5	Below pH 5.0, decomposition of organic detritus will be severely impaired. Autochthonous and allochthonous debris will accumulate rapidly. Most fish species will be eliminated.
4.5 and below	Below pH 4.5 all of the above changes will be greatly exacerbated. Lower limit for many algal species.

Source: Department of Energy Report

plants, a lowered pH also results in the disappearance of bacteria that decompose dead organic matter (fungi take over this role). By the time the pH has been lowered to the 4 to 3 range, only a few types of plants and invertebrates can survive.

While increased acidity is clearly destructive to aquatic life, its effects on terrestrial plant and animal life are far less certain. Laboratory experiments with extremely strong concentrations of acid rain (pH 3 and below) have caused lesions on leaves of some types of food plants, but there never has been any reported natural occurrences of such damage in the United States. In addition, the experiments showed that damage occurs only under certain conditions, and such factors as acid content of rain, frequency and duration of exposure, intensity of rainfall, and even the size of raindrops all play a part in determining whether serious leaf damage develops.

Although no direct damage can be seen, experiments have shown acid rain can have an effect on yields of some types of food crops, but results are mixed. While a pH of less than 3.5 is considered unlikely in most agricultural regions, one study by the Brookhaven National Institute on Long Island, New York, reported soybean yields were 17 percent less than expected when exposed to acid rains of pH 3.3.

Foliar Injury to Plants Following Exposure to Simulated Acidic Precipitation

Plant	Effect	pH Value
Bean	Foliar aberrations, decrease in growth, foliar lesions, plasmolysis of cells, reduction in dry weight	2.5
Bean	Reduction in dry weight	4.0
Soybean	Necrotic lesions and chloritic areas in leaves	3.0
Bean, sunflower	Foliar lesions	3.1
Sunflower	Foliar lesions	3.4
Mustard greens	Reduction in dry weight	3.0
Broccoli	Reduction in dry weight	3.0
Radishes	Reduction in dry weight	3.0
Spinach	Foliar lesions, reduction in dry weight	3.0
Eastern white pine	Foliar lesions	2.3
Scots pine birch	Foliar necrotic spots	2.5
Yellow birch	Foliar lesions, decrease in growth	3.1
Oak	Bifacial necrosis	3.2
Hybrid poplar	Foliar lesions	3.4

Source: Department of Energy Report

Rainfall of pH 4.1 produced yields 10 percent under what was expected. Radishes, bushbeans, beets, carrots, broccoli, and corn were similarly affected by acid rainfall. While other studies have tended to corroborate reduced crop yields, at least one showed that cauliflower, cabbage, and green pea crops were largely unaffected by even concentrated acid rain.

There is also concern, but as yet only inconclusive evidence, about the ability of acid rain to leach out organic chemicals from the foliage of living plants. Studies have shown increased loss of organic carbon from the tree canopy as a result of exposure to acid rain, and experiments on bean plants revealed that losses of nutrients from the foliage rose as acid concentrations grew stronger. But some scientists believe that leaching only causes the affected plant to increase its nutrient uptake from the soil.

Scientists have also been concerned about the leaching of nutrients from surface soil, because this is where plants get the essential raw materials they need to survive. While soil is not nearly so sensitive to acid rains as lakes, studies of acidified soil have shown increased loss of sodium, potassium, and other substances through leaching. But acid rains do supply nitrates and sulfates, which are two major components of chemical fertilizer. Plants may also cancel out effects of leaching by recycling nutrients. They accomplish this by taking nutrients in through their root systems deep in the ground and by depositing nutrient-rich dead organic material on the soil surface. Further complicating the debate on acidification of soil is the contention the agricultural uses of fertilizer and other chemicals far outweigh the effects of acid rain.

A more serious problem in acidified soil is the fact that metallic elements are released from normally stable compounds by low pH. Some studies have found that a buildup of metals, especially aluminum, does occur and that it could reach levels that are toxic to plants.

The effect of acidification on soil microorganisms, many of which are necessary for supporting plant life, has been cited as an area of major concern, even though experiments to date have been inconclusive. Tests generally have been conducted over a short period of time in which soils were exposed to simulated rains with very high concentrations of acids. Even under these harsh conditions, however, effects on microbes varied from one soil to another. In one case, for instance, nitrogen fixation by microbes was inhibited at a given low pH, while in another—at the same pH—there was no change at all. Of potentially greater concern is the effect of toxic metals, released by acids in the soil, on soil microbes.

There are other documented and suspected effects of acid rain, but none approaches the magnitude of damage done to ecologically sensitive lakes. Trees, for example, may in some cases be harmed by exposure to high concentrations of acid rain, with some species especially vulnerable. Red spruce trees on Camels Hump in Vermont are being killed off by acid rains there, and studies of conifers in New Jersey's Pine Barrens have indicated a decline in growth rates for shortleaf, loblolly, and pitch pine.

Studies of the changes in tree growth in North America released in February of 1984 produced conclusive evidence, based on 7,000 tree-ring samples, of inhibited tree growth. Cautious scientists are unsure, however, whether acid rain alone is the cause, since other pollutants could be a factor in the decline in growth, which began as early as 1956. Still, some scientists believe that the release of toxic metals (aluminum and potassium) in acidified soils poses the threat to trees. These substances are thought to damage fine root systems of trees. Other scientists point to acid destruction of a fungus normally associated with the root systems of loblolly pine and red spruce. Not only had laboratory tests shown that acid killed the fungus, and that the fungus was needed for tree growth, but also, when Robert I. Bruck checked the tree population on the top of Mount Mitchell, the highest peak on the Eastern Seaboard, he found dead red spruce along with dead or dying fungus on the spruces' roots. (The problem with trees has generally been more serious at higher altitudes than nearer sea level.)

The potential for public health problems from acid rain also exists, but to date the studies that have been conducted have not confirmed speculations that freshwater supplies could be contaminated by high levels of metallic elements released from the soil by acid rain. While levels of aluminum, zinc, copper, manganese, nickel, mercury, and cadmium have been found to be higher in acidified lakes and streams, they remain below public health standards for drinking water.

Of more immediate concern in some locales in New England was the fact that slightly acidic water was leaching copper and lead from pipes in the public water supply system. The problem was easily corrected by treating the water, once the leaching was detected.

SPREADING IT AROUND

The most serious consequences of acid rain today are still confined to fairly localized regions. Why, then, did such a heated controversy over acid rain develop in the United States in the late 1970s?

The primary reason was the energy crisis and the decision by the United States government to reduce dependence on foreign oil imports by encouraging utilities and industrial users to switch from oil to coal. Though many questions about acid rain needed further study, the government's decision (however inescapable) brought matters to a head. This is because it was well known that, for a given energy output, burning coal produces far more sulfur and nitrogen oxides—the two main ingredients of acid rain—than oil. Simply increasing the number of electrical generating plants and industrial boilers using coal, not to mention any new facilities to meet increased needs of the future, meant thicker clouds of pollutants and even stronger acid concentrations in the rain.

On the other hand, the cost of preventing sulfur and nitrogen oxides from entering the atmosphere loomed as the major argument against cleaning up the air, especially considering the lack of certainty about many aspects of the problem. With the utility industry estimating that it would cost $100,000,000,000 over 25 years to install devices to clean up smokestack exhausts, the debate was on. Many studies were undertaken, and one important focus was on monitoring. The aim was to find out where the acid rain was, how it got there, and whether the problem was indeed getting worse.

Earlier studies of precipitation in the United States had provided only indications of a general pattern of the spread of acid rain. The earliest study was conducted in 1955–56; it included precipitation samples taken at 24 locations in the Eastern United States. Analysis of that data showed that a large part of the Northeast was already being subjected to acid rain. A government study of precipitation at 30 locales around the United States (seventeen in the East) was conducted from 1964 to 1966; it showed precipitation west of the Mississippi to have a pH generally above 6.5.

In what was often considered a pivotal study, scientists from 1972 to 1974 pieced together data from earlier studies and extrapolated a set of expected pH values for various monitoring sites. They then compared these with actual pH values obtained from 1972 to 1974. They found acid rain had spread over a still wider area and that rainfall had become more acidic. While this study was used to support claims that acid rain was indeed getting worse, it has been seriously challenged. For example, sampling before 1972 included only wet deposition, not wet and dry. This alone could account for increased acid deposition. In fact, some critics have charged that analysis of the data shows no evidence of trends toward lowered pH, or westward extension of the area affected by acid rain. Even findings at Hubbard Brook, New Hampshire, where acid rain does occur and monitoring has been going on continuously since the 1960s, were said in 1980 to indicate no clear, long-term trend toward increasing acidity.

But the broad pattern of the acid rain phenomena in the United States does seem fairly clear. In past decades, heavy industry in the Midwest (and more recently big power plants in the West) steadily increased oil and coal burning activities and pumped millions of tons of sulfur and nitrogen oxides into the atmosphere. The prevailing pattern of wind and weather in the United States is west to east; acid-forming pollutants were thus carried up into the Northeast (and into lower Canada). All along the way acid rain in relatively strong concentrations does occur, and for years it has washed down upon Pennsylvania, New York, and, to one degree or another, all the New England states.

Ironically, this spread of pollutants across whole regions seems to have been aided by a measure designed to alleviate local pollution problems—namely, tall smokestacks. Coal-burning power generating plants began resorting to unusually tall smokestacks in the 1960s as a means of dispersing pollutants over a wider area (without actually reducing emissions). The average height of the stacks in 1960 was just over 73 meters (240 feet). But by 1969 the average had jumped to over 182 meters (600 feet).

The largest emitter of sulfur dioxide in the United States is the General James M. Gavin plant of the Ohio Electric Company near Cheshire, Ohio. Right next to it is the fifteenth largest producer of the gas. Between them, the two plants produce more than 540,000 metric tons of sulfur dioxide each year. The smokestack on the Gavin plant, completed in 1975, is 336 meters (1,103 feet) high. Because it was started in 1971, the Gavin plant is not required to have any means of removing the sulfur dioxide from the gases released by the coal it burns.

What to Do About Acid Rain

The three government reports issued in June 1983 represented the culmination of efforts to analyze existing data and current studies on acid rain, with the aim of developing future government policy on the problem. The first report, issued early in June by a federal interagency task force, confirmed what was already widely held—that industrial pollution was the major cause of acid rain in the Eastern United States.

The White House Office of Science and Technology issued a preliminary summary of its findings a few weeks later. This report was largely concerned with the effects of acid rain and expressed deep concern over the possibility of killing off microorganisms in the soil. Noting that these microbes are essential to the food chain of all living things, the report went on to recommend strong and immediate action to reduce sulfur emissions into the atmosphere.

The National Academy of Sciences issued its report at the end of June. By far the most comprehensive of the studies, it came to a few important conclusions. First, it found that 90 to 95 percent of the acid rain in the Northeast was caused by industrial pollutants. Second, it said that a reduction in sulfur emissions from smokestacks would result in a directly proportional reduction in acid precipitation. Third, it pointed out the regional character of the acid rain problem and estimated that formation and deposition occurs over an area that is perhaps as large as 1,000,000 square kilometers (385,000 square miles).

Thus, the report left open the question of identifying the exact sources of pollutants and for that matter said that it was not yet possible to predict what areas would benefit from a reduction of emissions at any particular locale. The number of pollution sources and the mixing area itself were simply too large to be any more specific. Still, a number of controversial points in the acid rain debate had been settled, and this cleared the way for action to reduce sulfur dioxide emissions, the major contributor to acid rain.

By far the most effective—and most expensive—method to clean up smokestack emissions is flue-gas desulfurization. This method removes up to 90 percent of the sulfur dioxide by passing smokestack gases through a limestone slurry. Sulfur dioxide in this process is converted into gypsum (calcium sulfate) and another compound, calcium sulfite. The major problem with this approach is the cost of installing the equipment in existing coal-burning facilities. While new plants could easily be designed with flue-gas desulfurization units, retrofitting old plants can cost up to $300,000,000.

A cheaper and less effective method is the limestone injection multistage burner. Here limestone is introduced directly into the firebox while the coal is burning. It combines chemically with sulfur compounds present to become calcium sulfate, and thereby reduces emissions of sulfur dioxide by about 50 to 60 percent.

Two other methods are directed at the source of the problem—the sulfur content of the coal itself. One approach, coal washing, involves rinsing the coal before it is burned to remove soluble sulfates. The other approach would require utilities to use coal with low sulfur content during the summer season when acid rain is most serious.

Whatever method, or combination of them, is utilized to lower sulfur emissions, there is little doubt that it will be costly. Since utility companies will have to bear most of the cost, that will mean only one thing: the electricity bill is going to go up.

WORLD WAR II BAT BOMBS—AMERICAN MILITARY PLANNERS WERE ACTUALLY SERIOUS

In the desperate days that followed the Japanese attack on Pearl Harbor, the United States military was probably willing to consider just about any idea for new weapons. Then along came a Pennsylvania dental surgeon named Lytle S. Adams with a zany plan to use bats, armed with incendiary devices, to bomb Japan (and thereby bring the empire to its knees). President Franklin Roosevelt listened to the idea and approved further investigation; a respected zoologist teamed up with Dr. Adams; and the wartime National Defense Research Committee (NDRC) and the Army Chemical Warfare Service became involved in the project.

The plan, as it evolved, was to attach to the bats small napalm incendiary devices with fifteen-hour fuses. The bats would then be chilled to 10°C (50°F) to force them to hibernate. United States bombers would transport the hibernating bats, packed in special containers, over Japan. Once dropped from the plane, the containers were designed to fall slowly enough to allow the bats to warm up and fly away before hitting the ground. The bats would then make their way into houses and other buildings, and there gnaw through the retaining bands that held the bombs to their bodies. Hours later the bombs would go off, setting fire to the buildings.

Some sense of the fundamentally madcap nature of the project surfaced after the first real test of bat bombs at Muroc Lake, California, in 1943. The dummy bombs proved to be too heavy for the bats, and the slow descent container fell too fast, hitting the ground and injuring many of the still-sleeping bats. But that wasn't all. The few bats that had been armed with live incendiary bombs escaped from their cage on the ground at Muroc Lake airfield. Subsequently, a general's car, hangars, and other buildings at the airfield burst into flames, proving that at least the incendiary device was functioning perfectly.

By the end of 1943, the United States Navy had become involved in the project. At that point Dr. Adams called for a test over the southern California desert using some 10,000 bat bombs. Noting that all of southern California might be endangered, a Navy lieutenant refused to permit the test. After that flap, Dr. Adams was dropped from the project.

The Navy persevered, though, and after further tests came up with a study that showed the bat bombs to be very effective indeed. It said a plane carrying regular incendiary bombs could set up to only 400 fires, whereas one carrying bat bombs might start as many as 4,748 fires. Plans for large-scale production of bat bomb devices were even set up. But in March 1944, before work had begun, the Navy suddenly scrapped the project, citing the unpredictable nature of bats and the time needed to get the bat bombs airborne. Perhaps someone higher up decided porpoises would be more reliable.

MATHEMATICS

More than in any other subject, advances in current mathematics present difficulties for the nonmathematician. In fact, mathematicians often have trouble understanding the developments in mathematics outside of the specific field in which they are specialists. Often, the cause of the difficulty is not the quality of the thought being expressed in new research; it is simply that, more than other subjects, mathematics is highly sequential. To understand one new idea, you may have to have already understood a hundred or more previous ideas.

On the other hand, there are a number of important areas in mathematics that are accessible to nonmathematicians, especially if the reader has taken an amateur's interest in the subject in the past. One of these areas is number theory, the part of mathematics that deals with relations among the whole numbers. As it happens, during the early 1980s, the most important advances that related to practical or theoretical matters in mathematics were concerned with number theory in one form or another. That is not to say that we can present detailed proofs of these results in this book and expect them to be understood by most readers; but the meaning of the results proven in this area is generally accessible to the amateur mathematician at least.

Some of the outstanding problems in number theory were solved in the past few years. Other advances did not involve solving new problems, but involved reaching new plateaus in finding specific numbers. While mathematicians know, for example, that the number of primes is infinite, finding specific large primes or sequences of primes is still of great interest.

As a result of some of the advances in number theory and related work with codes, mathematicians in the early 1980s found their research suddenly of considerable interest to the National Security Agency and even the White House. Since most work in theoretical mathematics is not watched closely by the government, although applied mathematics may be in some cases, this development alarmed the mathematical community.

Scientists from various backgrounds have been working on the question of the origin of mathematical talent. More than other human forms of intelligence, mathematical talent seems to be something possessed by a few persons in the population, while mathematical competence is something that tends to go more closely with intelligence in other fields. Finding a biological explanation for mathematical talent became at least a possibility in the early 1980s.

A number of the new results in the early 1980s were intimately linked to computers. While no computer program has yet proved a new and original mathematical theorem (see "Mathematics in 1984," page 371), many of the important searches for particular types of numbers can be conducted more efficiently by far with a computer than any other way. In fact, in searching for very large numbers, it is difficult to believe that these computations could be accomplished in any way that does not use a computer.

Computers are not, strictly speaking, part of mathematics; but it seems convenient to include computers in mathematics because of the intimate relation computers and mathematics have always had. Before there was word processing, graphics, or games, computers were solving computational problems. Mathematicians, such as Alan Turing, John von Neumann, and John Kemeny, contributed extensively to the theory and practice of computation and machine science.

In the early 1980s, there were many significant advances in computers—technically, theoretically, and socially. The social impact was penetration of the office and home by personal computers. While the personal computers were less powerful and less advanced than their predecessors in many ways, the development of simple software that almost anyone could use without special training was a significant improvement over the giant computers that spoke only to and obeyed only data-processing specialists.

But the computer news that was the most exciting for many people was the increasing possibility that some form of true artificial intelligence would be developed. In fact, even though no system of artificial intelligence has evolved to where workers in the field would agree that it was really "intelligent," great progress was made in the early 1980s. As part of the "fall-out" from that effort, computers that can help people solve many new kinds of problems are in experimental stages. There is no doubt that computers in the future will be both smarter and easier to use as a result of the work on artificial intelligence—whether or not the goal of true artificial intelligence is actually reached.

1980

Finding the "Monster"

Group theory is one of the most important branches of mathematics. It deals with structures that have a few very common properties. These properties are as follows: (1) A combination of two members of a group is another member of the same group. (2) There is a member of the group, called the identity element, that can be combined with every other member without changing the other member. (3) For each member of a group there is another member that can be combined with the first to produce the identity element for the group. (4) Finally, if you combine three members of a group, you get the same result when you combine the first two with the last or the first one with the last two. If a structure has these four properties, it is a group.

Some common groups are the rotations of an equilateral triangle about its center, the integers $(0, +1, -1, +2, -2, \ldots$, and so forth) with the operation of addition, the positive rational numbers (fractions with positive integers for numerators and denominators) with the operation of multiplication, and the kinds of arrangements possible for a set of 5 objects. For example, if you consider the rational numbers with the operation of multiplication as a group, (1) multiplying any two numbers will give another rational number; (2) the number 1 can be multiplied by any other number without changing the number; (3) for each member of the group, such as $\frac{2}{3}$, there is another member—the *reciprocal*—such as $\frac{3}{2}$, so that the product of the number and its reciprocal is 1; and (4) equations such as $(\frac{1}{2} \times \frac{3}{4}) \times \frac{9}{1} = \frac{1}{2} \times (\frac{3}{4} \times \frac{9}{1})$ are always true. If you know the kind of group you are dealing with and the number of members it has, you know a great deal about how it will behave.

One important type of group is called a *finite simple group*. For the past 50 years, the classification of the finite simple groups, which are the building blocks of all groups, has preoccupied a number of mathematicians. The simple groups are to groups what elements are to chemistry in the sense that all the possible groups can be built from the simple groups. The finite simple groups have been put into eighteen different families, such as the cyclic groups or the alternating groups. There are 26 groups, however, that do not belong to any of the families. These are called *sporadic groups*.

Understanding all possible finite simple groups has come slowly, but the task is now essentially complete. One of the last steps was finding a particularly large finite group that had been nicknamed the "monster" by Bernd Fisher and John G. Thompson in 1974. The name was given because of the immense number of elements in the group.

The key to the monster came from examining the Leech lattice, a close-packing of spheres in 24-dimensional space that is almost certainly the densest sphere packing possible for that space. The Leech lattice was constructed in 1965 by John Leech at the University of Glasgow. In 1968, John Horton Conway (the inventor of the game of Sprouts) found the symmetry group of the Leech lattice, which contains

8,315,553,613,086,720,000, or $2^{22} \times 3^9 \times 5^4 \times 7^2 \times 11 \times 13 \times 23$ elements.

In January 1980 Robert L. Griess, Jr. of the University of Michigan used the Leech lattice to construct the "monster". The number of elements in the "monster" group that Griess found is 808,017,424,794,512,875,886,459,904,961,710,757,005,754,368,000, 000,000, which is equal to $2^{46} \times 3^{20} \times 5^9 \times 7^6 \times 11^2 \times 13^3 \times 17 \times 19 \times 23 \times 29 \times 31 \times 41 \times 47 \times 71$.

TURN OF THE CENTURY CONFUSION

As we approach the twenty-first century, the natural question is "When does it start?" Look at it this way: There was no year 0. In fact, Time Zero is a moment from which we begin measuring in either direction. Therefore, the first year in the Common Era or *Anno Domini* was 1 C.E. or A.D. 1, which ended 365 days after Time Zero. After 100 such years, by definition a century, the first century ended at midnight on December 31, 100 C.E. and the second century started on January, 1, 101 C.E.

Through the centuries the same pattern continued, with the century ending in a year ending in 00 and starting in a year ending in 01. Thus, the twenty-first century will start just past midnight on January 1, 2001 C.E. Happy new century!

On a related note, the twenty-first century will arrive a day later than many people expect. The year 2000 C.E. will have 366 days, not 365, even though 2000 ends in 00, and years ending in 00 have not been leap years for the past 300 years. However, to keep the calendar year in as close agreement as possible with the solar year, a leap day is not added in years ending with 00 three out of four times. The actual deviation of the solar year from the calendar year is about 6.25 hours, so each leap year corrects with an error of about an hour. The missing leap years in the 00 years correct for that hour, while the fourth 00 year has a leap year to correct that correction.

Furthermore, the twentieth century will last a few seconds longer than 100 years. Since 1972, periodic "leap seconds" have had to be added to the year to compensate for a slowdown in the earth's rotation. By 1983's end, a total of 13 such "leap seconds" had been added to the century.

Codes, Security, and Factoring

What would happen if a mathematician invented an unbreakable code?

Based on the record so far, another mathematician would find a way to break it. Along the way, the mathematician might find that there were some serious national security questions raised about what could be said in public about the code.

Mathematics and codes tend to go together naturally. One naive view of mathematics is that the whole subject is just a code that makes it easier to think about nature. Anyone who has ever read Poe's short story "The Gold Bug" will remember that mathematical ideas were used to break the code that was central to the story. When the Germans thought they had an unbreakable code in World War II, the British had a great advantage— because the mathematician Alan Turing broke the code. After Turing and his group had solved the problem, the most important thing the British could do to hold onto their advantage was to keep Turing's work secret. Otherwise the Germans would have stopped using the code, and the British might have lost the war. As a result, the British often had secret information on which they could not act, for fear of revealing that they had learned to decipher the German code.

A more contemporary, and still unfinished, story about codes includes a 1980 meeting at the White House in which a mathematician's research came very close to being classified as a national security risk. The story also involves a 1982 session over a few beers in Winnipeg in which a whole new method of factoring large numbers originated in a remark by a computer engineer. There are actually two strands to the story, however. One starts with three mathematicians who suggested a new kind of code in 1976, while the other begins with the first efforts to build machines to factor large numbers in the 1920's.

Factoring a whole number consists of finding smaller whole numbers that, when multiplied, give the original number. For example, 35 can be factored as 5×7 and 24 as $2 \times 2 \times 3$. Numbers that cannot be factored except trivially into themselves and 1 are called *prime*. Many theoretical and practical problems in mathematics and engineering require knowing the factors of numbers.

Small numbers are easily factored. For example, you can divide a number by the primes in order until you reach the largest prime smaller than the square root of the number. Unfortunately, as the numbers get bigger, the time required for performing this operation increases exponentially, which means that the work takes a long time and takes a longer time as the numbers get larger. In fact, even when more efficient methods of factoring are used, the time needed for factoring increases exponentially with the size of the number being factored. Thus, roughly speaking, for every 3 additional digits in a number, the amount of time needed to factor it by a particular method doubles. In the 1920's the first mechanical factoring machine was able to factor numbers up to 20 digits, considerably better than anyone could do by hand.

By way of comparison among methods, factoring the 58-digit number $2^{193} - 1$ by dividing by primes on the fastest computers now available would take 35,000 years of computer time. (Numbers of the form $a^n \pm 1$ are particularly important; also, if a is a small number and n is a large one, the numbers are often very hard to factor.) On the other hand, using the most efficient factoring method available in 1983, the number $2^{193} - 1$ could be factored in 38.3 minutes (it turns out to be 13821503 × 6165444023324834061655 × 14732265321145317331353282383; try that on your hand calculator).

When large computers became available, a group of mathematicians who regularly met to work on factoring numbers of the form $a^n \pm 1$ quickly began to use the new tool. The group was known as the Cunningham Project Table, after A.J. Cunningham who had worked on this problem around the turn of the century, and it met each year in Winnipeg to report on progress. After 15 years, the Cunningham Project people had gone about as far as they could apparently go. In 1982, they decided to proceed with a paper on the results of the project, which amounted to saying that numbers of only about 50 digits could be factored. In the meantime, they had compiled a list of the "Ten Most Wanted Numbers," numbers that they could not factor, often because the numbers had 60 digits or more. The American Mathematical Society published the results as a logical conclusion to the project; some considered it terminated.

What is the connection between factoring and codes? In 1976, three mathematicians proposed that, since large numbers were so difficult to factor, a code could be built that used the factors of a large number as its key. Then, even if the large number was made public, only someone who already knew the factors could decipher the message. In April, 1977, three computer scientists at the Massachusetts Institute of Technology developed an explicit and practical way to exploit this idea. The system is called a Public Key Code or RSA cryptography (after the initials of the developers—Ronald Rivest, Adi Shamir, and Leonard Adleman).

The RSA system was considerably improved in 1980 when Adleman, Robert Rumely, and Carl Pomerance developed a quick computer test of whether or not a number is prime. The Adleman-Rumely-Pomerance test can tell whether or not a fifty-digit number is prime within fifteen seconds. The Public-Key system could then be based on the difference in the ease of telling whether or not a given large number is prime and the difficulty of finding its factors if it is not. To begin with, the people using the code take advantage of the fast system to find two very large prime numbers. Typically, each of the prime numbers should have about fifty digits, which is just beyond the power that the Cunningham group thought a computer program could factor. The product of the two primes will then have about a hundred digits—far beyond the ability of the computer systems available in 1980 to factor. From the product, if you know the two original primes, it is easy to calculate the number of integers that do not have any common factors with the product; this number is called Euler's phi function of the product. (In fact, from the way the product was formed, Euler's phi function is just the product of 1 less than each of the two primes.) Finally, one of the many numbers that has common factors with the phi function is chosen. In fact, almost all numbers will have no common factors with the phi function. The number chosen should be smaller than the phi function. Now you are ready to begin coding. The secret information is either the two original primes or the phi function. The public information is the product and the small number that has no factors in common with the phi function.

Step 1 is to assign to each entity likely to receive a message a large product and a small number with no factors in common with the phi function of that product. Each potential recipient or sender gets a published list showing these two numbers. Privately, the individual phi functions are also given out. Step 2 is to translate the message into numbers according to any simple system: e.g. use 1 for A, 2 for B, or perhaps use the ASCII code common to microcomputers. Step 3 is to break the messages into long blocks of numbers of a convenient size. Step 4 is to use the two numbers for the intended recipient to translate those blocks into other numbers; specifically, to raise the number in each block to the power of the small number, divide it by the product, and take the remainder. The remainders for each block then become the message that is sent. (This would, of course, be nearly impossible by hand, but is an easy task for a computer.)

Now the recipient uses the knowledge of the phi function to decode the message. The message is again broken into blocks of the same size. From the phi function, it is easy to calculate a number so that if each block of numbers is raised to that power, and the power is divided by the original product, the result is the block that was originally encoded. Then the simple code can be used to retrieve the message in the form of letters.

Anyone can obtain all of the numbers involved in this system except for the two original primes or the phi function. However, without knowing either of these, it is almost impossible to decode the message, since it would take a giant computer, such as the superfast Cray, years to factor the product of the two primes. Remember that in 1982 the Cunningham Project mathematicians decided that they would have to give up on factoring numbers with more than about fifty digits and the product of the two primes is chosen to have about a hundred digits.

Although news of the RSA system reached the mathematics community quickly by word of mouth, the formal presentation was scheduled for a conference of the Institute of Electrical and Electronics Engineers. Shortly before the conference, however, a letter from a J. A. Meyer was received pointing out that the presentation might be in violation of Federal laws governing the export of sensitive material to foreign governments. On legal advice, the presentation of the RSA coding system was dropped from the conference. Later, an enterprising reporter established that J. A. Meyer worked for the National Security Agency (NSA), although the NSA denied that it was behind the letter.

It was already apparent that the NSA was determined to do what it could to control advances in coding theory. Its then director, Admiral Bobby Inman, had spoken out against publication of research on cryptography in 1979. In 1980, when Adleman's grant from the National Science Foundation was due to be renewed, he was informed that national security reasons precluded renewing the part of the grant that involved cryptography. A few days later, Admiral Inman called Adleman to offer NSA support for his research. Of course, it could then be classified.

The issue was resolved at the White House meeting shortly thereafter. Although the NSA agreed not to classify the work if it supported it, Adleman was able to get his original grant from the National Science Foundation renewed, which is what he preferred. Since then, there has been an uneasy truce between the NSA and the mathematical cryptography club. At least one professional organization has asked its members to submit all papers on cryptography to the NSA for review before publication, but compliance is voluntary.

Now, back to Winnipeg. In the fall of 1982, some members of the Cunningham Project were having a beer with Tony Warnock, an engineer for Cray Research. The topic of conversation was the impossibility of getting computers to factor numbers larger than fifty digits. Gus Simmons and Marvin Wunderlich were explaining to Warnock the technical problem involved, which is that every time the computer made a small change in a long array of numbers (a vector), it had to change the whole vector. Since the vectors involved are very long, this took a lot of time and effectively made the problem insoluble. Warnock said that such a change was no problem on the Cray, which has a different system for keeping track of vectors than the system used on other, slower computers. In fact, to make a small change in a vector, you need only change the small part and leave the rest untouched. Therefore, instead of taking a lot of time to make the changes, the Cray could make such changes very fast. Furthermore, the Cray is specifically designed for speed in all its operations.

No one had ever thought that the problem was in the way the computer worked. Mathematicians simply assumed that all computers worked the same, but that some were faster than others. While the additional speed of the Cray was well known, and would have helped in factoring somewhat, the difference in vector handling was not known to the mathematicians.

After he got back from Winnepeg, Gus Simmons got his hands on a Cray. With the help of Jim Davis and Diane Holdridge, he found that factoring numbers that had taken 100 computing hours on a normal computer could be done in less than two hours on the Cray. Davis developed methods that were even faster that also used the Cray's special capabilities.

Now that the idea of exploiting the capabilities of a specific computer design has been raised, Wunderlich is also working on the problem, but from a different angle. Some computers use parallel processing instead of performing one computation after another. Wunderlich realized that a parallel processing computer could perform many different trial divisions at once. He has tried out the idea on a couple of computers, but without really getting it going. However, soon he will be able to use the NSA's MPP (Massively Parallel Processor), which performs 16,384 calculations at a time. If Wunderlich can get a program to operate efficiently, he should be able to cut the factoring time to about half of what Davis's algorithm can do on the Cray.

Finally, another group of researchers is building the Georgia Cracker, a special-purpose computer designed specifically to do factoring. If it works, it will be nearly as fast as the NSA's MPP.

Where does this leave the Public-Key codes? In trouble, as long as you start with fifty-digit primes. However, it still appears that the system will work if you start with hundred-digit primes, since that would require the ability to factor numbers of around two hundred digits. At the end of 1983, that seemed to be beyond any reasonable capability of either mathematicians or computers.

1982

The Largest Known Prime Found

In 1982, a Cray Research computer in Chippewa Falls, Wisconsin took every second of the time it was not doing something else to work on a particularly difficult problem, looking for very large prime numbers. The seconds mounted up until, after about 600 hours of computer time, one Saturday the Cray suddenly announced "$2^{86243} - 1$ is prime." The Cray then proceeded to print out all 25,692 digits of the newly found prime number, the largest ever found.

(Prime numbers are natural numbers, such as 1, 2, 3, ..., that have no factors other than themselves and 1. Thus, 2, 3, 5, 7, 11, 13, 17, ... are prime, while 4, 6, 8, 9, 10, 12, 14, 15, 16, ... are not. By convention, 1 is not considered to be prime.)

Prime numbers of the size found in 1982 are far removed from the "simple" fifty-digit or hundred-digit primes used in Public Key Codes. (See the preceding article.) The 600 hours were spent by the Cray doing 80,000,000 multiplications and 80,000,000 additions each second. (The programmer, David Slowinski, used an algorithm developed in 1930, that does not require division, a great help in speed for computers as well as for people, since division is more time-consuming and difficult than the other elementary mathematical operations.)

Slowinski did not try to test every number in sequence. Instead, he based his success on having the computer examine just a single set of numbers, numbers of the form $2^n - 1$, where n is prime. Such numbers, a subset of the ones studied by the Cunningham Project Table (see preceding article), are called Mersenne numbers, after the 17th-century monk Marin Mersenne (1588–1648) who asserted that these numbers were prime if n was 2, 3, 5, 7, 13, 17, 19, 31, 67, 127, and 257, but were not prime if n was some other prime. Actually, Mersenne was wrong on both counts. (The history of mathematics shows that one does not have to be the first discoverer nor correct about the discovery to get it named for you.) Neither $2^{67} - 1$ nor $2^{257} - 1$ are prime; and there are other values of n, including 61, 89, and 107, that do produce prime Mersenne numbers when plugged into $2^n - 1$. In fact, the large prime that Slowinski's program uncovered is the 28th Mersenne prime. Previously, in 1979, he had found the 27th Mersenne prime, a pygmy of only 13,395 digits when compared with the 28th.

The key to finding such large prime numbers lies in the fact that in

1876, Edouard Lucas developed a comparatively fast method for determining whether or not a given Mersenne number is prime. This method, which was refined by Derrick Lehmer in 1930, only works for Mersenne numbers, but it makes it much more feasible to determine if a Mersenne number is prime than if some other number of 25,692 digits is prime.

At last report, early in 1983, Slowinski had his program running on several computers scattered around the country, each one working on a different part of the Mersenne sequence. Even so, the time required to check each successive Mersenne number increases exponentially (the time increases greatly as the numbers involved become larger). Therefore, it may be quite a few more computer hours before a larger prime number is found. That is, if such a prime exists.

Although Euclid proved conclusively more than 2000 years ago, in what is perhaps the most elegant of the early proofs in number theory, that there is no largest prime, it is not known whether or not there is a largest Mersenne prime. (In fact, knowing whether or not the sequence of Mersenne primes is infinite or not would settle some other unsolved problems in mathematics.) Since Slowinski's program is designed to work on Mersenne numbers only, and since it is possible that Slowinski has already discovered the largest possible Mersenne prime, the computers may not come up with another prime. On the other hand, somewhere in the country, one of the computers may suddenly announce a new largest known prime.

AND THE LONGEST π

The number π, which is best known as the ratio of the circumference of a circle to its diameter, has been calculated with increasing accuracy over the ages. In the Bible, the value is assumed to be 3, while the Egyptians progressed to a fraction near 3.16, the Greeks reached fractions as close as 3.14, and the Chinese found a fraction that was accurate to 3.141592. Since decimals were invented, the accuracy has been measured in the number of decimal places. In 1699, π was extended to 71 decimal places and in 1873 to 707 places. But the 1873 calculation, done of course by hand, was found in 1948 to have a mistake in the 528th place.

The next year, 1949, computers were brought in. In 70 hours of computer time, the first electronic digital computer found π to 2037 places. Since then, computers have calculated π to astonishing lengths.

The current record was set by a Japanese supercomputer (similar to the Cray) programmed by Yasumasa Kanada and Y. Tamura in 1982. They calculated π to 4,000,000 digits. The first ten digits of π are 3.141592653 and the last ten known are 3349203390, but, since π is an infinite, nonrepeating decimal, the calculation could be continued in the future.

The Ten Most Interesting Small Numbers

Mathematicians often relate the following "proof" that all numbers are interesting. Consider the least number that is not interesting. Wouldn't that be of some interest?

0	Zero was not originally recognized as a number. The first known reference to 0 was in India around A.D. 900. It was primarily used to hold a place in a number (so that 101 and 11 could be told apart). It is the only number that cannot be used as a divisor.
φ	The Golden Ratio was recognized by the Greeks as producing self-similar divisions of a line segment or self-similar rectangles. It is also the only number from which you can obtain the reciprocal by subtracting 1, since φ is 1.61803398... while the reciprocal, $1 \div \phi$ is 0.61803398... . It is also the limiting ratio of adjacent terms in the Fibonacci series.
e	To twenty decimal places e is 2.71828182845904523536 (but it, like π, is a non-terminating, non-repeating decimal). This number is most commonly first encountered as the base of the natural logarithms, but it shows up all over mathematics.
π	Although best known as the ratio of the circumference of a circle to its diameter, π shows up in mathematics (and in physics) even more than e does. For the decimal value of π see "And the Longest π," page 361.
Feigen-baum's number	If some mathematical relation is such that it alternates between critical values, hovering on the brink of catastrophe, the ratio of the successive critical values will get closer and closer to Feigenbaum's number, which is approximately 4.669201609102990097.
5	Five is interesting in so many ways that some have called it the most interesting number of them all. There are 5 Platonic solids. Equations with exponents of 5 or greater cannot always be solved by algebraic means. According to the Banach-Tarski theorem, a sphere can be disassembled into 5 pieces and be reassembled as a different size (with no overlaps or holes).
6	The smallest perfect number is 6. It is also the smallest number whose cube is itself the sum of three cubes: $6^3 = 3^3 + 4^3 + 5^3$.
9	Nine is interesting in the decimal system, although it would be much less interesting in some other method of writing numbers.
17	Seventeen is the only number that is the sum of four consecutive prime numbers: $17 = 2 + 3 + 5 + 7$, and is the number of possible different ways to cover a plane with symmetric figures.
39	This number is the smallest whole number for which mathematicians have found no interesting property.

1983

Putting Your Primes All in a Row

Much of the progress in mathematics involved prime numbers in one way or another in the early 1980s. One reason is that there are still many unsolved problems connected with prime numbers.

One such instance is the conjecture that there might be an infinite number of twin primes. Twin primes are primes that are exactly two units from each other, such as 11 and 13 or 17 and 19. As you look at larger and larger numbers, pairs of twin primes get rarer. Although it has been known since Euclid's time that there is no largest prime number, there might be a largest pair of twin primes. One result that suggests this is that the series of the reciprocals of the twin primes has a sum; that is, the sum of $\frac{1}{3} + \frac{1}{5} + \frac{1}{7} + \frac{1}{11} + \frac{1}{13} + \frac{1}{17} + \frac{1}{19} + \frac{1}{29} + \frac{1}{31} + \frac{1}{41} + \frac{1}{43} + \frac{1}{59} + \frac{1}{61} + \ldots$ and so forth, exists. The sum of the reciprocals of *all* the primes, however, does not exist. One possible reason that the sum of the reciprocals of the twin primes exists might be that there are only a finite number of twin primes; but there is no guarantee that this is the case.

Since twin primes have peculiar properties, mathematicians have also looked at related phenomena. For example, how many sequences of primes exist that are just one unit apart? Clearly, only one such sequence exists, 2, 3. What is the longest sequence of primes that are exactly two units apart? It is the sequence 3, 5, 7, the only prime triple. Finally, if you allow the primes to be separated by n units, where n is any whole number, what is the longest sequence you can find?

Here is an example of such a sequence that is easily found by inspection of a table of primes: The sequence 5, 11, 17, 23, 29 is five primes long and the constant difference between the primes is six.

Paul A. Pritchard of Cornell University adapted the trick of using the spare milliseconds in the Cornell computer to find such sequences of primes. After the computer scrounged 250 hours of running time, it came up with the current champion. This is a sequence of eighteen prime numbers with a constant difference. The sequence starts with 107,928,278,317. Each prime in the sequence is exactly 9,922,782,870 units from the one before it, so the eighteenth prime in the sequence is 276,615,587,107. Check it out.

Pritchard is not just trying to get into the Guinness *Book of Records*. Number theory predicts how the prime numbers will be distributed and Pritchard is finding sequences of primes to check the theory.

The Unexpected Proof of Mordell's Conjecture

Any mathematical result that is proved from basic principles and definitions may be called a *theorem*. According to one calculation, about 200,000 *new* mathematical theorems are proved each year. Most of these are of interest to a few mathematicians only, those working in the specialized branch of the subject that the theorem is about. And even most of these theorems are quickly forgotten. Once in a while, however, a theorem is proved that instantly excites everyone's imagination, even appearing in the popular science magazines or daily newspapers. Generally speaking, such a theorem is the proof of one of the well-known unsolved conjectures of mathematics. One such example would be the proof in the late 1970s of the four-color conjecture, which is now the theorem that any map on a plane or a sphere can be colored using just four different colors. Both the antiquity of a conjecture that is proved and how easily the result may be explained to nonmathematicians seem to be the main criteria for instant fame.

Yet, in 1983, a young, obscure German mathematician, Gerd Faltings, made news all over the world with the proof of a conjecture only sixty years old and so arcane that many mathematicians outside its special field would have difficulty understanding it. The reasons for Faltings' impact are that the conjecture (now established) is expected to have important corollaries in many widely-separated fields of mathematics; and also that almost no one who had worked on the conjecture expected the problem ever to be solved by anybody. In fact, as recently as 1979, the number theorist Paulo Ribenboim stated that a proof of the conjecture "seems, with good reason, very remote." Faltings' proof—not especially complicated as proofs today go and breaking little new mathematical ground in its methods—was a bolt out of the blue.

The conjecture involves the type of numbers that can be the solutions to certain classes of equations.

In the beginning of Western mathematics, only the natural numbers— 1, 2, 3, ..., and so forth—were known. Followers of the Greek mathematician and philosopher Pythagoras believed that these numbers were all that existed. However, they could accept the ratios of natural numbers, which we know today as fractions such as $\frac{1}{2}$, $\frac{3}{4}$, or $\frac{17}{100}$, since these quantities were derived directly from natural numbers. Although the early Greeks did not work with equations as we know them, their notions were based on geometric ideas that corresponded to the following correct idea about equations: Equations of the form $ax + b = c$, where a, b, and c are natural numbers and b is greater than c, can always be solved as a ratio of natural numbers. The Pythagoreans went further to believe that all equations should have such ratios for their solutions.

This confidence was shattered when the followers of Pythagoras proved that the equivalents of such simple equations as $x^2 = 5$ or $x^2 = 2$ had solutions that, although they were positive numbers, could not be expressed as ratios of natural numbers. Although the Greeks did not deal with negative numbers, the existence of negative solutions was never a problem. Equations such as $x + 5 = 3$ simply had no solution as far as the Greeks were concerned. But the geometric solution of $x^2 = 2$ is the diagonal of a square one unit on a side. The length of such a line clearly exists, but is not the ratio of natural numbers.

A millennium and a half later, after the negative numbers had been accepted, the ratios of two integers—positive or negative whole numbers or 0—came to occupy the place in mathematics of the ratios of two natural numbers. Such ratios are called *rational numbers*. Essentially, the rational numbers are all the fractions, positive or negative, that are equivalent to a fraction that has an integer for a numerator and a nonzero integer for a denominator.

The question of the existence of rational solutions to equations in which the numbers used in forming the equation are all rational turns out to be at the heart of many problems. If there is only a single variable (or "unknown," often represented by x) in an equation, the question of the existence of rational solutions was resolved fairly early. When equations had two variables, say x and y, the question of whether both x and y were rational at the same time proved to be more difficult to resolve. Still, by the end of the nineteenth century, the number of rational solutions had been determined for equations in two variables so long as the exponents, or powers, of the variables were less than 3. Such equations either have no nontrivial rational solutions or they have an infinite number of them. For example, the equation $y = 2x + 3$ has an infinite number of rational solutions; $x = 0$, $y = 3$; $x = \frac{1}{2}$, $y = 4$; $x = -\frac{1}{2}$, $y = 2$; and so forth. An equation such as $x^2 - y^2 = 1$ has only the trivial rational solution $x = \pm 1$, $y = 0$, since it is easy to see that no two integers have squares that are exactly one unit apart except for (positive or negative) 1 and 0. It is a little more difficult to see that this is the only rational solution, although this is the case also. However, the superficially similar equation $x^2 - 2y^2 = 1$ has an infinite number of nontrivial solutions, such as $x = 3$, $y = 2$; and $x = 99$, $y = 70$. With $x^2 - 4y^2 = 1$, you are back to the single trivial solution. As noted, the general problem of equations with the exponents less than 3 has been solved.

Equations in which the exponents can be 3 are a different matter. No one has a general solution to that problem. In some cases, they behave like equations with exponents of 2 and in other cases they behave like equations with higher exponents. These equations, which have graphs that are called *elliptic curves*, did show up, however, in the other remarkable proof of 1983 (See "Perhaps the Most Indirect Proof" on page 367).

In 1922, Lewis J. Mordell guessed that if the exponents of at least one of the two variables in an equation was 4 or greater, the equation would turn out to have only a finite number of rational solutions. This is in contrast to the infinite number of solutions for equations with exponents less than 3.

The approach to solving such problems is unexpected for people unfamiliar with the power of complex numbers in mathematics. Complex numbers are numbers formed by adding a real number (that is, a number that is either rational or irrational; essentially any number that can be represented by a decimal, including infinite decimals) and the product of a real number and the square root of negative 1. The complex numbers form a system all their own. Often it turns out to be easier to solve a problem for complex numbers than for real or for rational numbers. Finding the number of rational solutions to equations has proven to be one of those situations.

When an equation in two variables is graphed using real numbers for the variables, the graph is a line or a curve. If complex numbers are used, however, the graph is a surface. For equations whose exponents are less than 3, the surface is equivalent to a sphere; that is, it has no holes or handles. An elliptic equation, one with powers of 3, has a surface that is equivalent to a doughnut or coffee cup. It has a hole or a handle. If the exponents are greater than 3, there will be two or more such holes or handles. Mordell's original conjecture was that equations whose complex surfaces had two or more holes would have a finite number of rational solutions. Surfaces may have, in addition to holes, places where the surfaces are not smooth, such as sharp corners. To work mathematically with surfaces that are not smooth, mathematicians have to eliminate the "rough spots"—more correctly called *singularities*.

Faltings drew on a great deal of recent work in handling such equations in proving that Mordell had been right all along. In fact, he went slightly farther than Mordell to show that certain other classes of solutions also existed in finite numbers. Although Faltings' proof is not constructive—it does not give a method of finding any of the solutions for a particular equation—nor does it tell how many solutions there are for a particular equation, simply knowing that there are only a finite number of solutions to be found represents a huge step forward.

Perhaps the Most Indirect Proof

Karl Friedrich Gauss (1777–1855) is often cited as the greatest mathematician of all time, although his name is much less well known to the general public than comparable figures in other sciences. One reason for this may be that some of Gauss's best work was in the field of number theory, a part of mathematics that is known for being particularly devoid of applications—although Gauss developed the basis for the solution of many practical problems in other areas of mathematics.

Gauss was such a good mathematician that he rarely conceived of problems that he himself could not solve, but one of his few unsolved problems resisted all efforts until 1983. In 1983, Don Zagier of the University of Maryland and the Max Planck Institute and Benedict Gross of Brown University finally solved it. They succeeded by proving something about a kind of curve that implied something else totally unexpected about the curve, which in turn implied the solution to Gauss's problem about some rather peculiar classes of numbers.

Number theory (sometimes called *arithmetic* by mathematicians, although it is not the familiar arithmetic of elementary school) concerns the relations among the integers, numbers that are either whole numbers (such as 0, 1, 2, and 3) or the negatives of whole numbers. While the integers seem simpler than fractions and decimals, there are many interesting and surprising relationships to be found among them, making their study actually more complicated than the study of apparently more sophisticated sets of numbers. Many of the results of number theory concern numbers that can only be formed as products in one way—as the product of the number itself and 1. These numbers, such as 2, 3, 5, 7, 11, 37, 137, and 1439, are called *prime*. Forming a whole number that is not prime into a product of primes—called *factoring*—can be done in essentially one way. For example, $24 = 3 \times 8$ or 4×6, but 8, 4, and 6 are not primes. The product $2 \times 2 \times 2 \times 3$, or $2^3 \times 3$, is the only way to factor 24 into primes.

Mathematics deals with numbers beyond the whole numbers and their negatives as well. Most people are familiar with the numbers that mathematicians call the *real* numbers, which include all the numbers used in counting, the fractions, and the decimals—including infinite decimals. The complex numbers, numbers formed by adding multiples of the square root of negative 1 to real numbers, form another familiar set of numbers.

Normally a complex number is of the form $x + iy$ where x and y can be any real numbers and i is the square root of negative 1. Gaussian integers are just complex numbers in which both x and y are ordinary integers. It happens that many properties of ordinary integers are preserved in the Gaussian integers; for example, a number can only be factored in one unique way. Gauss also considered a generalization of the Gaussian integers in which x and y were still ordinary integers but i was replaced with the square root of any negative integer, not necessarily negative 1. (Because of various complication in this idea, x and y are allowed to be half integers in certain circumstances.) With this generalization, it was quickly shown that a number can be factored only in one way if the negative integer whose square root is used is either 1, 2, 3, 7, 11, 19, 43, 67, or 163. It proved to be difficult to find if additional negative integers would produce number systems with the unique factorization property.

In the 1950s an amateur mathematician finally showed that those nine negative integers were the only ones that would produce unique factorization (although it took another 20 years before mathematicians accepted his proof). Not all the questions were resolved, however.

Gauss had also suggested that the numbers for which unique factorization failed could be characterized by a number that he called the *class number*, which described how close to unique factorization a set of complex "integers" based on a particular square root of a negative integer was. The nine integers that produced unique factorization were all the ones that had a class number of 1. The mathematicians Harold Stark and Alan Baker found the similar set of integers for the class number 2. But Gauss's problem was as follows: Given any integer k, find the largest integer on which you can base the complex "integers" that will have the class number that has the designated value k.

This problem resisted analysis, although Dorian Goldfeld finally showed in the late 1970s that he could solve the problem if he had just one tool—a particular curve that had certain properties, one with an equation of the form $x^3 + y^3 = z^3$ (called an elliptic curve).

Specifically, Goldfeld called for a function associated with the curve that had an order of vanishing greater than or equal to 3. This was a far-fetched connection to begin with, but Gross and Zagier, who worked on the problem, had to make an equally far-fetched connection; they found that the order-of-vanishing problem could be solved by looking at points on the curve where the coordinates are all rational. Understand that these mathematical objects had not been shown to be connected previously—Gaussian integers were not expected to be related to elliptic curves and rational points. Eventually, Gross and Zagier put together a 300-page proof that the connections were there, solving Gauss's problem about a hundred years after it was proposed.

Hormones, Left-Handedness, and Mathematics

For several years feminist groups have been crusading for a changed perception of girls by mathematics teachers in school. It has long been known that a disproportionate number of boys score far above average on standard mathematics examinations. Some educators suspected a "reverse Pygmalian effect" was the cause of the difference; if teachers did not expect girls to do well, the girls would not do well.

Now there comes persuasive evidence that at least among the most gifted children, the difference between boys and girls may actually be traced to events long before birth. The idea originated with Dr. Norman Geschwind, a Harvard neurologist. He had been documenting certain correlations between apparently unrelated traits: left-handedness, immune system disorders, certain learning disabilities, myopia (nearsightedness), and certain kinds of giftedness. He attributed this constellation of traits to the influence of the male hormone testosterone during prenatal development. If there is a large amount of the hormone in the fetal environment, or if the fetus is unusually sensitive to it, Geschwind theorized, it might alter the functional anatomy of the brain. It has long been known that the different hemispheres of the brain undertake different types of tasks. The left hemisphere is associated with language, for example, while the right hemisphere is associated with shape recognition. The testosterone might make the right hemisphere rather than the left hemisphere the center for language processing and manual dexterity. This would result in a left-handed child. There would also be an increased incidence of immune-system illness because this too is linked with testosterone. Perhaps most interesting of all, the condition might also cause both an increased incidence of developmental disturbance leading to autism, dyslexia (inability to read), and stuttering. At the same time, it might lead to particular giftedness in mathematics, art, or music.

Two researchers at Johns Hopkins, Camilla Benbow and Julian Stanley, have now added considerable interesting evidence to support Geschwind's hypothesis. They have been studying the scores of a large group of thirteen-year-olds who take the College Board's Scholastic Aptitude Test (SAT). These children take the test hoping to join special mathematics programs for the gifted. When their first results were announced in 1980, they claimed that the test results showed four times as many mathematically gifted boys as girls. Predictably, the reaction of many educators and feminists was to criticize the study as somehow unconsciously selecting boys over girls.

Four years later, Benbow and Stanley had a lot more test results on hand, enough to present seemingly irrefutable evidence of the superiority of thirteen-year-old males to thirteen-year-old girls at the highest levels of mathematical ability.

The SAT is normally administered in the eleventh or twelfth grades. Most of the thirteen-year-olds were in the seventh grade when they took the test. The mathematics portion of the SAT is largely algebra and geometry, courses that almost none of the thirteen-year-olds had taken. Nevertheless, out of approximately 40,000 boys and girls who took the SAT in the national talent search during the four-year period, 280 of the children got scores of 700 or over, the equivalent to the 95th percentile for twelfth-grade boys. In other words, the 280 thirteen-year-olds scored higher on the test than 95% of the twelfth-grade boys who take the test, nearly all of whom have taken both algebra and geometry. Almost equal numbers of thirteen-year-old boys (19,883) and girls (19,937) took the test, but of the 280 highest scorers, 260 were boys and only 20 were girls, a ratio of 13 to 1.

Benbow and Stanley decided to test Geschwind's hypothesis about mathematics talent clustering with left-handedness, immune disorders, and myopia. The researchers got in touch with the high scorers. They found that 20 percent of the group was left-handed (about twice the normal incidence); 60 percent of the high-scorers reported allergies or asthma, both immune-system disorders; and 70 percent were nearsighted.

The students had also taken the verbal portion of the SAT. There was little difference in sex among the high scorers on the verbal portion. Checking high scorers on the verbal section of the SAT did not reveal as much left-handedness or immune-system disorders.

Although more boys than girls are dramatically talented in mathematics at age thirteen, presumably because there is more testosterone in their fetal environment, there are also more boys that score very poorly on the test. This is presumed to be caused by the same mechanism. The extra testosterone in the fetal environment of boys tends to lead to severe learning disabilities. Just as educators have noticed for generations that there are more gifted male mathematics students, the educators have also noticed that many forms of learning disability are found predominately among boys. Geschwind says, "It's a funny mechanism. At first it looks like you have to deliberately produce damage to produce giftedness."

Although more boys than girls are at both ends of the scale, it should be noted that the average scores for boys out of the nearly 40,000 children tested were only about 30 points higher than the average scores for the girls, a difference that could probably be accounted for in several ways.

1984

Update 1984: Mathematics

ARTIFICIAL INTELLIGENCE PRIZE

In the 1930s Kurt Gödel, Alonzo Church, and others established that machines could not be constructed that could prove all mathematical theorems. However, they did not rule out the possibility of a machine proving *some* mathematical theorems. In 1984, Woodrow W. Bledsoe of the University of Texas in Austin became the chairman of a committee that is determining the rules for a prize of $100,000 to be given to the person who defines a machine that can prove any significant mathematical theorem that has not previously been proved by a person.

CONNECTING THE DOTS

To the average person, a graph is either a bar graph, a line graph, or a circle graph, but to a topologist, a graph is a set of points connected by curves. Graph theory is related to the following puzzle: Given three dots on a sheet of paper labeled 1, 2, and 3 and 3 dots labeled A, B, and C, you are to connect each dot labeled with a numeral to each dot labeled with a letter without any of the connecting lines crossing. No matter where the dots are on the paper, all six connecting lines cannot be drawn if the paper is kept flat. An imaginative solution is to roll the paper into the twisted tube called a Möbius strip and then the dots can be connected. Another similar graph that cannot be formed consists of five points in a plane, each of which is supposed to be connected to all four of the others with no curves crossing.

In 1930, Kasimir Kuratowski showed that these two graphs determine whether or not a given set of connections can be made in the plane. If they are a part of a more complex graph, then the connections cannot be made, but if they are absent, the connections can be made. This solved the problem for the plane, but it obviously did not solve it for the Möbius strip.

The two graphs that cannot be connected form the list of minimal graphs for the plane for the problem of connecting points. In the 1960s Kurt Wagner proposed that for any property of graphs whatsoever and for any surface, the list of minimal graphs (those that by their presence or absence determine whether or not a graph has the property) is finite. In 1984 Neil Robertson and Paul Seymour determined that Wagner's conjecture is indeed true for most ordinary surfaces (technically, for all surfaces that have an upper bound to the number of handles on the surface).

Artificial Intelligence

Will the computer ever become smarter than the people who created it? What will be the result of developing a machine that can actually think—and that will view the universe in much the same way that humans do? Will the transformation of computers from mere number processors to possibly rational beings herald a new computer revolution that will signal enormous societal change?

These are a few of the questions being raised as we enter an exciting time in computer science—and, possibly, human history. This is a time when a great deal of speculation is surrounding a relatively new endeavor—artificial intelligence.

Artificial intelligence has several definitions, but essentially it is the quality which allows a machine to do something which we as people would recognize as intelligent. Machines—computers—that are intelligent would approach problem solving in the same way humans do and not as computers have traditionally done.

Artificial intelligence, or "AI," covers several fields—from pure computer science to psychology. It is even of concern to the business world because a computer system that is, to a degree, intelligent, can be used in many applications. But AI, a term coined by John McCarthy of M.I.T. in the early Fifties, is also the culmination of the human being's many attempts to recreate its own image.

Of course, artificial intelligence is still in its infancy; however, many nations are competing to be the first with the new "fifth generation" system. The Japanese claim that they will have these new systems by the 1990s. MITI, the Japanese ministry of international trade and industry, announced on April 14, 1982, that it was committing $1,000,000,000 to the development of the new system, which would have, among other abilities, the ability to translate between Japanese and English, to read, and to talk to people in ordinary language.

The term "fifth generation" for such a computer is based loosely upon the first four stages of computer development, each of which was a major technical advance—vacuum tubes, transistors, integrated circuits, and very large scale integrated circuits. "Fifth generation" describes the advance that will enable a computer to be intelligent. This term became popular in 1983 when a popular book, *The Fifth Generation* by artificial intelligence pioneer Edward A. Feigenbaum and Pamela McCorduck, argued strongly in favor of American development of such a computer.

The key moment when artificial intelligence was recognized as something that was of concern to more than a handful of specialists, however, came earlier, in August, 1980. The American Association for Artificial Intelligence had scheduled its first annual meeting for Stanford University, planning for an expected attendance of the several hundred specialists in the somewhat arcane field. Nearly a thousand participants showed up, however, including 40 reporters and many people who were just there with the intent of investing money in promising new developments. Artificial intelligence had spread beyond the interests of the few to embrace the many.

Most of the attention is still focused on the areas of research and the work place. Obviously, computers that could approach problems intelligently will be an incredible boon to scientific research. The National Research Council declared that artificial intelligence will have a tremendous impact. "It would surely create a new economics, a new sociology and a new history. If artificial intelligences can be created at all, there is little reason to believe that they could not lead swiftly to the constructions of super intelligences able to explore significant mathematical, scientific or engineering alternatives at a rate far exceeding human ability."

EXPERT SYSTEMS

One area of AI that is already well developed, at least in experimental models, is in "expert systems." Expert systems are programs that contain the knowledge as well as "heuristics," which are expert guesses, for a specific subject. These programs contain many thousands of "if/then" questions that enable the computer to solve specific problems by using the data it contains. The data, whether it is questions to ask in making a diagnosis of a disease or subtle clues that tell where oil may be found, was originally in a human mind. By putting this data on a computer, people who are less skilled than the "experts" can work with the computer to conclusions that match what the expert would have reached in person.

Two of the best-known experimental expert systems are INTERNIST-1 and CADUCEUS, both programs designed to assist physicians in making diagnoses. INTERNIST-1, which is "expert" in internal medicine was shown in 1983 to be nearly as accurate as the attending physicians in diagnosing cases selected from the *New England Journal of Medicine*, while CADUCEUS goes beyond internal medicine to have "expert" knowledge of physiology and anatomy. Another expert system, DELTA/CATS-1 has been used by The General Electric Company to aid in the maintenance of diesel locomotives. Since GE has only a single real expert in this field on its staff, DELTA/CATS-1 has been a significant money-saver for the company.

Only about 50 expert systems are in existence at this time. A major reason is because compiling them is extremely time consuming. An expert system generally requires more than 50,000 lines of programming and years of interviews of experts by the programmers.

Expert systems still represent only the tip of AI potential. They have not been designed to really "think." And they are totally dependent upon the data provided by the programmer. But they are another step towards a true thinking machine.

EDUCATION

Another field that already uses some artificial-intelligence techniques is education. A responsive tutor for teaching subtraction, DEBUGGY, is programmed to focus on the subskill that is causing a student to make mistakes, then correct it. While this technique has been successfully used by real expert teachers in the past, most classroom teachers are not equipped to tell from a student's wrong answers what is causing the error; DEBUGGY is.

Another program, less useful than DEBUGGY, but promising as another approach to education by way of artificial intelligence, is called WEST. WEST "watches" students play a board game called "How the West Was Won." Keeping track of how well the students are doing is just a matter of keeping statistics on each move. When WEST observes a pattern that is generally quite successful or generally not successful, the computer breaks into the game to give the student hints on how to improve performance.

MAN AND MACHINES

The machine/man relationship has always been a fascinating symbiosis. Man has created an incredible variety of mechanisms to perform a virtually limitless range of tasks. Machines are "work savers." They either do work that would have been performed manually or they enhance a person's ability to do a task. For example, a pulley can be used to lift a weight beyond the capacities of any single individual.

So, to a large degree, any machine is an extension of a human's ability. But for eons people have dreamed about a machine that would copy "humanity." These dreams extend far beyond the automatons, robots, and thinking computers of science fiction.

Greek mythology has several examples of machines built by the gods which are "intelligent." Hephaestus, the smith, created a host of machine workers to serve the gods. He also created a special man of metal. This was Talos, a bronze human with a liquid essence of life built in.

Real craftsmen from ancient Egypt to Renaissance Europe constructed mechanical marvels—clockwork mechanisms that could ape some human or animal functions. Leonardo da Vinci designed several mechanical wonders and searched for ways in which mechanical contraptions could copy the natural actions of birds. In the 19th Century, there was even a mechanical fraud. Johann Maelzel, an Austrian showman, introduced a wooden chessplaying automaton called the Turk. It was a great success in several countries—until it was discovered that the chess master was actually an expert assistant who sat inside the wooden figure and made the moves. However, his lucrative tours lasted for many years before the ultimate denouement.

The fact that the audiences accepted such a marvel is an indication of just how many wonders had already been accomplished. Even though chess playing computer programs that can consistently beat the top players have not yet been developed, the concept was not far fetched more than a century ago.

In the early 19th Century, an Englishman named Charles Babbage developed working models and plans for computation machines that presaged the development of the modern computer. His machines were based on complex systems of mechanical relays. Babbage planned to use a punched-card system developed for the automated weaving loom to instruct the machine, and he conceived the nature of programming his creations—all long before the use of electrical impulses made his work more than theoretically possible.

Babbage's most powerful machine, had he succeeded in building it, would have been essentially the first computer. But Babbage's work was forgotten, and later scientists had to recreate his basic innovations. During World War II, it became necessary to increase computing power for the war effort. Although they did not complete them until shortly after the war, scientists produced the first working digital computers. In the 1940s ENIAC was developed. ENIAC, which stands for Electronic Numerical Integrator and Calculator, was an impressive electronic marvel. Over three meters (ten feet) tall and 30 meters (100 feet) long, ENIAC could perform a multiplication in about three thousandths of a second.

Quite naturally, a machine which could solve a mathematical problem in a tiny fraction of the time a human could solve it, demanded respect. And to the general population, the computer was, and is, accorded a large degree of awe.

One only has to look at the wealth of science fiction written after the development of the first early computers to see the degree of respect—and fear—the general population accorded these new machines that were mysteriously more than machines.

Most of these science fiction stories automatically accord "intelligence" to even these early machines. And these fiction writers invented many examples of thinking machines. Some are good, some are evil. Isaac Asimov even developed laws of robotics that supposedly would allow for the coexistence of robots and people with a degree of harmony—with people as the superiors, of course. More recently, *Star Wars'* friendly robots, R2D2 and C3PO, have reinforced the notion of intelligent machines working as tireless servants of humanity.

The flip side of that coin can be seen in Arthur C. Clarke's *2001: A Space Odyssey*. This novel has HAL—a reasoning computer that ultimately threatens the lives of the two astronauts. Although Arthur Clarke denies it, it seems more than a coincidence that the letters H...A...L are the three letters in the alphabet that precede IBM—the company that represents computers to the world.

COMPUTERS AND ARTIFICIAL INTELLIGENCE

Computers present scientists with a mechanism that has the potential to mimic human thought processes. But the problem with attributing intelligence to these machines is the fact that the early computers—and this includes most computers in use today—weren't really copying human thought processes. Computers generally approach problems in a sequential fashion, using algorithms, while human problem solving is done in a heuristic manner. This is essentially a hit or miss fashion, problem solving by rule of thumb. Only the most likely avenues are considered. Heuristics allow humans to solve problems by selecting the most promising alternatives and exploring them more fully. One of the premises of the expert systems is that they will approach difficult problems more like a human.

Most algorithmic approaches treat all possibilities as equals. The result is that the algorithmic approach is more accurate, but far slower and not applicable in many circumstances to abstract problem solving.

While computers provide the mechanics that make artificial intelligence possible, the heart of AI lies in the set of instructions that the computer follows—the programs or software. This is an important point. It's at the heart of the distinction between mere supercomputers and true fifth-generation machines. A supercomputer is simply the fastest machine available. Supercomputers, such as the Cray 1, have extremely high processing rates and are vital in scientific research. Japan is also working hard on faster supercomputers in the Super Speed Computer project, which may surpass the Cray 1 within a few years.

Although a fast, powerful computer can be used as a base for a fifth generation system, the way the computer is instructed to solve a problem is the key. Scientists are hard at work around the world developing "super" programs which will give computers intelligent capabilities.

AI's Roots

Although people have dreamed of an intelligent machine for eons, AI really got its start in the late 1940s and early 1950s. Since computers were still in their infancy, a great deal of the original AI concepts were of a philosophic nature.

Alan Turing was one of the most important pioneers. He was a British scientist who explored some very important concepts that form the core of AI. In the Fifties he developed a test that laid the ground rules for discovering if a machine were intelligent. Basically, the test is constructed to determine if a machine could actually fool a human.

This is how he visualized the test: A game would be played with a man, a woman, and an interrogator, with the interrogator in a separate room, communicating with the other two by teletype. The object would be for the interrogator to guess which of the other two is the man and which is the woman by asking questions. One of the two is supposed to aid the interrogator as much as possible in making the correct guess, while the other one is supposed to steer the interrogator into guessing wrong. It was clearly Turing's notion that in such a situation, the interrogator would be unable to tell which was which.

What will happen when a machine takes the part of the player who is trying to steer the interrogator away from the correct answer in this game? Will the interrogator decide wrongly as often as when the game is played between a man and a woman? If so, then, according to this "Turing test," the machine is exhibiting intelligence.

Most workers in AI accept a form of the Turing test as a legitimate definition of artificial intelligence; that is, if a person questioning a remote entity by way of a terminal cannot tell whether the entity is a machine or a person, and if the entity actually is a machine, the machine passes the Turing test, and can be deemed intelligent.

CREATING "INTELLIGENCE"

A computer that could argue a human to a standstill in Turing's game could indeed be considered intelligent. But to what degree? Since Turing's time there have been a number of impressive programs that mimic humans quite successfully. Margaret Boden in her book *Artificial Intelligence and Natural Man* relates an incident which took place between a computer running a program called ELIZA and a sales Vice President of a computer company. ELIZA is programmed to simulate conversation of the sort a psychoanalyst might have with a client. The sales VP assumed that a person was operating the teletype.

VP: Do you think that I can use the computer this morning?

Teletype response: Why do you ask?

VP: I might be able to make some additional telecomp sales.

Teletype response: Why aren't you sure?

VP: My prospective customers have never seen a demonstration of this system.

Teletype response: Does this mean very much to you?

VP: Of course it does.

Teletype response: You seem very positive about that.

VP: This is ridiculous.

Teletype response: What do you mean this is ridiculous?

VP. Is there any reason why I can't use the computer now?

Teletype response: Tell me more about your prospective customers.

At this point the VP requested the person he thought he was speaking with to telephone him. When he received no answer, he called the computer room.

VP: Why are you being snotty to me?

Operator: What do you mean I'm being snotty to you?

Since this answer was consistent with the dialogue the Vice President just had with the computer, he was almost certainly extremely angry. But the ELIZA program, while clever, didn't reflect true reasoning ability on the part of the computer. It is a relatively simple program which takes clues from certain words in the sentences fed into it; then, depending on the clues, it either produces a given sentence or it takes a specified word from the sentence it has received and puts it into a stock sentence. There is no understanding behind the conversation. If the Vice President had suspected that he might be talking to a machine, then a few questions would have soon assured him that ELIZA was not actually a person.

ELIZA was originally developed at the Massachusetts Institute of Technology as an approach to artificial intelligence. The program or ones based on the same ideas are currently available in several different forms for personal computers.

Actually, clever as these programs are, they still don't pass the Turing test, which requires that a human not be able to tell the difference between a machine and another human.

However, although these programs aren't intelligent in themselves, they, and others like them, are valuable as first steps towards true intelligent programs.

In order to be considered truly intelligent, the programs, and, by extension, the computers, must do more than just simulate human response. They must actually "think" through the problem.

PROGRAMMING FOR INTELLIGENCE

Obviously, incredible hurdles hinder the programmer who tries to create a "thinking" program. Since, by definition, a thinking program will contain elements of human problem solving, the programmer must understand the human thought process.

Consider the problem of semantics—how to program a computer to understand the various shades of meaning in language. "The car screeched to a halt when the traffic cop raised his hand." The sentence makes perfect sense to anyone familiar with English. However, a computer is faced with a progression of difficulties. Why did the car "screech" to a halt. How did the policeman stop the car with his hand?

Therefore, in order to communicate with humans, and to understand various semantic structures, the computer must be programmed to follow "human" patterns. But a human has a wealth of experiences, of knowledge, to apply to a variety of situations. In order for a computer to be intelligent, it must apply certain standards of common sense—and be able to make inferences.

This also means that computers must be able to learn. A computer that cannot learn must depend upon the knowledge and foresight of the programmer to cover all possible contingencies. And a program that is totally dependent upon the programmer's input, cannot be intelligent in itself. The ability to take experiences from one set of occurrences and apply them to a totally different set of problems is a cornerstone of intelligence.

Programmers have been struggling with making computers learn for years. And they have been successful to a degree. One program developed by Terry Winograd called SHRDLU was one of the first to "learn" on a limited basis. (SHRDLU comes from the compositor's use of ETAOIN SHRDLU, the first through the twelfth most commonly used letters in English, as a filler when no specific words are intended.) Although SHRDLU's ability to learn new words is essential to its operation, the program was designed mainly to develop concepts related to natural-language use by a computer.

SHRDLU operates in a theoretical world in which there are blocks, cubes and pyramids. These can be displayed in various positions on a computer monitor. The program is asked about relationships among the different figures and asked to perform certain actions by typing sentences into the computer. For example, it may be asked to stack a block on top of a cube.

One of the most impressive features of this program is its ability to grasp new concepts within its mythical world. For example, the concept of ownership can be learned. The programmer can claim to "own" types of physical objects within SHRDLU, say a red block and a blue cube. From the point that ownership was claimed, the program will "remember" that fact and act accordingly in its actions.

This is just one simple example of a program learning. There have been a number of other programs designed to learn since the introduction of SHRDLU. One interesting area is in developing chess programs. Most programs have followed normal computer search patterns. However, for the computer to look ahead just three moves would necessitate a search of a possible 1,800,000,000 alternative possibilities.

In order to play more as a human plays, the programs have to follow certain conditions and only pursue the most promising lines of offense and defense. Even so, computers have not yet challenged the top ranks of human experts—although they are well into expert rankings.

THE FUTURE OF ARTIFICIAL INTELLIGENCE

One concern at this time is the consideration of exactly what is intelligent. Over the years, various programs have achieved a degree of success in developing programs that are "intelligent." First steps have been taken in developing usable programs that allow machines to recognize characters or speech patterns. To an extent, these programs make inferences and learn.

Impressive strides have been made in the field of artificial intelligence. Already, computer systems have been programmed to accomplish surprisingly human tasks. For example, computers can now read words, "smell" fragrances, and even help doctors diagnose illnesses.

The office is another area that will benefit from AI advances. The Japanese are working on a speech recognition program which will allow an individual to operate a typewriter by merely speaking into a console. On a related line, machinery already exists that is able to read typed characters and even produce written reports.

Artificial intelligence has raised some serious questions, however, especially in connection with the related field of robotics. Here machines are often seen to be in competition with humans, actually replacing human workers.

Today, relatively stupid robots can do such work as painting, welding, and wiring. They can also deliver mail and even help with surgery. They can do the work faster, more effectively, and much cheaper than human labor. The question is what happens when robots become more intelligent, thanks to advances in AI?

Supporters of robotics claim these machines would only do those jobs humans find boring or jobs that are particularly hazardous, such as working in a nuclear reactor. But the fact is that within a few years, robots will have replaced several hundred thousand humans on simple assembly line operations. And what happens when robots are provided with superior programming? Will they continue to be a major force for human development or will they just put more people out of work?

CONCLUSION

Most computer scientists believe that the next computer "revolution" will be centered around machines that think. If so, the impact of AI on society will probably be significant. On one hand, AI will provide us with a good map of the warrens of the human mind. And on the other hand, we will introduce another type of intelligence into our world—an intelligence which should be capable of making incredible scientific, business, and societal advances—if it is managed properly.

The ultimate question, however, is this: Will we create something which will think as we think? Will we create a machine that can not only compose a beautiful sonata, but also enjoy listening to the music?

The next few decades will almost certainly witness some extremely important advances in computers, memory and programming techniques. It is, to say the least, an exciting time.

BETTING ON THE FIRST COMPUTER

Although Charles Babbage designed the first actual computer in 1842, his machine, called The Analytical Engine, was never built. Today, we know about its principles largely from the writings of Babbage's enthusiastic and brilliant supporter, Lady Ada Lovelace, Lord Byron's daughter. Because Lady Ada is credited with being the first computer programmer by some, the Defense Department has named its powerful standardized programming language ADA in her honor.

Although there were many unsolved technical problems involved in building a working computer that operated with switches instead of electron tubes or transistors, the principal reasons The Analytical Engine did not get built were Babbage's habit of redesigning it every time he had a new idea (which was quite often) and a lack of money. Babbage hoped for government financing, but the British government had advanced £17,000 to Babbage already for an earlier idea, which Babbage had abandoned without completing when he thought of The Analytical Engine, and they were not about to grant additional funds. Eventually, Lady Ada and Babbage devised a system for winning bets in horse racing as a mode of funding, a system that severely depleted the resources they could devote to the project.

LIFE IMITATES ART

A famous (among mathematicians and physicists) spoof of the 1930s was a list of ways that a mathematician or a physicist might capture a lion in the Sahara. (Sample way: The Schrödinger Method: At any given moment there is a positive probability that a lion has entered the cage. Sit down and wait for the lion.) Local police in Arizona recently asked mathematicians at the University of Arizona to develop a mathematical method for finding lost persons in the Sonoran Desert. The mathematical theory was developed and is used to determine the probability that a lost person is in a particular sector of the desert.

MEDICINE

No field of science, except possibly the closely related science of biology, saw such impressive gains in the early 1980s as medicine. Fundamental strides were taken in resolving the mysteries of cancer and heart disease, the two leading killers in the United States. Progress was made on developing vaccines for malaria, hepatitis B, herpes, cholera, and chicken pox. The science of genetics began to emerge from the theoretical phase into the practical, with potential treatments for sickle-cell anemia and thalassemia and with the location of a chromosomal marker for Huntington's disease. Transplants became increasingly practical, including transplants of pancreatic cells to cure diabetes, of livers and other organs, of bone marrow, and even of tissues into the brain. Furthermore, approval of the immune-system-suppressing drug cyclosporine in 1983 means that transplants will be safer than ever in the future. In addition, an artificial heart supported a human being for 112 days.

The 1980s opened with approval of the first American clinic to use new techniques of fertilization outside the body in helping couples solve their infertility problems. If a problem develops after a successful start for the fetus and before the birth of the baby, new techniques of fetal surgery are now available to help resolve it. At the same time, there is a continuing interest in suppressing fertility when it is not wanted.

A better understanding of the relationship between diet and health was another continuing trend in the 1980s. Because everyone eats, diet-related stories continue to be big news. The stories were concerned with how foods are preserved (by possibly cancer-causing nitrates); controversial ways to lose weight; possible dangers of salt, fats, or artificial sweeteners; and even the latest word on how much a person of a given height and build should weigh for good health. In line with this interest, *The Science Almanac* provides a comprehensive set of tables to help you in applying the new information to your own diet, as well as a list of some of the best recent books on diet and health.

At a more fundamental level of medical technique, gene splicing resulted in the production of useful human proteins that were grown in vats of bacteria or yeast. The development of monoclonal antibodies and nuclear magnetic resonance provided entirely new ways to diagnose serious diseases. Studies of the chemicals in the brain began to reveal the complex interactions involved in learning and in body control. Some clinical applications of this new knowledge are already available for treating recognized brain disorders and even conditions, such as depression, that once were thought to be nonorganic.

One result of all this activity is not surprising. The life expectancy for people in the United States continues to increase—although most biologists believe that the *maximum* life expectancy is unchanged. The increases come from lifting the *average* life expectancy. One result of this increase is that diseases of the aging are undergoing considerable study. Another result may be the increased importance of such diseases as cancer and heart disease.

One of the major areas in which progress was especially rapid, both in theory and in application, was immunology. Human beings have been manipulating the immune system, with very little understanding of what they were doing, at least since the eighteenth century, when young children were exposed to smallpox in hopes of protection against the disease in later life. Today, the basic mechanisms of the immune system are being revealed one by one, showing it to be an amazingly complex system and the center of the cell's interaction with its environment. Actually, so many different components of the immune system work together in such a complex way that it is not surprising that sometimes things go wrong, producing autoimmune diseases.

Understanding the immune system has turned out to be central to successful organ transplants, treating allergies and cancer, the development of protection against viruses, the diagnosis of diseases, and a host of other medical areas, some of which are just beginning to emerge. The progress made in the early 1980s is impressive, but it is clear that it is only a beginning.

There was a cloud over medicine in the early 1980s despite these great advances. Acquired Immune Deficiency Syndrome, or AIDS, appeared, seemingly from nowhere. Unlike any communicable disease seen before, it resisted massive efforts to halt it, being invariably fatal. Its cause was not even known until 1984, when an intensive effort located a virus that attacks part of the immune system. AIDS appeared just as doctors were beginning to understand immunity.

1980

U.S. Approves First "Test-tube" Clinic

Following on the heels of the successful fertilization in England of a human egg outside the body, the first U.S. clinic to use this new technique was approved in January 1980. The new fertilization technique produces what have been nicknamed "test-tube babies" by the general public. More scientifically, the technique, pioneered by British Drs. Steptoe and Edwards, should be called *in vitro* (literally, "in glass") fertilization, especially since the actual fertilization takes place in a small laboratory dish, called a Petri dish, not a test tube.

Approval of the U.S. clinic, at Eastern Virginia Medical School in Norfolk, Virginia, was opposed by various groups on religious or philosophical grounds, but for many childless couples, the clinic provided new hope. At the time the approval was finally granted, almost 25,000 childless couples had already contacted the proposed clinic.

The "test-tube" approach is used when both prospective parents are fertile, but the woman's blocked or missing oviducts prevent an egg and sperm from meeting to produce a pregnancy. Ripe eggs are surgically removed from the infertile woman's ovaries. The eggs are then placed with the husband's sperm in a petri dish. If fertilization occurs, the physicians can detect it by microscopic examination, since the fertilized eggs begin dividing. At the proper time, the fertilized egg is placed in the woman's uterus, which has been hormonally prepared to accept it.

An ethical problem that has been difficult to resolve is what to do with the excess eggs and, if any, the excess fertilized eggs. Fertilized human eggs would be very desirable for various research purposes, but they could also be viewed as life forms that should be treated with respect, since they are incipient human beings. Ethical guidelines have been set up that, in fact, allow the fertilized eggs to be used for research for a few divisions only. After the specified time, the fertilized eggs must be destroyed. Despite these concerns, the U. S. commission looking into the ethical questions was disbanded just as the problems became a reality and was not reconstituted, at least not by the time this book went to press.

A spokesperson for the Eastern Virginia Medical School indicated that the *in vitro* technique would be used only with couples for whom no other method of fertility enhancement could produce pregnancy. The clinic estimated that it would be able to treat about 50 couples during the first year of operation.

Interferon Produced in Lab

The 1980s opened with the announcement that interferon, an immune factor made naturally by the human body, had been "synthesized" in the laboratory by the new process called the recombinant DNA technique. The breakthrough was announced in a cascade of publications and proclamations, most dramatically by Charles Weissmann of the University of Geneva at a seminar held at the Massachusetts Institute of Technology in January 1980.

Interferon had been discovered in 1957. Doctors had noticed in the 1930s that when a person had one virus disease, he or she seemed to be protected against catching any other virus disease. In 1957 Alick Isaacs and Jean Lindenmann showed that this effect was produced by a chemical released by cells infected by a virus. They named the substance *interferon*. There was some hope that interferon might be effective against a wide range of illnesses, from cancer to the common cold. But interferon was not available in sufficient quantity to be examined or tested. Only tiny amounts were available from infected human cells, and interferon is species-specific; interferon from animals is not effective in humans.

Actually, it was quickly realized that there was more than one kind of interferon, but only tiny amounts of mixtures of the different types could be obtained by standard means. The new method of obtaining pure interferon announced early in 1980 was not like the chemical synthesis that had been performed in the past. Instead, the process involved locating the genes in human cells that produce interferons. The genes were then copied into living bacteria, which produced the human interferons. Finally, the interferons were purified by the use of monoclonal antibodies, substances produced by fusing cancer cells with antibody-producing cells. The result was not *synthetic* interferon, but actual human interferon. (For a further discussion of the technical elements involved see BIOLOGY, "Gene-Splicing Used to Produce Interferon," page 146.)

Dr. Weissmann's research was sponsored by Biogen S.A., a research organization supported largely by a major pharmaceutical company. Biogen applied for a patent on the new process, and the pharmaceutical house Schering-Plough received worldwide licensing rights.

Others had been working feverishly along slightly different lines. Within weeks of Weissmann's announcement, the Recombinant DNA Advisory Committee of the National Institutes of Health received a host of proposals for laboratory production of a variety of natural substances from major genetic engineering firms.

By June, Genentech of South San Francisco announced that it had synthesized two different kinds of interferon: leukocyte interferon, which is produced by white blood cells, and fibroblast interferon, which is produced by connective tissues in the body. (Since then, these types have come to be known as alpha and beta interferon, respectively.)

A few months later, there were reports of the synthesis of another type of interferon, now called gamma-interferon, which might be far more powerful and effective. By November, at the First Annual Conference on Interferon Research, four companies had produced interferons from human genes implanted in bacteria. They reported that there were at least ten varieties of alpha interferon, more than two varieties of beta interferon, and an unknown number of varieties of the gamma type.

The next step was the announcement of faster and more productive ways of growing the interferons. In the early 1970s, small amounts of interferon were grown in human cell cultures, but by 1981, as a result of gene splicing, interferons were being grown in bacterial cultures, then in yeast cultures, which yielded thousands of times more. There were estimates that the market for interferon might be worth $375,000,000 by 1985 and some $1,400,000,000 by 1990, although interferons were still not commercially available for regular clinical use as of the middle of 1984.

While the race for production went on, however, the results of interferon tests against various diseases were mixed at best. Interferon was not a broad-range cure-all as had been hoped, although it did seem to be of some use against certain cancers, and against such more common ailments as the common cold and warts (both caused by viruses).

More important, research on interferon was leading to a whole family of other substances that promised to be more valuable. By 1983, gamma-interferon was classed among the lymphokines, a family that included the interleukins and other immune substances (see the article on immunity that begins on page 491). It began to appear that interferon, instead of being a single great discovery, would one day be seen as the precursor of a whole complex range of natural immune substances.

The genetic engineering companies were in the forefront of research on these new substances. It appeared that they would continue to be important in the unfolding saga of research, production, and distribution of the immune substances. Just how they would fit in with the publicly funded research scientists and the practicing medical community remained to be seen. (See also, "Experiment with Engineered Bacteria Stopped," page 173.)

Successful Treatment for Hodgkin's Disease

In a report in the May 1980 edition of *Annals of Internal Medicine*, Vincent DeVita and others underscored the vast improvements in the treatment of certain cancers during the preceding decade. One such cancer is Hodgkin's disease, which affects the lymph glands. At one time, Hodgkin's disease, a common form of cancer, was invariably fatal. Today, that grim picture has changed for many patients.

DeVita and his colleagues at the National Cancer Institute reported on a follow-up study of 198 patients who had been treated between 1964 and 1976 with a multidrug regimen called MOPP, which consists of mechlorethamine, vincristine, procarbazine, and prednisone. MOPP produces an immunosuppressant action apparently extremely effective against Hodgkin's.

In 1964 median survival after diagnosis of Hodgkin's disease was only two years. DeVita reported that of the 198 patients treated with MOPP after 1964, 158 (or 80 percent) experienced a complete remission after three months of treatment. Even more encouraging, more than half the total surveyed (107) had remained disease-free for ten years after treatment. DeVita hypothesizes that these patients can be considered "cured."

Still, from 20 to 25 percent of Hodgkin's disease victims were not helped significantly by MOPP. Another drug combination, ABVD (adriamycin, bleomycin, vinblastine, and dacarbazine), developed in the 1970s, however, cured nearly 60 percent of the patients resistant to MOPP. While ABVD had some unpleasant side effects, they were not nearly so drastic as those produced by MOPP. MOPP could lead to sterility and even to other forms of cancer.

But the real breakthrough came in patients who alternated between MOPP and ABVD. In one study, 92 percent experienced complete remissions and all the remaining patients were helped to some degree. While Hodgkin's disease can still be dangerous, the outlook for most patients is good.

Similar progress has been shown in the treatment of various other forms of cancer, although cancer remains the number two cause of death in the United States. The biggest gains have come, as in Hodgkin's, with chemotherapy, but surgery, radiotherapy, and improved detection are all involved in a revolution in cancer treatment. At the same time, fundamental understandings in the cause and progress of cancer offer hope of revolutionary new approaches to detecting and cure of the disease (see "Cancer Genes," page 180 and "Genes and Cancer," page 193).

Pancreatic Cell Transplants

Researchers from the Washington University Medical School in St. Louis, Missouri, reported that they had succeeded in transplanting the pancreatic islets of Langerhans of a rat to a mouse with induced diabetes, and that the islet cells appeared to help control the recipient's blood sugar level over a considerable period of time.

The significance of these findings to victims of diabetes may be incalculable. There are different types of diabetes, but in all cases, the fundamental problem is control of the metabolism of sugar. For many diabetics, the underlying cause of the disease is lack of production of insulin, the substance made by healthy pancreatic islets of Langerhans, that is essential to the body's metabolism of sugar. Diabetes is controlled today with restricted diet (sometimes combined with drugs to lower blood sugar) and, for patients who have insufficient insulin, with injections of insulin from swine or cattle or injections of human insulin produced by gene splicing. Insulin injections, although they allow diabetics to lead fairly normal lives, do have serious risks, and their control of blood sugar levels is crude at best. If transplanting islet cells to human diabetics becomes a common procedure, it will be far superior to insulin injections as a treatment for many diabetics.

Paul Lacey, Joseph Davie, and Edward Finke reported that they had succeeded in overcoming the tendency of the recipient's body to reject the foreign cells by taking two precautions. First, they removed the islets from the rat donor and incubated them for a week before transplantation. This incubation period did not harm the islets, but many of the accompanying rat white cells died. (Live white cells increase the risk of rejection by the recipient.) Second, they injected the recipient with antibodies both to rat and mouse white cells to further reduce the rejection process.

The pancreatic islets were then injected into the mouse's bloodstream. They lodged in the mouse's liver, where they were able to secrete insulin and to control the blood sugar level in the mouse. In one case, blood sugar was still under control 116 days after the transplant procedure.

Later in 1980, Franklin Lim and Anthony Sun announced that they had had some success in inoculating animals with pancreatic cells enclosed in tiny semipermeable membranes. Such membranes allow the insulin to be released by the cells, but do not admit the recipient's white cells, which would likely kill the foreign cells.

These two procedures and similar ones seem likely to provide new and improved treatment for diabetes in humans.

1981

Cholesterol Dangers

Within one week in January 1981, scientists released two ground-breaking studies of the health effects of cholesterol, the substance linked to the consumption of saturated fats (mainly animal fats). The first of the two studies, released on January 8, announced compelling evidence of a direct link between cholesterol-rich foods and the threat of early heart attack. The second study linked very low cholesterol levels with cancer of the colon in men.

Cholesterol is a normal, even essential, steroid alcohol produced by the liver, but excess amounts in the bloodstream have been suspected for years of contributing to heart and artery disease. The problem seems to come both from consumption of food high in cholesterol itself (found only in animals) and with saturated fats that cause cholesterol to be produced. Saturated fats include coconut oil and a few other plant oils but are largely found in animals.

Although the two studies appeared contradictory in identifying both high- and low-cholesterol levels as hazardous, they actually began to delimit a safe range for blood cholesterol levels. Within months of publication of the two research studies, experts at a National Institutes of Health conference were describing this safe range.

The cholesterol-heart disease study took place at Rush-Presbyterian-St. Luke's Medical Center in Chicago. Begun in 1957, it charted the diet of 1,900 randomly selected employees at a Western Electric plant near Chicago. Their diets were scored as low, moderate, or high in cholesterol derived from saturated fats. The researchers also recorded other pertinent factors such as age, weight, blood pressure, smoking habits, and alcohol consumption. Reinvestigation of the study group twenty years later, in 1977, found that those with the lowest intake of saturated fats had a 33 percent lower rate of coronary disease.

This statistical evidence corroborated scientists' belief that blood cholesterol levels are directly tied to the consumption of saturated fats and that cholesterol contributes to atherosclerosis, the clogging of the blood vessels associated with heart disease. Cholesterol in the blood causes this clogging action but scientists were not sure how this happened. They believed that the effect was caused either by changing the structure of fat-carrying blood cells or by altering the normal process of cell division.

The clue to what was actually happening was discovered in 1980. A strain of rabbits found by Yoshio Watanabe of Kobe University in Japan was known to develop high cholesterol levels and human-like coronary artery disease. Studies on the rabbits established that humans who have a genetic tendency toward high cholesterol levels develop coronary artery disease by the same process as the rabbits. In both cases, the number of receptors for low density lipoproteins (LDLs) is smaller than normal, leading to a build-up of LDLs. LDLs carry cholesterol to the cells of the body, where the cholesterol serves a variety of purposes. With the Wantanabe rabbits as experimental animals, scientists were able to develop by 1983 a drug treatment that is effective for people with this form of genetic heart disease.

It was not known at the time whether similar mechanisms are involved in nongenetic heart disease or whether the same treatment would be effective, but a later study of carefully matched men with high-cholesterol levels, with one group using drugs to lower cholesterol and another not, revealed about half as many deaths from heart attacks in the group using the drug.

The cholesterol-cancer study was conducted by the National Heart, Lung, and Blood Institute in Bethesda, Maryland. It found a link between extremely low blood cholesterol and the risk of cancer of the colon in men. The findings were based on an ongoing study of 5,200 residents of Framingham, Massachusetts, that dates back to 1948. Examination of the statistics suggested that men with low blood cholesterol had a rate of colon cancer three times that of the entire study group, including women.

The Chicago study linked heart disease with a blood cholesterol level of more than 250 mg per 100 ml of blood serum. The Framingham study linked cancer of the colon in men with a blood cholesterol level of less than 190 mg per 100 ml of blood serum. Thus the "safe" limits of cholesterol in the blood probably fall within these upper and lower limits.

Since dietary efforts to lower cholesterol can only lower or raise it by 15 percent, it is unlikely that a low-cholesterol diet would ever bring the level down low enough to incur serious risk of cancer.

BEERS FOR THE HEART

A study of 44 college students at Baylor College of Medicine involved sedentary students who consumed three beers a day for twenty days. The research showed that the booze raised levels of the high-density lipoprotein cholesterol (HDLC), which is thought to help prevent heart disease. The beer raised HDLC levels to the same as that of joggers who had not been drinking regularly. Joggers who tried the beer supplement to their diets did not gain any HDLC from the beer, however.

Dietary Sodium: A Growing American Health Concern

Sodium is an important mineral, playing a key role in maintaining proper blood volume and regulating pressure in the blood vessels. However, the body needs only a relatively small amount, far less than most Americans consume daily. Most sodium in the diet comes from the use of salt (sodium chloride). In 1983 evidence was found to suggest that the chloride in salt may also be a factor in disease. Confirmation of abusive salt consumption is revealed in such reports as that of the National Center of Health Statistics, which estimates the number of people in this country suffering from salt-related illnesses:

Disease	Approx. No. in U.S.	Disease	Approx. No. in U.S.
Angina	4,000,000	Emphysema	2,000,000
Arteriosclerosis	680,000	Hypertension	35,000,000
Arthritis	23,000,000	Kidney malfunction	6,000,000
Borderline hypertension	25,000,000	Rheumatic heart failure	2,000,000
Cardiovascular disease	40,000,000	Rheumatism	800,000
Diabetes	4,000,000		

In 1982, researchers at the University of Minnesota discovered that red blood cells from patients with essential hypertension (high blood pressure) absorbed sodium twice as fast as those from people with normal blood pressure. An experimental test could be used to predict those people who are at special risk from dietary sodium before they develop hypertension.

Medical and health professionals in the United States agree that a proper diet of the natural foods can adequately supply the amount of sodium required daily. However, with salt added in cooking, at the table, and in processed foods, daily sodium intake may be as high as six to ten times the needed amount. While most added sodium comes from salt (2,000 mg per teaspoon), many other common food ingredients and additives signal the presence of sodium: baking soda, also known as sodium bicarbonate or bicarbonate of soda (821 mg per teaspoon); sodium benzoate; sodium citrate; sodium alginate; sodium propionate; sodium sulfite; baking powder (370 mg sodium per teaspoon); brine used in pickling; disodium phosphate; sodium saccharin; and monosodium glutamate (750 mg sodium per teaspoon). A "safe and adequate" daily sodium intake, as recommended by the National Research Council, is about 1,100 to 3,300 mg for an adult. Some medical professionals, advocating salt restriction only where therapeutically indicated, point out that a low-salt diet could be harmful in diabetes, in those who exercise heavily, and in cases of anorexia and diarrhea.

Dietary guidelines set forth by the Departments of Agriculture and Health, Education and Welfare, and the Food and Nutrition Board of the National Research Council, as well as by other health professionals and consumer organizations, all stress the need to avoid excessive salt intake. Salt substitutes (containing potassium chloride) should be used under the supervision of a physician, as an excess of potassium can be as detrimental to health as too much sodium. Various food manufacturers are making available a greater number of low-sodium products. In 1984, the Food and Drug Administration started to require all foods that list their nutritional content to state the number of milligrams of sodium contained in 100 grams of the product, together with the number of milligrams in a serving.

Some High-Sodium Foods

Food	Portion	Approx. sodium (milligrams)
Fast Food Chicken Dinner	One dinner	2,243
Dill Pickle	One large	1,928
Garlic Salt	1 teaspoon	1,850
Manhattan-Style Clam Chowder	1 cup	1,808
Meat Tenderizer	1 teaspoon	1,750
Frozen Fish Dinner	One dinner	1,212
Beef Bouillon Made from Cube	1 cup	1.152
Canned Tomato Soup	1 cup	1,140
Chicken Noodle Soup	1 cup	1,107
Frozen Beef Pot Pie	One pie	1,093
Soy Sauce	1 tablespoon	1,029
Instant Chocolate Pudding	1 cup	972
MacDonald's Big Mac	One	963
Kentucky Fried Chicken	4 ounces	745
Fast Food Cheeseburger	One	709
Frozen Pizza	4 ounces	656
Nabisco Wheat Thins	1 ounce (16 crackers)	370
Bottled Italian Salad Dressing	1 tablespoon	362
Hostess Twinkies	One	241
Potato Chips	1 ounce	220
Pillsbury's Sugar Cookies	Three	210
Frozen Fruit Pie	⅛ pie	180
Salted Popcorn	1 cup	175
Barbecue Sauce	1 tablespoon	140

Toxic Shock Syndrome

On January 13, 1981, the Minnesota Department of Health announced the first documented link between toxic shock syndrome (TSS) and the use of high-absorbency tampons. Suspicions that such a link existed had already caused the recall of Rely tampons from the market on September 22, 1980.

TSS first gained national notice in 1980 when 819 cases, including 69 deaths, were reported to the U.S. Centers for Disease Control. The outbreak of this mysterious and dangerous disease reached its peak in August of 1980, when 119 cases were reported. Most occurred in women under 30 years of age in the states of the upper Midwest. The symptoms of TSS are fever, vomiting, diarrhea, and in extreme cases, a rash and extreme low blood pressure.

Cautions broadcast to women using high-absorbency tampons, and the recall of Rely, brought the number of reported cases down to 31 in December 1980. Tampon sales had declined by 20 percent.

The Minnesota study was based on observations of 80 victims of TSS and 160 unaffected women of comparable age in the states of Minnesota, Wisconsin, and Iowa over a period of nearly a year. Researchers reported that users of high-absorbency tampons ran a risk of TSS up to 30 times greater than women using smaller, less absorbent tampons.

A month after publication of the Minnesota study, researchers at the University of Washington in Seattle released the first study purporting to show the causes of TSS. Their report showed that nearly 80 percent of the TSS sufferers studied had a new strain of the bacteria *Staphylococcus aureus*, which carries a latent parasitic virus, or bacteriophage, that makes a highly poisonous toxin. The women studied also had used high-absorbency tampons. (Later research tended to discredit the influence of the virus infection in the staph bacteria.)

The exact means by which the tampons encourage the disease are not yet clear. But the research does suggest that the emergence of a new strain of *S. aureus* may be partly responsible for the 1980 epidemic of the disease and that high-absorbency tampons are a contributing factor rather than a primary cause.

Toxic shock reentered the news in 1984, when a few women using the newly licensed contraceptive sponge developed the syndrome. A study of the sponge-related TSS revealed that it occurred in women who had kept the sponge in place longer than was recommended, and stronger warnings on the package were suggested. (See also, "Progress on Toxic Shock," page 472.)

Malaria Vaccine

Researchers at New York University applied for a patent for a malaria vaccine, raising hopes for this urgently needed weapon against one of mankind's most persistent enemies. Malaria infects some 200,000,000 people a year, most of them in tropical regions, and it accounts for the death of 1,000,000 children each year in Africa alone. The patent represented the solution of a crucial scientific problem in creating the vaccine. But it soon appeared that nonscientific problems might postpone its use indefinitely.

Malaria has defied vaccine research because of the complexity of its spread in the human body. The infection is introduced by the bite of a carrier mosquito that injects the parasitic protozoan *Plasmodium* in the form of sporozoites. These cells migrate to the liver, where they establish the infection in the form of merozoites. The merozoites then occupy red blood cells, proliferate by cell division, and in six to twelve days produce the fever, chills, and other symptoms of the disease. Some merozoites then evolve into gametocytes, which can infect blood-sucking mosquitoes and create new carriers for the disease.

Each stage of the infection creates a different antigen, the surface protein that provokes antibodies in the human immune system. The antibodies needed to combat the sporozoid stage are useless against the merozoite state.

Ruth and Victor Nussenzweig of New York University (N.Y.U.) had been working since 1963 to isolate the key antigen of the initial sporozoite stage, the likeliest target of preventive inoculation. They accomplished this by laborious dissection of the salivary glands of mosquitoes.

The next requirement was a larger supply of antigens than could be obtained by dissection. This problem was solved by Nigel Godson, also of N.Y.U. He managed to isolate the gene encoding the antigen of the monkey strain of malaria by breaking up fragments of DNA and testing each against the antibody known to adhere to it. The details of Godson's technique are the key to the patent application filed by N.Y.U.

Genentech, a leading biological engineering firm, was prepared to develop the vaccine in exchange for an exclusive license for its eventual manufacture. Such licensing, however, required the consent of the public agencies that helped fund the N.Y.U. research. One of them, the World Health Organization (WHO), objected to Genentech's demand for exclusive rights. WHO stated that if the company finds marketing of the vaccine unprofitable, it might continue to hold the license while producing no vaccine or too little.

World Health Organization's Model List of Essential Drugs

Many of the developing nations must not only depend on foreign suppliers for their drug needs, but they are faced with making an appropriate selection from drug compendiums listing at least 1,500 to 2,000 products. The World Health Organization (WHO) Expert Committee on the Use of Essential Drugs has studied this problem and has composed a list of just over 200 active substances that are suitable for the primary health care of a majority of the population. Realizing the wide differences existing in the health care field in different countries, the Expert Committee suggests using the list as a model in providing essential drugs of good quality from which each country can make its own selection. The following list is extracted from WHO's Technical Report Series No. 641.

I. ANESTHETICS
1.1 *General anesthetics and oxygen* — ether, anesthetic; halothane; nitrous oxide; oxygen; thiopental
1.2 *Local anesthetics* — bupivacaine; lidocaine
2. ANALGESICS, ANTIYPYRETICS, NONSTEROIDAL ANTI-INFLAMMATORY DRUGS AND DRUGS USED TO TREAT GOUT   acetylsalicylic acid (aspirin); allopurinol; ibuprofen; indomethacin; paracetamol
3. ANALGESICS, NARCOTICS AND NARCOTIC ANTAGONISTS   morphine; naloxone; complementary drug: pethidine*
4. ANTIALLERGICS/ANTIHISTAMINES   chlorpheniramine
5. ANTIDOTES
5.l *General* — activated charcoal; ipecacuanha
5.2 *Specific* — atropine; deferoxamine; dimercaprol; sodium calcium edetate; sodium nitrite; sodium thiosulfate
6. ANTIEPILEPTICS   diazepam; ethosuximide; phenobarbital; phenytoin
7. ANTIINFECTIVE DRUGS
7.1 *Amebicides* — metronidazole
7.2 *Anthelmintic drugs (against parasitic worms)* — mebendazole; niclosamide; piperazine; tiabendazole
7.3 *Antibacterial drugs* — ampicillin; benzathine benzylpenicillin; benzylpenicillin; chloramphenicol; cloxacillin; erythromycin; gentamicin; metronidazole; phenoxymethylpenicillin; salazosulfapyridine; sulfadimidine; sulfamethoxazole + trimethoprim; tetracycline
7.4 *Antifilarial drugs* — diethylcarbamazine; suramin sodium
7.5 *Antileprosy drugs* — clofazimine; dapsone; rifampicin
7.6 *Antimalarials* — chloroquine; primaquine; pyrimethamine; quinine
7.7 *Antischistosomals* — metrifonate; niridazole; oxamniquine
7.8 *Antitrypanosomals* — melarsoprol; nifurtimox; pentamidine; suramin sodium
7.9 *Antituberculosis drugs* — ethambutol; isoniazid; rifampicin; streptomycin
7.10 *Leishmaniacides* — pentamidine; sodium stibogluconate
7.11 *Systemic antifungal drugs* — amphotericin B; griseofulvin; nystatin

8. ANTIMIGRAINE DRUGS   ergotamine
9. ANTINEOPLASTIC AND IMMUNOSUPPRESSIVE DRUGS   azathioprine;
   bleomycin; busulfan; calcium folinate; chlorambucil; cyclophosphamide;
   cytarabine; doxorubicin; fluorouracil; methotrexate; procarbazine; vincristine
10. ANTIPARKINSONISM DRUGS   levadopa; trihexyphenidyl
11. DRUGS AFFECTING THE BLOOD
11.1 *Antianemia drugs* — ferrous salt; folic acid; hydroxocobalamin
11.2 *Anticoagulants and antagonists* — heparin; phytomenadione; protamine
   sulfate; warfarin
12. BLOOD PRODUCTS AND BLOOD SUBSTITUTES
12.1 *Plasma substitute* — dextran 70
12.2 *Plasma fractions for specific uses* — albumin, human normal
13. CARDIOVASCULAR DRUGS
13.1 *Antianginal drugs* — glyceryl trinitrate; isosorbide dinitrate; propranolol
13.2 *Antiarrhythmic drugs* — lidocaine; procainamide; propranolol;
   complementary drug: quinidine*
13.3 *Antihypertensive drugs* — hydralazine; hydrochlorothiazide; propranolol;
   sodium nitroprusside; complementary drug: reserpine*
13.4 *Cardiac glycosides* — digoxin
13.5 *Drugs used in shock or anaphylaxis* — dopamine; epinephrine
14. DERMATOLOGICAL DRUGS
14.1 *Antiinfective drugs* — neomycin + bacitracin
14.2 *Antiinflammatory drugs* — betamethasone; hydrocortisone
14.3 *Astringents* — aluminum acetate
14.4 *Fungicides* — benzoic acid + salicylic acid; miconazole; nystatin
14.5 *Keratoplastic agents (used to soften hard scabs)* — coal tar; salicylic acid
14.6 *Scabicides and pediculicides* — benzyl benzoate; gamma benzene
   hexachloride
15. DIAGNOSTIC AGENTS   edrophonium; tuberculin, purified protein derivative
15.1 *Ophthalmic* — fluorescein
15.2 *Radiocontrast media* — adipiodone meglumine; barium sulfate; iopanoic
   acid; meglumine amidotrizoate; sodium amidotrizoate
16. DIURETICS   amiloride; furosemide; hydrochlorothiazide; mannitol
17. GASTROINTESTINAL DRUGS
17.1 *Antacids (nonsystemic)* — aluminum hydroxide; magnesium hydroxide
17.2 *Antiemetics* — promethazine
17.3 *Antihemorrhoidals* — local anesthetic + astringent and antiinflammatory
   drug
17.4 *Antispasmodics* — atropine
17.5 *Cathartics* — senna
17.6 *Diarrhea*
17.6.1 *Antidiarrheal* — codeine
17.6.2 *Replacement solution* — glucose (dextrose); oral rehydration salts (for
   glucose-salt solution); potassium chloride; sodium bicarbonate (baking
   soda); sodium chloride (table salt)
18. HORMONES
18.1 *Adrenal hormones and synthetic substitutes* — dexamethasone;
   hydrocortisone; prednisolone
18.2 *Androgens* — testosterone

18.3 *Estrogens* — ethinylestradiol

18.4 *Insulins* — compound insulin zinc suspension; insulin injection

18.5 *Oral contraceptives* — ethinylestradiol + levonorgestrel; ethinylestradiol + norethisterone

18.6 *Progestogens* — noresthisterone

18.7 *Thyroid hormones and antagonists* — levothyroxine; potassium iodide; propylthiouracil

19. IMMUNOLOGICALS

19.1 *Sera and immunoglobulins* — anti-D immunoglobulin (human); antirabies hyperimmune serum; antivenom sera; diphtheria antitoxin; immunoglobulin, human normal; tetanus antitoxin

19.2 *Vaccines*

19.2.2 *For specific groups of individuals* — influenza vaccine; meningococcal vaccine; rabies vaccine; typhoid vaccine; yellow fever vaccine

19.2.1 *For universal immunization* — BCG vaccine (dried); diphtheria-pertussis-tetanus vaccine; diphtheria-tetanus vaccine; measles vaccine; poliomyelitis (live attenuated); tetanus vaccine

20. MUSCLE RELAXANTS (PERIPHERALLY ACTING) AND CHOLINESTERASE INHIBITORS   neostigmine; suxamethonium; tubocurarine

21. OPHTHALMOLOGICAL PREPARATIONS

21.1 *Antiinfective* — silver nitrate; sulfacetamide

21.2 *Antiinflammatory* — hydrocortisone

21.3 *Local anesthetics* — tetracaine

21.4 *Miotics (agents which cause the pupil to contract)* — pilocarpine

21.5 *Mydriatics (agents which dilate the pupil)* — homatropine

21.6 *Systemic (agents affecting the eye, through administration by mouth)* — acetazolamide

22. OXYTOCICS (agents which stimulate the contraction of the uterus)   ergometrine; oxytocin

23. PERITONEAL DIALYSIS SOLUTION   intraperitoneal dialysis solution (of appropriate composition)

24. PSYCHOTHERAPEUTIC DRUGS   amitriptyline; chlorpromazine; diazepam; fluphenazine; haloperidol; lithium carbonate

25. DRUGS ACTING ON THE RESPIRATORY TRACT

25.1 *Antiasthmatic drugs* — aminophylline; epinephrine; salbutamol; complementary drug: ephedrine*

25.2 *Antitussives (for relieving coughs)* — codeine

26. SOLUTIONS CORRECTING WATER, ELECTROLYTE, AND ACID-BASE DISTURBANCES

26.1 *Oral* — oral rehydration salts (for glucose salt solution); potassium chloride

26.2 *Parenteral (by infusion)* — compound solution of sodium lactate; glucose; glucose with sodium chloride; potassium chloride; sodium bicarbonate; sodium chloride; water for injection

27. SURGICAL DISINFECTANTS   chlorhexidine; iodine

28. VITAMINS AND MINERALS   ascorbic acid; ergocalciferol; nicotinamide; pyridoxine; retinol; riboflavin; sodium fluoride; thiamine

---

*Suggested as possible alternatives when infectious organisms develop resistance to essential drugs, or in cases of rare disorders or exceptional circumstances.

Microwave Sickness

A Workmen's Compensation Board in New York State ruled on February 26, 1981, that the death of Bell Telephone employee Samuel Yannon in 1974 was caused by "microwave sickness." This ruling could make Yannon's widow eligible for benefits to pay the cost of her husband's long and painful illness. It also raised concern for thousands of other workers exposed to microwave radiation.

Yannon, a phone company employee since 1930, was assigned in 1954 to work on the 87th floor of the Empire State Building in New York City. His job was to monitor and tune television transmission signals. Television signals are weak microwave radiation, although they are much stronger near broadcast equipment, such as that at the top of the Empire State Building. About ten years later Yannon began complaining of sight and hearing problems. Gradually his condition worsened. He got cataracts in his eyes, began having balance problems (presumably from inner ear involvement), and showed signs of severe premature senility. He spent his last years in a mental hospital unable to recognize even his wife, and was a victim of severe physical wasting, losing nearly 100 pounds. He died there at the age of 63 in 1974.

The phone company opposed the widow's appeal to the compensation board, claiming that Yannon was a victim of Alzheimer's syndrome. But attending doctors testified that some characteristic symptoms of that disease were not present.

The board's finding in favor of Yannon also stirred concern about standards for microwave exposure. The federal standards for exposure to microwaves is 10 milliwatts per square centimeter, and testimony showed that Yannon had been exposed to only 1.5 milliwatts per square centimeter.

IT'S NOT THE SCREEN, IT'S THE JOB

The National Academy of Sciences reported in the summer of 1983 that 40 to 80 percent of all persons working at video display terminals (VDTs) report eyestrain or stress. The report went on to say that the VDTs themselves present no hazard to health. The cause of the distress is the jobs that involve VDTs. These jobs constitute a new kind of sweatshop in which workers are allowed little or no autonomy, are rated on the number of keystrokes they make per hour, and are forced to work in a regimented way.

VDTs remain under suspicion, however. Major studies are being sponsored by the National Institutes of Occupational Safety and Health on the effect of VDTs on pregnant women, and by IBM on physiological signs of visual fatigue.

Laetrile Report

The controversial drug laetrile, which caused a stir among victims of cancer in the late 1970s and which continues to be the basis of treatment in centers in Mexico and elsewhere, was the subject of a report sponsored by the National Cancer Institute and issued early in May 1981.

In carefully designed clinical tests carried out at four prestigious cancer treatment centers, the institute reported flatly that laetrile had proven itself totally "ineffective as treatment for cancer." The report dashed the last remote hopes that laetrile might be approved at any time for use in the United States.

The tests were begun in 1980 at the Mayo Clinic in Rochester, Minnesota; the UCLA Medical Center; the University of Arizona Health Sciences Center; and the Memorial Sloan-Kettering Cancer Center in New York City. Laetrile was administered in all to 178 patients with advanced cancers, the people most likely to resort to laetrile. The report states that the drug failed on all four of the claims made by its proponents. It failed to promote regression of cancer, it failed to extend life, it failed to improve symptoms of the disease, and it failed to contribute to an increase in weight or physical activity by the patients.

The statistics showed that of the 156 patients available for full evaluation, 90 percent showed progression of the disease three months after the beginning of treatment; 80 percent died within eight months. These statistics are about what would be expected with no treatment at all. Ten weeks after treatment, only 5 percent of the patients themselves reported any improvement and 3 percent reported either a weight gain or increased physical activity. Researchers said that these last statistics are comparable to reports received on administration of placebos (pills with no medical effect).

Laetrile is manufactured from apricot pits and contains an ingredient that the body converts to cyanide, a potent poison. The research program did not document negative effects of the drug, but such effects are likely at high dosages.

Despite the lack of evidence that laetrile was of use in treating cancer, clinics outside the United States continued to offer the drug, along with other treatments not permitted in the United States or deemed ineffective by American physicians. Patients from the United States continue to travel to Mexico or other countries to get treatment when American doctors offer little hope. The appeal of such clinics did not even seem to be tarnished when well-known cancer victims, such as actor Steve McQueen, died in one.

Hepatitis Vaccine

The Food and Drug Administration (FDA) gave its approval to a new vaccine effective against hepatitis B on November 16, 1981. Hepatitis is a debilitating liver disease; the B strain infects up to 100,000 persons in the United States each year, with a mortality rate of 2 percent. The disease is even more serious worldwide, infecting more than 500,000,000 people, most in less developed regions. Furthermore, it is suspected of causing liver cancer. The FDA announcement launched the first new viral vaccine in a decade and the first one ever produced entirely from human blood.

The hepatitis vaccine is produced by collecting blood samples from persons who have been infected by the disease. Scientists at Merck Institute for Therapeutic Research in Pennsylvania developed a technique to isolate the antigen, or surface protein, that identifies the disease-causing agent. This protein, called hepatitis B surface antigen, stimulates antibody production by the human immune system. Once the antigen is separated from the virus causing the disease, it can be injected as vaccine, prompting the body of the recipient to develop immunity without contracting the infection.

Most other vaccines have been made by a process of "attenuation." Cells containing the target disease and antigen have been passed through the system of other animal hosts where they mutate to survive in these animals. They then are retrieved in the mutated form, which retains the antigen, but without the strain of the disease that is infectious to humans.

Hepatitis B, however, infects no animals except humans and chimpanzees. In the procedure developed by the Merck labs, attenuation is accomplished by a technique of separating antigens from disease-causing viruses that requires more than a year of processing. As a result, the vaccine will cost about $100 for three injections, limiting its initial use in the United States to people considered to be at high risk.

There are many critics of vaccines derived from human blood without mutation in another animal host. They fear that the vaccine might contain other viruses infectious to humans. Both this concern and the high cost of the present method of production have led researchers to explore other ways of producing the vaccine. An alternative is to produce the hepatitis B surface antigen by genetic engineering and grow it in yeast cells. The first successful clinical trial of this vaccine was announced June 1, 1984. Another is to synthesize the antigen from peptides that are the molecular equivalent of the natural protein. This second process has not yet produced an antigen of sufficient strength to create immunity in human recipients, however.

Report on Nitrites

The National Academy of Sciences issued a 600-page report in December 1981 reporting its findings on the hazards of nitrites and nitrates in foodstuffs. Nitrites (and to a lesser extent nitrates) can be converted by the body into nitrosamines, a class of known cancer-causing substances. Fear had been expressed by scientists that some cancers might be caused or at least promoted by the use of nitrites and nitrates in food processing.

The researchers reported that the major source of nitrites in the diet is processed meats, to which nitrite is added to prevent food spoilage, particularly the growth of the deadly botulism bacterium. Nitrates, on the other hand, are found most plentifully in fresh vegetables, where they occur naturally along with vitamins C and E, both of which appear to inhibit the conversion of nitrates to nitrosamines in the body.

Dietary intake represents only a minor portion of nitrite and nitrate exposure, however. Much larger dosages are encountered by cigarette smokers and by workers in the rubber, leather, and rocket fuel industries. Another serious concern is the high level of these compounds in the drinking water in agricultural regions where use of nitrogen-based fertilizers is heavy.

The academy recommended the cautious reduction of nitrites in processed meats and research into other means of preventing spoilage. A later report, issued in 1982, suggested that when nitrites are used in food processing, vitamins C and E should be added as well to inhibit nitrosamine production in the body. The vitamins should be especially useful in bacon, the major source of nitrites in the American diet, particularly vitamin E, which is fat-soluble. However, because fat-soluble vitamins also tend to accumulate in human fat, a person who ate a great deal of vitamin-E-laced bacon might be in danger of toxic accumulations of that vitamin.

SOMETHING GOOD ABOUT COLAS

While millions of parents try to limit their children's intake of cola drinks, medical science has found that such drinks may benefit children who live in old apartment houses with peeling paint or near areas of high automobile traffic. Such children have been found to have high levels of poisonous lead in their bodies. Bruce J. Aungst of the DuPont Company and Ho-Leung Fung of the State University of New York at Buffalo report that cola significantly reduces the absorption of lead paint or other lead contaminants in rats. It is not actually known that the same effect would be found in children.

Fetal Surgery

The months of 1981 were filled with new reports of advances in the treatment of unborn infants. Doctors concerned with the protection and health of the fetus had only recently learned to intervene in fetal development by administering massive doses of vitamins or other substances to the mothers during troubled pregnancies. Doctors had also experimented with putting certain substances directly into the amniotic fluid.

In June two doctors at Mount Sinai School of Medicine in New York reported on the first known case of "selective abortion." It was discovered that one of a pair of fetal twins was suffering from Down's syndrome, a birth defect that almost always leads to severe retardation. At the urging of the children's parents, the doctors used a syringe to draw blood from the heart of the affected child, thus ending its life. The surviving twin was born four months later and was healthy and normal at seven months of age.

Thomas Kerenyi and Usha Chitkara, the physicians, emphasized the ethical questions arising in such a procedure. In an August issue of the *Journal of the American Medical Association*, Michael Harrison, Mitchell Golbus, and Roy Filly, pioneers in fetal intervention research at the University of California at San Francisco, reviewed the development of new techniques in the field and urged that a special ethical review board be established by the American Medical Association to oversee new procedures and to establish ethical guidelines.

In September Dr. William H. Clewall of the University of Colorado Health Sciences Center at Denver reported still another dramatic advance. He had succeeded in implanting a small valve in the head of a fetus diagnosed as having hydrocephalus, a defect that usually leads to brain damage before birth. The valve allowed the normal circulation of the infant's cerebro-spinal fluid and kept fluid pressure on the brain within normal limits, thus eliminating or greatly reducing the risk of brain damage. The child came to term and was apparently normal and healthy.

In December the three doctors in San Francisco reported still another advance. They actually operated on a fetus outside of the womb to repair a urethral blockage that was causing kidney damage. The fetus was in its twenty-first week of development. The mother carried it to term and it was delivered normally, but it died soon after birth from the kidney damage it had suffered prior to the operation.

Further advances in techniques for improving the survival of fetuses and for protecting them from otherwise damaging deficiencies or defects seem likely in the years to come.

American Test-tube Baby

On December 28, 1981, the first "test-tube baby" in the United States was delivered at Norfolk (Virginia) General Hospital. The child, named Elizabeth Jordan Carr, was born to Judith and Roger Carr of Westminster, Massachusetts.

In a procedure first successfully accomplished in Britain in 1977, an ovum from Mrs. Carr's ovary was surgically extracted and united with sperm from Mr. Carr in a laboratory dish. Soon after conception occurred, the tiny fetus was implanted in Mrs. Carr's uterus, where it attached itself and grew normally to term.

The process, called *in vitro* fertilization, allows a woman whose Fallopian tubes are blocked to give birth to her own child. Normally, the ovum travels from the ovaries down the Fallopian tube to the uterus, where conception occurs.

By June of 1982, the test-tube baby procedure was well on its way to becoming routine. Doctors reported nineteen babies already delivered (twelve in Britain, six in Australia, and one in the U.S.) and over 150 women presently carrying babies who were conceived outside the womb.

Medical publications and meetings continue to explore the ethical implications of the new procedure, which has raised questions among some scientists and in some religious groups.

NEW CHILDBIRTH OPTIONS

Women who want a more natural way to give birth can now opt for a birthing chair. Equipped with leg rests and seat-level handles for gripping, the chair can be raised, lowered, and tilted backward. Its vertical position allows a woman to use her abdominal muscles more efficiently and to take advantage of gravity. Furthermore, a mother may view the birth of her baby directly. A major disadvantage of the birthing chair is that a woman must be transferred from the labor bed to the chair and, once in it, as it is currently designed, she is not able to adjust the seat-level handles by herself.

Another option is the birthing room, combining a home-like atmosphere with the safety of a hospital environment. The woman, admitted to the birthing room along with her husband or any other support person desired, labors, delivers, and recovers in the same bed, but this is far from the old-fashioned birth at home. This birthing bed is especially constructed to incorporate features of a birthing chair and a labor-delivery bed: add stirrups and the bed becomes a delivery table; raise the upper section to a vertical position and drop the lower section to form a footrest; or detach it completely and the bed is converted into a comfortable chair.

1982

Smoking Report

In 1982, the U.S. Surgeon General's report on the health of the nation was stronger than ever before in detailing the dangers of tobacco, especially cigarette smoking. In the fourteenth report since 1964, Surgeon General C. Everett Koop concluded, "Cigarette smoking is clearly identified as the chief preventable cause of death in our society and the most important public health issue of our time."

The Koop report asserted that smoking accounted for approximately 340,000 deaths per year in the United States. Of these, approximately 129,000 were from cancers caused by smoking. The report noted that lung cancer mortality had increased sixfold since 1950 to reach an estimated 111,000 deaths in 1982, and that 85 percent of these deaths were attributable to smoking. The report also echoed the warnings of private researchers that there are probable links between smoking and cancers of the larynx, esophagus, stomach, pancreas, kidney, bladder, and cervix. Lung cancer remains the most serious form of cancer associated with smoking because of its high incidence and resistance to treatment. (Heart disease caused by smoking accounts for more deaths, however.) Only one in ten victims of lung cancer survives five years beyond the diagnosis of the disease. Surgeon General Koop told the nation that if it were not for the dramatic increase in lung cancer, "the overall cancer mortality would have fallen."

The Koop report also went beyond earlier government reports in associating smoking directly with a variety of other health problems, including heart ailments and chronic lung diseases such as emphysema. Smoking by pregnant women, the report said, has been clearly linked with miscarriage, premature delivery, and birth defects among the young. The report also cited dangers in tobacco consumed in pipes, cigars, and snuff and acknowledged the danger to nonsmokers as a result of smoky rooms.

Among the results of warnings such as the Koop report have been the spread of no-smoking areas in public places, including no-smoking areas in restaurants. A move to eliminate all smoking on short commercial airline flights, however, failed in 1984.

Also in 1984, however, the United States congress acted to stiffen warnings on cigarette packages and in cigarette advertising. Among the new warnings were statements that linked smoking to cancer, heart disease, and emphysema specifically and also an alert to pregnant women not to smoke.

Life Expectancies

Life expectancy figures are often quoted, but frequently in contexts that suggest that they are poorly understood by the persons doing the quoting. For example, when it is stated that the average lifespan in 1900 was 48 years, a writer may go on to imply that few people in 1900 lived to be much beyond 50 or so. Yet, if you examine the biographies of people born in 1850, you will find that many of them lived well into the twentieth century. The confusion arises because the average lifespan includes deaths in childhood, extremely prevalent in 1900. If a person survived childhood, he or she had a life expectancy that was only 6 to 9 years less than the life expectancy of people living today.

For most people, a more useful figure than the average life expectancy at birth is the average remaining life expectancy at a specific age. It is not surprising that the average remaining expectancy for people who have lived to 75 is around 10 years; people who have lived to be 75 have already shown a significant amount of toughness.

However, average life expectancy from birth is a useful figure in comparing national populations, as it tends to reflect differences in deaths at early ages caused by poor medical care or poor sanitation. A nation with a low average life expectancy from birth knows that it needs to make improvements.

Expectation of Life at Birth

Year	Total	Total Male	Female	Total	White Male	Female	Black and Other Total	Male	Female
1920	54.1	53.6	54.6	54.9	54.4	55.6	45.3	45.5	45.2
1930	59.7	58.1	61.6	61.4	59.7	62.5	48.1	47.3	49.2
1940	62.9	60.8	65.2	64.2	62.1	66.6	53.1	51.5	54.9
1950	68.2	65.6	71.1	69.1	66.5	72.2	60.8	59.1	62.9
1955	69.6	66.7	72.8	70.5	67.4	73.7	63.7	61.4	66.1
1960	69.7	66.6	73.1	70.6	67.4	74.1	63.6	61.1	66.3
1965	70.2	66.8	73.7	71.0	67.6	74.7	64.1	61.1	67.4
1970	70.9	67.1	74.8	71.7	68.0	75.6	65.3	61.3	69.4
1975	72.6	68.8	76.6	73.4	69.5	77.3	68.0	63.7	72.4
1980	73.7	70.0	77.5	74.4	70.7	78.1	69.5	65.3	73.6
1981*	74.1	70.3	77.9	74.7	71.0	78.5	70.3	66.1	74.5
1982*	74.5	70.8	78.2	75.1	71.4	78.7	70.9	66.5	75.2

* Preliminary

Source: U.S. National Center for Health Statistics

Average Remaining Life Expectancy in the United States

Age in 1980	Total	Expectation of life in years			
		White		Black	
		Male	Female	Male	Female
At birth	73.7	70.7	78.1	63.7	72.3
3	71.8	68.7	76.0	62.3	70.9
5	69.8	66.8	74.0	60.5	69.0
8	66.9	63.8	71.1	57.6	66.0
10	64.9	61.9	69.1	55.6	64.1
12	63.0	59.9	67.1	53.6	62.1
15	60.0	57.0	64.2	50.7	59.2
18	57.2	54.2	61.3	47.9	56.2
20	55.3	52.4	59.4	46.0	54.3
23	52.5	49.6	56.5	43.4	51.4
25	50.7	47.8	54.5	41.7	49.5
28	47.9	45.1	51.6	39.1	46.7
30	46.0	43.2	49.7	37.4	44.8
33	43.2	40.4	46.8	34.9	42.1
35	41.3	38.6	44.9	33.2	40.2
38	38.5	35.8	42.0	30.8	37.5
40	36.7	34.0	40.1	29.1	35.7
42	34.8	32.1	38.2	27.5	33.9
45	32.1	29.4	35.4	25.2	31.4
48	29.5	26.8	32.7	23.0	28.9
50	27.8	25.2	30.9	21.6	27.3
53	25.3	22.7	28.2	19.7	24.9
55	23.7	21.2	26.5	18.4	23.4
58	21.4	18.9	24.0	16.6	21.3
60	19.9	17.5	22.4	15.5	19.8
63	17.8	15.4	20.0	13.9	17.8
65	16.4	14.2	18.5	12.9	16.5
70	13.2	11.3	14.8	10.5	13.4
75	10.4	8.8	11.5	8.3	10.7
80	7.9	6.7	8.6	6.3	8.2
85 and over	5.9	5.0	6.3	4.5	6.1

Source: U.S. National Center for Health Statistics

Baby Doe

On April 15, 1982, a baby known only as Baby Doe (to protect the privacy of his parents) died at the age of six days.

Baby Doe had been born in Bloomington, Indiana, on April 9 with a physical deformity—an incomplete esophagus, making normal nourishment impossible—and with Down's syndrome or congenital mongolism, a handicap that almost always leads to severe mental retardation. With new neonatal procedures, the doctors could attach the incomplete esophagus to the baby's stomach and thus avert starvation, but they were powerless against Down's syndrome, which is caused by a chromosomal defect. Baby Doe's parents, upheld by local courts, declined to authorize the operation, and the infant died.

News of this event reached President Ronald Reagan and he professed shock and disapproval. In May the Department of Health and Human Services (HHS) issued a regulation warning hospitals that "failure to feed or care for handicapped infants" might threaten the hospital's federal funding. HHS also instituted a 24-hour "hotline" on which hospital employees could report cases like that of Baby Doe.

The American Hospital Association and the American Academy of Pediatrics both sued HHS over what they perceived to be federal infringement on physicians' and parents' responsibilities. In April 1983, federal judge Gerhard Gesell (related to the Gesells well known for their research in child development) ruled that the HHS regulation and hotline were "ill-considered."

Other confrontations on this issue continued in 1983 and 1984. One developed around an infant called Baby Jane Doe, who was born October 11, 1983, on Long Island, New York. She suffered from spina bifida, an open spinal cord, hydrocephalus, and other grave birth defects. When her parents decided against elaborate corrective surgery, HHS challenged the decision in a court case that focused on access to the infant's medical records. The parents were upheld by a New York State Court of Appeals decision that called their action within "the most private and most precious responsibility vested in parents for the care and nurture of their children."

Medical associations insist that they are committed in principle to upholding the sacredness of life, and the AMA suit against the federal government's interference was upheld by a federal district court in May 1984. On the same day, a Bronx, New York district judge decided to order an operation, opposed by parents, on a baby with problems similar to the original Baby Doe, including Down's syndrome.

Lasers for Common Eye Ailment

One of the leading causes of blindness in the United States is senile macular degeneration (SMD), a condition that afflicts a large number of elderly people. It causes the degeneration of the macula, the part of the retina of the eye responsible for the reception of central (as opposed to peripheral) vision, thus leading to a particularly disabling form of blindness.

In a major study published in the June 1982 issue of the *Archives of Ophthalmology*, researchers at the National Eye Institute announced that laser treatment had proven extremely effective in one of the two major forms of the disease. Between 5 and 20 percent of the cases of SMD are "wet," meaning that the damage to the macula is caused by leakage from blood vessels in the retina itself. It is in these cases that laser therapy seals off the leaking blood vessels without damaging surrounding tissue and thus saves the patient's central vision.

Lasers produce beams of coherent light of high intensity. As such, they can transmit energy through transparent materials, such as the lens and inner fluid of the eye, and deliver the energy to a small opaque spot, such as the retina. The amount of the energy, which is released as heat upon striking the retina, can be precisely controlled by the surgeon. This laser-generated heat has been used to reattach detached retinas as well as to seal blood vessels in cases of SMD.

In a two-year study, institute researchers compared the results of laser treatment with conventional treatment. After eighteen months, only 25 percent of those in the study group had lost significant vision, compared with 60 percent in the control group.

A spokesman estimated that the new treatment might benefit as many as 1,000,000 patients in ten years and that it could save the U.S. government $2,500,000,000 in tax funds that would otherwise be paid as benefits to those blinded by this form of SMD.

By the early 1980s, in fact, the laser was gradually replacing the surgeon's knife routinely in many forms of operations. Not only can lasers be used to seal blood vessels that are already leaking, but also they can be used to make incisions. The incisions are smoother than those made by a scalpel and also seal the leakage of blood at the same time. While eye operations of various types became the first proving-ground for medical lasers, their use has spread to other parts of the body, especially areas where bleeding could cause damage.

(For another medical application of lasers, unrelated to surgery, see "New Pain Treatments," page 449.)

Reye's Syndrome Warning

The U.S. Department of Health and Human Services (HHS) took the unusual step in 1982 of announcing plans to seek a warning label on aspirin, alerting users that in children the drug may cause the dangerous disease called Reye's syndrome. The announcement on June 4 provoked immediate objections from the aspirin industry because it preceded any action by the Food and Drug Administration or any conclusive scientific evidence of a link between the medication and the disease.

Reye's syndrome is a rare but frequently fatal transformation of a viral infection, such as flu or chicken pox, into seizures progressing to coma and death. A quarter of the children affected die of liver and brain degeneration. Many of those who survive suffer permanent brain damage. The disease seems to affect children between five and sixteen years of age.

The HHS announcement suggests that the syndrome is "iatrogenic"—caused by medical treatment; in this case, the aspirin frequently given to children with viral fevers. No link has been suggested between the syndrome and acetaminophen, the other commonly prescribed analgesic.

Concern about a possible link between the syndrome and aspirin came after a survey by the Centers for Disease Control revealed that 137 of 144 Reye's cases examined included early administration of aspirin for fever. In the week after the HHS announcement, four additional independent studies were released. Two supported the hypothesis of a link between the medication and the disease; one found no such link; and one was inconclusive. In the absence of any hypothesis of how aspirin might transform a viral disease into Reye's syndrome, most medical researchers in 1982 remained skeptical.

In November 1982 Secretary Schweiker of HHS announced that he was delaying the order for a warning label on aspirin. He said at a press conference that the original research findings had been called into question and that it would take additional time and study to determine if there is indeed a link between aspirin and the disease. He had received a recommendation from the American Academy of Pediatrics requesting further study before ordering the warning label. But doctors on the academy's infectious diseases committee protested the delay and suggested that it was only in response to strong pressure from the manufacturers of aspirin.

It should be noted, also, that various studies have shown that fever is one of the principal ways that the body fights disease. Unless the fever is so high as to be life threatening, taking aspirin to reduce the fever may actually prolong the illness.

The year after Secretary Schweiker announced the delay in labeling to await further research, laboratory findings in two separate studies suggested that the link between aspirin and Reye's syndrome was not coincidental.

In January 1983 scientists at the University of Michigan reported some evidence that a link between aspirin and Reye's syndrome might indeed exist. They reported that twelve of sixteen ferrets given aspirin and inoculated with flu virus died of symptoms resembling those of Reye's syndrome. The animals were found to have excess ammonia in their blood, a finding compatible with what has been found in biopsies of human Reye's victims. But the role of aspirin in producing such a chemical imbalance remains unknown.

Later in the year, Kwan-sa You of Duke University Medical Center demonstrated that aspirin solutions cause swelling of the small bodies in cells called mitochondria. Mitochondria process the energy reactions in the cell and are essential to life. The structural deformation of mitochondria in liver cells from victims of Reye's syndrome is essentially the same as the transformation You observed in cell cultures. It is known that aspirin in normal individuals is metabolized by the liver into a chemical that does not cause such swellings. Therefore, You speculated, it is possible that children who develop Reye's syndrome lack some essential enzyme involved in the conversion of aspirin in the liver to a harmless form.

You's suggestion focuses on one of the many medical mysteries connected with disease: Why, when exposed to the same environmental stimulus, do some people get a disease and others not? In the case of Reye's syndrome, many children take aspirin for fevers, but very few develop Reye's syndrome.

Despite the research, 1984 saw no warning labels on aspirin.

There was, however, a new development that year in the over-the-counter painkiller field. The prescription drug ibuprofen was approved for nonprescription sale, and two manufacturers of aspirin and acetaminophen moved swiftly to offer it to the public. Ibuprofen was criticized by some doctors for its possible side effects, especially on those persons with high blood pressure, but HHS declared that warning labels on the drug were adequate. The labels state that ibuprofen should not be taken without consulting a physician, but do not spell out which illnesses might preclude the use of ibuprofen. A spokesman for the HHS pointed out that if aspirin were to undergo the same kind of review ibuprofen had undergone, it is doubtful that the powerful drug would be approved for nonprescription use.

Starch Blockers

On July 1, 1982, the U.S. Food and Drug Administration (FDA) asked manufacturers to halt distribution of "starch blockers," a popular diet aid invented in 1981 that was just gaining wide notice and popularity. By July 1982 it was estimated that over 1,000,000 starch-blocker tablets had been sold. The FDA charged that the pills cause serious side effects and that their long-term effects had not been determined. Consumers of the pills had reported confusion, nausea, diarrhea, vomiting, and flatulence. Researchers speculated that these symptoms were caused by the passage of unusually large amounts of undigested starch to the intestines. Bacteria in the large intestine presumably cause the starch to ferment, resulting in gas and painful cramps.

The FDA also expressed concern about the long-term effects of starch blockers, including dietary deficiencies caused by the incomplete digestion of starches, interaction with prescription drugs, and possible harmful effects on such groups as diabetics and pregnant women.

The concept of starch blockers was developed in 1981 by Notre Dame biochemist J. John Marshall. He isolated a protein in kidney bean extract that slows the action of the enzyme alpha amylase, which breaks down starch in digestion. According to manufacturers, one starch blocker pill is able to prevent digestion of up to 100 grams (3.5 ounces) of starch, the equivalent of about 400 Calories. The undigested starch would be excreted, allowing dieters to eat starchy foods such as bread, potatoes, and pasta without digesting their full caloric value.

Manufacturers of the new diet aid disputed the jurisdiction of the FDA in ending sale of their product. They claimed that starch blockers are a natural food supplement rather than a drug. The FDA, however, defines a drug as "any substance which affects any function of the body," and under that broad definition exercised its power against starch blockers pending further investigation.

The dispute went to the federal courts, where it was settled in the FDA's favor. In September some states seized remaining shelf stock of the pills, and in October the FDA outlawed any future manufacture of starch blockers.

In December 1982 a study by George W. Bo-Linn and associates at Baylor University Medical Center put the cap on the story. Volunteers who ate spaghetti dinners with and without taking starch blockers excreted the same amount of Calories. In other words, discounting the side effects, starch blockers did not work as a diet aid in any case.

Nutritive Values of Foods

Food and Amount	Grams	Energy Cal-ories	Protein Grams	Fat Grams	Carbo-hydrate Grams	Calcium Milli-grams	Phos-phorus Milli-grams	Iron Milli-grams	Potas-sium Milli-grams	Vitamin A Int. Units	Ascorbic Acid Milli-grams
Dairy Products											
Cheese, natural:											
Blue, 1 oz.	28	100	6	8	1	150	110	0,1	73	200	0
Camembert, 1 oz.	38	115	8	9	Tr.	147	132	.1	71	350	0
Cheddar, 1 oz.	28	115	7	9	Tr.	204	145	.2	28	300	0
Cottage, creamed, 4% fat, 1 c.	225	235	28	10	6	135	297	.3	190	370	Tr.
Cottage, 1% fat, 1 c.	226	165	28	2	6	138	302	.3	193	80	Tr.
Cream, 1 oz.	28	100	2	10	1	23	30	.3	34	400	0
Mozarella, whole milk, 1 oz.	28	90	6	7	1	163	117	.1	21	260	0
Parmesan, grated, 1 Tb.	5	25	2	2	Tr.	69	40	Tr.	5	40	0
Provolone, 1 oz.	28	100	7	8	1	214	141	.1	39	230	0
Ricotta, whole milk, 1 c.	246	430	28	32	7	509	389	.9	257	1,210	0
Ricotta, part skim, 1 c.	246	340	28	19	13	669	449	1.1	308	1,060	0
Swiss, 1 oz.	28	105	8	8	1	272	171	Tr.	31	240	0
Cheese, pasteurized process:											
American, 1 oz.	28	105	6	9	Tr.	174	211	.1	46	340	0
Swiss, 1 oz.	28	95	7	7	1	219	216	.2	61	230	0
spread, American, 1 oz.	28	80	5	6	2	159	202	.1	69	220	0
Cream, sweet:											
Light, table, 1 c.	240	470	6	46	9	231	192	.1	292	1,730	2
Heavy, whipping, 1 c.	238	820	5	88	7	154	149	.1	179	3,500	1
Whipped, pressurized, 1 c.	60	155	2	13	7	61	54	Tr.	88	550	0

Food and Amount	Grams	Energy Calories	Protein Grams	Fat Grams	Carbohydrate Grams	Calcium Milligrams	Phosphorus Milligrams	Iron Milligrams	Potassium Milligrams	Vitamin A Int. Units	Ascorbic Acid Milligrams
Cream, sour, 1 c.	230	495	7	48	10	268	195	.1	331	1,820	2
Cream, imitation:											
Creamer, powdered, 1 tsp.	2	10	Tr.	1	1	Tr.	8	Tr.	16	Tr.	0
Topping, frozen, 1 c.	75	240	1	19	17	5	6	.1	14	650	0
Milk, fluid:											
Whole, 3.3% fat, 1 c.	244	150	8	8	11	291	228	.1	370	310	2
Lowfat, 2%, 1 c.	244	120	8	5	12	297	232	.1	377	500	2
Lowfat, 1%, 1 c.	244	100	8	3	12	300	235	.1	381	500	2
Skim, 1 c.	245	85	8	Tr.	12	302	247	.1	406	500	2
Buttermilk, 1 c.	245	100	8	2	12	285	219	.1	371	80	2
Milk, canned:											
Evaporated, whole, 1 c.	252	340	17	19	25	657	510	.5	764	610	5
Condensed, sweetened, 1 c.	306	980	24	27	166	868	775	.6	1,136	1,000	8
Milk Beverages:											
Chocolate Reg., 1 c.	250	210	8	8	26	280	251	.6	417	300	2
Shake, choc., 10 oz.	300	355	9	8	63	396	378	.9	672	260	0
Milk desserts:											
Ice cream, reg., 1 c.	133	270	5	14	32	176	134	.1	257	540	1
Ice Milk, 1 c.	131	185	5	6	29	176	129	.1	265	210	1
Sherbet, 1 c.	193	270	2	4	59	103	74	.3	198	190	4
Custard, baked, 1 c.	265	305	14	15	29	297	310	1.1	387	930	1
Pudding, cooked from mix, 1 c.	260	320	9	8	59	265	247	.8	354	340	2
Yogurt:											
fruit-flavored, 8 oz.	227	230	10	3	42	343	269	.2	439	120	1
Plain, 8 oz.	227	145	12	4	16	415	326	.2	531	150	2

Eggs

Eggs:

	Weight (g)	Calories									
Raw, whole, 1	50	80	6	6	1	28	90	1.0	65	260	0
Raw, white, 1	33	15	3	Tr.	Tr.	4	4	Tr.	45	0	0
Raw, yolk, 1	17	65	3	6	Tr.	26	86	.9	15	310	0
Fried in butter, 1	46	85	5	6	1	26	80	.9	58	2900	0

Fats, Oils, Related Products

Butter, 1 Tbs.	14	100	Tr.	12	Tr.	3	3	Tr.	4	430	0
Butter, whipped, 1 Tbs.	9	65	Tr.	8	Tr.	2	2	Tr.	2	290	0
Vegetable shortening, 1 c.	200	1,770	0	200	0	0	0	0	0	—	0
Lard, 1 c.	205	1,850	0	205	0	0	0	0	0	0	0
Margarine, 1 Tbs.	14	100	Tr.	12	Tr.	3	3	Tr.	4	470	0
Margarine, soft, 8 oz.	227	1,635	1	184	Tr.	53	52	.4	59	7,500	0

Oils:

Corn, 1 Tbs.	14	120	0	14	0	0	0	0	0	—	0
Olive, 1 Tbs.	14	120	0	14	0	0	0	0	0	—	0

Salad dressings:

Blue cheese, regular, 1 Tbs.	15	75	1	8	1	12	11	Tr.	6	30	Tr.
Blue Cheese, lo-cal, 1 Tbs.	16	10	Tr.	1	1	10	8	Tr.	5	30	Tr.
French, regular, 1 Tbs.	16	65	6	3	2	2	1	13	—	—	—
French, lo-cal, 1 Tbs.	16	15	Tr.	1	2	2	2	.1	13	—	—
Italian, Regular, 1 Tbs.	15	85	Tr.	9	1	2	1	Tr.	2	Tr.	—
Italian, lo-cal, 1 Tbs.	15	10	Tr.	1	Tr.	Tr.	—	Tr.	2	Tr.	—
Mayonnaise, 1 Tbs.	14	100	Tr.	11	Tr.	3	4	.1	5	40	—

Fish, Shellfish, Meat, Poultry, Related Products

Food and Amount	Grams	Energy Cal-ories	Protein Grams	Fat Grams	Carbo-hydrate Grams	Calcium Milli-grams	Phos-phorus Milli-grams	Iron Milli-grams	Potas-sium Milli-grams	Vitamin A Int. Units	Ascorbic Acid Milli-grams
Fish and shellfish:											
Bluefish, baked, 3 oz.	85	135	22	4	0	25	244	0.6	—	40	—
Clams, raw, 3 oz.	85	65	11	1	2	59	138	5.2	154	90	8
Clams, canned, 3 oz.	85	45	7	1	2	47	116	3.5	119	—	—
Crabmeat, canned, 1 c.	135	135	24	3	1	61	246	1.1	149	—	—
Fish stick, breaded, 1	28	50	5	3	2	3	47	.1	—	0	—
Haddock, breaded, fried, 3 oz.	85	140	17	5	5	34	210	1.0	296	—	2
Oysters, raw, 1 c.	240	160	20	4	8	226	343	13.2	290	740	—
Salmon, canned, 3 oz.	85	120	17	5	0	167	243	.7	307	60	—
Sardines, canned in oil, 3 oz.	85	175	20	9	0	372	424	2.5	502	190	—
Scallops, breaded, fried, 6	90	175	16	8	9	—	—	—	—	—	—
Shrimp, canned, 3 oz.	85	100	21	1	1	98	224	2.6	104	50	—
Shrimp, fried, 3 oz.	85	190	17	9	9	61	162	1.7	195	—	—
Tuna, canned, 3 oz.	85	170	24	7	0	7	199	1.6	—	70	—
Meat and meat products:											
Bacon, fried, 2 pcs.	15	85	4	8	Tr.	2	34	.5	35	0	—
Beef, braised, 3 oz.	85	245	23	16	0	10	114	2.9	184	30	—
Beef, ground lean, 3 oz.	85	185	23	10	0	10	196	3.0	261	20	—
Beef, roast, lean, 3 oz.	85	135	26	3	0	11	210	3.2	278	Tr.	—
Beefsteak, broiled, 3 oz.	85	330	20	27	0	9	162	2.5	220	50	—
Chili w/beans, canned, 1 c.	255	340	19	16	31	82	321	4.3	594	150	—
Corned Beef, canned, 3 oz.	85	185	22	10	0	17	90	3.7	—	—	—
Corned beef hash, 1 c.	220	400	19	25	24	29	147	4.4	440	—	—
Dried Beef, chipped, 2½ oz.	71	145	24	4	0	14	287	3.6	142	—	0

Food											
Ham, boiled, 1 oz.	28	65	5	5	0	3	47	.8	—	0	—
Ham, roasted, 3 oz.	85	245	18	19	0	8	146	2.2	199	0	—
Lamb Chop, broiled, 3 oz.	85	360	18	32	0	8	139	1.0	200	—	—
Lamb, roast leg, 3 oz.	85	235	22	16	0	9	177	1.4	241	—	—
Liver, beef, fried, 3 oz.	85	195	22	9	5	9	405	7.5	323	45,390	23
Pork chop, broiled, 2.7 oz.	78	305	19	25	0	9	209	2.7	216	0	—
Pork roast, 3 oz.	85	310	21	24	0	9	218	3.7	233	0	—
Veal, broiled cutlet, 3 oz.	85	185	23	9	0	9	196	2.7	258	—	—
Veal, roast, 3 oz.	85	230	23	14	0	10	211	2.9	259	—	—
Sausages and Luncheon Meats:											
Bologna, 1 oz.	28	85	3	8	Tr.	2	36	.5	65	—	—
Braunschweiger, 1 oz.	28	90	4	8	1	3	69	1.7	—	1,850	—
Brown/serve sausage, 1	17	70	3	6	Tr.	—	—	—	—	—	—
Deviled ham, canned, 1 Tbs.	13	45	2	4	0	1	12	.3	—	0	0
Frankfurter, cooked, 1	56	170	7	15	1	3	57	.8	—	—	—
Salami, dry, 1 slice	10	45	2	4	Tr.	1	28	.4	—	—	—
Salami, cooked, 1 oz.	28	90	5	7	Tr.	3	57	.7	—	—	—
Vienna sausage, 1	16	40	2	3	Tr.	1	24	.3	—	—	—
Poultry:											
Chicken, breast fried, 3.3 oz.	79	160	26	5	1	9	218	1.3	—	70	—
Chicken, leg fried, 2 oz.	38	90	12	4	Tr.	6	89	.9	—	50	—
Chicken, half broiled, 10 oz.	176	240	42	7	0	16	355	2.0	483	160	3
Chicken, canned, 3 oz.	85	170	18	10	0	18	210	1.3	117	200	—
Ch. chow mein, canned, 1 c.	250	95	7	Tr.	18	45	35	1.3	418	150	13
Turkey, roast, 3 oz.	85	160	27	5	0	7	213	1.5	312	—	—

Fruits and Fruit Products

Food and Amount	Grams	Energy Calories	Protein Grams	Fat Grams	Carbohydrate Grams	Calcium Milligrams	Phosphorus Milligrams	Iron Milligrams	Potassium Milligrams	Vitamin A Int. Units	Ascorbic Acid Milligrams
Apple, medium, 1	138	80	Tr.	1	20	10	14	.4	152	120	6
Apple juice, 1 c.	248	120	Tr.	Tr.	30	15	22	1.5	250	—	2
Applesauce, unsweetened, 1 c.	244	100	Tr.	Tr.	26	10	12	1.2	190	100	2
Apricots, raw, 3	107	55	1	Tr.	14	18	25	.5	301	2,890	11
Apricots, canned, 1 c.	258	220	2	Tr.	57	28	39	.8	604	4,490	10
Apricots, dried, 1 c.	130	340	7	1	86	87	140	7.2	1,273	14,170	16
Apricot nectar, canned, 1 c.	251	145	1	Tr.	37	23	30	.5	379	2,380	36
Avocado, 1	216	370	5	37	13	22	91	1.3	1,303	630	30
Banana, 1	119	100	1	Tr.	26	10	31	.8	440	230	12
Blackberries, 1 c.	144	85	2	1	19	46	27	1.3	245	290	30
Blueberries, 1 c.	145	90	1	1	22	22	19	1.5	117	150	20
Cherries, sour, canned, 1 c.	244	105	2	Tr.	26	37	32	.7	317	1,660	12
Cherries, sweet, raw, 10	68	45	1	Tr.	12	15	13	.3	129	70	7
Cranberry sauce, canned, 1 c.	277	405	Tr.	1	104	17	11	.6	83	60	6
Dates, pitted, 10	80	220	2	Tr.	58	47	50	2.4	518	40	0
Fruit cocktail, canned, 1 c.	255	195	1	Tr.	50	23	31	1.0	411	360	5
Grapefruit, medium, ½	241	50	1	Tr.	13	20	20	.5	166	540	44
Grapefruit juice, 1 c.	247	100	1	Tr.	24	20	35	1.0	400	20	84
Grapes, seedless, 10	50	35	Tr.	Tr.	9	6	10	.2	87	50	2
Grapejuice, canned, 1 c.	253	165	1	Tr.	42	28	30	.8	293	—	Tr.
Grape drink, canned, 1 c.	250	135	Tr.	Tr.	35	8	10	.3	88	—	—
Lemon, peeled, 1	74	20	1	Tr.	6	19	12	.4	102	10	39

Food	Grams	Calories	Protein (g)	Fat (g)	Carb. (g)	Calcium (mg)	Phos. (mg)	Iron (mg)	Potassium (mg)	Vit. A (IU)	Vit. C (mg)
Lemonade, conc., 1 can	219	425	Tr.	Tr.	112	9	13	.4	153	40	66
Melon, cantaloupe, ½	477	80	2	Tr.	20	38	44	1.1	682	9,240	90
Melon, honeydew, 1/10	226	50	1	Tr.	11	21	24	.6	374	60	34
Orange, raw, 1	131	65	1	Tr.	16	54	26	.5	263	260	66
Orange juice, 1 c.	248	110	2	Tr.	26	27	42	.5	496	500	124
Peach, pitted, 1	100	40	1	Tr.	10	9	19	.5	202	1,330	7
Peaches, syrup pack, 1 c.	256	200	1	Tr.	51	10	31	.8	333	1,100	8
Peaches, water pack, 1 c.	244	75	1	Tr.	20	10	32	.7	334	1,100	7
Pear, Bartlett, 1	164	100	1	1	25	13	18	.5	213	30	7
Pears, syrup pack, 1 c.	255	195	1	1	50	13	18	.5	214	10	3
Pineapple, raw, 1 c.	155	80	1	Tr.	21	26	12	.8	226	110	26
Pineapple, syrup pack, 1 c.	255	190	1	Tr.	49	28	13	.8	245	130	18
Pineapple juice, 1 c.	250	140	1	Tr.	34	38	23	.8	373	130	80
Plum, medium, 1	66	30	Tr.	Tr.	8	8	12	.3	112	160	4
Prunes, cooked, 1 c.	250	255	2	1	67	51	79	3.8	695	1,590	2
Prune juice, 1 c.	256	195	1	Tr.	49	36	51	1.8	602	—	5
Raisins, 1 c.	145	420	4	Tr.	112	90	146	5.1	1,106	30	1
Raspberries, raw, 1 c.	123	70	1	1	17	27	27	1.1	207	160	31
Strawberries, raw, 1 c.	149	55	1	1	13	31	31	1.5	244	90	88
Watermelon, 4 by 8 in. wedge	926	110	2	1	27	30	43	2.1	426	2,510	30
Grain Products											
Bagel, egg, 1	55	165	6	2	28	9	43	1.2	41	30	0
Barley, uncooked, 1 c.	200	700	16	2	158	32	378	4.0	320	0	0
Biscuit, homemade, 1	28	105	2	5	13	34	49	.4	33	Tr.	Tr.
Breadcrumbs, grated, 1 c.	100	390	13	5	73	122	141	3.6	152	Tr.	Tr.

417

Food and Amount	Grams	Energy Calories	Protein Grams	Fat Grams	Carbohydrate Grams	Calcium Milligrams	Phosphorus Milligrams	Iron Milligrams	Potassium Milligrams	Vitamin A Int. Units	Ascorbic Acid Milligrams
Breads:											
French, 1 slice	35	100	3	1	19	15	30	.8	32	Tr.	Tr.
Raisin, 1 slice	25	65	2	1	13	18	22	.6	58	Tr.	Tr.
Rye, 1 slice	25	60	2	Tr.	13	19	37	.5	36	0	0
Pumpernickel, 1 slice	32	80	3	Tr.	17	27	73	.8	145	0	0
White, soft, 1 slice	25	70	2	1	13	21	24	.6	26	Tr.	Tr.
Breakfast cereals:											
Oatmeal, 1 c.	240	130	5	2	23	22	137	1.4	146	0	0
Bran flakes, (40%), 1 c.	35	105	4	1	28	19	125	5.6	137	1,540	0
Corn flakes, plain, 1 c.	25	95	2	Tr.	21	V	9	V	30	V	13
Corn flakes, sugared, 1 c.	40	155	2	Tr.	37	1	10	V	27	1,760	21
Oats, puffed, sugared, 1 c.	25	100	3	1	19	44	102	4.0	—	1,100	13
Wheat, puffed, plain, 1 c.	15	55	2	Tr.	12	4	48	.6	51	0	0
Wheat, shredded, 1 biscuit	25	90	2	1	20	11	97	.9	87	0	0
Wheat germ, toasted, 1 Tbs.	6	25	2	1	3	3	70	.5	57	10	1
Cakes:											
Angelfood, 1/12	53	135	3	Tr.	32	50	63	.2	32	0	0
Cupcake, choc. icing, 1	36	130	2	5	21	47	71	.4	42	60	Tr.
Cake, choc. w/icing, 1/16	69	235	3	8	40	41	72	1.0	90	100	Tr.
Gingerbread, 3 in. sq. piece	63	175	2	4	32	57	63	.9	173	Tr.	Tr.
Boston cream pie, 1/12	69	210	3	6	34	46	70	.7	61	140	Tr.
Fruitcake, 1 slice	15	55	1	2	9	11	17	.4	74	20	Tr.
Poundcake, 1 slice	33	160	2	10	16	6	24	.5	20	80	0
Spongecake, 1/12	66	195	5	4	36	20	74	1.1	57	300	Tr.
Brownie, homemade, 1	20	95	1	6	10	8	30	.4	38	40	Tr.

Cookies:

Choc. chip, 4	42	200	2	9	29	16	48	1.0	56	50	Tr.
Fig bars, 4	56	200	2	3	42	44	34	1.0	111	60	Tr.
Gingersnaps, 4	28	90	2	2	22	20	13	.7	129	20	0
Sandwich, 4	40	200	2	9	28	10	96	.7	15	0	0
Vanilla wafers, 10	40	185	2	6	30	16	25	.6	29	50	0
Cornmeal, 1 c.	122	435	11	5	90	24	312	2.9	346	620	0
Crackers, graham, 2	14	55	1	1	10	6	21	.5	55	0	0
Crackers, saltines, 4	11	50	1	1	8	2	10	.5	13	0	0
Doughnut, cake type, 1	25	100	1	5	13	10	48	.4	23	20	Tr.
Doughnut, glazed, 1	50	205	3	11	22	16	33	.6	34	25	0
Macaroni, cooked, 1 c.	130	190	7	1	39	14	85	1.4	103	0	0
Macaroni & cheese, canned, 1 c.	240	230	9	10	26	199	182	1.0	139	260	Tr.
Muffin, blueberry, 1	40	110	3	4	17	34	53	.6	46	90	Tr.
Muffin, bran, 1	40	105	3	4	17	57	162	1.5	172	90	Tr.
Muffin, corn, 1	40	125	3	4	19	42	68	.7	54	120	Tr.
Noodles, egg, cooked, 1 c.	160	200	7	2	37	16	94	1.4	70	110	0
Noodles, chow mein, 1 c.	45	220	6	11	26	—	70	—	—	—	0
Pancake, from mix, 1	27	60	2	2	9	58	70	.3	42	70	Tr.

Pies:

Apple, 1/7	135	345	3	15	51	11	30	.9	108	40	2
Banana cream, 1/7	130	285	6	12	40	86	107	1.0	264	330	1
Blueberry, 1/7	135	325	3	15	47	15	31	1.4	88	40	4
Cherry, 1/7	135	350	4	15	52	19	34	.9	142	590	Tr.
Custard, 1/7	130	285	8	14	30	125	147	1.2	178	300	0

Food and Amount		Energy	Protein	Fat	Carbohydrate	Calcium	Phosphorus	Iron	Potassium	Vitamin A	Ascorbic Acid
	Grams	Calories	Grams	Grams	Grams	Milligrams	Milligrams	Milligrams	Milligrams	Int. Units	Milligrams
Lemon meringue ½	120	305	4	12	45	17	59	1.0	60	200	4
Mince, ½	135	365	3	16	56	38	51	1.9	240	Tr.	1
Peach, ½	135	345	3	14	52	14	39	1.2	201	990	4
Pecan, ½	118	495	6	27	61	55	122	3.7	145	190	Tr.
Pumpkin, ½	130	275	5	15	32	66	90	1.0	208	3,210	Tr.
Pizza, cheese, ⅛	60	145	6	4	22	86	89	1.1	67	230	4
Popcorn, oil & salt, 1 c.	9	40	1	2	5	1	19	.2	—	—	0
Pretzel sticks, 10	3	10	Tr.	Tr.	2	1	4	Tr.	4	0	0
Rice, instant, 1 c.	165	180	4	Tr.	40	5	31	1.3	—	0	0
Rice, long grain, 1 c.	205	225	4	Tr.	50	21	57	1.8	57	0	0
Roll, brown/serve, 1	26	85	2	2	14	20	23	.5	25	Tr.	Tr.
Roll, hamburger, 1	40	120	3	2	21	30	34	.8	38	Tr.	Tr.
Spaghetti, 1 c.	130	190	7	1	39	14	85	1.4	103	0	0
Spaghetti w/sauce, canned, 1 c.	250	190	6	2	39	40	88	2.8	303	930	10
Spaghetti w/meatballs, 1 c.	248	330	19	12	39	124	236	3.7	665	1,590	22
Toaster pastry, 1	50	200	3	6	36	54	67	1.9	74	500	—
Waffle, from mix, 1	75	205	7	8	27	179	257	1.0	146	170	Tr.
Legumes (Dry), Nuts, Seeds, Related Products											
Almonds, slivered, 1 c.	115	690	21	62	22	269	580	5.4	889	0	Tr.
Beans, dry:											
Great northern, cooked, 1 c.	180	210	14	1	38	90	266	4.9	749	0	0
White, w/pork & sauce, 1 c.	255	310	16	7	48	138	235	4.6	536	330	5

Food											
Red kidney, 1 c.	255	230	15	1	42	74	278	4.6	673	10	—
Lima, 1 c.	190	260	16	1	49	55	293	5.9	1,163	—		
Blackeye peas, 1 c.	250	190	13	1	35	43	238	3.3	573	30		
Cashews, roasted in oil, 1 c.	140	785	24	64	41	53	522	5.3	650	140		
Coconut, grated, 1 c.	80	275	3	28	8	10	76	1.4	205	0	2
Lentils, cooked, 1 c.	200	210	16	Tr.	39	50	238	4.2	498	40	0
Peanuts, roasted in oil, 1 c.	144	840	37	72	27	107	577	3.0	971	—	0
Peanut butter, 1 Tbs.	16	95	4	8	3	9	61	.3	100	—		
Peas, split, cooked, 1 c.	200	230	16	1	42	22	178	3.4	592	80	0
Pecans, chopped, 1 c.	118	810	11	84	17	86	341	2.8	712	150	2
Walnuts, English, 1 c.	120	780	18	77	19	119	456	3.7	540	40	2

Sugars and Sweets

Candy:

Food											
Caramel, 1 oz.	28	115	1	3	22	42	35	.4	54	Tr.	Tr.
Chocolate, milk, 1 oz.	28	145	2	9	16	65	65	.3	109	80	Tr.
Chocolate coated peanuts, 1 oz.	28	160	5	12	11	33	84	.4	143	Tr.	Tr.
Fudge, 1 oz.	28	115	1	3	21	22	24	.3	42	Tr.	Tr.
Marshmallows, 1 oz.	28	90	1	Tr.	23	5	2	.5	2	0	0
Jams and preserves, 1 Tbs.	20	55	Tr.	Tr.	14	4	2	.2	18	Tr.	Tr.
Jellies, 1 Tbs.	18	50	Tr.	Tr.	13	4	1	.3	14	Tr.	1
Syrup, choc. fudge, 2 Tbs.	38	125	2	5	20	48	60	.5	107	60	Tr.
Syrup, Molasses, light, 1 Tbs.	20	50	—	—	13	33	9	.9	183	—		

Sugars:

Food											
Brown, 1 c.	220	820	0	0	212	187	42	7.5	757	0	0
White, granulated, 1 Tbs.	12	45	0	0	12	0	0	Tr.	Tr.	0	0
Powdered, 1 c.	100	385	0	0	100	0	0	.1	3	0	0

Vegetable and Vegetable Products

Food and Amount	Grams	Energy Cal-ories	Protein Grams	Fat Grams	Carbo-hydrate Grams	Calcium Milli-grams	Phos-phorus Milli-grams	Iron Milli-grams	Potas-sium Milli-grams	Vitamin A Int. Units	Ascorbic Acid Milli-grams
Asparagus, spears, 4	60	10	1	Tr.	2	13	30	.4	110	540	16
Beans, lima, cooked, 1 c.	170	170	10	Tr.	32	34	153	2.9	724	390	29
Beans, green, cooked, 1 c.	125	30	2	Tr.	7	63	46	.8	189	680	15
Beans, green, canned, 1 c.	135	30	2	Tr.	7	61	34	2.0	128	630	5
Beans, yellow, cooked, 1 c.	125	30	2	Tr.	6	63	46	.8	189	290	16
Bean sprouts, raw, 1 c.	105	35	4	Tr.	7	20	67	1.4	234	20	20
Beets, cooked, 1 c.	170	55	2	Tr.	12	24	39	.9	354	30	10
Broccoli, cooked, 1 stalk	180	45	6	1	8	158	112	1.4	481	4,500	162
Broccoli, chopped, 1 c.	185	50	5	1	9	100	104	1.3	392	4,810	105
Brussels sprouts, 1 c.	155	55	7	1	10	50	112	1.7	423	810	135
Cabbage, raw, shredded, 1 c.	70	15	1	Tr.	4	34	20	0.3	163	90	33
Cabbage, cooked, 1 c.	145	30	2	Tr.	6	64	29	.4	236	190	48
Carrot, raw, 1	72	30	1	Tr.	7	27	26	.5	246	7,930	6
Carrots, cooked, 1 c.	155	50	1	Tr.	11	51	48	.9	344	16,280	9
Carrots, canned, 1 c.	155	45	1	Tr.	10	47	34	1.1	186	23,250	3
Cauliflower, raw, 1 c.	115	31	3	Tr.	6	29	64	1.3	339	70	90
Cauliflower, cooked, 1 c.	180	30	3	Tr.	6	31	68	.9	373	50	74
Celery, raw, 1 stalk	40	5	Tr.	Tr.	2	16	11	.1	136	110	4
Collards, cooked, 1 c.	190	65	7	1	10	357	99	1.5	498	14,820	144
Corn, cooked, 1 ear	140	70	2	1	16	2	69	.5	151	310	7
Corn, canned, 1 c.	210	175	5	1	43	6	153	1.1	204	740	11
Corn, canned, cream, 1 c.	256	210	5	2	51	8	143	1.5	248	840	13

Food											
Cucumber, raw, 7 slices	28	5	Tr.	Tr.	1	7	8	.3	45	70	3
Lettuce, Boston, 1 head	220	25	2	Tr.	4	57	42	3.3	430	1,580	13
Lettuce, iceberg, 1 head	567	70	5	1	16	108	118	2.7	943	1,780	32
Mushrooms, raw, 1 c.	70	20	2	Tr.	3	4	81	.6	290	Tr.	2
Okra, cooked, 10 pods	106	30	2	Tr.	6	98	43	.5	184	520	21
Onions, chopped, 1 c.	170	65	3	Tr.	15	46	61	.9	267	Tr.	17
Peas, canned, 1 c.	170	150	8	1	29	44	129	3.2	163	1,170	14
Peas, frozen, cooked, 1 c.	160	110	8	Tr.	19	30	138	3.0	216	960	21
Peppers, sweet, 1 pod	74	15	1	Tr.	4	7	16	.5	157	310	94
Potato, baked, 1	156	145	4	Tr.	33	14	101	1.1	782	Tr.	31
Potato, boiled, 1	135	90	3	Tr.	20	8	57	.7	385	Tr.	22
Potato, french fries, 10	50	135	2	7	18	8	56	.7	427	Tr.	11
Potato, mashed, 1 c.	210	135	4	2	27	50	103	.8	548	40	21
Potato chips, 10	20	115	1	8	10	8	28	.4	226	Tr.	3
Potato salad, 1 c.	250	250	7	7	41	80	160	1.5	798	350	28
Pumpkin, canned, 1 c.	245	80	2	1	19	61	64	1.0	588	15,680	12
Radishes, raw, 4	18	5	Tr.	Tr.	1	5	6	.2	58	Tr.	5
Sauerkraut, canned, 1 c.	235	40	2	Tr.	9	85	42	1.2	329	120	33
Spinach, raw, 1 c.	55	15	2	Tr.	2	51	28	1.7	259	4,460	28
Spinach, cooked, 1 c.	180	40	5	1	6	167	68	4.0	583	14,580	50
Squash, cooked, 1 c.	210	30	2	Tr.	7	53	53	.8	296	820	21
Sweet potato, boiled, 1	151	170	3	1	40	48	71	1.1	367	11,940	26
Tomato, raw, 1	135	25	1	Tr.	6	16	33	.6	300	1,110	28
Tomatoes, canned, 1 c.	241	50	2	Tr.	10	14	46	1.2	523	2,170	41
Tomato catsup, 1 Tbs.	15	15	Tr.	Tr.	4	3	8	.1	54	210	2
Tomato juice, 1 c.	243	45	2	Tr.	10	17	44	2.2	552	1,940	39
Turnips, cooked, 1 c.	155	35	1	Tr.	8	54	37	.6	291	Tr.	34

Food and Amount	Energy		Protein	Fat	Carbo-hydrate	Calcium	Phos-phorus	Iron	Potas-sium	Vitamin A	Ascorbic Acid
	Grams	Cal-ories	Grams	Grams	Grams	Milli-grams	Milli-grams	Milli-grams	Milli-grams	Int. Units	Milli-grams
Miscellaneous Items											
Barbecue sauce, 1 c.	250	230	4	17	20	53	50	2.0	435	900	13
Beverages:											
Beer, 12 oz.	360	150	1	0	14	18	108	Tr.	90	—	—
Wine, dessert, 3½ oz.	103	140	Tr.	0	8	8	—	—	77	—	—
Wine, table, 3½ oz.	102	85	Tr.	0	4	9	10	.4	94	—	—
Cola, 12 oz.	369	145	0	0	37	—	—	—	—	0	0
Fruit-flavored soda, 12 oz.	372	170	0	0	45	—	—	—	—	0	0
Gelatin dessert, 1 c.	240	140	4	0	34	—	—	—	—	—	—
Mustard, prepared, 1 tsp.	5	5	Tr.	Tr.	Tr.	4	4	.1	7	—	—
Olives, green, 2 giant	16	15	Tr.	2	Tr.	8	2	.2	7	40	—
Pickles:											
Dill, medium, 1	65	5	Tr.	Tr.	1	17	14	.7	130	70	4
Sweet, small, 1	15	20	Tr.	Tr.	5	2	2	.2	—	10	1
Relish, sweet, 1 Tbs.	15	20	Tr.	Tr.	5	3	2	.1	—	—	—
Popsicle, 1	95	70	0	0	18	0	—	Tr.	—	0	0
Soup, canned:											
Cream of chicken, 1 c.	245	180	7	10	15	172	152	0.5	260	610	2
Tomato, 1 c.	250	175	7	7	23	168	155	.8	418	1,200	15
Bean with pork, 1 c.	250	170	8	6	22	63	128	2.3	395	650	3
Beef broth, 1 c.	240	30	5	0	3	Tr.	31	.5	130	Tr.	—
Clam chowder, Manhattan, 1 c.	245	80	2	3	12	34	47	1.0	184	880	—
Vegetable beef, 1 c.	245	80	5	2	10	12	49	.7	162	2,700	—

Code: Tr. – Trace; V – varies with brand.

Source: U. S. Department of Agriculture

Vitamins: Essential Nutrients for Good Health

Fat-Soluble Vitamins

Vitamin/ Year Discovered	Chief Metabolic Function	Effects of Deficiency	Recommended Dietary Allowance*
Vitamin A 1913	Essential for normal growth and development; for normal function of epithelial cells and normal development of teeth and bones. Prevents night blindness.	Retarded growth. Reduced resistance to infection. Abnormal function of gastrointestinal, genitourinary, and respiratory tracts due to altered epithelial membranes. Interferes with production of "night purple."	Micromilligrams of retinol equivalents† Infants: 400; Children, 4–6 yr.: 500 Males: 1,000 Females: 800; Pregnant: 1,000; Lactating: 1,200
Vitamin D 1925	Regulates absorption of calcium and phosphorus from the intestinal tract. Affords antirachitic activity.	Interferes with utilization of calcium and phosphorus in bone and teeth formation. Development of bone disease, rickets, caries.	Micromilligrams Infants and children: 10. Teenagers and young adults: 7.5 Adults, 23 and up: 5; Pregnant: 10; Lactating: 10
Vitamin E 1936	Protects tissues, cell membranes, and Vitamin A against peroxidation. Helps strengthen red blood cells.	Decreased red blood cell resistance to rupture.	Milligrams alpha-tocopherol equivalent Infants 6–12 mo.: 4; Children 4–6 yr.: 6 Adult males: 10 Adult females: 8; Pregnant: 10; Lactating: 11

Vitamin/ Year Discovered	Chief Metabolic Function	Effects of Deficiency	Recommended Dietary Allowance*
Vitamin K 1935	Essential for formation of normal amounts of prothrombin and blood coagulation.	Diminished blood clotting time. Increased incidence of hemorrhages.	Not established. Synthesized by intestinal bacteria in adequate amounts. Dietary deficiency is rare.

Water-Soluble Vitamins

Vitamin/ Year Discovered	Chief Metabolic Function	Effects of Deficiency	Recommended Dietary Allowance*
Vitamin B_1 (Thiamin) 1936	An important aid in carbohydrate metabolism. Needed for proper functioning of the digestive tract and nervous system.	Loss of appetite. Impaired digestion of starches and sugars. Various nervous disorders, beriberi. Loss of muscle coordination.	Milligrams Infants 6–12 mo.: 0.5; Children 4–6 yr.: 0.9 Adult males: 1.4 Adult females: 1.1; Pregnant: 1.5; Lactating: 1.6
Vitamin B_2 (Riboflavin) 1935	Needed in formation of certain enzymes and in cellular oxidation. Prevents inflammation of oral mucous membranes and the tongue.	Impaired growth, lassitude and weakness. Causes cheilosis or glossitis. May result in photophobia and cataracts.	Milligrams Infants 6–12 mo.: 0.6; Children 4–6 yr.: 1.7 Adult males: 1.7 Females: 1.3; Pregnant: 1.6; Lactating: 1.8
Niacin (Nicotinic Acid) 1867	Important, as the component of two enzymes, in glycolysis, tissue respiration, and fat synthesis. The antipellagra principle of B complex.	Results in pellagra, gastrointestinal and mental disturbances.	Milligrams Infants 6–12 mo.: 8; Children 4–6 yr.: 11 Adult males: 18 Adult females: 14; Pregnant: 16; Lactating: 19

Vitamin	Function	Deficiency Symptoms	Recommended Daily Allowance*
Vitamin B6 (Pyridoxine) 1934	Acts, as do other B vitamins, to break down protein, carbohydrate, and fat. Acts as a catalyst in the formation of niacin from tryptophan.	Increased irritability, convulsions, and peripheral neuritis. Anorexia, nausea and vomiting.	Milligrams Infants 6–12 mo.: 0.6; Children 4–6 yr.: 1.3 Adult males: 2.2 Adult females: 2.0; Pregnant: 2.6; Lactating: 2.5
Vitamin B12 (Cyanocobalamin) 1948	Essential for development of red blood cells. Required for maintenance of skin, nerve tissues, bone and muscles.	Results in pernicious anemia. Weakness, fatigue, smooth tongue, sore and cracked lips.	Micromilligrams Infants 6–12 mo.: 1.5; Children 4–6 yr.: 2.5 Adult males: 3.0 Adult females: 3.0; Pregnant: 4.0; Lactating: 4.0
Vitamin C (Ascorbic Acid) 1919	Needed to form the cementing substance, collagen, in various tissues (skin, dentine, cartilage, and bone matrix). Assists in wound healing; bone fractures.	Lowered resistance to infections, susceptibility to dental cavities, pyorrhea, and bleeding gums. Delayed wound healing. Specific treatment for scurvy.	Milligrams Infants 6–12 mo.: 35; Children 4–6 yr.: 45 Adults: 60; Pregnant: 80; Lactating: 100
Folic Acid (Folacin) 1946	Needed for manufacture of nucleic acid, essential in several synthesis processes. Needed for normal functioning of hematopoietic system.	Megaloblastic (pernicious) anemia; sprue.	Micromilligrams Infants 6–12 mo.: 45; Children 4–6 yr.: 200 Adult males: 400 Adult females: 400; Pregnant: 800; Lactating: 500

*1980 Revised Recommended Dietary Allowances.

‡One retinol equivalent: 1 mg retinol or 6 mg B-carotene.

Genital Herpes Epidemic

The August 2, 1982, edition of *Time* magazine devoted its cover story to the epidemic of genital herpes sweeping the United States in the late 1970s and early 1980s. Although the illness had received considerable previous publicity both in general and specialized publications, the *Time* story seemed to symbolize its seriousness.

There were an estimated 20,000,000 cases of genital herpes in the United States in 1982, of which 400,000 cases were new that year. The disease is not severe and scarcely ever life-threatening to adults, but it is also not curable; it is life-threatening to newborns whose mothers have herpes. Between outbreaks in an individual patient, generally consisting of sores—painful in early attacks—in the genital region, the herpes virus retreats to nerve cells unreachable by the immune system. In this hide-away, the virus can remain indefinitely in a nonactive state. At some time, however, in most victims, the virus will reactivate and cause another round of sores. Then it will return to the nerve cells. Among the most serious effects of the disease in adults are psychological problems.

Genital herpes is spread by contact with the sores in their active phase, when they are "shedding" viruses. Since the virus normally attacks the lower portion of the body, most transmission is by way of sexual contact. However, it has been shown that the virus can survive in hot tubs for more than four hours when it lodges on the plastic walls of the tubs. There is some evidence that the infection has been transmitted in health-club hot tubs, causing some health clubs to close. Because the most common method of transmission is sexual, herpes has helped change patterns of sexual behavior.

Genital herpes is caused by one form—*Herpes simplex 2*—of a puzzling family of common viruses. Those who have caught the disease suffer painful sores around the genital area for periods ranging from a week to several months. A new drug, acyclovir, approved by the Food and Drug Administration in April 1982, can be moderately effective in reducing the severity of the first attack. But acyclovir does nothing to cure the disease itself, and the victim is subject to later outbreaks against which acyclovir will be much less effective. Initial approval of acyclovir was as a topical drug to be applied to the affected area, but in 1984 doctors reported that an oral form of acyclovir was more effective in controlling the disease.

When the immediate first outbreak of the disease is controlled, whether by drug therapy or by the body's own immune system, the virus retreats to the sacral ganglia, a center of nerve cells near the spinal cord in the lower back. There, safe from the challenge of the human immune system, it can live indefinitely. It may emerge at any time afterward to cause another active phase of the disease. One of the great mysteries is what triggers these recurring outbreaks. Victims report that the virus is responsive to various kinds of stress and that it seems to "know" the most inopportune time to strike again.

The outbreak of genital herpes has caused increased attention to the whole herpes family of viruses. *Herpes simplex l* is the cause of common cold sores on the mouth and face. Sufferers of this ailment, like those of genital herpes, have the virus in their systems at all times. During inactive periods, it stays in the trigeminal ganglia, the cluster of nerves connecting the face and the brain; like the genital form it recurs unpredictably. Facial herpes is transmitted by contact also. It normally appears on the upper half of the body. However, about 10 percent of herpes l cases actually occur on the lower half of the body, where they cause genital herpes. Likewise, about 10 percent of facial herpes cases involve herpes 2.

A cousin of the simplex viruses is *Herpes varicella zoster*, the virus that causes chicken pox, a highly contagious disease that appears most often among children. Apparently this virus, too, is never finally expelled from the body, but in most people it seems never to cause further disease symptoms. In a small minority, however, it can reappear in adult life as the painful disease called shingles, which causes a blistering rash on various parts of the body. Children can actually catch chicken pox from contact with a person who has shingles. But for reasons unknown, an adult cannot get shingles from contact with a victim of chicken pox.

There was progress in this area in 1984, also. In Japan, an effective vaccine against chicken pox was announced. Some scientists were concerned, however, about mass inoculations, since it might be possible for the vaccine to cause shingles or other complications years after.

Still another member of the herpes family is the cytomegalovirus, usually called CMV. It has been estimated that perhaps 80 percent of the U.S. population has been infected with this virus and that nearly 100 percent of homosexual males have had it. CMV seems generally harmless but can be deadly to patients whose immune systems have been depressed by drug treatment or disease. It is known to cause birth defects when the mother has an active case of CMV disease.

The most mysterious member of the family is the Epstein-Barr virus. It has been shown to be the cause of mononucleosis, the "kissing disease" common among high school and college students. But it is also known to have a direct relation to Burkitt's lymphoma, a type of cancer seen almost exclusively in parts of Africa, and to another form of pharyngeal cancer found among people in parts of China. How one virus can have such widely varying effects has not been determined. There is some evidence that *Herpes simplex* can also cause cancer.

Herpes viruses seem to be among the most "intelligent" of all viruses. They establish a long-standing relationship with their human host, causing complaints not serious enough to kill and yet serious enough to cause real health problems. In sufferers of genital herpes, the most disturbing effects are those on sexual performance and on the highly psychological process of courting. The most serious effects, however, may be the potential dangers to newborns of infected mothers.

Modern childbirth techniques, including Caesarean section, can reduce the risk of death or permanent damage to the newborn, but normal deliveries to a woman with active herpes are associated with a high infant mortality rate and a high incidence of retardation and blindness. In December 1983, the Federal Centers for Disease Control announced that the epidemic of herpes had resulted in large numbers of herpes-infected newborns. Half of those infected died, while the remainder were physically or mentally impaired. In Seattle, where a major study took place, the rate of herpes-infected newborns in 1982 was ten times the rate in the 1960s. Although nationwide statistics are not available, estimates are that about 500 cases of herpes-infected infants were born in 1983.

Fortunately, childbirth during the inactive phase of the disease does not have these risks. Doctors advise, however, that an expectant mother with herpes be checked for signs of onset of the active disease often during the last months of pregnancy. In the meantime, researchers are trying to develop a test that would reveal herpes in its inactive phase.

Late in 1983 researchers announced progress in developing a vaccine against herpes (see "Recombinant Herpes Vaccine," page 481). The vaccine might protect the many who have not yet fallen victim to genital herpes, but it would apparently present no new hope to those who have already contracted the disease.

THE RULE OF 86

Many medicine labels inform the user to store below 86 degrees Fahrenheit. Why 86? The answer is that the *United States Pharmacopeia*, the standard reference book of drugs, declares that standard room temperature for storage is between 15 and 30 degrees Celsius. Thirty degrees Celsius, a nice round number, translates to 86 degrees Fahrenheit.

Oraflex Withdrawn

On August 5, 1982, Secretary of Health and Human Services Richard S. Schweiker made an extraordinary announcement that Eli Lilly and Co., a major pharmaceutical house, had agreed to withdraw its new and much publicized arthritis drug, Oraflex, from the market.

The immediate cause of the withdrawal was a report earlier that same day that health authorities in Britain had recalled the drug. The British statement said that Oraflex (also known as Opren and by its generic name benoxaprofen) had been responsible for 61 deaths and that there had been reports of 3,500 bad reactions, especially among the elderly.

Health advocates outside the United States government had been opposing Oraflex for some months as more and more cases in which side effects were severe began to surface. Although full evidence was lacking, there were suggestions that the drug might cause toxic effects to the gastrointestinal tract, the liver, bone marrow, skin, eyes, and nails.

Lilly had heavily promoted the new drug. Oraflex has many of the desirable properties of aspirin, but it can be prescribed for less frequent administration. Lilly also claimed that the new drug would not have some of aspirin's undesirable side effects.

When the first reports of death among elderly people taking the maximum dosage of Oraflex surfaced in Britain in May, Lilly advised doctors not to prescribe large doses to older patients. Although Lilly also revised its labeling at that time in an effort to keep the drug on the market, Oraflex was withdrawn from the United States market only three months after the Food and Drug Administration (FDA) had approved it. One of the reasons that this action was so extraordinary is that the approval procedures of the FDA are supposed to be designed to screen out any drugs with serious unwanted side effects.

NIACIN INTOXICATION STRIKES BAGEL FANS

Some fourteen guests at a party in New York City on April 27, 1983, came down with a strange ailment after eating some pumpernickel bagels. They got an itchy rash and reported a generally unpleasant "warm" sensation. Investigators concluded that they were victims of *niacin intoxication*, brought on by excessive doses of the vitamin niacin. It was discovered that a poorly labeled box of niacin at the bakery had been dumped into the pumpernickel flour to enrich it. Each bagel contained 190 milligrams of niacin, as opposed to the recommended daily allowance of 13 milligrams.

Side Effects of Popular Prescription Drugs

Product Name/Use	Generic/Chemical Name	Possible Side Effects
ALDOMET® Antihypertensive	Methyldopa	Drowsiness, headache, nasal congestion, dry mouth, dizziness, diarrhea, nausea, numbness of extremities, skin rash, breast enlargement, impotence, jaundice, anemia.
ALDORIL® Diuretic/antihypertensive	Hydrochlorothiazide/ methyldopa	Frequent urination, muscle weakness or cramps, headache, dizziness, skin rash, nausea, vomiting, reduced appetite, diarrhea, blurred vision, dry mouth, nightmares, depression, jaundice, breast enlargement, impotence, elevated blood sugar, elevated uric acid.
AMOXIL® Antibiotic	Amoxicillin	Nausea, vomiting, diarrhea, irritation of mouth and tongue, rash, itching, fever, sore throat, joint pain, swollen glands, rectal and vaginal itching.
ATIVAN® Tranquilizer	Lorazepam	Dizziness, weakness, headache, disorientation, depression, nausea, sleep disturbances, agitation, eye function disturbance.
BENADRYL® Antihistamine	Diphenhydramine hydrochloride	Drowsiness, dizziness, blurred vision, upset stomach, difficult urination, dry mouth and throat, headache, loss of appetite, skin rash, fast heartbeat.
CECLOR® Antibiotic	Cephalosporin	Nausea, vomiting, diarrhea, stomach cramps, rash, itching, joint pain, fever, swollen glands.
CLINORIL® Anti-inflammatory agent	Sulindac	Abdominal pain, cramps, dyspepsia, nausea, vomiting, diarrhea, constipation, flatulence, anorexia, rash, itching, dizziness, headache, nervousness, depression.
DALMANE® Hypnotic	Flurazepam hydrochloride	Dizziness, lightheadedness, unsteadiness, drowsiness, headache, constipation, diarrhea, heartburn, nausea, vomiting, blurred vision, shortness of breath, dry mouth, skin rash.

Product Name/Use	Generic/Chemical Name	Possible Side Effects
DARVOCET-N® Analgesic	Propoxyphene napsylate/acetaminophen	Nausea, vomiting, constipation, drowsiness, dizziness, lightheadedness, euphoria, itching, skin rash, blurred vision, headache.
DIABINESE® Antidiabetic agent	Chlorpropamide	Anorexia, diarrhea, nausea, vomiting, weakness, headache, heartburn, loss of appetite, skin rash, jaundice, low blood sugar, anemia, low grade fever, sore throat.
DILANTIN® Anticonvulsant	Phenytoin sodium	Swollen or bleeding gums, nausea, vomiting, blurred vision, skin rash, slurred speech, mental confusion, joint pain, dizziness, insomnia, headache.
DIMETAPP® Antihistamine	Brompheniramine maleate; phenylephrine hydrochloride; phenylpropanolamine hydrochloride	Dry mouth, drowsiness, dizziness, lassitude, giddiness, headache, blurred vision, headache, loss of appetite, skin rash, nausea, diarrhea.
DYAZIDE® Diuretic/anti-hypertensive	Triamterene hydrochlorothiazide	Frequent urination, nausea, vomiting, diarrhea, skin rash, headache, weakness, dizziness, muscle cramps, dry mouth, constipation.
E-MYCIN® Antibiotic	Erythromycin	Nausea, vomiting, diarrhea, skin rash, rectal and genital itching, sore mouth or tongue.
E.E.S.® Antibiotic	Erythromycin	Abdominal cramping, nausea, vomiting, diarrhea, skin rashes, rectal and genital itching, sore mouth or tongue.
EMPIRIN® w/codeine Analgesic	Aspirin (acetylsalicylic acid)/codeine phosphate	Dizziness, lightheadedness, euphoria, confusion, headache, nausea, constipation, vomiting, sweating, skin rash, itching.
HYDRODIURIL® Diuretic/anti-hypertensive	Hydrochlorothiazide	Frequent urination, muscle weakness, cramps, lightheadedness, skin rashes, nausea, vomiting, reduced appetite, stomach cramps, diarrhea, headache, blurred vision.

433

Product Name/Use	Generic/Chemical Name	Possible Side Effects
HYGROTON® Diuretic/anti-hypertensive	Chlorthalidone	Frequent urination, dry mouth, anorexia, gastric irritation, nausea, vomiting, cramping, diarrhea, constipation, jaundice, dizziness, vertigo.
INDERAL® Beta-adrenergic blocking agent	Propranolol hydrochloride	Lightheadedness, insomnia, weakness, fatigue, mood swings, depression, hallucinations, vomiting, diarrhea, constipation, slow heart rate, tingling of fingers, hair loss.
INDOCIN® Anti-inflammatory agent	Indomethacin	Headache, indigestion, heartburn, nausea, vomiting, diarrhea, bloating, constipation, hearing loss, blurred vision, anemia, asthma.
ISORDIL® Antianginal agent	Isosorbide dinitrate	Dizziness, weakness, flushing, headache, restlessness, nausea, vomiting, palpitations, low blood pressure, rash.
KEFLEX® Antibiotic	Cephalexin	Diarrhea, nausea, vomiting, skin rash, itching, joint pains, fever, swollen glands, dizziness, fatigue, headache, vaginitis and vaginal discharge.
LANOXIN® Cardiotonic	Digoxin	Anorexia, nausea, vomiting, diarrhea, blurred vision, headache, weakness, apathy, palpitations, slow heart rate.
LASIX® Diuretic/antihypertensive	Furosemide	Frequent urination, weakness, muscle cramps, lightheadedness, dizziness, nausea, vomiting, blurred vision, ringing in ears, anemia, skin rash.
LO/OVRAL® Contraceptive	Norgestrel ethinyl estradiol	Abdominal cramps, bloating, breakthrough bleeding, change in menstrual flow, fluid retention, breast tenderness or enlargement, weight loss or gain, rash, depression, headache, jaundice.
LOPRESSOR® Antihypertensive/beta-blocker	Metoprolol tartrate	Fatigue, dizziness, depression, headache, insomnia, wheezing, cold extremities, palpitations, heart failure, diarrhea, nausea, constipation.

Product Name/Use	Generic/Chemical Name	Possible Side Effects
MELLARIL® Tranquilizer	Thioridazine	Dry mouth, blurred vision, constipation, dizziness, nasal congestion, fast heartbeat, difficult urination, impotence, swollen breasts.
MINIPRESS® Antihypertensive	Prazosin hydrochloride	Dizziness, headache, drowsiness, weakness, palpitations, fatigue, nausea, vomiting, diarrhea, constipation, abdominal pain, nervousness, vertigo, depression.
MONISTAT-7® Antifungal agent	Miconazole nitrate 2%	Vulvovaginal burning, itching, irritation, pelvic cramps, hives, headache.
NAPROSYN® Anti-inflammatory agent	Naproxen	Constipation, heartburn, abdominal pain, nausea, diarrhea, headache, dizziness, drowsiness, lightheadedness, pounding heart, sweating, skin eruptions.
NORINYL® Contraceptive	Norgestrel ethinyl estradiol	Breast tenderness or enlargement, weight gain or loss, nausea, vomiting, spotting, fluid retention, rash, depression, headache.
OMNIPEN® Antibiotic	Ampicillin	Diarrhea, nausea, vomiting, black tongue, cough, difficult breathing, fever, mouth and throat irritation, rash, rectal and vaginal itching.
ORTHO-NOVUM® Contraceptive	Norethindrone mestranol	Abdominal cramps, bloating, breakthrough bleeding, spotting, dysmenorrhea, breast tenderness or enlargement, weight loss or gain, migraine, rash, depression.
PERSANTINE® Vasodilator	Dipyridamole	Headache, dizziness, flushing, nausea, diarrhea, stomach irritation.
PREMARIN® Estrogen replacement therapy	Conjugated estrogens	Nausea, vomiting, abdominal cramps, bloating, breast tenderness or enlargement, breakthrough bleeding, migraine, headache, dizziness, depression, skin rash, blood clots in the legs.

435

Product Name/Use	Generic/Chemical Name	Possible Side Effects
SLOW-K® Potassium supplement	Potassium chloride	Nausea, vomiting, abdominal discomfort, diarrhea, obstruction, bleeding, ulceration, perforation.
SYNTHROID® Thyroid replacement therapy	Levothyroxine sodium	Rapid heart rate, weight loss, chest pain, tremor, headache, diarrhea, nervousness, insomnia, sweating, heat intolerance.
TAGAMET® Antiulcer agent	Cimetidine	Diarrhea, dizziness, rash, muscular pain, blood disturbances, increase in breast size, may prolong effects of other medication.
THEO-DUR® Bronchodilator	Theophylline (anhydrous)	Nausea, vomiting, stomach pain, diarrhea, headache, irritability, restlessness, insomnia, convulsions, palpitations, flushing, high blood pressure.
TIMOPTIC® Antiglaucomatous agent	Timolol maleate	Ocular irritation, conjunctivitis, blepharitis, keratitis, localized and generalized rash, slight reduction in resting heart rate.
TRANXENE® Tranquilizer	Clorazepate dipotassium	Drowsiness, dizziness, nervousness, headache, fatigue, irritability, insomnia, dry mouth, blurred vision, mental confusion, skin rash, depression.
TYLENOL® w/codeine Antipyretic	Acetaminophen/codeine phosphate	Lightheadedness, dizziness, euphoria, dysphoria, nausea, vomiting, constipation, rash, itching, breathing and heart difficulties.
VALIUM® Tranquilizer	Diazepam	Drowsiness, dizziness, weakness, dry mouth, headache, mental confusion, depression, skin rash, itching, insomnia, nervousness, irritability, jaundice, difficult urination.
ZYLOPRIM® Antigout agent	Allopurinol	Skin rash, nausea, vomiting, diarrhea, abdominal pain, drowsiness, blood disorders, kidney or liver damage, hair loss, visual disturbances.

Nuclear Magnetic Resonance

At a fall meeting in 1982 of a group called Neuroradiologicum in Washington, D.C., reports were received on new techniques of seeing inside the human body. The most talked about of the new techniques is called NMR, which stands for nuclear magnetic resonance. NMR was originally developed for use on humans in the early 1970s, but it did not come into clinical use until the early 1980s. NMR takes its place with the venerable X-ray and with the newer and remarkably successful computerized axial tomography (CAT) systems, and offers some great advantages over each.

NMR uses no radioactive substances or penetrating radiation and requires no invasive procedures (such as introducing foreign substances into the bloodstream). The patient is placed in an extremely strong magnetic field, which detects the hydrogen nuclei in the water molecules of the body. Hydrogen atoms that are in normal cells produce a different signal from the hydrogen atoms in cancer cells. The technique can also tell nerves affected by multiple sclerosis from those that do not have the disease, and detect the motion of blood through the heart or the brain. By following the metabolic path of specific chemicals, such as glucose, NMR can examine precisely what the metabolism of the chemical is in a given patient. (For more technical details on NMR, see page 227.)

Nuclear magnetic resonance imaging is apparently safe for all but the few patients who might have metal appliances (valves, plates, artificial joints) in the affected part of the body. The equipment is relatively expensive ($500,000 to $4,000,000), but is already being installed in large hospitals.

NMR will be competing with a number of other new imaging techniques, in addition to CAT scans, each with its own acronym. Positron emission transaxial tomography (PET) uses positron-emitting isotopes, and may have more research applications than clinical ones. Single photon emission computed tomography (SPECT) is an extremely sensitive imaging technique that can actually be used to measure blood flow in small vessels in the brain. Digital subtraction angiography (DSA) uses conventional X-rays, but converts images to digital information for processing by a computer.

One other technique that uses no radioactive substances and is noninvasive is ultrasound imaging. This may become especially useful because it can provide immediate images on a screen monitor. It has already shown itself useful in some brain studies and procedures and may one day be used during delicate operations to tell the surgeon more than can be seen with the naked eye about the tissues being examined.

Modern Diagnostic Tests

Here is a brief review of modern selected procedures that help establish the cause and nature of disease.

Test	Purpose	Procedure
Amniocentesis	Determination of the health of the fetus. Detection of many genetic defects and some hereditary disorders. Generally used in pregnant women over 35 years of age.	Insertion of a hollow needle through the abdominal wall into the uterus to remove a small amount of fluid from the amniotic sac. The placenta is located by ultrasound.
Biopsy	Examination of tissue samples and cells to determine abnormalities. Indicated when cancer, or certain liver or kidney diseases, are suspected.	Removal of a small piece of tissue for microscopic examination. Specific procedure depends upon body organ being studied.
Bone marrow	Examination of specimen from bone interior to determine red and white blood cell balance and presence of iron.	A small amount of bone marrow is withdrawn with a syringe and special needle, usually from hip bone or sternum.
Bronchoscopy	Inspection of the interior of the tracheobronchial tree through a bronchoscope. Helpful in diagnosing upper bronchial cancer, obtaining biopsy specimens, and removing obstructions. Indicated in cases of severe, prolonged cough or coughing up of blood, and for an undiagnosed abnormality on chest X-ray.	A fiberoptic (a flexible material made of glass or plastic that transmits light along its course by reflecting it from the side or wall of the fiber) bronchoscope is passed from the mouth through the trachea and into the bronchi. Sedatives are administered before the test. A local or general anesthetic may also be administered.
Cholecysto-graphy	Examination of the gallbladder by X-ray study. Useful in detecting gallstones or gallbladder dysfunction.	A radiopaque substance or contrast medium is given to make the gallbladder more visible. Then an X-ray is taken to determine the presence of gallstones. To test for gallbladder dysfunction, a fatty meal is eaten and another X-ray is taken.

Test	Purpose	Procedure
Colonoscopy	Inspection of the colon above the rectum; obtaining tissue samples for biopsy; removal of polyps. Used after repeated blood stool tests have been positive to look for suspected colon cancer or serious colon disorders.	Insertion of an endoscope into the anus, which is then threaded the entire length of the large intestine. Prior to test, the colon is cleansed by enemas, and a sedative is given.
Computerized Axial Tomography (CAT scan)	Imaging of internal structures, especially the brain. Works only for parts of the body that can be completely immobilized. Useful in diagnosing brain disorders and injuries, liver or pancreatic tumors.	Numerous parallel X-ray beams are projected through the body from various angles. These images are converted by a computer into cross-sectional views. An opaque dye is sometimes used as a contrast medium. Needs lower level of radiation than conventional X-rays.
Culdoscopy	Endoscopic examination of the female internal organs and pelvic cavity. Used in cases of suspected endometriosis (abnormal uterine tissue growth outside the uterus), infertility, and other disorders involving pelvic organs.	An endoscope (a flexible fiberoptic tube permitting visualization of a hollow organ) is inserted into the pelvic cavity through an incision in the vaginal wall.
Cystoscopy	Examination of the inside of the bladder and urethra with a cystoscope. Indicated when there is a suspicion of bladder or urinary tract cancer or polyps.	A fiberoptic cystoscope or tube is passed through the urethra into the bladder. Sedatives are given before the test. Usually a hospital procedure performed under local or general anesthesia.
Echocardio-graphy	To record the size, motion, and composition of various cardiac structures. Used in heart valve problems that are difficult to diagnose.	Use of ultrasound produces an image or photograph of an organ or tissue. Ultrasonic echoes are recorded as they strike tissues of different densities.
Fluoroscopy	Examination of inner parts of the body by means of the fluoroscope. Upper body: X-ray visualization of	For an upper gastrointestinal (GI) series, a patient is given a barium drink. X-rays and fluoroscopic studies are then

Test	Purpose	Procedure
	esophagus, stomach, and duodenum to detect ulcers. Indicated when such symptoms as indigestion, stomach pain, nausea, and loss of appetite become persistent. Lower body: X-ray visualization of large intestine. Useful in detecting diverticulosis (weakened outpouches in the colon wall) and colon tumors.	taken. For a lower GI series, barium mixture is fed into the empty colon via a tube inserted into the anus. While barium is being fed in, fluoroscopy (the projection of moving X-rays onto a special screen) is undertaken to look at the intestinal lining. Regular X-rays are taken after the large intestine is filled.
Gastroscopy	Examination of the stomach and abdominal cavity. Collection of biopsy specimens and photographing of organs. Used in cases where there is a suspicion of stomach ulcers with upper gastrointestinal tract bleeding of unknown origin or unexplained abnormalities.	This procedure calls for a flexible fiberoptic endoscope to be threaded from the mouth through the esophagus into the stomach and upper intestine. Air may be blown through the tube for better visualization. Either a local anesthetic or mild sedative is administered.
Hysterosalpingo-graphy	X-ray studies of the uterus and oviducts. Used as an aid in diagnosing infertility after other tests have shown a woman to be ovulating and the male to have normal sperm count.	After injection of a radiopaque material or dye into the uterus, X-rays are taken of the uterus and Fallopian tubes.
Nuclear Magnetic Resonance (NMR)	A new technique, NMR imaging is becoming available at large medical centers. It promises many of the advantages of the X-ray CAT and other scans without the hazard of ionizing radiation.	Various atoms in the body, acting like tiny magnets, can be caused to emit weak radiowaves as a result of a combined magnetic field and radio stimulation. These waves, computer detected and analyzed, project an image on a TV monitor.
Positron Emission Tomography (PET)	A scanner that makes images from radioactive isotopes introduced into the patient's body. These images are deciphered by the use of computers.	Specially prepared chemicals, containing radioactive atoms that emit positrons, are injected or inhaled. A scanning device detects these radioactive isotopes within the body.

Test	Purpose	Procedure
Pyelography	X-ray examination of the renal pelvis and ureter. Indicated where there is recurrent kidney or bladder infection, or with severe symptoms of prostate enlargement; also, in certain cases of hypertension.	A dye, containing iodine, is injected into the arm. X-rays are taken to determine the size and functioning of the kidney or, in men, of the prostate gland.
Radioisotope scanning	The study of the function and condition of internal organs, vessels, or body fluids. Used in patients with symptoms of various serious diseases. There are specific scans for virtually all the body organs.	A radioactive substance is injected into the blood and concentrates in the organ under study. The presence and location of these substances in the body is detected by special apparatus.
Sigmoidoscopy	Inspection, through a speculum, of the interior of the sigmoid colon. Indicated for the detection of abnormalities in the lower 12 inches of bowel.	A tubular speculum is inserted through the anus to examine the sigmoid flexure (the lower part of the descending colon, shaped like the letter S).
Spinal Tap	Lumbar puncture. Used in cases of suspected meningitis, and in certain types of spinal cord or brain damage.	A needle is injected at the juncture between the third and fourth vertebrae to extract spinal fluid for laboratory analysis.
Thermography	A process for measuring temperature by means of a thermograph. The technique has been used to study blood flow in limbs and to detect cancer.	A regional temperature map of a body or organ is obtained with thermographic imaging. An infrared camera scans the body and the computer codes body temperature.
Ultrasonography	Use of ultrasound to produce an image or photograph of an organ or tissue. Used where there is a suspicion of thyroid or gallbladder dysfunction, and in hard-to-diagnose cases of heart and valve problems.	A transducer, placed over the area to be examined, sends ultrasonic waves into the body. The ultrasonic echoes are recorded by the transducer as they strike tissues of different densities, and are converted into visual images.

Genetic Cures for Blood Diseases

Late in 1982 independent teams of doctors simultaneously announced the first successes in treating diseases by means of genetic manipulation. The diseases that responded to this revolutionary treatment were beta-thalassemia and sickle cell anemia. Both are inherited and were previously incurable blood disorders caused by imbalances in hemoglobin.

Three patients suffering from beta-thalassemia were treated in the new fashion by Timothy Ley, Arthur Nienhuis, and their associates at the National Heart, Lung and Blood Institute in Bethesda, Maryland. They worked with Joseph DeSimone, Paul Heller, and their team from the University of Illinois College of Medicine, who developed the treatment in animals.

Beta-thalassemia is caused by a genetic difference that produces defects in two primary kinds of hemoglobin, alpha globins and beta globins. When either of these two globins are not synthesized by the body in sufficient supply, hemoglobin cannot be synthesized properly either, and small, ineffective red blood cells result. Doctors have known this much about the disease for years, but have had no means of bringing the blood chemistry back into a healthy balance.

Recent research suggested that the key to the problem might be a third type of red cell—gamma globin—which is normally active only before a person's birth. In the early stages of life, gamma globins supervise the production of hemoglobin, but once they have launched this production, they become inactive. The molecules become coated with molecular methyl groups that keep them from acting. Scientists reasoned that a healthy blood balance might be regained if they could reawaken the gamma globins.

A tool for accomplishing that soon presented itself—a drug called 5-azacytidine that strips the methyl groups and restores the gamma globins to activity. Tests on baboons showed that the drug did boost gamma globin activity and that it did supervise the production of healthy red cells. Then the three beta-thalassemia patients were given the drug to combat their beta globin deficiencies. As the doctors hoped, the gammas returned the blood of the patients to a healthy balance between alpha and beta globins. However, the good results were actually puzzling to the researchers. Improvement occurred faster than the drug could possibly have stripped the methyl groups from the genes and the genes to have then been expressed in new cells. The reason for this was still unknown at the start of 1984, although there were several competing theories.

Further research suggests that the removal of the methyl groups does occur, and that the genes for gamma globin are then expressed. There seems to be, however, some other immediate effect that precedes the genetic change. Then, the newly expressed genes continue to produce gamma globins, prolonging the effects of the drug for a few weeks after each dose.

The important advance is that the drug appears to have altered the role of a body cell whose activity (and lack of it) had been genetically determined. The uncertainty about the actual cause of the change, however, led researchers at other institutions to try different drugs that have some of the same properties as 5-azacytidine, but which were not expected to reduce methylization of genes. The drugs also produced higher levels of gamma globin in monkeys and baboons. Thus, the treatment involving 5-azacytidine may have been discovered by accident.

In sickle cell anemia, which most often strikes people of African descent or, to a lesser degree, people of Mediterranean ancestry, the theory behind the use of 5-azacytidine was different. The beta globin is synthesized, but produces cells with a misshapen "sickle-shaped" form. Such sickle cells are not efficient carriers of oxygen. The disease is progressive because the sickle cells are gradually produced in larger numbers. At the same time as the experiments on beta-thalassemia were conducted, two patients at the National Heart, Lung and Blood Institute and another at the Illinois University College of Medicine received treatment with 5-azacytidine, based in part on the work of George Dover and a team from Johns Hopkins Medical School as well as on the work of DeSimone, Heller, and their coworkers. Doctors hoped that because sickle cell patients have some cells that produce normal beta globin, the reawakened gamma globin might stimulate only healthy beta globin. The hope was realized, and all three patients improved. Again, however, the recovery was more rapid than expected, so the exact cause is unknown.

These attempts to address serious blood diseases by chemically altering the agents of blood reproduction were especially encouraging because earlier experiments—seeking a cure by bone marrow transplants—were not clear successes.

Treatments of the blood diseases in 1980 by Dr. Martin Cline of UCLA Medical School raised an outcry because the procedure used had not been tried first on animals. Furthermore, Cline's approach was not successful, although it caused the patients no harm. Cline was reprimanded and grant money was taken away from him. It appeared that the scandal might set back research on inherited blood diseases for years. But the announcement of the success of the 1982 drug treatment, which had been thoroughly tested in baboons first, appeared to reverse the mood of discouragement.

A Chance for Jamie

During the fall of 1982, newspapers and television brought us the story of Jamie Fiske, a baby born with an incurable liver disease called biliary atresia. Surgery was attempted and was unsuccessful. The only hope Jamie had for survival was a liver transplant, and she was admitted to the University of Minnesota Hospital, which specializes in such surgery. Because of her small size and because of the necessary tissue matching, the chances of finding a suitable liver for transplant seemed small indeed.

We might never have heard of Jamie Fiske—there are scores of other similar babies—had it not been for the determination of her father, Charles, who was an administrator at the Boston University School of Medicine and who knew a thing or two about the medical world. He asked for and received permission to speak briefly to the American Academy of Pediatrics at its convention late in October, and he used the chance to plead for his daughter. Newsmen at the convention picked up the story and soon he was on television and in newspapers from coast to coast, letting people know how urgent the need for a suitable organ had become.

As a result of his public appearances, the parents of a ten-month-old child who died in an auto accident in Salt Lake City, Utah, offered their child's liver for transplant. Tissue types matched, and the transplant was accomplished on November 5, 1982.

The case brought up troubling questions in the medical profession and among students of public policy as well. At present, there is no clear system for matching prospective recipients of transplanted organs with organs that become available. No one could blame Charles Fiske for his energy and perseverance in seeking a suitable donor. But if every prospective transplant recipient had to take his case to the newspapers, how would medical men decide who "deserves" to get the first available organ?

Journalists themselves would not be a very reliable medium of communication in any case, since after the third or fourth such story, audiences would tire of the subject and reporters would seek some new kind of human interest.

The Fiske case called forth editorials and letters to the major medical journals and to newspapers around the country. Some suggested that a national computerized registry of transplant resources be established. Others even suggested a lottery. As transplants become more common, the issue will have to be resolved, and it will take the cooperation not only of doctors but of a general public sensitive to the difficult questions of life and death that choosing a recipient necessarily involves.

First Artificial Heart

A team of doctors headed by William DeVries made medical history at the Utah Medical Center on December 2, 1982, when they implanted a mechanical heart into Barney Clark, a retired 62-year-old dentist from Seattle, Washington. The device, called a Jarvik-7 after its inventor Robert Jarvik, "beat" more than 13,000,000 times while keeping Clark alive from December 2, 1982, until March 23, 1983, when he died of massive organ and tissue failure.

The polyurethane, Dacron, and Velcro Jarvik-7 replaced the two lower chambers of Clark's natural heart, the ventricles that pump blood to the lungs and force it into the arteries. It was attached to the two upper chambers of Clark's own heart, the auricles that receive blood from the veins and from the lungs. The plastic ventricles were opened and closed to simulate heart action by an external air compressor connected to Jarvik-7 by hoses. It was designed to pump a minimum of 7 liters (8 quarts) of blood each minute.

Clark, who had suffered for years from severe heart insufficiency caused by cardiomyopathy, would have had only hours or days to live if the artificial heart had not been implanted in a seven-hour operation. During his 112 days of added life, Clark rebounded again and again from serious setbacks, but he could not finally overcome the severe debilitation of his years of illness. He underwent additional surgery for lung complications, then was forced back into the operating room for a procedure to end severe nosebleeds. Other crises were precipitated by brain seizures, pneumonia, kidney failure, and severe depression.

The only crisis precipitated by the Jarvik-7 itself happened twelve days after its implantation. Clark's blood pressure plunged to a dangerous low, and he was rushed into the operating room where exploratory surgery revealed a cracked mitral valve, the artificial link between the plastic left ventricle and the natural left auricle. A spare valve was used to replace the cracked one.

After sixteen weeks, Clark's lungs deteriorated to the point where they could no longer supply adequate oxygen to his tissues. In a fatal spiral of complications, the oxygen deprivation led to collapsed artery walls that closed the circulatory system to the blood steadily pumped by the Jarvik-7. Barney Clark was pronounced dead on March 23, 1983, after several hours without neurological response. The artificial heart continued to beat regularly until it was switched off by an unidentified member of the medical team.

Doctors at the center said in the weeks after Clark's death that the experimental implantation had been a success. They suggested that if the heart had been implanted in a healthier person, the survival time might have been greatly lengthened. Despite success, however, the routine implantation of artificial hearts seems to be years in the future, pending further research. Barney Clark's 112 days were the first human example of the work yet to be done.

For nearly a year after Clark died, authorization to conduct another experimental implant of an artificial heart was withheld from the Utah team. Early in 1984, however, permission to try again was granted. Once again, as they had before operating on Barney Clark, the team began the search for the right candidate. There were many volunteers, despite the problems Clark had faced. This time doctors were hoping to find a patient who did not have as many other physical difficulties as Clark had.

WHAT'S NATURAL?

Developers of a new rotary pump for artificial hearts have been taking lots of kidding about the heart that will deliver no pulse. The rotary pump, which is much simpler than those now used in artificial hearts for humans, has kept a calf alive for 99 days without ill effects. When asked if the rotary pump was not just too different from the thump-thump of the natural heart, engineer Gordon Jacobs of the Cleveland Clinic replied, "Well, Boeing 747s don't flap their wings, either."

BLUE PEOPLE

In and around Troublesome Creek, Kentucky, for the past 162 years, a number of the inhabitants have had blue skin. This genetic anomaly is the result of a recessive gene brought to the region six generations ago by Martin Fugate, a French-born orphan, and by his locally born wife, Elizabeth Smith. In fact, the local name for the blue people is "the blue Fugates." The blueness is caused by the absence of the enzyme diaphorase, an enzyme that turns the blue protein methemoglobin into the red protein hemoglobin. Left to itself, hemoglobin gradually converts to methemoglobin. Despite the defect, the blue people produced by the inbreeding common in isolated villages are normally quite healthy—although somewhat odd to the eye.

Dr. Madison Cawein, however, introduced the blue people to a cure for the condition, pills of methylene blue that must be taken daily to turn the skin back to pink. Methylene blue works by providing an electron donor. The body uses this as an alternative approach to converting methemoglobin to hemoglobin. But the blue people picture methylene blue as working another way. As the dye is excreted, it turns the urine blue, so the blue Fugates say, "I can just see that old blue running out of my body."

Second Heart Attack Risk Factors

Each year 1,500,000 people in the United States will suffer heart attacks, and one-third will not survive. Of the survivors, another 100,000 will succumb to a second heart attack within a two-year period. It was a desire to reduce this high mortality rate that led Dr. Arthur Moss and his colleagues at the University of Rochester School of Medicine and Dentistry to undertake an investigation of signs that could signal a second attack. Heart attack victims were studied for residual damage, and a detailed record was kept of survivors and nonsurvivors within a two-year period following the initial attack. Four signs proved to be predictors of early mortality.

Pulmonary rales. Normally, the heart contracts to pump blood to the tissues, but a damaged heart may not contract sufficiently after a heart attack. Blood is not pumped out of the heart at a fast enough rate. Consequently, blood backs up and the lungs become filled with fluid, resulting in pulmonary rales. (A *rale* is an abnormal sound that occurs during breathing. If the sound comes from the lungs, it is a pulmonary rale.) This condition, readily detected on stethoscopic examination, made another heart attack three times more probable.

Ejection fraction. The ability of the heart to pump blood is lessened when a portion of the heart muscle dies. To get an accurate measure of the fraction of the blood squeezed out with each beat, some of a patient's red blood cells were tagged with a trace of radioactivity and their movement was observed with a special machine. If the resulting so-called ejection fraction was poor, the patient was almost two and a half times at greater risk of incurring a second, fatal heart attack.

Shortness of breath. This risk factor was not a sign stemming from the initial heart attack, but rather a preexisting symptom showing up in the medical history of the patient. It doubled the risk of another heart attack.

Disrupted tempo. The fourth risk factor was determined by a study of 24-hour tape recordings of patients' heartbeats. Those who had extra heartbeats, a fast heart rate, or abnormal rhythms had a one and a half times greater risk of fatality than those whose hearts showed no disruption.

A class of drugs—the beta blockers, which are often used to control blood pressure—seem to prevent second heart attacks. Research data are being amassed in a large-scale, ongoing four-year study that will involve 2,000 patients at 25 hospitals for the purpose of an in-depth evaluation of this type of drug therapy.

1983

Depo-Provera

A fifteen-year battle over Depo-Provera, an injectable, long-acting contraceptive, went before an extraordinary board hearing of the Food and Drug Administration (FDA) early in 1983, but the hearings ended inconclusively, leaving the contraceptive still unapproved and the final decision up to FDA Commissioner Arthur Hayes.

The hearings pitted the manufacturer of the drug, the Upjohn Company, against consumer advocates, notably the Health Research Group. In the middle was the FDA, seemingly tied in knots by its own procedures. The stakes are high—Upjohn has already realized $25,000,000 in profits from the drug even without FDA approval, and ultimate profits could be many times that amount. In league with the company are the World Health Organization and 80 nations from around the world that are seeking shipments of the contraceptive for use in family planning programs. The drug cannot be shipped from the United States to other countries without FDA approval.

Depo-Provera is a progestogen, a synthetic acetate of the hormone progesterone, which is active in uterine acceptance of an ovum for fertilization and in the production of milk by the mammary glands. Its contraceptive property is that a single injection of the drug can halt ovulation in a human female for up to three months. Thus Depo-Provera is at least as effective as oral contraceptives, and it is both less costly and far less complex to administer than the pill.

Experiments mandated by the FDA in the 1970s linked progesterone with breast tumors in two of a group of sixteen beagles and with endometrial, or uterine, cancer in two of a group of 52 rhesus monkeys. The World Health Organization argues that these results on a handful of animals are insignificant against its own evidence—successful clinical trials with 11,000 women, and regular nonregulated use in places like Thailand, where 86,000 women have been injected with Depo-Provera since 1965.

The FDA's own rules tie its decision to the animal tests, which makes it difficult for approval to be given in the face of negative results. Upjohn dismisses the tests as meaningless because the drug response of the chosen animals has been proven different from that of humans. But opponents of the drug bristle at the suggestion that any drug should be approved for use if it has been demonstrated to be carcinogenic in animal tests.

By the middle of 1984, no decision had been reached.

New Pain Treatments

Miracle pain cures are the stock in trade of medical charlatans and hence are always the targets of skeptics in the medical profession. But within a single month in 1983, debunkers of offbeat analgesics were confronted with evidence of the effectiveness of two unorthodox pain treatments. The first is a new technological application, the "cold laser." The second is the ancient Chinese remedy, acupuncture. In both cases, although the details of how the processes work are not well understood, the treatments seem to use the body's own chemical painkillers to achieve their effects. (See "The Intriguing Neurotransmitters," page 451.)

On February 7, 1983, Judith Walker of the Walker Pain Institute in Los Angeles reported on cold lasers to a seminar sponsored by the National Center for Devices and Radiological Health, a part of the Food and Drug Administration (FDA) of the federal government. She announced that more than 70 percent of a test group treated with cold lasers were "pain free without additional treatment."

On her way to the seminar, however, Walker learned that her use of cold lasers was illegal. Experimental use of the lasers is permitted by the FDA only if either the FDA or an institution sponsoring research has granted an "investigational-device exemption." Of a dozen groups testing the cold-laser technique, only one was known to have received the exemption, although other institutions might have granted the exemption without notifying the FDA. The FDA did not plan any prosecutions, however.

Walker's study focused on more than two dozen chronic sufferers of trigeminal neuralgia, which causes intermittent episodes of excruciating pain in the face and neck, and of osteoarthritis, a painful bone disease. Treatment consisted of 30 sessions of doses from a 1-milliwatt helium-neon laser, called a cold laser because it transmits relatively little energy compared with the "hot" lasers used in surgical procedures. Unlike the effects of lasers that are used to cauterize blood vessels or cut tissues, irradiation from the cold laser is so subtle that patients remain unaware that they are being "zapped" at all. The beam is directed at two key nerve centers near each ankle and each wrist. Laser stimulation of these sites— sites long known to acupuncturists—has the effect of stimulating 90 percent of the cortical nerves in the whole body according to advocates of the method. Walker administered the treatments over ten weeks and followed the study group for six months.

Another believer in cold lasers is Wolfgang Bauermeister, who operates out of another Los Angeles pain center. Dr. Bauermeister has treated patients with pains ranging from migraine headaches to stomach aches with a reported 85 percent success rate. He believes that the lasers release the body's own natural painkillers, the endorphins.

The extreme low level of radiation produced by the laser forces the FDA to ponder two questions: first, whether there is any danger in the experimental treatment and second, whether the treatments have any desirable effects. Its division of risk assessment has yet to be convinced that the devices accomplish anything at all, and so is requiring special approval of commercial applications in order to protect consumers from fraud. Because the cold laser treatments are already becoming popular, however, the agency is unlikely to impose a ban on them. Enthusiastic approval by the agency is equally unlikely.

The ancient Oriental art of stimulating those same nerve centers with needles rather than lasers also gained new advocates in 1983. In March 1983 Per E. Hansen and John H. Hansen of Denmark reported complete patient preference for acupuncture over other analgesics for relief of trigeminal neuralgia. In a test group of sixteen facial pain sufferers, the Hansens gave half of the patients acupuncture and half bogus needle stimulations slightly displaced from the nerve centers. In all cases the needles were inserted into the same general hand and foot areas and left in place for about fifteen minutes. Those receiving true acupuncture reported long-lasting freedom from neuralgia attacks. Those receiving the acupuncture "placebo" did not, thus lending new experimental credence to one of the most ancient of treatments for pain.

The study of pain itself has become more important in recent years. Doctors have realized that pain—the symptom—can sometimes become a major contributor to stress that worsens the disease. Pain-treatment clinics that specialize in various methods—traditional American medicine as well as the exotic—are becoming a fixture on the medical landscape, and general practitioners are also making more efforts to eliminate pain.

HIGH-VOLTAGE HIGHS

A research group at the University of Salford, England, has observed that electric currents stimulate the body to produce "natural opiates" such as beta-endorphin and met-enkephalin. These normal products of the body are not really opiates, but they affect the same receptors that opiates do, producing similar effects. The disturbing thing about these findings is that just standing near a high-voltage source can produce currents like the ones used in the investigation, or even higher ones. It is possible that persons exposed to such frequencies for a long time could experience withdrawal symptoms on changing their environment.

The Intriguing Neurotransmitters

It was once believed that only one or two chemicals were active in the transmission of nerve impulses, and that they were perhaps active in the process of thought itself, since thought results from connections between nerves. Chemicals that transmit messages from or to nerves are called neurotransmitters. In more recent years, not only were the neurotransmitters found in various new chemicals extracted from the brain and from nerve cells, but it was also discovered that many long-familiar substances, such as insulin, also had roles to play in the central nervous system. In 1984, a list of neurotransmitters (compiled by Dorothy Krieger of Mt. Sinai Medical Center) looked like this:

Years	Neurotransmitters Described
1920s	acetylcholine and epinephrine
1940s	norepinephrine
1950s	various amino acids (GABA, glutamic acid, aspartic acid) and dopamine
1960s	substance P
1970	serotonin
1974	"hypothalamic-releasing hormones"—thyrotropin-releasing hormone (TRH), luteinizing hormone-releasing hormone (LHRH), and somatostatin (SRIF)
1975	enkephalins and endorphins
1976	vasoactive intestinal polypeptide (VIP) and cholecystokinin (CCK)
1977	adrenocorticotropic hormone (ACTH)
1978	other pituitary hormones, insulin, vasopressin, oxytocin, and angiotensin
1979	glucagon
1981	corticotropin-releasing hormone
1982	growth hormone-releasing hormone

Very small amounts of serotonin are found in the neural tissue of vertebrates, mollusks, and arthropods. Serotonin is the poison in the sting of a hornet and in the venom of some spiders. Dopamine is currently widely used for the treatment of shock.

Disturbances in neurotransmitters are a major factor in certain neurological diseases. For example, schizophrenia may involve a disturbance of serotonin metabolism, injury to neurons that produce dopamine almost certainly causes Parkinson's disease, and malfunction in GABA-containing neuronal systems may be implicated in epilepsy.

Medical Research Scandal

On February 16, 1983, the National Institutes of Health (NIH) released a final report on the case of Dr. John Roland Darsee, a case that has raised serious questions about the quality of medical research in the United States.

Darsee was a brilliant research doctor whose first well-publicized research on the treatment of heart disease with drugs was carried on at Emory University Medical School in Atlanta. In 1979 he was appointed to a research position at Harvard Medical School, and his admirers predicted that he would soon be appointed to a prestigious chair at the school.

The first questions were raised in early 1981. In May of that year colleagues of Dr. Darsee caught him in an obvious falsification of data. The Medical School made its own in-house investigation of the matter, dragging it on through the year, and not reporting it to the NIH, whose research funds were being used, until December 1981. The director of the Cardiology Research Lab, Dr. Eugene Braunwald, was later accused of countenancing a cover-up.

The NIH, after its own investigation under the direction of distinguished outside physicians, barred Darsee from receiving any NIH grants for ten years, asked Harvard for a refund of $122,000, and put Dr. Braunwald's Cardiology Lab on a one-year probation. Dr. Darsee had already left Harvard and had taken up a hospital position in Schenectady, New York.

Three months after the NIH report, Emory University announced that it had discovered widespread instances of falsified records in Dr. Darsee's research there between 1974 and 1979. In all, at Harvard and Emory, Dr. Darsee had been author or co-author of 116 research reports in major medical publications. Just how his research findings can be corrected is a problem that haunts other medical researchers. The list of articles that cite Darsee's findings covers some 26 feet of computer printout paper.

If the Darsee case were an isolated instance, it would soon be forgotten. But there is evidence that similar falsifications have been going on at other major medical centers. In the early months of 1983, other instances of the same behavior surfaced at Mount Sinai in New York and at Boston University.

Why did it happen? What does it mean? The NIH investigators reported about the Harvard lab that work there had "a hurried pace and an emphasis on productivity." Younger doctors had "limited interaction with senior scientists," meaning that Darsee's patently unrealistic results were not discovered for many months.

New Light on Emphysema?

An apparent breakthrough in the pathology of emphysema has opened the way for experiments on possible treatment of the disease. The new findings on chemical changes in the disease process were announced on March 22, 1983, by Ines Mandl of the Columbia University College of Physicians and Surgeons. By the end of the year, however, other scientists cast some doubt on the exact mechanisms involved.

For some years scientists have understood the basic effects of emphysema, a chronic lung disease that affects 11,000,000 Americans and kills about 56,000 each year. The chief characteristic of the disease is deterioration and rupture of the air sacs, or alveoli, in the lungs. The damage is caused by the failure of elastin, a fibrous protein in the alveoli that gives the cells their elasticity. The elastin proteins are chemically eroded by an enzyme called elastase, which breaks the fibrous proteins down into smaller peptides.

The body normally controls elastase with an "elastin protector" called alpha$_1$ antitrypsin. Individuals with a genetic deficiency of this protector have long been known to be predisposed to emphysema. Other individuals develop emphysema when exposed to cigarette smoke.

Mandl's experiments with induced emphysema in hamsters exposed to cigarette smoke managed to uncover the manner in which cigarette smoke causes the disease. She found that cigarette smoke introduced into the lungs of 100 hamsters oxidized the elastin protector, rendering it useless. Moreover, the smoke also caused a greater than normal production of elastase by the white blood cells. Thus the smoke both dispersed the protective substance and encouraged greater production of the antagonist. Furthermore, cigarette smoke reduced the supply of an enzyme needed in rebuilding damaged lung tissue.

Other researchers had reported similar results in humans, although the mechanism involved was not known. However, studies designed to test specifically the activity of the alpha$_1$ antitrypsin taken from the lungs of smokers and nonsmokers revealed contradictory findings. All studies showed that cigarette smoking contributed to too much elastase activity, but the evidence that oxidation of the elastin protector occurred in humans was lacking.

Emphysema treatment still might be improved by direct administration of the protector substance alpha$_1$ antitrypsin. One further possibility is injection of antioxidants to protect the body's own elastase protector, but research has yet to confirm that this would actually be effective.

Solution to Lyme Disease Mystery

In the early 1970s, pediatricians around Old Lyme, Connecticut, began to see alarming symptoms in a number of their young patients. The sickness began with a rash, fever, chills, and muscle aches. But in a fair number of cases, after the first symptoms came really serious joint pain and inflammation—apparently a kind of juvenile arthritis. If left untreated, the disease could do permanent damage to joint tissue. Knowing how to treat it posed the big question. By 1975 the syndrome was recognized and was christened Lyme disease after the town.

The pattern of the disease suggested that there must be an infectious agent involved. Because Old Lyme is a rural area, it seemed plausible that an insect or similar carrying agent might be involved. By 1978 a new tick, which received the name *Ixodes dammini*, was isolated in the area, and presumed to be somehow involved. By 1980 it had been established that the treatment of victims with penicillin during the early stages was effective in reducing the risk of arthritic symptoms and damage. This discovery suggested that the source of the infection must be a bacterium. Then in 1982 a bacterial candidate was isolated in tissue of the *I. dammini* tick. It was of the spirochete family best known for causing syphilis, but this variety had not been previously identified. It was called *I. dammini* after its host the tick.

In 1983 research teams from Yale University and the State University of New York at Stony Brook announced that the connection of the new spirochete with the disease had been established. One team found the bacterium itself in three sufferers of the disease and found antibodies to it in more than 90 percent of 135 victims. The other isolated both bacterium and antibody in two patients. These discoveries are sufficient for most researchers to conclude that the *I. dammini* spirochete is the cause of Lyme disease. In the classical Koch test for establishing a particular organism as the cause of a particular disease, the organism must also be grown in a pure culture and then used to infect experimental animals. Researchers have not yet carried out this program.

Questions still remain. For example, it is not clear just how the tick transmits the spirochete to human victims when it bites them. It is also odd that the bacterium itself was not isolated in more than a few patients. There remains the possibility that the disease itself is a complex one caused more by an autoimmune reaction to the bacterium than by the organism itself. But in a few short years, progress in understanding the new ailment has been rapid.

Report on Bone Marrow Transplants

The May 13, 1983, issue of the *Journal of the American Medical Association* reported on the increasing use of bone marrow transplants as a treatment for aplastic anemia and certain types of leukemia. Both diseases drastically reduce the ability of the body to manufacture white blood cells, an essential part of the immune system (see "Our New Understanding of Immunity," page 491). Most such cells are made by specialized cells in bone marrow. In leukemia, the cancer cells in the bone marrow may produce modified white cells that are ineffective even when they are made in sufficient numbers. Bone marrow transplants can also be a lifesaver for children who are born without a functioning immune system, and for victims of thalassemia, a blood disorder. In all these cases, it is the bone marrow as the "nursery" for blood cells that is the effective source of the disease, even though the symptoms appear in the cells and antibodies circulating in the bloodstream.

The first attempts at bone marrow transplants were made in the late 1960s, but at that time doctors knew so little about the immune system that the transplants (which were used only as a last resort) were almost always unsuccessful. Because the marrow is the source of most of the cells that form the immune system, the "foreign" marrow produced cells that attacked the body of the recipient, a condition known as graft-versus-host disease. However, scientists reasoned that, just as there are different blood types, there are also different tissue types. In the late 1970s, as knowledge of tissue typing increased, more and more of the transplants proved successful. However, bone marrow transplants still usually require a donor who is closely related to the patient. An identical twin is the ideal donor.

The bone marrow procedure seems scarcely like a transplant at all. The marrow cells are taken from the anesthetized donor by a suction device similar to a hypodermic needle. No incision is required. The cells are then processed and dripped into the blood of the recipient intravenously. The cells find their way into the bones, carried there by the bloodstream.

The key to the process is the elaborate preparation of the patient. All recipients are given massive doses of immunosuppressant drugs to forestall rejection of the foreign tissue. Leukemia patients also receive total-body irradiation to reduce the leukemic cells that are destroying them. The dosages of radiation might be fatal if not followed up with the transplant of the marrow cells.

Depending on the type of anemia or leukemia, short-term survival rates range from 15 to 80 percent. The marrow cells, by a process still not understood, seek out the bones of the recipient, attach themselves, and begin to help produce white blood cells. One of the more promising recent developments is the success of a new immunosuppressant called cyclosporine, which suppresses immunity reactions to the transplant but does not appear to inhibit the body's defenses against bacterial infection (see "Cyclosporine Approved," page 479).

Another recent improvement in the technique is the use of monoclonal antibodies (see "Monoclonal Antibodies," page 459) combined with a powerful poison to destroy T-cells, the blood cells in the graft that are the main factors in the rejection process. The poisons used are diphtheria toxin or ricin, a plant poison derived from castor beans. Both are among the most toxic substances known because they act like catalysts; that is, a single molecule of the toxin can inactivate crucial parts of a cell's machinery without being changed itself. Thus, one molecule of the poison that is transported into the cell by a monoclonal antibody will result in the death of the cell. The purpose of the monoclonal antibody is to direct the poison to the desired type of cell, so that other cells are not affected. The technique is not perfected, so it cannot be used in a living person, but it can be used outside the body. Then the poison is removed from the marrow and the marrow can be safely given to the patient. Another approach, still in the research stage also, is growing the stem cells that produce the T-cells in the laboratory, and then just transplanting the stem cells.

With these new techniques, it may be possible to extend bone marrow transplants even to recipients for whom a donor of a suitable tissue type cannot be found and to people with sickle cell anemia, who are currently treated by other less drastic means.

In 1984 a famous case came to an end when a bone marrow transplant was tried on David, the "bubble boy." David was born with a defective immune system. Because his brother had previously died from the same condition, he was placed in a germ-free "bubble" immediately after a sterile Caesarean birth. This action saved his life, but he was condemned to stay in the bubble all his life. As David grew older, it was clear that a bone marrow transplant might help him, but a suitable donor could not be found. With the new techniques available, however, the doctors attending David decided to try anyway, using bone marrow taken from his sister.

The transplant did not take. Also, although the cause is not known, David became ill for the first time in his life. While ill, David was set free from his bubble. He lived only fifteen days after that, dying at age twelve. An autopsy revealed no graft-versus-host disease, but the actual cause of death remains unknown. David's last name was never revealed publicly.

Hair and Violence

The report of a part-time researcher named William Walsh on May 15, 1983, to a gathering at the Schizophrenia Foundation, part of the New Jersey Brain Bio Center, sent out waves to many different disciplines. Mr. Walsh, employed by day at the Argonne National Laboratory near Chicago, spends his spare time working with prisoners. Over the years he became obsessed with the possible influence of body chemistry on criminality. Finally, in 1983, he presented a report of great elegance and persuasiveness on his analysis of human hair in relation to violence.

Why hair? Because it has been discovered that hair cut from near the scalp has about 200 times the concentration of trace elements to be found in blood, and the "storage" of these elements in hair is relatively permanent. Trace elements are minor constituents of the body, although they often have a great effect on certain functions. A small amount of a given element may be needed to form a particular hormone or neurotransmitter, for example. In most cases, amounts of the element beyond what are needed act as poisons. Also, a deficiency of the element means that the necessary hormone or neurotransmitter cannot be made in sufficient amounts.

The major finding of Walsh's study is that there is a fairly clear *chemical* profile of two different types of violent personalities. Measuring the trace elements in the hair of 24 matched pairs of violent and nonviolent siblings and, in a second study, of 96 extremely violent men and 96 controls matched by age, race, socio-economic level, and size of resident city, Walsh found that there were clear differences in the hair samples of the violent siblings and the violent men. Furthermore, the violent men broke clearly into two types. The first type Walsh calls episodic violent; men in this group are subject to sudden fits of murderous anger and are prone to periodic violence. Like all of those in the violent group, they showed increased levels of such elements as lead, cadmium, and calcium, and abnormally low levels of lithium, zinc, and cobalt. They were distinguished from the second type of violent men by also showing low levels of copper and high levels of sodium.

Walsh calls the second violent type antisocial. Although not so impulsive as the first group, these men are thought to make up the majority of career criminals. They seem relatively impervious to "correction," returning again and again to violent behavior. Unlike the first group, they have high levels of copper and low levels of sodium in their hair samples.

Walsh's findings caused considerable comment in the legal community. The suggestion that there may be a chemical basis to violent behavior raises the question of whether a given criminal is responsible for his or her acts or is merely the victim of a chemical imbalance. It seems certain that Walsh's study could be the basis of a persuasive defense of a murderer in the hands of a determined lawyer.

On the less controversial side, the study suggests that there may be ways to treat violence-prone individuals to correct or at least minimize any dangerous chemical imbalance. This may even be accomplished through a simple change of diet. Experiments in several state prison systems suggest that reducing or eliminating prisoners' intake of refined carbohydrates (white sugar; and white flour, pasta, and rice) reduces violent episodes in those prisons. Refined carbohydrates are associated with a high level of cadmium in hair samples—one of the elements Walsh found uniformly high in both violence-prone types.

Walsh suggests that as techniques of analysis grow more sophisticated, hair analysis may become a routine medical test as common as the typing of blood. He says there is already some evidence that the trace element profiles of the hair of normal persons seems to break them into six types, presumably based on differences in metabolism. Such results could have important implications in studying disease patterns, especially if it became apparent that certain metabolic types were more subject to certain ailments. Further study is needed, however, on connecting the trace elements to the chemicals in the body that are affected by their presence or absence.

Hair analysis is a technique that fell into disrepute in the 1960s and 1970s, when many quacks offered personality profiles based on bogus hair analyses. In the health-conscious 1980s, mail-order hair analysts offered to solve nutritional problems if a sample of hair was sent to them. Skeptics who sent in doll's hair or dog's hair got just as useful analyses as those who sent in their own hair; all were worthless. A mail-order laboratory cannot adjust for the shampoos or other chemicals that have been used on the hair. In many cases, the laboratories provide analyses for substances that are not retained in hair (such as vitamins), or use faulty experimental techniques. While the results of a dedicated scientist such as Walsh may have some significance, the variables are too great to be applied in any consistent way to a large population until better laboratory techniques become routinely available.

There is hope on that front, however. Researchers have developed a standardized protocol that includes everything from the kind of scissors to be used (the wrong ones can contaminate the sample) to the way the sample should be washed. Also, a four-year study of what the elements are in "normal" hair is now available.

Monoclonal Antibodies

The bloodstream is filled with tens of thousands of slightly different chemicals, each "trained" to attack a very specific protein. These chemicals are the antibodies, the first defense against such foreign invaders as viruses and bacteria. In a "lock-and-key" fashion, each type of antibody is designed to attach itself to the specific protein, a process called *binding*. If only there were a way to provide more antibodies of a specific type, these antibodies would have a wide range of uses in diagnosis, research, and perhaps even treatment of disease. Since 1975 such a source of specific antibodies has existed. The products of that source are called *monoclonal antibodies*.

By mid-1983 monoclonal antibodies were coming into wide use in a variety of medical areas. Several processes for producing these antibodies had been developed in rapid succession. Various modifications on the basic idea made it possible to produce significant quantities of an antibody of the same specificity; that is, a large amount of an antibody that would bind to a specific protein.

Monoclonal antibodies are produced by causing the fusion of human cancer cells and cells from another animal in such a way that the results are cancer-like cells called hybridomas. These hybridomas can be induced to multiply in the laboratory as would some cancer cells. Each hybridoma produces genetic copies, or *clones*, of itself, which accounts for the name *monoclonal*. As the hybridomas multiply, they also produce antibodies. For a particular line of hybridomas, derived from the same original fused cell, all the antibodies have the same makeup.

The similarity has proved valuable in a number of medical procedures. Since millions of different hybridomas are initially formed, it is possible to select the genetic lines that produce useful antibodies and to discard the rest (see also "Human Hybridomas," page 148). The greatest hope for monoclonal antibodies is that one of the genetic lines may be effective in treating previously incurable diseases, especially cancer. That hope persists with good reason, although no one expects progress along these lines to be easy. In the meantime, the first routine uses of the antibodies have been somewhat more prosaic, although various experimental uses have continued to demonstrate the great potential of this method.

Monoclonal antibodies' most immediate use has been in improving a number of diagnostic procedures. For example, a pregnancy test using the antibodies is much more sensitive than older tests and can be conclusive as little as ten days after conception, even before delayed menstruation provides a clue. Tests have also been devised for hard-to-diagnose disorders such as chlamydia, a venereal disease that may afflict millions. Prior to the monoclonal antibody test, only the most severe cases could be diagnosed with certainty. The same laboratory that developed the test for chlamydia has also created a quick test for gonorrhea and herpes, reducing the time for diagnoses from three to six days to half an hour. Since all three diseases are hard to detect in women and have serious effects on newborns whose mothers have one of the diseases, the new test may help prevent many birth defects. These tests were beginning to reach the marketplace in 1984, the first commercially available products of monoclonal antibodies.

A second use has been in imaging—in seeking points in the body where infection or other disease is present. By producing an antibody sensitive to a certain cancer, for example, and tagging it with a radioactive marker, researchers can put the antibody into the patient's bloodstream and "see" what part or parts of the body it goes to. A similar procedure might be used to seek the main site of a nonspecific infection.

Imaging and purification with monoclonal antibodies has become an essential tool of the research biologist. In work with cells or tissue outside of the body, fluorescent dyes are often combined with the antibodies instead of radioactive elements. In this way, researchers can pinpoint the exact site of a reaction in a tissue or, with the aid of a microscope, within a cell. Along with recombinant DNA techniques, monoclonal antibodies have become the heart of current research into the basic processes of life. In fact, the two are being used effectively together. Proteins produced by bacteria that have been engineered using recombinant DNA techniques are then purified using monoclonal antibodies.

Researchers have also used the antibodies combined with ricin, a potent poison from the castor bean plant, or with diphtheria toxin to remove specific cells from bone marrow before transplanting the marrow to patients with leukemia. By providing selective immunosuppressant effects in transplant patients, the need for transplants to come from closely related donors is eliminated. (See also "Report on Bone Marrow Transplants," page 455). The antibodies have also been used to suppress immune reactions in patients after kidney transplants.

Perhaps the method of attaching poison to the antibodies can be used in time to attack malignant tumors. For now, however, the risk of the poison reaching healthy cells of the same type by one route or another is still too great. In the meantime, however, intensive research into these new substances has already proved monoclonal antibodies to be one of the most promising new tools of both the doctor and the research biologist.

The Chemistry of Depression

Researchers reporting to a National Institute of Mental Health seminar on stress in July 1983 outlined some of the complex chemical interactions in the body that cause chronic depression, one of the most prevalent and debilitating of all mental afflictions.

These new findings were made possible by the production of the natural substance corticotropin-releasing factor (CRF) in the laboratories of the Salk Institute in La Jolla, California. After extracting the substance from the brains of sheep, researchers were able to study its properties.

The researchers discovered that when CRF is administered to normal human subjects, it produces symptoms of acute depression. Then, through a series of more complicated experiments, they traced the action of this substance.

When people encounter a stressful situation or event, the brain apparently sends a signal to the hypothalamus to produce more CRF. The CRF triggers the release of a variety of other brain chemicals, including adrenocorticotropic hormone (ACTH), which is produced by the pituitary gland. The ACTH triggers the release of cortisol and epinephrine from the adrenal glands. Cortisol mobilizes the body to encounter danger in what is often called the "fight or flight" state. When it reaches the brain, however, it also has the effect of limiting the further release of ACTH, thus creating a "loop" and limiting the reaction.

In a rare ailment called Cushing's disease, this mechanism does not work properly, and the ACTH-cortisol loop runs wild. Some defect in the same system may also be responsible for chronic depression in many people. Those who are depressed appear to have high levels of CRF, which in turn keeps levels of both ACTH and cortisol relatively high. Researchers reason that, in effect, depressed people may be living in a permanent state of high stress, which can be exhausting and debilitating. Other symptoms of depression—drowsiness and lethargy—may be explained by the fact that other chemicals triggered by CRF are the natural opiates.

The oversupply of CRF may be caused by early conditioning, such as chronic stress in childhood. In effect, the body switches on to meet a stressful challenge and then never quite switches off even after the cause of the stress has disappeared, eventually causing symptoms of depression. The connection between psychological factors and chemical ones may be extremely intricate. But some understanding of the chemical mechanisms involved may eventually lead to effective chemical treatment for chronic depression.

Commonly Abused Drugs

Drug Group	Drug	Drug Effects	Withdrawal Effects
Narcotic analgesics, or pain relievers	Heroin Codeine Meperidine (Demerol) Morphine Methadone ((Dolophine), synthetic chemical with morphine-like action Opium	Affect nerve cells of the brain and spine. Taken by mouth, smoked, or injected, causes drowsiness, euphoria, mood changes, and mental clouding. Reduces the ability to feel pain. Lessens hunger, thirst, and sex drives.	Symptoms peak within 48-72 hours. Pains develop in back, arms, and legs. Withdrawal begins 4-6 hours after last use. "Cold turkey" withdrawal (sudden stopping of heavy drug use) causes chills, shaking, sweating; muscle aches and abdominal pain; vomiting and diarrhea.
Depressants, or anti-anxiety agents	Barbiturates (pentobarbital, secobarbital) Methaqualone (Quaaludes) Minor tranquilizers (Valium, Librium, Tranxene) Alcohol	Have a general depressant effect on the central nervous system. Moderately large doses produce drowsiness, euphoria, poor muscle coordination, and impaired judgment and speech. Overdosage is a real danger when barbiturates are combined with alcohol.	Symptoms start within 24 hours and include anxiety, agitation, nausea and vomiting, sweating, fast heart beat, tremors, and cramps. Symptoms peak at 2 to 3 days; there may be convulsions, delirium, and hallucinations.

Stimulants, or Psychoactive drugs	Amphetamines (when injected, called "speed") Cocaine	Act on the central nervous system to produce increased alertness and activity. In some persons, even small and infrequent doses can incur reactions ranging from restlessness and anxiety to convulsions and coma. Heavy, frequent doses can result in brain damage. Death follows injected overdose. Chronic users may become depressed, or paranoid with heavy doses.	There is a strong craving for another dose. Depression, cramps, sleeplessness, apathy, irritability, and mental confusion can occur. Cocaine withdrawal is accompanied by depression, apathy, and insomnia.
Hallucinogens, or psychedelics	LSD (D-lysergic acid diethylamide) PCP (phencycline) Mescaline (peyote cactus)	Bring about changes in thought, self-awareness, sensations, and emotions. Physically, changes range from dilated pupils and dry mouth to tremors and a rise in blood pressure. Psychological effects are more dramatic: "hearing" color and "seeing" smell; time and space stretch or shrink; there is a sense of detachment. Reactions to higher doses range from speech and vision changes to delirium and fear of death. PCP intoxication can produce violent and bizarre behavior that can result in serious harm to users or to other persons.	Control over normal thought processes may be lost for long periods of time. User may become engulfed by negative emotions: anxiety and depression, or may experience a break from reality that can last a few days or go on for months. Heavy users may incur organic brain damage.
Cannabis, or THC	Marijuana THC (delta-9-tetrahydro-cannabinol) Hashish	Produce reddening of the eyes and an increased heart rate. As use continues, sight, hearing, and touch become enhanced. Short-term memory may be impaired.	After heavy, regular use, withdrawal symptoms include insomnia, restlessness, nausea and loss of appetite. Continued long-term use can lead to emotional and psychological dependence.

463

Aspartame

The new artificial sweetener aspartame, most familiar from the brand name version NutraSweet, was approved in July 1983 as an additive to soft drinks. Leading soft drink manufacturers began using it almost immediately. This raised questions that had first been publicized in 1981, when the Food and Drug Administration approved aspartame to be sold as a sweetener directly to consumers.

In a letter to the *New England Journal of Medicine,* one researcher warned that even in fairly small doses, aspartame may have some of the qualities of the mood-changing drugs and that its action is greatly strengthened when it is consumed with carbohydrates—a strong likelihood with soft drinks.

Aspartame increases levels of a natural substance called phenylalanine, which in turn appears to stimulate production of the catecholamines, which are known to affect appetite, sleep, and mood changes in the brain. The researcher warned that aspartame-sweetened beverages might be dangerous to people suffering from Parkinsonism or even from more common ailments, such as insomnia.

The Coca-Cola company claims that the new sweetener is perfectly safe, but they announced that as a precaution they were blending aspartame with saccharin in their beverages, thus reducing the dosage of aspartame. Saccharin does not meet the Food and Drug Administration's rules for use as a food additive, because it has been found to cause cancer in animals. Congress has exempted saccharin from the rules, however, so its use is permitted. Some people think that a mixture of aspartame and saccharin is even more dubious than pure aspartame. At least there is no evidence yet that aspartame causes cancer. Other soft drink companies may or may not follow Coca-Cola in using the aspartame-saccharin mixture. The advantage of aspartame over saccharin is that it does not leave the metallic aftertaste that some people notice with saccharin.

CANDY IS DANDY

Research conducted by Michael W. Yogman and Steven H. Zeisel of Boston's Children's Hospital Medical Center and Boston University School of Medicine has led to the conclusion that a mother's eating a candy bar before nursing will help the baby fall asleep more quickly. Although often thought of as a source of quick energy, the sugar in candy actually promotes sleep. Eating a candy bar raises the sugar content of the mother's milk.

Some Known or Suspected Carcinogens

Chemicals	Source	Site: Target Organs and Tissues
Naturally occurring		
Aflatoxins; Mycotoxins (formed by molds)	Shelled peanuts and peanut products, tree nuts, grains, cottonseed meal and oil; Meat, eggs, and milk of livestock consuming aflatoxin-contaminated feed.	Liver, stomach, kidney.
Industrial Chemicals		
Composed of chemicals or chemical mixtures for which the major source of exposure results from industrial activity (uses, manufacture or production).		
Arsenic	Glass, ceramics, paints, dyes, pesticides, hair tonics and dyes.	Lungs, bladder, skin, larynx.
Asbestos fibers/dust (Actinolite, amosite, anthophyllite, crocidolite, chrysotile, and tremolite)	Automobile brake linings; patching compounds, spackling, hand-held hair dryers, thermal clothing, ironing board covers, electric irons and stoves, gas fireplaces, home insulation.	Lungs, chest or abdominal cavities (mesothelioma—malignant tumors of the chest or abdominal cavities), gastrointestinal tract (esophagus, stomach, and large intestine).
Auramine	Dye for paper, textiles, leather.	Bladder, pancreas.
Benzene	Chemicals, plastics, detergents, paint removers.	Aplastic anemia and/or leukemia and/or thrombocytopenia.

465

Chemicals	Source	Site: Target Organs and Tissues
Benzidrine and its salts	Aniline dyes, rubber, plastics, printing inks, fireproofing of textiles.	Bladder, liver.
Beryllium	Metal alloys.	Lungs.
Bis (chloromethyl) ether (BCME) and chloromethyl methyl ether (CMME)	Organic chemicals.	Lungs.
2-Naphthylamine	Dyes and paints	Bladder.
Polychlorinated biphenyls/PCBs	Transformers and electrical equipment, paper, dyes, herbicides, lacquers, plastics, resins, textile flame proofers, wood preservatives.	Toxic to the nerves and have adverse skin and liver effects. Birth defects, cancer, and immunological defects in laboratory animals.
Soots, tars, and oils (mixtures of aromatic hydrocarbons)	Shale oils, coal tars, soot, creosote oils, and cutting oils.	Skin, scrotum, lungs, bladder.
Vinyl chloride	Plastics, aerosols.	Angiosarcoma of liver, lungs, brain, blood system.
Industrial Processes		
Cadmium and cadmium compounds	Electroplating, alloys, soldering, plastic stabilizers, nickel-cadmium batteries, fungicides. In phosphorus and pigments for television tubes, inks, artists' colors, glass, ceramics, textiles, and paper.	Prostate, respiratory tract.

466

Chromium and chromium compounds	Alloys, protective coatings on metal, on magnetic tapes, and as pigments for paints, cement, paper, rubber, composition floor covering, and other materials.	Lungs, nasal sinuses.
Nickel and nickel compounds	Metal alloys and plating, batteries, spark plugs, ceramic glazes, paint pigments.	Lungs, nasal passages.
Hematite (red iron ore, bloodstone, ferric oxide)	Paint pigments, face rouge.	Lungs, skin.

Pharmaceuticals

Heaviest exposure to pharmaceutical chemicals occurs among patients being medically treated.

Chloramphenicol	Prescribed as a broad-spectrum antibiotic for treatment of bacterial and rickettsial infections in humans and nonfood-producing animals.	Serious and fatal blood abnormalities (aplastic anemia terminating in leukemia).
Cyclophosphamide Cytoxan (related to nitrogen mustards)	Approved synthetic anticancer drug. Interferes with the growth of susceptible neoplasms and some normal tissue to a degree.	Development of secondary malignancies; urinary-bladder cancers.
Diethylstilbestrol (DES) Stilbestrol	A synthetic estrogen used for treatment of menopausal and postmenopausal symptoms and breast engorgement following delivery. Also used for prostate and other cancers. Used in animal feed to promote growth.	Development of vaginal and cervical cancer in some daughters of women exposed to relatively large doses during pregnancy.
Melphalan 1-phenylalamine mustard	Approved prescription drug for treatment of two types of cancer: multiple myeloma and cancer of the ovary; also of some investigational value in other types of cancer.	Development of secondary cancer; leukemia

Chemicals	Source	Site: Target Organs and Tissues
Oxymetholone Anadrol, adroyd, anapolon	A synthetic hormone with testosterone-like action. Only approved use has been treatment of postmenopausal osteoporosis. Illegally taken by athletes to enhance muscle strength.	Liver damage and cancer; leukemia.
Phenacetin Acetophenetidin	Combined with other drug substances in over-the-counter remedies for pain and fever.	Kidney, urinary tract, nasal passages.
Phenytoin Diphenylhydantoin	A widely used prescription pharmaceutical for control of epilepsy. Also used as an intermediate in manufacturing textiles, lubricants, epoxy resins, and certain plastics.	Malignant lymphomas.

Radiation

Ionizing radiation (nonmedical)	Corresponds to alpha, beta, gamma rays; neutrons, and X-rays. Used widely in industry, nuclear-power reactors, and medical/dental X-rays. Also used to treat cancer.	Leukemia, epithelial tumors, thyroid.
Ionizing radiation (medical, ultraviolet)	Chiefly medical and dental X-rays. Dangers arise from ignoring proper precautions, or taking X-rays when not needed or too often.	Skin.
Radioactive ores and metals	Inhalation of radon (radioactive gas produced by natural decay of uranium and radium) with total exposure to radiation. Inhalation of radioactive dusts, damaging areas where deposited (nose, throat, and lungs).	Lungs.
Radioluminescent paint/radium	Luminescent watch dials. Use of radium in medical and laboratory procedures.	Bone sarcoma, leukemia and lymphomas, colon cancer.

Mysterious TAF

Early in 1980, Judah Folkman of Harvard Medical School, together with a team of researchers from various Boston institutions, announced the discovery of a substance Folkman called the *tumor angiogenesis factor*, or TAF for short. The substance is secreted by tumor cells and it succeeds in persuading the host body to grow blood vessels into the tumor and nourish it. Soon the tumor is better supplied with blood than the tissues around it. This allows the tumor to grow rapidly and is one of the factors that eventually kill the host (the person or animal afflicted with the cancer).

In an August 1983 article in the British medical journal *Lancet*, Shant Kumar of Christie Hospital in Manchester, England, and a team of scientists reported that TAF had been encountered in quite a different situation. It appears that a heart damaged by a heart attack secretes TAF. The TAF encourages the body to grow new blood vessels to repair or go around the tissue damage caused by the attack. It has long been known that blood vessels will soon form in an area that has been damaged, but only recently has the reason for this become somewhat clear. The TAF causes the walls of veins to develop leaks. As cells from the walls leak out, they head toward the source of the TAF. Along the way, they form themselves into tubes, which eventually become new blood vessels.

The appearance of tumor angiogenesis factor in normal repair work on the body gives TAF quite a different face. And it suggests one of the basic facts about cancer that is often not taken into account. The processes that cause a cell to become cancerous are generally the same processes that occur naturally in the body. Something, however, turns these processes on at the wrong time and then fails to turn them off. A life-saving response to injury becomes life-threatening when it is turned on in the absence of injury and continues to operate continuously.

Dr. Folkman called the news about TAF's role in heart attacks "very exciting and important." He and his colleagues had worked on the tendencies of tumors to promote blood-vessel formation for many years without receiving much recognition. In the same week as the newly found effects of TAF were announced, however, Folkman and his colleagues published an article of their own reporting that heparin (a widely used anticlotting drug) and cortisone together inhibit the action of TAF. Furthermore, they were able to prove that the inhibition of TAF resulted in halting the growth of many solid tumors, although some tumors were unaffected. The drug combination also stopped metastases of cancers and even caused tumors to disappear completely.

Like many discoveries, there was an element of the accidental in finding the heparin-cortisone combination. Heparin by itself promotes angiogenesis (the development of new blood vessels), but sometimes the effects of heparin are hard to see because the drug causes inflammation of the tissues being studied. This was a problem to the Folkman group who were using heparin in an effort to understand how and why angiogenesis occurs. Cortisone reduces inflammation, so a heparin-cortisone combination was expected to promote angiogenesis and make the results more visible. Surprisingly, all angiogenesis stopped. In detailed experiments, Folkman and his group found that fragments of heparin, fragments that lacked the anticlotting properties, combined with high doses of cortisone cured most types of solid cancers in mice. The heparin could even be administered orally, since the digestive system broke the heparin into fragments that possessed the anti-TAF activity. At first, the mice died from the same kinds of opportunistic infections that occur in AIDS. The cortisone was suppressing the immune system in the mice. However, by carefully protecting them from infection, the mice lived long enough to have their cancers cured by the new drug treatment.

The combination of new insights into TAF and the effects of heparin may cause reevaluation of post-heart attack treatment. Heparin, used to prevent clots in the circulatory system after a heart attack, may at the same time slow the body's repair work by inhibiting the action of TAF (but only if cortisone or certain related steroids are also present).

In September word of another inhibitor of TAF appeared. Anne Lee and Robert Langer, who had been working with Folkman, tested shark cartilage extract, which proved to be a potent inhibitor. Langer had shown that some substance in cartilage from calves inhibited TAF, but there was not enough cartilage in calves nor enough of the inhibitor in the cartilage. Reasoning that sharks are very large and have skeletons composed almost entirely of cartilage, Lee and Langer obtained some 6.1-meter (7-foot) basking sharks that weighed about 409 kilograms (900 pounds). Extract from the sharks' cartilage stopped tumor growth in the corneas of rabbits. Shark cartilage, therefore, may prove to be a significant new treatment for cancer. It is known that sharks are unusually free of cancers of the type that affect mammals or bony fish.

Any new treatment for cancer based on inhibiting TAF is still years away from clinical trials on humans. For one thing, none of the inhibitors has been identified chemically. In fact, the chemical composition of mysterious TAF itself is not known. Nevertheless, the work on TAF so far suggests that important advances can be made.

Do We Dream to Forget?

Not all science is made in laboratory experiments. Some of the most productive ideas may be speculation based on known facts. In the August 1983 issue of *Nature*, the famous scientist Francis Crick, co-discoverer of the DNA molecule, and Graeme Mitchison speculated on the physiology of dreams.

Crick and Mitchison base their theories on recent research into REM (rapid eye movement) sleep, the state in which most dreaming occurs. They also took into account studies of the cortex of the brain and even computer simulations of learning.

The two scientists note that the cortex contains many local networks of neurons. Since there are so many connections, Crick and Mitchison think that some of them may be random or accidental. At the end of a busy day, the number of connections may be so great that the unwanted ones get in the way of the wanted memories or learning paths. Crick and Mitchison suggest that REM dreaming is a means of erasing the variety of coincidental connections the brain has made during the day. In effect, the cortex is clearing or "debugging" itself of superfluous associations. Most of what we dream never comes to consciousness, and Crick and Mitchison suggest that this is as it should be, since the dreams may well be, in computer terms, "garbage." Although REM sleep seems to improve memory, the theory here is that it does so by eliminating impediments to the memories we really want to store.

Some evidence cited in favor of the new theory is that REM sleep occurs in animals, newborns, and even fetuses. Theories of dreaming that are based on psychological processes fail to explain REM sleep in these cases (although current research into animal thought and development of newborns also suggests that there may be more going on psychologically in these groups than once believed). If the clearing system for unwanted connections were to fail, the two authors suggest, the failure might produce such aberrations as fantasies, obsessional behavior, hallucinations, and perhaps even symptoms of schizophrenia. This new theory implies considerable disagreement with conventional psychoanalytic theories since it suggests that the problems in mental illness may be caused by an inability to *forget* rather than by an inability to remember.

Crick needn't worry about his scientific reputation, which is secure from his earlier work. Yet if his dream theory were shown to be true and useful, he could become even more famous for it than for his pioneering work on DNA.

Progress on Toxic Shock

The bacterial gene apparently responsible for producing the virulent toxin in toxic shock syndrome (TSS) (see also page 392) was announced on October 19, 1983, by Dr. Richard Novick, director of the Public Health Research Institute of the City of New York; the results of his large research group's efforts were published in the journal *Nature*. In addition to the staff from the New York institute, the researchers included leading specialists on TSS from the Universities of Wisconsin and Minnesota.

Location of the gene effectively proves that TSS is not caused by a virus that infects staph bacteria, as some researchers had previously believed. These researchers suggested that only the virus-infected staph produces the toxin, but the new research shows that the toxin is produced by the action of a gene normally present in the bacterium.

The isolation of the gene should allow the production of pure toxin under laboratory conditions for further testing. The first experiments are expected to prove conclusively that the toxin from the bacterium *Staphylococcus aureus* is indeed the cause of TSS. Another important benefit of the discovery should be the development of a simple blood test for women of childbearing age, who are at highest risk of contracting the ailment. Most women have antibodies to the staph bacterium in their system, but some small percentage apparently do not and so are defenseless.

It is possible that a vaccine against TSS might also be developed, but this is considered doubtful because it is not common enough to warrant widespread inoculation. The Centers for Disease Control have reported 2,204 cases from the first identification of the disease in the 1970s through the first half of 1983. There had been 103 deaths.

DIET AND NUTRITION: STRONG SUSPECTS IN CANCER

The National Academy of Sciences estimates that 60 percent of the cancers in women, and 40 percent in men, are due to diet, and some researchers think this is far too conservative. In the past, the implication of diet in cancer centered mainly on food additives, pesticide residues, and other contaminants. Latest findings of the National Academy of Sciences suggest that the issue of diet is more likely to involve the kinds and amounts of foods that we eat. Recommended dietary modifications include the following: Eat little or no salt-cured, salt-pickled, or smoked foods. Eat less fat, as well as fewer fatty meats and dairy products. Eat more whole grains and fiber-rich foods. Eat more fruit and vegetables, especially in the cabbage family. Drink alcohol in moderation only.

Athletes and Steroids

The quadrennial Pan American games, under way at Caracas, Venezuela, were disrupted on August 22, 1983, when four weight lifters were stripped of their medals because of evidence that they had been using anabolic steroids. These substances, either synthetic chemicals that resemble the male hormone testosterone or testosterone itself, are believed by many athletes to help the body manufacture muscle tissue and perhaps improve physical stamina. To provide these benefits, if they exist, athletes take the steroids in steady doses over a period of months.

The steroid issue was not a new one. It has been an open secret for some years that many world-class athletes in events requiring strength or great stamina use steroids. The sudden wave of disqualifications in Caracas was caused by the use of a new and extremely sensitive test for the drug. Checking urine samples using mass spectrometry and gas chromatography, officials claimed to be able to detect any steroid use within the past year or more. In practice, the tests measured the amount of testosterone in the athletes' urine.

By the end of the games, 23 medals had been surrendered because of the drug test results, and dozens of other athletes, including twelve Americans, withdrew from the games prior to their events rather than submit to the tests. Athletes who were disqualified in Caracas were expected to be banned from the 1984 Olympic Games in Los Angeles as well.

Technically, steroids are a large class of chemical compounds that include a number of substances produced naturally in the body, including cholesterol. For athletes, however, the word *steroids* refers only to testosterone or its analogs. The presence of more testosterone in men than in women has been shown to be the cause of men having larger and stronger muscles than women. Therefore, the athletes believe that taking additional steroids will produce thicker and stronger muscle fibers, making larger and stronger muscles.

There were several issues involved in the dispute.

The first was the reliability of the new tests themselves. Some athletes who refused to submit to the tests did so on the grounds that the tests were so sensitive they might detect minuscule amounts of steroids or other banned substances in eye drops, cold medicines, and other common products. They would rather withdraw from competition than take a chance of being unjustly accused of cheating.

A second issue concerned the regulations themselves. The International Olympic Committee requires that competitors be free of any trace of more than 100 drugs. If their tests for all of these become as sensitive as the one for steroids, it might become possible for officials to disqualify almost any athlete, since most athletes occasionally take drugs for pain or other legitimate reasons.

The third, and in many ways the most serious, issue had to do with the side effects of steroids. In females, steroids cause growth of facial hair and other male characteristics, and may have other unknown effects on the body. In men, the substance may cause a halt in the natural production of testosterone and produce atrophy of the testicles; it may also be linked to enlargement of and cancer of the prostate and to atherosclerosis with concomitant heart disease. In both sexes, steroids are associated with tumors of the liver, the organ that metabolizes them. Many doctors have stated that they would not prescribe steroids to anyone who had the hormones in normal supply because of these side effects.

Finally, there is a serious question about whether taking steroids actually produces the effects that athletes are seeking. There are perhaps an equal number of research studies showing that people taking steroids are stronger and studies that show no significant difference. Although muscle fibers look thicker under a microscope in people who take steroids, careful studies show that the apparent gain is caused by water retention in the cells. In fact, the entire weight gain caused by steroids is apparently due to water.

In some sports, of course, extra weight is useful for its own sake. At least one female weight lifter, Jan Todd, decided she would rather gain weight by eating than by taking steroids. After she had gained 36 kilograms (80 pounds), she set the world record in the squat.

There is another side effect of steroids that may help some athletic performances. Testosterone not only causes males to be more muscular than females, it also causes them to be more aggressive. In some sports, for example football, aggressiveness might be considered desirable. Some athletes on steroids, however, may become so aggressive that they are too short-tempered to obey coaches and so confident in their abilities that they injure themselves. (See the related articles "Effects of Fetal Exposure to Male Hormones," page 154; "Understanding Masculine and Feminine Behavior," page 183; and "Hormones, Left-Handedness, and Mathematics," page 369.)

Steroids have legitimate medical uses, so their manufacture cannot be banned completely. As long as one athlete believes that another has beaten him or her because of steroid use, however, athletes will continue to take chances with the drugs.

1983 Metropolitan Height and Weight Tables

Compared with figures as revised in 1959, this latest update on height and weight tables issued by the Metropolitan Life Insurance Company shows that adults can now weigh more than previously thought and still be considered healthy. There is an average increase allowed by the 1983 table in weight of 13 pounds for short men and 10 pounds for short women; 7 pounds for men of medium height, and 3 pounds for women of medium height; and 2 pounds for tall men, with 3 pounds allowed for tall women.

These tables first appeared in 1942, with a revision in 1959. The new tables are based on the insurance industry's largest survey yet of the association between weight and death rates. Four million people were involved over a period of 22 years. These tables serve as a reference for doctors and insurance companies in determining correct body weight of an adult. The measurements in the table indicate the weight "average" healthy Americans should strive for, and do not cover people who are ill or those with unusual conditions.

There is no clear explanation why rising weights are having fewer adverse effects on lifespan, although it is thought that the increased health consciousness of many Americans may be responsible for lowering the risk factors in certain diseases. People are more aware of exercise, lowering cholesterol levels, not smoking cigarettes, and curtailing salt intake. In itself, therefore, overweight may not now contribute to the extent it once did in such diseases as diabetes, hypertension, and heart impairment, since the added pounds may include healthy muscle or polyunsaturated, rather than saturated, fats.

Not everyone is happy with the new tables. The American Heart Association, for example, is opposed to raising the recommended weights, although doctors who have investigated the relationship between weight and longevity agree with the new ranges.

These tables apply to people from the ages of 25 to 59. The ranges given correspond with the weights of people who tended to live longer than people whose weights fell above or below the weights in the given range. While these weights may not be the ones that correspond to fashion at a given time, they are the ones that your doctor or your insurance examiner will expect you to attain and keep. The ranges of weights are stated in pounds, according to body frame size and the wearing of indoor clothing (five pounds of clothing for men and three pounds for women; heights include shoes with 1-inch heels).

Men

Height		Small Frame	Medium Frame	Large Frame
Feet	Inches			
5	2	128–134	131–141	138–150
5	3	130–136	133–143	140–153
5	4	132–138	135–145	142–156
5	5	134–140	137–148	144–160
5	6	136–142	139–151	146–164
5	7	138–145	142–154	149–168
5	8	140–148	145–157	152–172
5	9	142–151	148–160	155–176
5	10	144–154	151–163	158–180
5	11	146–157	154–166	161–184
6	0	149–160	157–170	164–188
6	1	152–164	160–174	168–192
6	2	155–168	164–178	172–197
6	3	158–172	167–182	176–202
6	4	162–176	171–187	181–207

Women

Height		Small Frame	Medium Frame	Large Frame
Feet	Inches			
4	10	102–111	109–121	118–131
4	11	103–113	111–123	120–134
5	0	104–115	113–126	122–137
5	1	106–118	115–129	125–140
5	2	108–121	118–132	128–143
5	3	111–124	121–135	131–147
5	4	114–127	124–138	134–151
5	5	117–130	127–141	137–155
5	6	120–133	130–144	140–159
5	7	123–136	133–147	143–163
5	8	126–139	136–150	146–167
5	9	129–142	139–153	149–170
5	10	132–145	142–156	152–173
5	11	135–148	145–159	155–176
6	0	138–151	148–162	158–179

Courtesy of the Metropolitan Life Insurance Company

FOR AN APPROXIMATION OF YOUR FRAME SIZE

Extend your arm and bend the forearm upward at a 90 degree angle. Keep fingers straight and turn the inside of your wrist toward your body. If you have a caliper, use it to measure the space between the two prominent bones on either side of your elbow. Without a caliper, place thumb and index finger of your other hand on these two bones. Measure the space between your fingers against a ruler or tape measure. Compare it with these tables that list elbow measurements for *medium-framed* men and women. Measurements lower than those listed indicate you have a small frame. Higher measurements indicate a large frame.

Men

Height in 1" heels	Elbow breadth
5'2"–5'3"	2-½"–2-⅞"
5'4"–5'7"	2-⅝"–2-⅞"
5'8"–5'11"	2-¾"–3"
6'0"–6'3"	2-¾"–3-⅛"
6'4"	2-⅞"–3-¼"

Women

Height in 1" heels	Elbow breadth
4'10"–4'11"	2-¼"–2-½"
5'0"–5'3"	2-¼"–2-½"
5'4"–5'7"	2-⅜"–2-⅝"
5'8"–5'll"	2-⅜"–2-⅝"
6'0"	2-½"–2-¾"

STRAWBERRIES A LA CROTONALDEHYDE

A ripe, freshly picked strawberry contains (besides many other things) acetone, acetaldehyde, methyl burrate, ethyl caproate, hexyl acetate, methanol, acrolein, and crotonaldehyde. These are natural substances, and yet every one is poisonous! Methanol may readily be recognized as wood alcohol, but crotonaldehyde may be less familiar as the chemical used in a "Mickey Finn," the famed knockout drink of the 1920s and 1930s. The knockout, however, sometimes lasted a lifetime if too much crotonaldehyde was slipped into the drink, since too large an amount of either methonal or crotonaldehyde could prove deadly.

Fresh food is better than food that has aged, however. As the inviting dish of strawberries sits on the table, for example, the methanol is being oxidized to formaldehyde, the same process that, when it occurs in the stomach, accounts for methanol's poisonous effects.

Chemistry of Suicide

For centuries, doctors and concerned laymen have assumed that suicide is an act brought on by a person's environment. Statistical studies show, for example, that suicide rates rise and fall with economic indicators and with unemployment rates.

Recently, however, researchers have discovered that there may be some chemical difference in those most apt to attempt—and to succeed in—doing away with themselves. The clue came unexpectedly in the mid-1970s from a study by psychiatrist Marie Asberg at the Karolinska Hospital in Stockholm, Sweden. She and her associates were measuring various trace substances in the spinal fluids of acutely depressed patients. They found that more than a third of the 68 patients had abnormally low levels of 5HIAA (5-hydroxyindoleacetic acid). The substance is known to be a product of the normal breakdown of serotonin, one of the important neurotransmitters, or messenger substances, in the brain; it seems to have a special effect on moods.

Some months after the original discovery, one of the depressed patients who had a low 5HIAA level committed suicide. One of the researchers asked casually why all those low-5HIAA patients kill themselves. The comment took the others by surprise, but a review of other patients who measured low in the substance did show that they were at greater risk of suicide than depressed patients with normal 5HIAA levels.

Since then, studies in the United States have at least partially verified the Swedish findings. The most recent study by Herman van Praag at Albert Einstein College of Medicine discovered a similar correlation. It has also been discovered that there is some correlation between low 5HIAA levels and impulsive, aggressive, and violent behavior. The director of intramural research at the National Institutes of Health comments that, "suicide occurs at the intersection of aggression and depression," and it is apparently that intersection that 5HIAA levels measure.

Researchers are now looking for a simpler way to test for 5HIAA levels. A spinal tap is too elaborate a procedure to use on large groups of people. One possibility is that 5HIAA levels are reflected in blood platelets, which seem to take up serotonin in the same way as neurons in the brain do. If a simpler test can be found, doctors would have a way of helping them to judge the relative risk of suicide among acutely depressed people. Such a simple screening test could alert doctors to high-risk patients and save many lives. (See also "The Chemistry of Depression," page 461.)

Cyclosporine Approved

Early in September 1983, the Food and Drug Administration (FDA) approved cyclosporine, a drug about which there had been exciting reports during the four years that it had been used on an experimental basis.

The drug first went under the name cyclosporin (without the "e") and then cyclosporin A, but finally settled down as cyclosporine. Sandoz Pharmaceuticals will manufacture it under the name Sandimmune.

Cyclosporine is the latest in a long line of immunosuppressant drugs. These substances are used to suppress the normal immune reaction of the body to foreign tissue during transplant operations and to treat, experimentally, such immune diseases as multiple sclerosis and rheumatoid arthritis. They have also been found helpful in treating certain types of cancer. The problem with most of the immunosuppressants is that even as they reduce the unwanted actions of the immune system they also render the recipient nearly defenseless against infection. The constant threat of infection and other unpleasant but less serious side effects have made them difficult to use.

Cyclosporine has been hailed as a miracle drug because it appears to have a selective effect on the immune system. It has been used in bone marrow transplants, for example, with great success because it suppresses the body's reaction to the foreign marrow cells but leaves it free to fight many common infections. This kind of selective action makes it preferable to the older immunosuppressants in many cases. One major concern about cyclosporine, however is its cost. After a transplant, cyclosporine must be taken for the rest of the recipient's life at a current cost of $4,000 per year.

Research in 1983 also showed that cyclosporine is effective in treating uveitis, a major cause of blindness. In this disease, the body's T-cells, part of the immune system, attack a sugar protein produced by the disease in the retina of the eye, eventually causing blindness. The disease had been treated in the past with conventional immunosuppressants, but now cyclosporine appears to be able to restrain the T-cells from their abnormal attack while leaving resistance to infection more or less unaffected.

Finally, there is considerable hope that cyclosporine will also prove to be effective against autoimmune diseases, such as multiple sclerosis, although its effectiveness in this area is yet to be established.

The process of approval of the drug by the FDA illustrates the pains the agency goes to in evaluating a drug. The Sandoz Company submitted 123 volumes of information and test results late in 1982; using a "streamlined" approval process, the FDA announced its decision only nine months later.

First Human Embryo Transfer

John Buster and Maria Bustillo, doctors at the Harbor-UCLA Medical Center in Torrance, California, reported in October 1983 on the first known embryo transfer in human beings. Earlier, writing in the July issue of the British medical journal *Lancet*, Dr. Buster had announced success in two cases with the following procedure:

Using sperm from the husband of a woman who wanted a child but was unable to produce the required egg, doctors artificially inseminated a woman donor. When conception occurred, the doctors removed the tiny embryo from the donor's womb during the first week of pregnancy and implanted it in the womb of the wife in the hope that it would attach itself and grow to maturity in the normal way. In two cases out of 25 this procedure was successful, and Dr. Bustillo reported in October that one of the two pregnancies had reached seven months and the other a few weeks less. Both pregnancies resulted in healthy, normal children early in 1984.

The babies will not be genetically related to the mother, and yet they will have been carried and nourished by her for the full nine months before birth. The procedure is suitable in cases where a woman wanting a child has blocked Fallopian tubes that cannot be repaired by surgery. The technique was adopted from one successfully used in tens of thousands of cases with cattle, although typically cattle breeders use frozen embryos.

These cases suggest new and complicated questions for family courts and other legal institutions in deciding whose children these transferred embryos really are. Another legal complication is that a corporation backed the experiments hoping to patent the method.

The possible legal nightmare became a reality in 1984 when a wealthy couple with frozen fertilized ova awaiting transplants in Australia was killed in an airplane crash. A further complication was that the ova had been fertilized by artificial insemination. The Australian government set up an ethics committee to determine the status of frozen fertilized ova and similar problems arising from the new technology.

Fertility aids are already big business. (See also "U.S. Approves First Test-tube Clinic," page 383, and "American Test-tube Baby," page 402.) Treatment with drugs that cause ova to mature can cost as much as $500 a month, for example, and does not work in all cases. While the embryo transplant requires careful timing (the donor's and recipient's menstrual cycles must be matched), the corporation funding the project, Fertility and Genetics Research Inc., hopes to develop a large pool of donors, thus providing matches for almost any potential recipient.

Recombinant Herpes Vaccine

Researchers announced success in November 1983 in using recombinant DNA techniques to create a new vaccine for the herpes simplex virus. The "carrier" is the venerable smallpox or *vaccinia* vaccine, the first effective preventive treatment for a communicable disease and the source of the word "vaccine" itself. The *vaccinia* virus is the cause of cowpox, which is related to smallpox but not so severe a threat to humans. The human immune system, faced with the cowpox organism, mounts an immune attack that produces immunity to both cowpox and smallpox. Smallpox was completely eliminated as an infectious threat in the human population in less than 200 years, thanks to the vaccine and effective quarantine procedures.

Vaccinia is a excellent virus for immunization on a large scale. Although *vaccinia* can cause severe reactions in some people, these reactions are both quite rare and also well understood by physicians. In addition to the safety factor of smallpox vaccine, known from worldwide use, the vaccine is more easily transportable than some vaccines, which may require refrigeration, a problem when used in large-scale applications. *Vaccinia* can be freeze-dried and still remain effective. Also, the techniques for making the smallpox vaccine are well known, and it can be administered with a pin prick. Finally, the virus is large (for a virus), which makes it easier to work with and opens the possibility of including more than one form of immunity in the same vaccine.

By splicing genes from the herpes virus into *vaccinia*, technicians have "taught" it to produce proteins resembling the parent herpes virus. This enriched vaccine alerts the immune system of a patient not only to the cowpox-smallpox viruses, but to the threat of herpes as well. The resulting immune response creates immunity to herpes. In tests with animals, rabbits inoculated with the new vaccine survived doses of live herpes virus that would otherwise be fatal.

Similar vaccines based on smallpox vaccine enriched with genes from other infectious threats such as hepatitis B and influenza, have also shown promise. Although there is an effective vaccine now against hepatitis B, it is so expensive that it is used only for people at especially high risk.

In fact, the addition of genes from several viruses to smallpox vaccine may eventually create a powerful multiprotective vaccine against a whole range of illnesses. But further research and testing are required before the new vaccines will become practical; and no vaccine will offer comfort to those who have already contracted the long-lasting herpes virus.

Researchers Locate Gene for Huntington's Disease

Huntington's disease is a hereditary disease of the central nervous system. It usually strikes young adults, causing them gradual loss of motor control, and intellectual and emotional deterioration. The ailment is progressive and results in death, usually ten or even twenty years after the appearance of the first symptoms. In the United States it is best known as the disease from which the famous folk singer Woody Guthrie died, and is sometimes known as "Woody Guthrie's disease."

One of the clearest facts about Huntington's disease is its pattern of inheritance. Since it is carried by a dominant gene, a child with one parent who carries the gene for Huntington's disease has an even chance of inheriting the gene and contracting the disease as an adult. Children of the disease's victim thus know from an early age that they too may suffer and die of it. But because it does not manifest itself until adulthood, the children have no way of knowing before adulthood whether they have the deadly gene.

However, in the November 17, 1983 issue of the journal *Nature*, a group of medical researchers representing many cooperating institutions announced that they had identified the relative location of the Huntington's gene, marking a remarkable step forward both in understanding the disease and in being able to predict, even before birth, whether a given individual is a carrier of the gene.

The discovery is a story of medical detection that involves relatively new technical skills in searching for genes and an international information-gathering effort that had its greatest success on the shores of Lake Maracaibo in Venezuela.

The clear pattern of inheritance suggests strongly that the disease is passed on by a single gene or by a combination of genes so close together that the combination is always inherited as a unit. Some years ago, the Hereditary Disease Foundation began preparing the family tree of American families with a history of the disease. More recently, the foundation learned of a community in a remote part of Venezuela where Huntington's disease is common. Apparently, a woman living there in the early 1800s, whose father was a European sailor, was the carrier who brought the disease to the region. Today, among her descendants, there are more than 100 sufferers of the disease and more than 1,000 children who have a 50 percent chance or a 25 percent chance of inheriting it (those with a 25 percent chance are children of a parent who may or may not have inherited the gene).

Nancy Wexler, the president of the foundation, organized a team of medical researchers to visit the community, which lives in houses on stilts over the waters of Lake Maracaibo. They collected blood and skin samples from 570 individuals. The samples were rushed to Massachusetts General Hospital in Boston for analysis in the laboratory of Dr. James Gusella.

Gusella had been studying samples from American families carrying Huntington's disease, and he believed that the gene in question was in chromosome four. The new samples from Venezuela confirmed the weak correlation he had already established, and he was able to pinpoint the location of the specific gene by correlating the incidence of a restriction enzyme at a particular point on the chromosome. Apparently, those who have the Huntington's disease gene also respond to this particular restriction enzyme, which cuts into the DNA of the chromosome only at a specific point. The location of the gene has been narrowed to an area on the chromosome that may hold as many as 100 genes out of the millions in the whole chromosomal system.

The most immediate result of these findings is the development of a clinical test that will predict accurately whether a child or even a fetus has inherited the Huntington's gene. In order for the test to work, however, blood samples from three generations (subject, parent, grandparent) should be available. A more serious problem is the ethical question that arises from the power to predict. Children of Huntington's victims have often nursed their parent through the long and agonizing course of the disease and are likely to have very mixed feelings about knowing whether they too will suffer. Researchers already have predictive material on a number of individuals both in the Venezuelan families and in U.S. families. This information is held under strict rules of medical confidence.

There is still much work to do, and researchers predict that it may be five or ten years before the exact gene implicated in the disease is isolated. But they remain confident that this breakthrough will be only the first.

WHY DRUGS GET ORPHANED

The Food and Drug Administration estimates that it costs $7,300,000 to bring a new drug to the market, while pharmaceutical manufacturers figure the cost at more than ten times as much. Most drug firms say they need an annual population of about 2,000,000 users to make a profit. Thus, a drug used to treat, say, Parkinson's disease, which has only about 1,500,000 victims, can exist and be "orphaned" because no one wants to make it. Furthermore, if the drug is made directly from plant or animal sources, it cannot be patented, which further reduces its profit potential.

AIDS

The early 1980s saw a series of medical "miracles" in genetic engineering and virology that seemed to promise a cure for virtually any communicable disease. But that climate of optimism was shaken in 1982 by an epidemic of a new, mystifying, and incurable ailment which acquired the name AIDS, standing for Acquired Immune Deficiency Syndrome.

By December 19, 1983, the Centers for Disease Control had recorded 3,000 fully documented cases of the disease, 43 percent of them fatal. Panic about contagion was widespread for two reasons. First, AIDS is not a "normal" communicable disease that produces recognizable symptoms and stimulates an immune reaction from the body. Instead, it is a syndrome that destroys the body's natural ability to fight disease and so leaves its victims to die of respiratory infections, viral diseases, or cancers. It is believed to be invariably fatal.

Second, AIDS was first reported exclusively among homosexual men, provoking prejudice against homosexuality, including the suggestion that victims of the disease "deserved" it for their immoral behavior. Then, when there were fragmentary reports of the disease developing among nonhomosexuals, the sense of fear and suspicion of victims—and all other known homosexuals—increased. Some victims of the disease were evicted from their homes, fired from their jobs, and even neglected by frightened attendants in hospitals.

Scientists have managed to alleviate some of the fear by defining more precisely who it strikes and how it is presumably transmitted. The important facts are that it seems to be transmitted only through intimate contact, through contaminated hypodermic needles or blood, or at birth from an infected parent. It cannot be passed by casual contact or by sneezes or coughing as are common viral infections. In April 1984, however, it was found that AIDS is caused by a virus. The virus has been named HTLV-III by Robert Gallo of the National Cancer Institute, who identified it. A similar, and probably the same, virus, named LAV, was also found in AIDS victims by Luc Montagnier of the Pasteur Institute in Paris.

The first AIDS cases were diagnosed by physicians in San Francisco and New York early in 1981. In San Francisco, a young male homosexual arrived at a hospital with a severe fungal infection and little immunological defense against any disease. He died by the end of the year of *Pneumocystis carinyii* (PCP), a lung disease thought to be long extinct in the United States. In New York, a dermatologist encountered two cases of Kaposi's sarcoma (KS) in one week, both in homosexual men. The disease was so rare that he only recognized it from descriptions he had read in old textbooks. Information on these apparently unrelated cases was exchanged between the two cities, and soon other physicians were noting an inexplicable rash of PCP and KS among gay men.

The Centers for Disease Control were also alerted to the appearance of these rare diseases. Their studies showed that the link between the cases was that the patients each had a severe immune deficiency, and further research suggested that the same type of deficiency might be responsible for other life-threatening illnesses in the gay community. By late 1981, 500 cases had been diagnosed; they included 27 women and 60 apparently heterosexual men. In each of these cases, the victims were ravaged by one or several of the opportunistic infections already known to afflict those with immune deficiencies.

Researchers still lack any certain knowledge of how the syndrome is acquired, but they are learning more about how it works. The immune system relies on white blood cells called lymphocytes to counteract various infections. Among these are the T-cells, which have many functions in fighting disease. Among the types of T-cells are "helper" cells that help mobilize other cells in the immune system to fight invaders, and "suppressor" cells that keep the immune system from running out of control. In normal individuals, helper cells outnumber suppressor cells by about two to one. But AIDS victims were found to have abnormally low counts of helper cells and proportionately high counts of suppressor cells. This imbalance seems capable of shutting down normal immune responses and inviting attack by common infectious agents that are everywhere but do not affect people with normal immune reactions.

Even with this guide, however, AIDS remained very difficult to diagnose in its early stages. Researchers finally found a definitive (and fairly uncomplicated) test in mid-1984. Finally they had a way to indicate the presence of the syndrome before the victim is besieged by debilitating infections.

More than 70 percent of the diagnosed cases have occurred in homosexual men who reported extreme sexual promiscuity (1,000 or more partners). This strongly suggests that the disease can be transmitted by intimate contact, although under just what circumstances continues to be a mystery. Some women have also been infected by sexual intercourse with AIDS victims.

Another 17 percent of the victims were intravenous drug users known to have shared needles with others. This suggests that the syndrome can also be transmitted by the use of contaminated needles or in the transfusion of contaminated blood. The case for this link with the blood is further strengthened by the fact that another 1 percent of the victims are hemophiliacs, who require frequent transfusions from pools of plasma that may be infected.

Another group reported at first to be at high risk were recent immigrants from Haiti. They represent about 5 percent of known cases. This apparent anomaly prompted theories about the possible origin of the disease in Haiti or in another developing country with poor health facilities. But increasing doubt was cast on the significance of the "Haitian connection," and none of the theories was verified.

Other groups, each with 1 percent of the cases, were persons receiving blood transfusions and persons having heterosexual relations with members of other risk groups. The most puzzling group of all was the 4 percent of the AIDS victims who apparently do not belong in any of the groups at high risk. They included infants and women as well as heterosexual men.

An effective treatment for AIDS victims is proving more elusive than its cause. At Memorial Sloan-Kettering Cancer Center in New York, doctors had some success in controlling AIDS-related cancers with interferon and interleukin-2, both substances produced by white blood cells. But the results are far from definitive. Some cancers in AIDS patients have responded to conventional chemotherapy.

No diagnosed AIDS patient, however, has ever regained immunological defenses. None of the treatments attempted, including bone marrow transplants and massive blood transfusions, has restored the necessary T-cell function. For the present, the AIDS victim who survives an initial onslaught of disease must live in constant fear of later life-threatening infections.

Doctors candidly admit that research into AIDS is helping rewrite medical text books, and it may provide important information on the development and operation of the healthy immune system. But for those suffering from the disease—or those who fear they may contract it—there is still little encouragement. "People are likely to ask what they can do to help themselves," says one doctor from Houston, "and you are forced to say, 'I don't know'."

For fiscal year 1984, the National Institutes of Health has granted $7,900,000 to seek an answer to the AIDS problem, and private institutions have appropriated still more. The number of cases reported and documented continues to rise rather steeply, and medical researchers would be especially pleased to be able to dispel the cloud of fear and anger that the AIDS epidemic has caused.

Specified Reportable Diseases

Disease	1950	1955	1960	1965	1970	1975	1977	1978	1979	1980
Amebiasis	4,568	N.A.	3,424	2,768	2,888	2,775	3,044	3,937	4,107	5,271
Aseptic meningitis	N.A.	3,348	1,593	2,329	6,480	4,475	4,789	6,573	8,754	8,028
Botulism[1]	20	16	12	19	12	20	129	105	45	89
Brucellosis (undulant fever)	3,510	1,444	751	262	213	310	232	179	215	183
Chickenpox (1,000)	N.A.	N.A.	N.A.	N.A.	N.A.	154.2	188.4	154.1	199.1	190.9
Diphtheria	5,796	1,984	918	164	435	307	84	76	59	3
Encephalitis:										
Primary infectious[2]	1,135	2,166	2,341	1,722	1,580	4,064	1,414	1,351	1,504	1,216
Post infectious[2]	N.A.	N.A.	N.A.	981	370	237	119	78	84	38
Hepatitis (1,000)										
Serum (1,000)	2.8	32.0	41.7	33.9	8.3	13.1	16.8	15.0	15.5	19.0
Infectious (1,000)	N.A.	N.A.	N.A.	N.A.	56.8	35.9	31.2	29.5	30.4	29.1
Unspecified (1,000)	N.A.	N.A.	N.A.	N.A.	N.A.	7.2	8.6	8.8	10.5	11.9
Leprosy	44	75	54	96	129	162	151	168	185	223
Leptospirosis	30	24	53	84	47	93	71	110	94	85
Malaria	2,184	522	72	147	3,051	373	547	731	894	2,062
Measles (1,000)	319.1	555.2	441.7	261.9	47.4	24.4	57.3	26.9	13.6	13.5
Meningococcal infections	3,788	3,455	2,259	3,040	2,505	1,478	1,828	2,505	2,724	2,840
Mumps (1,000)	N.A.	N.A.	N.A.	N.A.	105.0	59.6	21.4	16.8	14.2	8.6
Pertussis[3] (1,000)	120.7	62.8	14.8	6.8	4.2	1.7	2.2	2.1	1.6	1.7
Plague	3	0	2	8	13	20	18	12	13	18
Poliomyelitis, acute	33,300	28,985	3,190	72	33	8	18	15	34	9
Psittacosis	26	334	113	60	35	49	94	140	137	124

Disease	1950	1955	1960	1965	1970	1975	1977	1978	1979	1980
Rabies, animal	7,901	5,799	3,567	4,574	3,224	2,627	3,130	3,280	5,119	6,421
Rabies, human	18	4	2	2	3	2	1	4	4	0
Rheumatic fever, acute[4]	N.A.	N.A.	N.A.	N.A.	N.A.	N.A.	N.A.	851	629	432
Rubella[5] (1,000)	N.A.	N.A.	N.A.	N.A.	56.6	16.7	20.4	18.3	11.8	3.9
Salmonellosis[6] (1,000)	1.2	5.4	6.9	17.2	22.1	22.6	27.8	29.4	33.1	33.7
Shigellosis[7] (1,000)	23.4	13.9	12.5	11.0	13.8	16.6	16.1	19.5	20.1	19.0
Streptococcal sore throat, scarlet fever (1,000)	64.5	147.5	315.2	395.2	433.4	[8]330.8	[8]423.1	[8]397.0	[8]465.4	[8]370.0
Tetanus	486	462	368	300	148	102	87	86	81	95
Trichinosis	327	264	160	199	109	252	143	67	157	131
Tuberculosis[9] (1,000)	N.A.	76.2	55.5	49.0	37.1	34.0	30.1	28.5	27.7	27.7
Tularemia	927	584	390	264	172	129	165	141	196	234
Typhoid fever	2,484	1,704	816	454	346	375	398	505	528	510
Typhus fever: Flea-borne (endemic-murine)	685	135	68	28	27	44	75	46	69	81
Tick-borne (Rocky Mt. spotted fever)	464	295	204	281	380	844	1,153	1,063	1,070	1,163
Venereal disease (civilian cases): Gonorrhea (1,000)	287	236	259	325	600	1,000	1,002	1,013	1,004	1,004
Syphilis (1,000)	218	122	122	113	91	80	65	65	67	69
Other (1,000)	8.2	3.9	2.8	2.0	2.2	1.1	0.9	0.9	1.2	1.0

Code: N.A. Not available. [1]Beginning in 1975, includes foodborne, infant, and unspecified cases. [2]Data reported for 1975-1980 reflect new diagnostic categories. [3]Whooping cough. [4]Based on reports from states: 37 in 1960 and 1980, 36 in 1965 and 1975, 38 in 1970, 40 in 1978, and 39 in 1977 and 1979. [5]German measles. [6]Excludes typhoid fever. [7]Bacillary dysentery. [8]Based on reports from states: 41 in 1975, 1979, and 1980, 42 in 1977, and 40 in 1978. [9]Beginning 1960, newly reported active cases. New diagnostic standards introduced in 1975. Source: U.S. National Center for Health Statistics and U.S. Centers for Disease Control.

1984

Update 1984: Medicine

HEART DISEASE REDUCED BY LOWERING CHOLESTEROL

A massive, ten-year study involving 3,806 middle-aged men with high blood cholesterol levels established that the use of a drug, cholestyramine, reduced heart attacks, especially fatal heart attacks. Basil M. Rifkind of the National Heart, Lung, and Blood Institute, the director of the study, interpreted the results to mean that lowered cholesterol through diet, with drugs for those that need them, should be a goal for all men, women, and children. This interpretation is based upon evidence that the relationship between cholesterol and heart disease is progressive; the higher the blood cholesterol, the greater the risk of heart disease. In fact, the study group concluded that a 25-percent reduction in blood cholesterol would result in 50-percent reduction in heart-attack risk.

All the men in the study were placed on a lowered cholesterol diet that, by itself, reduced blood cholesterol by 8.5 percent. The group taking cholestyramine achieved an average reduction of 13.4 percent, which resulted in 19 percent fewer heart attacks and 24 percent fewer fatal heart attacks than the control group using diet alone. Use of the drug presented several problems. It does not taste very good to most people, must be taken six times a day, costs $150 per month, does not produce sufficient results without diet changes, and causes such unpleasant—but not dangerous—side effects that 27 percent of the men dropped out rather than continue taking it.

DIET AND CANCER

There was a lot of talk in 1984, but few certain conclusions, about the connection between diet and cancer. A conference convened at the end of February, for example, looked at the relationship between vitamin A and cancer in humans. Vitamin A has been shown to help prevent certain kinds of cancers in animals. The conferees reported on a number of studies of vitamin A and cancer in humans, however, that showed no relationship. A slight relationship was found between carotene, a precursor of vitamin A, and cancer prevention. It may be that some other substance in high-carotene vegetables is the effective agent, since carotene levels in the blood do not correlate with cancer risk.

Despite these ambiguous results, both the American Cancer Society (ACS) and the National Cancer Institute elected to promote diet as a means of cancer reduction.

VACCINES IN THE NEWS

Work in both the United States and Japan in 1984 brought forth a number of new vaccines, at least to the experimental stages.

Whooping Cough   Whooping cough, or pertussis, is a dangerous childhood disease that can be prevented by vaccinating children with ground-up, inactivated pertussis bacteria. Sometimes, however, some of the pieces of the vaccine mixture cause serious, long-term side effects in children receiving the vaccine. A Japanese team reported in January that they had determined which parts of the bacterium could be used to produce immunity while reducing the risk of side effects. A vaccine produced using just those parts was tested on 5,000 children with no detectable side effects and levels of immunity comparable to or better than the older type.

Flu   A team from the National Institutes of Health, the University of Maryland, and the University of Rochester (New York) developed an experimental live flu vaccine based on an innocuous version of the flu virus found by John Maassab of the University of Michigan. The new vaccine, which is administered as a nasal spray, uses the core of the innocuous virus and the outer coat of a disease-causing virus. Since it is the core that reproduces in cells, causing the disease, and the coat that produces the antibody reaction leading to immunity, the new virus seems to be both safe and effective.

Chicken Pox   Although chicken pox is not serious for most children, it can be very dangerous for adults and children with leukemia. It was announced at the end of May that an experimental vaccine against the disease passed its first major U.S. test with flying colors. Not enough time has elapsed since the test, however, to be certain that the immunity is sufficiently long lasting for general use.

Measles   Measles vaccine has been around for some time now, but measles does not seem to be going away. In fact, cases of measles were up 50% at the start of 1984. The measles virus was successful in locating unvaccinated people or that portion of the population for which the vaccine had not taken. Statistics like those for measles led the American College of Physicians to launch a campaign directed toward the nation's doctors to encourage them to promote more vaccinations.

Our New Understanding of Immunity

Perhaps the most rapidly growing and exciting area in medicine in the early 1980s has been the field of immunology. As with many life processes, immunity has turned out to involve more different approaches and more complex interrelations than previously thought possible. While a simple explanation of *what* happens when a "foreign" protein enters the blood has been accepted for years, *how* the body's immune system functions against that protein is only now becoming understood.

For centuries the way in which the body fought disease was a complete mystery, and the only use made of the immune system as an ally against disease was more or less accidental. Then, with the scientific revolution of the nineteenth century, deliberate attempts to work for or against the immune system started—and, in fact, the very concept of an immune system arose gradually. Theories were developed to account for the protection the immune system offered against disease. Finally, in the early 1980s, medical researchers began to explore this system, much in the same way space scientists have begun to explore the moon and the nearby planets. Amazing new methods and equipment have made it possible for the first time to produce hard information rather than abstract speculation. In the case of immunology, this new knowledge is already leading to major breakthroughs in the treatment of many different diseases.

HISTORY

Folk ideas about how to protect yourself from disease probably go back to the first humans. However, the scientific story of immunology begins in the 1700s when smallpox was epidemic throughout the Western world. People noticed that an individual could get smallpox only once, and they sought to control when they got the disease and how severe a case they incurred by "inoculating" themselves with the disease. Since it had been observed that smallpox could be spread by contaminated clothing or objects, children were deliberately made to sleep in sheets from the beds of smallpox victims. Children were thought to be somewhat less likely to die from the disease than adults. If the child was sufficiently strong and the sheets were from a weak strain of the virus, the child got a light case of smallpox and was immune thereafter.

Edward Jenner in England knew that milkmaids seemed more or less immune to smallpox. He theorized that because they contracted the less serious but apparently related disease called cowpox, they developed an immunity to smallpox. If he could induce cowpox in other people, perhaps they would develop the immunity as well. As we know, his reasoning turned out to be correct, and he devised the first vaccine in 1796, which contained an attenuated virus from the animal form of the disease. Although there were great battles throughout Europe and America about the effectiveness and safety of the vaccine, eventually large parts of the population were vaccinated. By the end of the 1970s, the last pockets of smallpox on Earth were eradicated and the World Health Organization declared the disease conquered once and for all.

The next great stride in immunology—although it was not thought of by that name—was the development of antiseptic techniques in surgery in the early part of the nineteenth century. Surgery had been very risky before that time because infection usually set in and the patient would die from the incision itself. Joseph Lister and others theorized that small "germs" got into the incision and caused terrible carnage there, killing the patient even if the operation had been otherwise successful. There were great battles between Lister's supporters and more conservative surgeons over the effectiveness of such antisepsis, but antiseptic surgery eventually won out ànd it has saved millions of lives. From the standpoint of immunology, the important fact is that although the immune system mobilizes against any foreign invader of the body, massive invasions of bacteria through a break in the skin present almost insuperable odds. The foreign microbes have the advantage, multiplying too quickly for the immune system to cope with them.

The germ theory developed through the late 1800s, thanks to the work of Louis Pasteur, Robert Koch, and others. Pasteur set the theory of vaccination on a firmer scientific basis and contributed to the development of safe and effective vaccinations for illnesses other than smallpox. At the same time, physicians made the crucial connection between the spread of infection and unsanitary conditions in water and food supplies. The resulting public health measures, by limiting human exposure to some of the most virulent diseases, have probably saved more lives than all medical treatments combined. But still the role of the body in fighting outside invaders was little understood.

Despite this lack of fundamental understanding, during the early 1900s considerable progress was made in fighting communicable diseases. Public health and sanitation laws eradicated some diseases, and others were conquered by the development of new vaccines (diphtheria, whooping cough). In the 1930s medical researchers discovered the first antibiotics—medications that specifically attack the bacteria that cause many severe infections. These "miracle drugs" took much of the terror out of diseases such as pneumonia, which had previously been potent killers, especially among the very young and the very old, although antibiotics worked directly on bacteria, not on the immune system. In the 1950s a polio vaccine was perfected. Since then vaccines for measles and mumps have virtually wiped out what had been common, yet hazardous, childhood diseases.

MODERN IMMUNOLOGY

In the meantime, some progress was being made toward connecting the experimental work on vaccines with a theory to explain why vaccines work. The progress began in another area of medicine altogether—blood transfusions. Around 1900 a French doctor named Paul Ehrlich learned that one body will also mobilize against foreign material from another body. When he transfused a pint of goat's blood into a sheep, he discovered that some mechanism in the sheep's blood rapidly broke down or "lysed" the foreign blood, in effect destroying it. He then tried giving one sheep blood taken from another sheep. In most cases, the same lysing process took place, but he discovered certain pairs of sheep in which one could tolerate the blood of the other. He found that blood extracted from one sheep and later retransfused was never destroyed. Somehow the body recognized its own blood.

This was the beginning of blood typing, but it also raised fascinating questions about the body's immune mechanisms. The body could recognize and reject tissue from different species and often even foreign tissue from the same species, but how did it do this? And how did it recognize its own tissue so as *not* to reject it?

In the twentieth century, a theory gradually developed to answer such questions. This theory, which was eventually substantiated by molecular biology, is as follows. Living tissues contain substances called *antigens*—markers that will alert another body that the tissue is foreign—and yet the immune system in a given body is somehow made or "trained" not to be set off by its own body's antigens. This makes possible retransfusion of blood, for example, and skin grafts from one part of the body to another. When foreign antigens are present, however, molecules called *antibodies* or special cells attack the bearers of the antigens.

Despite all this progress, the way in which the immune system works against a host of foreign invaders was still not well understood. But beginning in the 1950s, serious research began to reveal some secrets. Among the early victories, Peter Medawar and his colleagues at Oxford University developed a way to breach the rejection mechanism in animals. By injecting a small amount of foreign tissue from one strain of mouse into a newborn mouse from another strain, Medawar found that when the recipient mouse had grown up, it would accept extensive skin grafts from the original donor without rejecting them. This suggested that in early infancy the body is still at work making distinctions between "self" substances and foreign substances. It "recognized" some code in the skin grafts as part of "self" because of the earlier injection.

In the 1960s the field began to grow and change with bewildering speed. The discovery of the structure of the DNA molecule, one of life's principal building blocks, and the development of many new research techniques, gave immunologists tools to study just how the immune system works. As in many research efforts, the progress toward understanding has been intermittent, and along the way there have been many half-wrong hypotheses to confuse both the researchers themselves and the physicians interested in the practical clinical results.

Auto-immune Diseases

Although early emphasis was on understanding and using the immune system, some researchers also began to look toward immunology to explain a number of diseases not obviously caused by any outside agent. In these diseases, some part of the body appears to be attacked not by infectious disease, but by the immune system itself. Could the immune system be a culprit as well as a vital defense system?

It now seems clear that at least some of these diseases, including rheumatoid arthritis and systemic lupus erithematosis (SLE), have strong connections to a malfunction in the immune system. They are often called *auto-immune diseases*, suggesting an immune reaction against the self. Some researchers have theorized that multiple sclerosis may also be classed among these diseases, but the evidence is still not conclusive. Late in 1983 a group of researchers from various institutions across the United States established that juvenile diabetes also involves a reaction against the body's own cells, in this case the cells that produce insulin. (It should be pointed out that in none of these cases is the exact relationship between immune system and disease clear. For example, rheumatoid arthritis is also sometimes classified as a metabolic disorder since there seems also to be some malfunction in the way a victim digests and makes use of nourishment.)

Another commonly acknowledged connection is between the immune system and allergies. Allergies are among the most puzzling of ailments, but one way of defining them is as an excessive or in some way "mistaken" overreaction of the immune system to a substance. In most cases, allergies are only minor irritants to those who have them, but in other cases, as with acute penicillin allergy, allergies can be life-threatening. Asthma is a serious disease that is also basically an allergic reaction.

Genetics and Emotions

Studies of the auto-immune diseases have brought up several other interesting connections. The first is the observation that a disposition to diseases such as rheumatoid arthritis, asthma, and hay fever seems to run in families. This has caused immunologists to turn to the field of genetics, where there have been major advances in recent years. Geneticists have been successful in identifying the relationship of certain genes to certain predispositions for specific diseases. Work is now proceeding in clarifying the exact relationship between inheritance and the abilities and eccentricities of a person's immune system. In the future, physicians may be able to "type" individuals by genetic pattern and predict predisposition to certain immune-system-related ailments.

A second factor, which complicates the picture immensely, is the apparent psychosomatic element in many immune-related diseases. For example, surveys suggest that rheumatoid arthritis is far more common among divorced women than among married women. Temperament and emotional state are also known to have a strong influence on asthma and many allergies. These connections could suggest that emotional disturbance has a role in the development of disease or in acute attacks of disease. But this has not been proven, and the reverse may be true: the chemical imbalances associated with a disease may actually cause emotional instability. For asthma, for example, this seems most clearly to be the case. Wide-ranging studies in brain chemistry are discovering similarities between chemicals that act as messengers in the brain, affecting both mood and behavior, and chemicals produced by the immune system, such as interferon. This connection may eventually shed new light on the mind-body connection, but at present, the link is still elusive and poorly understood.

TRANSPLANTS

Immunology has also played an essential part in the new medical specialty of organ transplants. The first form of transplant that was clearly understood was blood transfusions. Until blood typing was developed by a succession of researchers, starting with Landsteiner, Jansky, and Moss in 1901, transfusion was too risky to attempt. By World War II, blood transfusions were commonplace.

Early efforts at transplanting tissues other than blood failed miserably when the recipient's body rejected the foreign tissue. Efforts were made to find tissue types that corresponded to the by then well-known blood types. These efforts led to serious research into tissue compatibility and tissue typing.

Because the immune system is responsible for the rejection of noncompatible organs, tissue typing is actually a study of the line a human organism draws between itself and all "other" tissue. We have already seen that skin transplants from one part of a single body to another are not rejected. Similarly, grafts or transplants between identical twins are always successful. Unfortunately, very few of the critically ill patients who might benefit from an organ transplant have an identical twin. In most transplants to date, the organ of a brother or sister, who shares a similar (though not quite identical) genetic inheritance, have been most successful.

Even when tissue types are nearly identical, however, rejection may still be a problem. Here the use of a wide range of immunosuppressant drugs has been essential. Researchers have discovered many compounds that seem to suppress the natural response of the immune system to foreign tissue, and these compounds have been used not only in transplant cases, but also as a treatment for acute cases of rheumatoid arthritis, SLE, and other possible auto-immune diseases.

The danger of these drugs, of course, is that they suppress not only the "mistaken" reactions of the immune system, but also those reactions that may be necessary to survival. Patients being treated with such drugs (or those whose immune systems have been damaged by radiation treatment for cancer) can be highly susceptible to disease, even to infections that normal people have strong natural immunity to. A serious search for substances that are selective in suppressing immunity has been under way for some years, and the approval of cyclosporine in 1983 signaled the arrival of some such selective drugs.

IMMUNE DEFICIENCY DISEASES

The effects of immunosuppressant drugs are mimicked in a life-threatening way by a variety of immune deficiency diseases. These have been brought to public attention in the 1980s by the epidemic of the disease called AIDS (Acquired Immune Deficiency Syndrome) primarily among male homosexuals and intravenous drug users. The contagious nature of the disease suggested that there is some infectious agent responsible for its onset, something that can be passed from person to person through intimate contact or the use of contaminated hypodermic needles. In 1984 the agent was found to be a virus. Much attention continues to be given to the effects of the syndrome on the immune system. Victims lose the ability to fight off a wide range of infections and so they contract Kaposi's sarcoma, an otherwise rare cancer, or a variety of opportunistic infections—those that an active immune system would fight off without trouble.

Another kind of immune deficiency is not acquired but is discovered as a defect at birth. Children born with severely deficient immune systems must be kept in a sterile (organism-free) environment because almost any infection could be fatal. Because the cells these children lack are made in the bone marrow, doctors have attempted bone marrow transplants as a way of creating a working immune system. In cases where tissue compatibility is no obstacle, these transplants have apparently prolonged the lives of the victims. This encouraging development makes it possible that other immune deficiency diseases might respond to like treatment.

THE CELLS

The greatest advances in immunology, however, have sprung from basic research on how the immune system works. Advanced research techniques have allowed doctors to study the various cells that are responsible for protecting the body from infection and other dangers and to isolate the substances these cells generate. These two breakthroughs promise future success in treating a variety of diseases still considered incurable today. But as with all basic research, it is still impossible to know just when—or how—our new knowledge will change medical theory and practice.

Although the body has many protections against disease, including the skin and the system of cilia that expel dust and other particles from the lungs, the immune system is that part of the body's defense found in the blood. Blood is a living tissue consisting of several types of cells, generally classified as red or white, embedded in a matrix of water, chemicals, and small particles called platelets. The cells in the blood are produced primarily by the bone marrow. The importance of bone marrow to the body can be observed by how well it is protected. Only the marrow and the brain are completely encased in bone, the safest place in the body.

Elie Metchnikoff established around the turn of the century that white cells in the blood are principal agents in the immune system. In fact, he observed certain white blood cells actively consuming disease-causing bacteria. On further investigation, it appeared that there were many different kinds of white cells and that somehow the labor of repelling foreign invaders was divided among them. The main cell types identified so far are:

Killer Cells: Short-lived white cells that are first to make contact with an invading organism and attack it, killing themselves in the process.

Macrophages: large white cells that arrive at the site of an infection after the first killers have made contact with the invader. The macrophages secrete enzymes that "digest" both dead invader and killer cells, helping to clean up the site of infection. But they also play an important role in mobilizing the larger system by fixing invaders' antigens on their own surfaces and "presenting" them to other cells (the lymphocytes) for identification.

Lymphocytes: These white blood cells are apparently the center of the immune system against infectious agents. They develop from "stem cells" that are produced in the bone marrow, and they are all *antigen-specific*. This means that there is a group of lymphocytes for every disease and other infectious or foreign organism that the body is ever likely to encounter. When the signal goes out that there is an invasion by a certain agent, lymphocytes specific to that agent come to the scene, where they begin to multiply and to mobilize to kill the attackers.

By processes not yet understood, lymphocytes mature into several subgroups. The groups so far identified include B-cells and three classes of T-cells.

B-cells are the principal manufacturers of antibodies—substances that are able to kill invaders in the bloodstream. Antibodies are made of a substance called immunoglobulin (IG), but there are at least five different types, each with different functions. B-cells are especially effective against a wide range of bacterial infections, but their antibodies are not effective against invaders such as viruses, which actually get inside cells.

T-cells also develop from the common bone marrow stem cell, being differentiated from B-cells after emerging from the marrow. It has been discovered that they develop under the influence of the thymus gland, a small organ just behind the breastbone. (The "B" in *B-cells* is for "bone marrow," while the "T" in *T-cells* is for "thymus.") The thymus appears to be most active and important in childhood; after a person reaches the age of twelve, the thymus begins to shrink. Some of its functions may be taken over by the spleen and the liver. The T-cells appear to be the most complicated of the lymphocytes and major efforts to understand them did not succeed until 1984. Three subgroups have been distinguished.

T-effector cells (called T_E cells) are the most active of the three types. They produce lymphokines (see below), which attract other lymphocytes and macrophages to the scene of the infection. They can also operate directly against a particular foreign cell. Unlike their cousins, the B-cells, they are effective against viruses and other invaders that actually take over body cells.

T-helpers (or T_H) cells, on the other hand, are important in coordinating the attack of the various white blood cells in combatting the foreign organisms or substance. They are a necessary part of the process by which B-cells are mobilized to make antibodies against the invader, and they may be important to the function of T_E cells as well.

T-suppressor (or T_S) cells are the third type, and they serve the paradoxical function of monitoring and limiting the action of the system. Without suppressor cells to provide a control, the immune system may overreact, and the absence of these cells may be related to certain auto-immune diseases in which the body seems to attack itself by mistake.

These descriptions are sketchy because more is being learned each day about these cells and the complex relationship they have to each other. Before our understanding is complete, other kinds of lymphocytes may be discovered, playing still other roles, but the general outlines are at last becoming clear.

THE LYMPHOKINES

Perhaps the most interesting area of research in recent years has been in identifying, studying, and learning to produce the *lymphokines*, the substances created naturally by various white blood cells to activate and control the immune system. Thanks to the development of recombinant DNA technology, these newly discovered substances can be synthesized outside the human body and produced in amounts great enough to allow widespread testing of their properties.

Interferon (later renamed interferon alpha), which was first synthesized in 1980 (see "Gene Splicing Used to Produce Interferon," page 146, and "Interferon Produced in Lab," page 384) was the first of this family of substances to receive widespread testing. Interferon beta was synthesized somewhat later. Neither of these substances is actually produced by the lymphocytes, but they appear to play some part in the whole immune process.

Immune interferon, now called *interferon gamma*, which is produced by T-cells, has begun to overshadow its cousins and has, in fact, come to be considered a prototype of this new class of substances. Interferon gamma appears to enhance the effectiveness of killer cells, macrophages, and T-cells in killing foreign cells. It also is important in the process by which a macrophage "fixes" a foreign antigen on its surface and presents it to a B- or T-cell for recognition. This recognition of the antigen is essential to the proper activation of the whole immune system.

Even more intriguing are the lymphokines called interleukin-1 and interleukin-2. As its number suggests, interleukin-1 was discovered first. It is produced by macrophages and possibly by other cells in the body, and its early name suggests at least one of its important functions. It was called "lymphocyte activating factor," or LAF, because it activates the T-cells and may stimulate T-cells to produce interleukin-2. Interleukin-1 also is important in promoting inflammation in the neighborhood of an infection and encouraging the production of enzymes that will help clean up the debris caused by the deaths of both invader organisms and white cells in the battle.

Interleukin-2 may be an even more significant discovery. It is made exclusively by T-cells, and its effect is to trigger the rapid reproduction of many more T-cells. Thus, interleukin-2 is the essential T-cell growth factor. In addition, interleukin-2 stimulates T-cells to produce another growth factor that favors the rapid reproduction of B-cells. All of these characteristics suggest that interleukin-2 may turn out to be a major new approach to handling disease, rather than the interferons.

In 1983 interleukin-2 was being used experimentally to treat victims of AIDS, the acquired immune deficiency that has struck homosexual men and intravenous drug users in epidemic numbers. Those who suffer from AIDS have a serious shortage of T-helper cells, and preliminary reports suggest that interleukin-2 improves the performance of the remaining T-cells in fighting infections. Researchers speculate that it may mimic a lymphokine that would be produced by a healthy T-helper population in a well person.

Several other lymphokines have been discovered, but much less is known about them. Two of them, called *tumor necrosis factor* (TNF) and lymphotoxin, appear to have important jobs in protecting the body from various kinds of cancers. When these substances have been fully analyzed and synthesized for production, they may offer new hope in treating malignancies.

There will certainly be other discoveries in the lymphokine field in the near future, and as a result, scientists should be able to piece together the elaborate workings of the immune system. Just how these new discoveries will affect the treatment of diseases cannot be determined yet, but immunologists are confident that greater understanding coupled with new technologies, such as recombinant DNA techniques, will eventually result in effective treatment for diseases now considered incurable.

PROSPECTS AND PROBLEMS

The prospects for further advances in immunology seem immense. Immunological research touches several other disciplines in which major new discoveries are being made. For example, recent studies suggest that cancer cells have some ability to "fool" the immune system into not mobilizing. Why the cancer cells are able to do this in some cases and not in others is a matter of some debate, and the answers may lie both in genetic factors still not well understood and in environmental factors. If the immune system could be "unfooled," it might be persuaded to take a major part in the internal war against the multiplying cancer cells. Thus research is proceeding not only to identify medications or treatments that directly affect a cancer, but also to identify those medications and treatments that might modify and restore the power of the body's own immune response.

Knowing how to turn the immune system on and off is, of course, vital to increasing the success of organ transplants. Similarly, since treatments for cancer often suppress the immune system, better understanding of immunity will help fight cancer by permitting immunity-suppressing treatments to be used with less risk to the patient from infectious disease.

Immunology is a word usually used to refer to a particular medical specialty, but interest in the powers of the body to resist outside challenges extends well beyond the small circle of technically trained immunologists. Not only are most practicing doctors faced with immunological questions in many guises, but non-professionals also seek to understand and use the new insights in the field.

Perhaps the most notable example in the early 1980s of this nonprofessional and nontechnical interest is that of Norman Cousins, a prominent magazine editor and writer. When he was stricken with a rare and little understood disease, his physicians put him in the hospital and warned that his condition was incurable and might lead to an early death. After realizing that his doctors understood little about the ailment and that hospital treatment was affecting it in no meaningful way, Cousins checked out of the hospital and began treating himself. He concluded that a large part of his problem had to do with attitude, so he scheduled certain periods of the day for laughter (by watching old Laurel and Hardy films), which he believed might be better medicine than any the pharmacy could offer. He also arranged for treatment with vitamins under a doctor's reluctant supervision. Whatever the cause, Cousins contradicted his doctors' prognoses and got better. He then wrote a magazine article, later made into a best-selling book called *The Anatomy of an Illness*.

Cousins' book and his general approach are one of a number of so-called holistic approaches to medical care espoused by a growing number of non-professionals. Some holistic practitioners concentrate on diet and exercise as a means to health and a barrier to sickness. Others stress the psychological aspects of health.

Immunological researchers take little direct note of such unorthodox views, occupying themselves with trying to understand the extremely complex and amazing chemical system that protects the body. Yet some professional research suggests that holistic views may soon need to be taken into account. The immune system may be sensitive to the presence of a number of trace elements and vitamins. Research into brain chemistry is providing detailed analyses of messenger chemicals affecting mood and behavior that are startlingly similar to substances produced by the immune system. Thus the relationship between attitude and physical well-being may be more direct than previously thought. Doctors studying stress report that too much stress has a direct negative effect on immune response. As physicians studying physical health begin to consider emotions as a cause of sickness, a complementary trend can be seen in psychiatry, where there is strong interest in the chemical causes of mental illness and distress. The search for health in the body may eventually lead directly to the mind, while the search for health in the mind may lead to the study of the chemistry of the body.

Even as immunology clears up certain mysteries about the operation of the body, it may increase appreciation for the interrelatedness of modern medicine. Specialists in many disciplines may discover that they are all seeking the same mysterious substance or process, and discoveries in one field may solve major problems in another apparently unrelated field. The new discoveries may also increase our respect for the wholeness of the human body and mind and for their interrelatedness.

Fourteen Useful Medical Self-Help Books

CANCER AND NUTRITION
Charles B. Simone (McGraw Hill Book Co., New York: 1983). The author, a full-time cancer researcher, outlines adjustments in lifestyle and eating habits that may help avoid the development of cancer.

THE COMPLETE DR. SALK, AN A TO Z GUIDE TO RAISING YOUR CHILD
Lee Salk (NAL Books, New American Library, New York: 1983). A comprehensive guide on child rearing that will delight readers with its specific yet sensitively expressed viewpoints.

THE COMPLETE HEALTH CLUB HANDBOOK
John Dietrich and Susan Waggoner (Simon and Schuster, New York: 1983). The authors cover the subject of fitness facilities in detail, discussing everything from health and racquet clubs to weight training centers.

CONSUMER DRUG DIGEST 1982
American Society of Hospital Pharmacists. (Facts on File, Inc., New York: 1982). This reference work is intended to help consumers, as patients, to be better able to participate in their own health care by understanding basic facts about physician-prescribed medications.

DIET AGAINST DISEASE
Alice Martin and Frances Tenenbaum (Houghton Mifflin Company, Boston: 1980). This three-part book fully explains the connection between diet and disease, interprets available scientific data for diet modification, and shows how to act on this advice through the purchase and preparation of food.

DO-IT-YOURSELF MEDICAL TESTING
Cathey Pinckney and Edward R. Pinckney (Facts on File, Inc., New York: 1983). The authors outline more than 160 medical tests that can be done at home. The book is intended to aid individuals in taking an active part, together with health professionals, in obtaining proper health care.

THE ESSENTIAL GUIDE TO NONPRESCRIPTION DRUGS
David R. Zimmerman (Harper and Row, New York: 1983). This book evaluates more than 1,000 drug substances and 1,000 brand-name medical products.

FOODS FOR HEALTHY KIDS
Lendon S. Smith (McGraw Hill Book Co., New York: 1981). Proper food programs are presented for children from birth to puberty. Special attention is given to such physical and behavioral problems as sleep disturbances, allergies, mood swings, tension, and hyperactivity.

THE HOME ALTERNATIVE TO HOSPITALS AND NURSING HOMES
Mara B. Covell (Rawson Associates, New York: 1983). A comprehensive guide, containing many illustrations and checklists, that explains in simple language how one can care for an ailing loved one at home.

NUTRITION HANDBOOK
Edmund S. Nasset, (3rd ed.) (Barnes and Noble Books, division of Harper and Row, New York: 1982). A readable and updated pocket guide on all aspects of nutrition, presented by an expert in the field.

OVER FIFTY-FIVE, HEALTHY AND ALIVE, A HEALTH RESOURCE FOR THE COMING OF AGE
H. Pizer (ed.) (Van Nostrand Reinhold Co., New York: 1983). This book addresses itself to the changing attitudes of Western society toward people 55 years and older. Subject matter includes discussion on how our bodies change, how to be a wise consumer, and new ways of looking at family and friends, marriage, and sexuality.

PREPARING FOR CHILDBIRTH, A COUPLE'S MANUAL
Adair Sirota (Contemporary Books, Inc., Chicago: 1983). The author provides fundamental information on the process and techniques of childbirth, stressing the involvement of couples during pregnancy.

PRESCRIPTION DRUGS
Thomas A. Gossel, Donald W. Stansloski, and the editors of *Consumers Guide 1982* (Beekman House, distributed by Crown Publishers, New York: 1982). Two clinical pharmacologists and the editors of *Consumers Guide* have joined forces in providing information on choosing a physician or pharmacist, understanding a doctor's prescription, and knowing what specific drugs do.

WHERE DOES IT HURT? A GUIDE TO SYMPTOMS AND ILLNESSES
Susan C. Pescar and Christine A. Nelson (Facts on File Publications, New York, 1983). The authors have provided a correlation between common symptoms and common illnesses that also discusses treatment and severity of the illnesses.

The Nobel Prize for Physiology or Medicine

Date	Name (Nationality)	Achievement
1901	Emil von Behring (Germany)	Discovery of diphtheria antitoxin.
1902	Sir Ronald Ross (England)	Work on malaria infections.
1903	Niels Ryberg Finsen (Denmark)	Light ray treatment of skin disease.
1904	Ivan P. Pavlov (Russia)	Study of physiology of digestion.
1905	Robert Koch (Germany)	Tuberculosis research.
1906	Camillo Golgi (Italy) Santiago Ramon y Cajal (Spain)	Study of structure of nervous system and of nerve tissue.
1907	Charles L.A. Laveran (France)	Discovery of the role of protozoa in disease generation.
1908	Paul Ehrlich (Germany) Elie Metchnikoff (Russia)	Pioneering research in the mechanics of immunology.
1909	Emil Theodor Kocher (Switzerland)	Work on the thyroid gland.
1910	Albrecht Kossel (Germany)	Study of nucleic acids .
191⁻	Allvar Gullstrand (Sweden)	Work on refraction of light in the eye.
1912	Alexis Carrel (France)	Vascular grafting of blood vessels.
1913	Charles Robert Richet (France)	Work on anaphylaxis allergy.
1914	Robert Bárány (Austria)	Inner ear function and pathology.
1915	No award	——————
1916	No award	——————
1917	No award	——————
1918	No award	——————
1919	Jules Bordet (Belgium)	Studies in immunology.
1920	Shack August Krogh (Denmark)	Discovery of motor mechanism of blood capillaries.
1921	No award	——————
1922	Archibald V. Hill (England) Otto Meyerhof (Germany)	Archibald's discovery of muscle heat production and Meyerhof's of oxygen-lactic acid metabolism.
1923	Sir Frederick Banting (Canada) John J.R. Macleod (England)	Discovery of insulin.
1924	Willem Einthoven (Netherlands)	Invention of electrocardiograph.
1925	No award	——————
1926	Johannes Fibiger (Denmark)	Discovery of Spiroptera carcinoma.
1927	J. Wagner von Jauregg (Austria)	Fever treatment, with malaria inoculation, of some paralyses.

Date	Name (Nationality)	Achievement
1928	Charles Nicolle (France)	Research on typhus.
1929	Christiaan Eijkman (Netherlands) Sir Frederick G. Hopkins (England)	Work with vitamins.
1930	Karl Landsteiner (U.S.)	Definition of four human blood groups.
1931	Otto H. Warburg (Germany)	Discovery of respiratory enzymes.
1932	Sir Charles Sherrington (England) Edgar D. Adrian (U.S.)	Multiple discoveries in the function of neurons.
1933	Thomas H. Morgan (U.S.)	Discovery of relation of chromosomes to heredity.
1934	George R. Minot (U.S.) William P. Murphy (U.S.) George H. Whipple (U.S.)	Discovery and development of liver treatment for anemia.
1935	Hans Spemann (Germany)	Discovery of the "organizer effect" in embryonic development.
1936	Sir Henry Dale (England) Otto Loewi (Austria)	Work on chemical transmission of nerve impulses.
1937	Albert Szent-Gyorgyi (Hungary)	Study of vitamin C and respiration.
1938	Corneille Heymans (Belgium)	Discoveries in respiratory regulation.
1939	Gerhard Domagk (Germany)	Discovery of first sulfa drug, Prontosil (was not allowed to accept award until after WWII).
1940	No award	————
1941	No award	————
1942	No award	————
1943	Henrik Dam (Denmark) Edward A. Doisy (U.S.)	Dam for discovery and Doisy for determining composition of vitamin K.
1944	Joseph Erlanger (U.S.) Herbert Spencer Gasser (U.S.)	Work on different functions of a single nerve fiber.
1945	Sir Alexander Fleming (England) Sir Howard W. Florey (England) Ernst Boris Chain (England)	Discovery of penicillin and research into its value as a weapon against infectious disease.
1946	Hermann J. Muller (U.S.)	Discovery of X-ray mutation of genes.
1947	Carl F. Cori (U.S.) Gerty T. Cori (U.S.) Bernardo A. Houssay (Argentina)	Coris for work on the metabolism of glycogen, Houssay for pituitary study.
1948	Paul Müller (Switzerland)	Discovery of effect of DDT on insects.

Date	Name (Nationality)	Achievement
1949	Walter Rudolf Hess (Switzerland) Antonio E. Moniz (Portugal)	Hess for middle brain function, Moniz for prefrontal lobotomy.
1950	Philip S. Hench (U.S.) Edward C. Kendall (U.S.) Tadeusz Reichstein (Switzerland)	Discovery of cortisone and other hormones of the adrenal cortex and their functions.
1951	Max Theiler (S. Africa)	17-D yellow fever vaccine.
1952	Selman A. Waksman (U.S.)	Discovery of streptomycin.
1953	Fritz A. Lipmann (U.S.) Hans Adolph Krebs (England)	Discovery by Lipmann of coenzyme A and by Krebs of citric acid cycle.
1954	John F. Enders (U.S.) Thomas H. Weller (U.S.) Frederick C. Robbins (U.S.)	Discovery of a method for cultivating viruses in tissue culture.
1955	Hugo Theorell (Sweden)	Study of oxidation enzymes.
1956	Werner Forssmann (Germany) Dickinson Richards (U.S.) André F. Cournand (U.S.)	Use of catheter for study of the interior of the heart and circulatory system.
1957	Daniel Bovet (Italy)	Discovery of antihistamines and work with curare.
1958	George Wells Beadle (U.S.) Edward Lawrie Tatum (U.S.) Joshua Lederberg (U.S.)	Beadle and Tatum for relating genes to enzymes, Lederberg for genetic recombination.
1959	Severo Ochoa (U.S.) Arthur Kornberg (U.S.)	Artificial production of nucleic acids with enzymes.
1960	Sir Macfarlane Burnet (Australia) Peter Brian Medawar (England)	Study of immunity reactions to tissue transplants.
1961	Georg von Békésy (U.S.)	Study of auditory mechanisms.
1962	Francis H.C. Crick (England) James D. Watson (U.S.) Maurice Wilkins (England)	Determination of the molecular structure of DNA.
1963	Sir John C. Eccles (Australia) Alan Lloyd Hodgkin (England) Andrew F. Huxley (England)	Study of mechanism of transmission of neural impulses.
1964	Konrad Bloch (U.S.) Feodor Lynen (Germany)	Work on cholesterol and fatty acid metabolism.
1965	Francois Jacob (France) Andre Lwoff (France) Jacques Monod (France)	Studies and discoveries on the regulatory activities of genes.
1966	Charles B. Huggins (U.S.) Francis Peyton Rous (U.S.)	Research on causes and treatment of cancer.
1967	Haldan K. Hartline (U.S.) George Wald (U.S.) Ragnar A. Granit (Sweden)	Advanced discoveries in the physiology and the chemistry of the human eye.

Date	Name (Nationality)	Achievement
1968	Robert Holley (U.S.) H. Gobind Khorana (U.S.) Marshall W. Nirenberg (U.S.)	Understanding and deciphering the genetic code that determines cell function.
1969	Max Delbruck (U.S.) Alfred D. Hershey (U.S.) Salvador E. Luria (U.S.)	Discoveries in the workings and reproduction of viruses.
1970	Julius Axelrod (U.S.) Ulf von Euler (Sweden) Sir Bernard Katz (England)	Discoveries in the chemical transmission of nerve impulses.
1971	Earl W. Sutherland, Jr. (U.S.)	Discovery of cyclic AMP.
1972	Gerald M. Edelman (U.S.) Rodney R. Porter (England)	Determination of the chemical structure of antibodies.
1973	Karl von Frisch (Germany) Konrad Lorenz (Germany) Nikolaas Tinbergen (Netherlands)	Study of individual and social behavior patterns of animal species.
1974	Albert Claude (U.S.) George E. Palade (U.S.) C. Rene de Duve (Belgium)	Advancement of cell biology, electron microscopy, and structural knowledge of cells.
1975	David Baltimore (U.S.) Howard M. Temin (U.S.) Renato Dulbecco (U.S.)	Discovery of reverse transcriptase and work with the interaction between viruses and host cells.
1976	Baruch S. Blumberg (U.S.) D. Carleton Gajdusek (U.S.)	Identification of hepatitis antigen and slow viruses.
1977	Rosalyn S. Yalow (U.S.) Roger C.L. Guillemin (U.S.) Andrew V. Schally (U.S.)	Advancements in the synthesis and the measurement of hormones.
1978	Daniel Nathans (U.S.) Hamilton O. Smith (U.S.) Werner Arber (Switzerland)	Discovery and use of restriction enzymes for DNA.
1979	Allan McLeod Cormack (U.S.) Godfrey N. Hounsfield (England)	Invention of computed axial tomography, or CAT scan.
1980	George D. Snell (U.S.) Baruj Benacerraf (U.S.) Jean Dausset (France)	Discovery of the role of antigens in organ transplants.
1981	Roger W. Sperry (U.S.) David H. Hubel (U.S.) Torsten N. Wiesel (U.S.)	Studies on the organization and local functions of brain areas.
1982	John R. Vane (U.S.) Sune K. Bergstrom (Sweden) Bengt I. Samuelsson (Sweden)	Studies on formation and function of prostaglandins, hormone-like substances that combat disease.
1983	Barbara McClintock (U.S.)	Discovery of mobile genes in chromosomes of corn.

PHYSICS

While there was progress in many different branches of physics in the beginning of the 1980s, the most dramatic progress involved the deeper understanding of the basic structure of the universe that comes from particle physics (although the most practical results concerned such solid-state developments as new chips or such technological improvements as new lasers).

While there are already more than a hundred subatomic particles known, so that finding new ones usually causes little stir, certain particles are more interesting than others. Thus, the discovery of a fifth quark at the beginning of the 1980s was a major event, as was the discovery in 1983 of predicted particles that prove one of the new unifying theories of physics. Understanding the importance of such discoveries, however, is not so easy as understanding the importance of, say, a new vaccine or the discovery of possible planets in formation. Therefore, it might be helpful to read this section through from the beginning.

The early 1980s in physics were marked almost as much by what was not found as by what was. As of 1984, the evidence for the decay of the proton, the charge of the quark, magnetic monopoles, anamalons, and the weakening of the gravitational force was short of convincing, although in varying degrees.

One of the salient facts about physics in the 1980s is that it has become very expensive. The main technological effort, for example, is toward the control of fusion—the principal hope for clean, renewable energy of the future; but even experimental fusion reactors cost enormous sums of money to build, operate, or staff. Even more expensive are the new particle accelerators. Such accelerators have grown in shortly over 30 years from a cyclotron that was less than a foot across to facilities that are kilometers long. Now American physicists are planning to build a particle accelerator that will be larger in area than many cities. The Desertron, as it has been nicknamed, will be one of the largest structures ever built by humans.

1980

Naked Beauty or Bare Bottom Found

Sometime after World War II physicists began to change their way of giving names to theoretical ideas. Before then, new ideas were given titles such as "special relativity theory" or "neutrons." A precursor of the new kinds of names came in 1953 when Murray Gell-Mann and Kazuhiko Nishijima decided to name one of the properties of subatomic particles "strangeness." Gell-Mann accelerated the trend in 1961 by calling his group-theoretic way of explaining the properties of particles "The Eight-Fold Way," a takeoff on a basic doctrine of Buddhism. Gell-Mann's crazy names finally reached the consciousness of the general public in 1964 when he described the particles involved in the next stage of his thinking as "quarks," after a word coined by James Joyce in *Finnegans Wake*:

—three quarks for Muster Mark!
Sure he hasn't got much of a bark
And sure any he has it's all beside the mark.

Although the names used currently for the original three quarks are only a little peculiar—*up, down,* and *strange*—they also were known at one time as *vanilla, chocolate,* and *strawberry.* This accounts for the different kinds of quarks being called different flavors of quark. When Sheldon Glashow found that he needed a fourth quark in 1974, he called it, with *strange* still in mind, the "charmed" quark, or *charm.* As two more quarks came to be required in 1978 and following years, it was no longer surprising that some physicists would call the new ones *beauty* and *truth.* However, at this point, more prosaic imaginations prevailed, so most physicists called beauty *bottom* and truth *top,* analogous to the two original quarks, *down* and *up.*

In any case, when physicists searched for the existence of the last two quarks to be named, they had an interesting choice as to how to describe their work to outsiders. Since the object was to find the beauty or bottom quark in undisguised form, they could choose to say they were looking for either "bare bottom" or "naked beauty." The search was serious, even if the names were playful.

The preliminary discovery was of a new particle, not a new theory. The particle was the upsilon meson, found by Leon Lederman and his group at the Fermi National Accelerator Laboratory in 1977 while they were looking for muons. It was quickly realized that the unexpected particle could be fit into the general scheme that had been developed to explain the heavier particles by postulating a new quark. The new quark was bottom or beauty, depending on the eye of the beholder. However, the bottom quark was hidden in the upsilon meson, since it was combined with its own antiquark (an antiquark is just like the original quark, but it has the opposite charge; bottom has a charge of $+\frac{1}{3}$, while antibottom has a charge of $-\frac{1}{3}$). The quark theory was designed to explain why there are so many different kinds of particles that participate in an exchange of forces called the strong interaction. Collectively, these particles are known as hadrons, although they come in two forms—baryons (made from three quarks) and mesons (made from a quark and its antiquark). Since there are only six leptons, the other constituent of matter, symmetry demanded six quarks.

The theory predicted that at a slightly higher energy, the upsilon meson would decay into "beauty-flavored" mesons, that is, mesons from which the properties of the beauty quark could be detected. Thus, although upsilon represented hidden beauty, it appeared that at a little higher energy than the upsilon the physicists would be able to find naked beauty—also known as bare bottom.

That is just what happened in 1980. A group consisting of Nariman B. Mistry, Ronald A. Poling, Edward H. Thorndike, and others used a new particle accelerator at Cornell University to bump the energy of the upsilon particle up to where it decayed into a new meson, the B meson, which in turn enabled the group to study the bottom quark almost directly. This allowed the group to make some preliminary determinations about important questions, such as the decay mode and decay time of the bottom quark. There were many questions still unanswered, however. (See also "Update 1984: Physics" for the related story of the discovery of the top quark.)

WILL THERE BE A "BIG WOOPS"?

As particle-accelerators become more powerful, some physicists are becoming concerned about the possibility of a "Big Woops." This would occur if some current models of the universe are correct about the cosmos having arisen out of a false vacuum. If this model is correct, a very high-energy collision might cause the entire universe to collapse suddenly into a true vacuum—that is, nothingness. As Archibald MacLeish predicted in his sonnet "The End of the World," everything would become "Quite unexpectedly...nothing, nothing, nothing at all." Woops!

Princeton scientists Piet Hut and Martin Rees say that we should not worry. Cosmic ray collisions have already produced collisions of higher energies than any planned in particle accelerators for the foreseeable future. If there were to be a "Big Woops" caused by particle collisions, it would have already occurred.

1981

Do Protons Decay?

When anything decays, it turns into something that is part of itself in some way, according to the ancient Greek idea that the whole is equal to the sum of its parts. Thus, there must be irreducible parts that cannot decay into anything, since they are the end products. That is philosophy.

A photon is a particle with no charge and no mass. It cannot decay into anything else, since it has no parts. An electron is a particle with a negative charge of 1 and a small amount of mass. If it were not for the charge, the electron could decay into photons (or into neutrinos, other particles that have either no mass or very little mass; the jury is still out on the neutrino's mass). None of these particles carries a charge, however. A negative charge of less than 1 occurs in quarks, but the electron is not made from quarks; so the electron, like the photon, cannot decay. (An electron can disappear into energy, however, if its charge is canceled by the positive charge of a positron. But this is not decay. *Decay* means that the particle falls apart by itself.) That is physics.

Protons have a lot more mass than electrons do; they also have a positive charge of 1. What prevents a proton from decaying into, say, quarks? Of course, if protons did this with any frequency, the universe would vanish, since all atoms are made from combinations of protons, neutrons, and electrons.

In 1938 E. C. G. Stueckelberg suggested that protons do not decay because they possess the smallest amount of some quality. This quality is always conserved and is not possessed by any particles lighter than protons. Eugene P. Wigner made the same suggestion independently in 1949, and eventually the idea became dogma. The quality became known as the *baryon number*, since it was something that only heavy particles could have (*baros* means "heavy" in Greek). The concept neatly divided the world of particles into the leptons (from the Greek for "light") such as electrons and muons, which did not have the baryon number, and the baryons, such as protons and neutrons, with a few other classes of particles left over.

Unfortunately, aside from keeping the proton from decaying, there appeared to be no reason for the baryon number to exist. It did not connect with any other known force or situation. Nevertheless, since protons did not noticeably decay, most physicists accepted the baryon number without question for about 30 years.

Some physicists, however, were skeptical. As early as 1954, experiments were begun to search for protons decaying. If the concept of the baryon number had as its sole *raison d'etre* keeping protons from falling apart, it was as sound physics as saying that protons are held together by angels.

In the 1970s a variety of new unifications of particle physics were proposed. These theories proposed that leptons, baryons, and other particles (excluding gravitational particles) could all be viewed as different manifestations of a single entity. Thus, the most far-reaching of these theories came to be called Grand Unification Theories, or GUTs. The GUTs also generally found that the baryon number was not a viable idea. In fact, using the GUTs, it was possible to predict approximately how long it would take for a proton to decay.

The most carefully refined prediction, reached in the early 1980s, was that in 10,000,000,000,000,000,000,000,000,000,000 years, on the average, a given proton would decay. (The number is usually shown in scientific notation as 10^{31}, which is the same as 1 followed by 31 zeros.) This time is a lot longer than the universe has existed. Since the decay is "on the average," however, it means that if you put 10^{31} protons together in one place and wait a year, there is a good chance that one of those protons will decay. In 1981 as many as nine different experiments were either ongoing or getting started based on this idea.

The experiment that most physicists thought had a chance of success was in the Morton salt mine in Ohio. It used the protons in 10,000 tons of water, monitored for proton decay by looking for radiation that would be emitted as a result of the decay. There are enough protons in 10,000 tons of water that, if the proton decayed on the average of one decay in 10^{31} years, 300 decays in a year should be observed in the Morton salt mine.

However, the first group to claim witnessing a proton decay was a team of Indian and Japanese physicists who said they saw a proton decay in the Kolar gold fields in India. Other physicists doubted this, however, and the Kolar claim was disallowed. Some detectors in Japan and Europe also recorded instances of possible proton decay, but they were not convincing either. What everyone was waiting for was some results from the Morton salt mine.

At the end of 1983, such results had not been announced. This could mean that protons do not decay. On the other hand, the calculation of 10^{31} years could be too low. If the average time were 10^{33} years, for example, the Morton salt mine might not have been observed for enough time to spot its first decay.

The Charge of the Quark

One of the mysteries of particle physics is why an electron has exactly the same negative charge as a proton's positive charge. In fact, all charged elementary particles ever found have exactly the same amount of charge (either positive or negative) or none. Although the reason for this is a mystery, the exact equivalence of charge is seen as a blessing, since electromagnetic forces would mess up the universe if things were arranged differently. If, for example, the proton's charge were slightly more than the electron's, then atoms would all be positively charged. Thus, they would repel each other. Since the electromagnetic force is stronger than the bonds that hold matter together, matter would fly apart. No stars, nor anything else, would form, since the electromagnetic force is also stronger than gravity. Thus, the equality of the charges on protons and electrons is essential, if inexplicable by today's theories.

When quarks were proposed by Murray Gell-Mann in 1964, he suggested that one of the ways that they are different from other particles is that they have charges that are either one-third or two-thirds of the fundamental unit of charge that is found on the electron and proton. In fact, finding the existence of a particle with such a charge would be considered the first absolute proof that quarks exist. (Other experiments that have "found" quarks have actually located particles predicted by the theory of quarks, not quarks themselves.)

Because free quarks have never been unambiguously found, however, it has not been possible to determine their charge directly. Quarks appear to occur only in combinations with each other that always have exactly 0 or 1 unit of charge. In fact, until 1981, it was difficult to show that quarks have any charge at all.

Early in 1981, however, physicists working in Hamburg, Germany, were reported to have found evidence for quarks having charge in experiments involving colliding electrons (with negative charge) and positrons (the same particles as electrons, but with positive charge). The evidence was not terribly conclusive, however, and did not show the amount of charge on each quark.

Later in the year, better evidence was reported from the Super Proton Synchrotron in Geneva, Switzerland. Although the physicists involved were reluctant to make claims, they were finding particles with fractional charge in an experiment that consisted of shooting neutrinos at a target of 23 tons of lead. According to current theory, a fractional charge could come only from a quark.

1982

Magnetic Monopoles

In the nineteenth century the remarkable symmetry between electricity and magnetism was already a well-known phenomenon. Like electric charges repel each other and opposite charges attract each other in much the same way as the like poles of two magnets repel each other and the opposite poles attract each other. However, in one situation the symmetry between electric and magnetic phenomena seemed incomplete: negative and positive electric point charges can exist in isolation (for example, in the form of protons or electrons), but magnetic poles are always found in pairs. Any magnetic particle, however small, contains a distinct "north" and "south" pole forming a magnetic dipole. By cutting the magnet in two one obtains two new dipoles.

Because the idea of symmetry in physics is an important one, scientists have been searching for a particle that would complete the symmetry between electric and magnetic phenomena; that is, the magnetic monopole, a particle that would carry only one magnetic pole, either the north pole or the south pole.

During the early 1970s two theoretical physicists, A.M. Polyakov of the Landau Institute for Theoretical Physics near Moscow, and Gerard 't Hooft at the University of Utrecht in the Netherlands, developed independently unification schemes that would unify the strong, the electromagnetic, and the weak forces. These schemes, also called Grand Unification Theories or GUTs, state that at extreme high temperatures and energies, such as those that reigned during the earliest moments of the universe, perfect symmetry exists and the differences between the strong, electromagnetic, and weak forces vanish. These theories require the existence of magnetic monopoles, simply called GUT monopoles, that would be about 10,000,000,000,000,000 times the mass of a proton. This is an extraordinary mass for a subatomic particle. Such a particle would have a mass about the same as that of a paramecium or ameba. The GUT monopoles would have been created during the first instants of the creation of the universe and should still be around.

No agreement exists on the number of monopoles and on where they should be located in the universe. In 1981 it was estimated that one monopole would pass through an area of 1 square centimeter (0.15 square inch) every 10,000,000,000,000,000 seconds, which is once in 280,000,000 years.

One way to detect a particle is to observe the track it makes when it passes through photographic emulsion, basically the same substance that is used to coat film for the purpose of detecting light. In 1975 scientists at the University of California at Berkeley and the University of Houston discovered an anomalously heavy track recorded in a stack of photographic emulsions and plastic sheets exposed to cosmic rays in a balloon at high altitude. The scientists claimed that this was a typical trace of a monopole; however, because of problems with the experimental setup, the possibility existed that the trace could have been produced by the nucleus of a heavy atom, and the discovery found little acceptance among physicists.

Aside from waiting for an unlikely event to be found in routine apparatus, one can use equipment that is specifically developed to watch for magnetic monopoles. This approach in the search for magnetic monopoles is based on the idea that a sudden change of the magnetic flux through a conducting ring, caused by a passing monopole, would induce a sudden change in the current through the ring. Until recently an instrument able to measure these small changes did not exist. By connecting a superconducting niobium loop to a SQUID (Superconducting Quantum Interference Device), Blas Cabrera at Stanford University obtained an instrument that would be sensitive enough to register the sudden change in magnetic flux caused by a passing magnetic monopole.

In May 1982 Cabrera announced that during 200 hours of operation of his apparatus, one sudden change in the current in the superconducting loop was detected that could have been caused by a passing magnetic monopole. Cabrera and other researchers are presently operating superconducting loops of larger areas, oriented along different axes, in order to increase the chance of detection. Thus far no further findings of monopoles with superconducting loops have been reported.

In February of 1984, however, S.N. Anderson and six other researchers from the University of Washington at Seattle and Eastern Washington University in Cheney discovered five events that may have been caused by magnetic monopoles. Their approach was different. Unlike the idea of flying detectors high in the sky, they put their detectors far underground. Thus, cosmic rays consisting of heavy atomic nuclei would not be present to confuse the picture.

In emulsions placed deep under the ground in the Homestake Mine in South Dakota they found tracks of heavy nuclei that had disintegrated into three fragments (ternary fission). As ternary fission is extremely rare, the researchers surmised that the fragments may have been caused by the presence of monopoles.

1983

Two W's and a Z

Physicists persist in believing that, as Albert Einstein remarked, "Subtle is the Lord,...but he is not malicious." From the time of the Greeks, who explained the diversity of nature in terms of the four elements of fire, air, earth, and water, to Einstein himself, who labored unsuccessfully for years in trying to show that gravity and electromagnetism are two aspects of the same force, there has been a continuing drive to find simplicity in the midst of complexity.

There have been successes and setbacks. From the welter of millions of compounds, it was found that there were only a few tens of elements making up all the compounds. As the number of elements grew to nearly 100, another breakthrough explained the elements in terms of simple structures made of (originally) just three different particles, the proton, the neutron, and the electron. This simple scheme soon gave way to the discovery of a confusing array of other particles, which, although they were not constituents of ordinary matter, needed to be explained. Once again the list was shortened by showing that most of the more than 100 particles could be explained as combinations of quarks. At first there were three quarks, but soon it was seen that six were required.

On a related front, however, there seems to be genuine simplicity developing. Only four basic forces are known in nature. These are (1) gravitation, the force of attraction between masses; (2) the electromagnetic force, such as the force between electric charges; (3) the strong force (also called color force), a short range force that holds the protons and neutrons together in the atomic nucleus; and (4) the weak force, responsible for slow nuclear processes, such as radioactive decay. These four forces, also called fundamental interactions, appear to have little in common, and there seems to be no reason why four different forces are needed. Consequently, physicists have been searching for an underlying unity among them, and with some success.

Theories unifying fundamental forces are called unification schemes. Theories that would unify the strong, weak, and electromagnetic interactions are called Grand Unification Theories or GUTs (see Do Protons Decay?, page 510, and Magnetic Monopoles, page 513).

A first step toward the unification of fundamental forces was taken by the Japanese theoretical physicist Hideki Yukawa in 1935. He introduced the idea that exchange particles, which he called *mesons*, would act as mediators of the strong force between protons and neutrons in the atomic nucleus. Yukawa's theory was experimentally verified in 1947 when the pi meson (also called the pion) was discovered in experiments with cosmic rays.

According to present theories, the other fundamental forces are also mediated by exchange particles (called vector bosons), and the masses of these particles should be inversely proportional to the range of the force. The exchange particle transmitting the force of gravity is the still hypothetical graviton, a particle that would have zero mass since gravitation has an infinite range. Electromagnetic forces are also infinite in range and are mediated by photons, which have no mass.

The strong force is a short-range force and consequently the pi meson should have a mass; it was found to be 15 percent of that of the proton. Because the range of the weak force is even shorter, theories predicted that it should be carried by particles with masses much larger than the proton mass.

A more precise insight into the nature of the particles transmitting the weak force became possible when Steven Weinberg, Abdus Salam, and Sheldon Lee Glashow succeeded in formulating a theory (called the electroweak theory) that unified the weak force and the electromagnetic force. By using a mathematical technique known as gauge symmetry, they were able to predict that three particles would mediate the weak forces in atomic nuclei: a W^+ particle with charge $+1$, and a W^- particle with charge -1, both with a mass of 81 GeV, which is 80 times the mass of a proton, and a neutral particle of even greater mass, the Z° with 93.8 GeV. (Masses of subatomic particles are given in terms of their energy, which Einstein showed was equivalent to mass. An energy of 1 GeV, or giga-electron-volt, is 1,000,000,000 electron volts. One electron-volt is equal to 0.0000000000016 ergs, or, in scientific notation, 1.6×10^{-12} ergs. In the United States the term BeV—for billion-electron-volts—is sometimes used instead of GeV.)

In 1973 experiments showed that in some weak interactions particles do not undergo a change in charge. This is only possible when the exchanged particles are neutral. The exchange of neutral particles is called neutral weak currents; these proved the existence of the Z° particle.

Although this confirmation was sufficient to lead to Weinberg, Salam, and Glashow's being awarded the Nobel Prize in Physics in 1979, there was not universal belief in the electroweak theory. For one thing, none of the predicted particles had ever been observed. Until a particle is found experimentally, there is always the possibility of an unpleasant surprise.

Until 1982 the direct observation of W and Z particles was not possible because even in the largest particle accelerators, fixed targets were used and most of the collision energy was transferred to translation energy of the target particles. For example, when protons accelerated to 270 GeV (270,000,000,000 electron volts) in the Super Proton Synchrotron (SPS) at CERN (*Centre Européen de Recherche Nucléaire* near Geneva, Switzerland) hit stationary targets, only 30 GeV (30,000,000,000 electron volts) became available for the production of new particles. However, if particles are accelerated in such a way that they would collide head-on, only a small part of the collision energy would be lost into translation energy, and virtually all the liberated energy would be transformed into new particles.

Under the impetus of the Italian scientist Carlo Rubbia, the SPS at CERN was converted into a huge proton-antiproton collider. The protons and antiprotons were both injected into the same ring and accelerated in paths opposite to each other; the energy available for the production of new particles was increased about fifteen times to 540 GeV (540,000,000,000 electron volts). In two locations of the ring the particles collide head-on. In these locations are placed the two large detectors, the UA-1, a general purpose 2,000-ton detector developed by a group of 134 physicists headed by Rubbia, and the UA-2, a smaller (200-ton) but more specialized detector developed by a group of 60 physicists under the leadership of Pierre Darriulat.

Neither detector can detect the W and Z particles directly. However, the W and Z particles have short lifetimes, and their presence can be inferred from the different decay products, energies, and directions. The W^+ decays into a positron and a neutrino, while the W^- decays into an electron and a neutrino. The Z° particle can decay into an electron and a positron. As the neutrinos cannot be detected, scientists at CERN concentrated in their search on the electrons (or positrons) that would carry about half the energy of the decaying W-particle (the rest of the energy is carried away by the neutrino), and on electron-positron pairs produced by the disintegration of Z° particles. The positron and the electron would then each carry about half of the energy of the Z° particle. The detectors are equipped with calorimeters, devices that measure the energy of the created charged particles.

During runs of the Super Proton Antiproton Synchrotron at the end of 1982 and during 1983, a total of 90 W-particles and twelve Z-particles were recorded. The electroweak theory developed by Weinberg, Glashow, and Salam was confirmed convincingly. The 1979 Nobel Prize had not been given to the wrong people.

Facts about Subatomic Particles

There are more than 100 different subatomic particles that have been located at one time or another, although many of them have extremely short lifetimes, and only a few of them are constituents of ordinary matter. Some particles apparently have infinite lifetimes, and so are called *stable*.

Although the following list is incomplete, it includes the particles that are involved in recent and important theories of the structure of matter.

A few definitions may help. MeV is the abbreviation for million-electron-volt, a measure of energy that is also used for mass, since mass and energy are equivalent. A lifetime shown as 2.2×10^{-6} seconds is the same as 0.0000022 seconds, where the number of zeros is the same as the negative exponent—if you count the zero in front of the decimal point. *Spin* is a number associated with the angular momentum of particles.

Name	Charge	Mass in Mev	Average Lifetime in Seconds	Spin
Vector Bosons				
Photon	0	0	Stable	1
Positive W	+1	81000	10^{-20}	1
Negative W	−1	81000	10^{-20}	1
Neutral Z	0	93800	10^{-20}	1
Higgs particle	0	1000000	N.A.	0
Graviton	0	0	Stable	0
Leptons				
Neutrino	0	0*	Stable*	½
Antineutrino	0	0*	Stable*	½
Neutrino (muon type)	0	0*	Stable*	½
Antineutrino (muon type)	0	0*	Stable*	½
Neutrino (tau type)	0	0*	Stable*	½
Antineutrino (tau type)	0	0*	Stable*	½
Electron	−1	0.511	Stable	½
Positron	+1	0.511	Stable	½
Muon	−1	105.7	2.2×10^{-6}	½
Antimuon	+1	105.7	2.2×10^{-6}	½
Tau	+1	1750		½
Antitau	−1	1750		½
Mesons				
Positive pi	+1	139.6	2.6×10^{-8}	0
Negative pi	−1	189.6	2.6×10^{-8}	0
Neutral pi	0	135	8×10^{-17}	0

Name	Charge	Mass in MeV	Average Lifetime in Seconds	Spin
Positive K	+1	493.7	1.2×10^{-8}	0
K-zero-short	0	497.7	9×10^{-11}	0
K-zero-long	0	497.7	5.2×10^{-8}	0
Negative K	−1	493.7	1.2×10^{-8}	0
Rho	0, ±1	773	N.A.	1
Omega	0	783	N.A.	1
Eta	0	2980	N.A.	0
Psi		3098	N.A.	1
B meson		1228	N.A.	1
D meson		1286	N.A.	1
Baryons				
Proton	+1	938.3	Stable†	½
Antiproton	−1	938.3	Stable†	½
Neutron	0	939.6	9.18×10^2	½
Antineutron	0	939.6	9.19×10^2	½
Lambda hyperon	0	1115.6	3×10^{-10}	½
Lambda antihyperon	0	1115.6	3×10^{-10}	½
Positive sigma	+1	1189.4	8×10^{-9}	½
Neutral sigma	0	1192.5	10^{-11}	½
Negative sigma	−1	1195.3	1.5×10^{-10}	½
Neutral xi	0	1315	3×10^{-10}	½
Negative xi	−1	1321	1.7×10^{-19}	½
Omega	−1	1672	1.3×10^{-10}	3/2
Quarks				
Up	+⅔	1.0	Stable	½
Down	−⅓	3.0	Stable	½
Strange	−⅓	102.2	Stable	½
Charm	+⅔	1530	N.A.	½
Top	+⅔	4600	N.A.	½
Bottom	−⅓	4700	$<5 \times 10^{-12}$	½
Gluons				
Red to blue	0	0	Stable	1
Red to green	0	0	Stable	1
Green to red	0	0	Stable	1
Green to blue	0	0	Stable	1
Blue to red	0	0	Stable	1
Blue to green	0	0	Stable	1
Neutral (1)	0	0	Stable	1
Neutral (2)	0	0	Stable	1

* Some theories suggest a very small mass and a "decay" of one kind of neutrino into another.
† Some theories suggest a proton decay on the order of 10^{31} years.

Quantized Hall Effects

When an electric current passes through a conductor placed in a magnetic field that is perpendicular to the direction of the electric current, a potential develops across the conductor. This potential, called the Hall voltage, is caused by the deflection of electrons by the Lorentz force, a force perpendicular to the magnetic field and to the direction of motion of the electrons. The net effect is that more free electrons are found on the edges of the conductor as well as more at one end caused by the original current. The difference in electron concentration along the edges can be measured as a small voltage across the conductor.

In 1980 Klaus von Klitzing and other researchers experimented with metal-oxide-semiconductor-field-effect transistors (MOSFETs) placed in strong magnetic fields at temperatures close to absolute zero. A MOSFET consists of a layer of metal and of silicon, separated by a thin layer of insulating silicon dioxide. If a voltage is applied between the metal and the silicon, electrons are free to move only in a very narrow region of the silicon next to the silicon dioxide layer. Because the electrons can move only in directions parallel to the silicon dioxide layer, they are said to form a two-dimensional electron gas. During routine measurements the researchers discovered two surprising phenomena: (1) The Hall resistance (defined as the Hall voltage divided by the current through the conductor) is not a smooth function of the electron density in the conductor, but exhibits several "plateaus" in which the Hall resistance is nearly independent of the electron density in the conductor; and (2) the electric resistance of the conductor drops to nearly zero when the Hall resistance reaches a plateau.

More specifically, they found that the plateaus correspond to values in which the Hall resistance equals the product of an integer with the ratio of two fundamental physical constants, the constant of Planck h and the square of the charge of the electron. Integers are whole numbers, such as 1, 2, 3, and so forth. The presence of only integral values means that the Hall effect for a two-dimensional electron gas is quantized.

Quantum effects have been observed since the beginning of the century, but most of them are statistical in nature. That is, a particular behavior, such as the radiation emitted by a black body, the energy of an electron produced by shining light on a metal, or the detection of electron beams through a slit, exhibits a behavior at the macroscopic level that can best be explained if only integral values are allowed for certain parameters at the level of subatomic particles. At the macroscopic level, nearly everything behaves at least superficially as if continuous changes take place, rather than "quantum jumps." The Hall effect in thin layers is one of the very few directly observable examples of such quantum behavior. Typically, the macroscopic quantum behavior of any system takes place only at temperatures near absolute zero. (For more information on quantum effects see "Seeing One Atom Evaporate," page 524.)

Roughly stated, it is now believed that the quantization of the Hall resistance is caused by the discrete energy levels, called Landau levels, occupied by electrons moving in a magnetic field.

To understand the Hall effect better, first consider the motion of free electrons in a two-dimensional conductor in the absence of magnetic and electric fields. The free electrons move in straight paths with random speeds and directions, and there is no net current of electrons in any direction. If you apply a magnetic field perpendicular to the plane of motion of the electrons, each electron changes its straight path into a circular one because of the Lorentz force, which acts perpendicularly to the direction of motion of each electron. All the free electrons move now in circular paths; the centrifugal forces acting on the electrons are counterbalanced by the Lorentz forces in exactly the same way as in the motion of electrons in a synchrotron. The radii of these orbits are proportional to the speed of the electrons.

If you now apply an electric field, the electrons, contrary to what is expected, not only move in the direction of the electric field, but also experience a drift perpendicular to the electric field. The force of the electric field on the electrons causes the circular orbits to drift in a direction perpendicular to the electric field. For each circle, the applied current pushes the electron for half its orbit and retards it for the other half, deforming the orbit continually in the same direction. This deformation produces the drift that is responsible for the build-up of the Hall voltage. The net result is an electric current across the conductor, perpendicular to the applied electric field. In the direction of the electric field, the conductor acts as an insulator. Perpendicular to the electric field, the conductor acts as an ideal conductor because an electric current exists in the absence of an electric field component in that direction. In turn, the Hall voltage sets up a new electric field in the direction across the conductor and the circular electron orbits now start drifting in the direction of the electric field initially set up by the voltage source, resulting in a current through the conductor.

Two additional factors have to be considered that describe the behavior of electrons in the two-dimensional conductor: the quantization of the energy levels of the orbiting electrons and the scattering of electrons as the cause of electric resistance.

Just as electrons circling an atom can only occupy certain fixed energy levels, electrons moving in orbits in the two-dimensional electron gas can occupy only certain fixed energy levels, termed Landau levels. Just as the Pauli exclusion principle allows only two electrons with opposite spin to occupy the same energy level in an atom, only a fixed number of electrons per unit area are allowed to occupy the same Landau level. (The number of electrons per unit area that occupy a same Landau level is called the degeneracy of that Landau level.) Consequently, the lowest Landau level can only accommodate a fixed number of electrons. Other electrons have to be accommodated in the consecutive Landau levels, and generally the highest Landau level will be only partially filled.

The second factor to take into account is the scattering of electrons in the conductor. It is the reason that a normal conductor offers resistance to electric current: electrons collide with atoms in the conductor and exchange energy with them. The result of a collision is that the electron jumps to a lower (or higher) energy state. The amount of energy exchanged depends on the energy of the atoms, or the temperature of the conductor. At low temperatures, the amounts of exchanged energy are much lower than the energy differences between the Landau levels, and electrons cannot jump from one Landau level to another; they can only be scattered from one orbit to another in the same Landau level.

If the electron density in the conductor becomes such that a certain number of Landau levels are completely filled up with electrons, while the higher Landau levels are empty, there is no possibility for electrons to be scattered. They cannot be scattered into other orbits in the same Landau level because the level is completely filled, and the next Landau level is out of reach. As no scattering can take place, the electric resistance vanishes.

In real systems, so-called localized electrons that are trapped in the conductor occupy energy levels that are distributed in between the Landau levels. These electrons do not contribute to the current, but are responsible for the smoothing of the Hall resistance curve.

In 1983, beside the integral values, however, fractional values of the ratio were observed : $\frac{1}{3}$, $\frac{2}{3}$, $\frac{2}{5}$, $\frac{3}{5}$, $\frac{2}{7}$, and so forth, instead of integers. The denominators of the fractions are always odd. A complete theoretical explanation for these fractional values has not yet been found, although Robert B. Laughlin of Lawrence Livermore National Laboratory has explained some values as the result of "quantum liquids" formed by the electrons. In many ways, the appearance of the fractionally quantized values is more startling than the integral values, for physicists have come to expect integral quantum phenomena, at least for particles.

Anomalons

When heavy nuclei are accelerated to relativistic speeds (speeds close to the velocity of light) in a heavy-ion accelerator, such as the Bevalac at Lawrence Berkeley Laboratory in Berkeley, California, and collide with atomic nuclei in a fixed target, large numbers of nuclear fragments are produced. These fragments are scattered in all directions in the material; some of the fragments collide again with other nuclei and break apart in even smaller fragments. These fragments can again collide with other nuclei in the target.

The distance over which a nuclear fragment travels through matter before colliding with an atomic nucleus is called the mean free path. For each fragment with a given atomic number (the number of protons in the particle), the length of the mean free path is a well-defined quantity that can be measured.

In a number of experiments physicists have been studying the mean free paths of nuclear fragments produced by the interaction of nuclei in stacks of photographic emulsions (the emulsion serves at the same time as target and as particle detector). The produced tracks can easily be followed with a microscope, and mean free paths have been determined for a large number of nuclear fragments.

From most of the experiments to date an unexpected result has emerged: a certain percentage (estimates range between 3 and 6 percent) of the fragments have shorter mean free paths than their "normal" counterparts, and therefore they are termed "anomalons." Anomalons seem to interact more easily with matter and behave very much as particles with a much larger interactive cross section. For example, magnesium nuclei that become anomalons can exhibit cross sections the size of lead nuclei.

Scientists have not been able to come up with a satisfactory theoretical explanation for anomalons, but several theories have been proposed. In the most commonly accepted explanation, anomalons consist of so-called "quark-gluon plasma": masses of quarks and gluons (the particles that keep the quarks together) in which the separate identities of protons and neutrons have disappeared. A second explanation is that anomalons have an increased reaction cross section because of an extraordinary nuclear configuration acquired during the collision with the target nucleus. On the other hand, skeptical physicists think that anomalons are an artifact of the detection medium or of the statistics used. The most careful analysis done to date was presented at the Second Workshop on Anomalons in the summer of 1983. The conclusion: Anomalons exist (probably).

Seeing One Atom Evaporate

Two of the most remarkable substances around are ordinary water and liquid helium-4 (the most common isotope of helium). Life could not exist without water and its unusual properties; and it often seems as if modern physics could not exist without helium-4 and *its* unusual properties. Helium-4 is the most common isotope of helium, but it is specified here, since helium-3, the other isotope, does not have the properties that are required for many special effects. In particular, liquid helium-4, at temperatures close to absolute zero, is useful for demonstrating quantum effects at a level where they can be detected directly. (See also "Quantized Hall Effects," page 520.)

The first effect that made the quantum seem real instead of simply a mathematical fiction was the photoelectric effect. As explained by Albert Einstein in 1905, the photoelectric effect is caused by packets of light waves, called photons, that behave like particles. When a photon strikes the surface of a metal, if the photon has a certain threshold energy, it will eject an electron. If light is viewed simply as waves, the number of electrons ejected would vary with the intensity of the light; but when seen as particles, the number of electrons ejected and their energy depend on the frequency of the light, since the frequency reflects the amount of energy in each particle. It is the latter effect that is observed, although classical physics would predict the former. Einstein's quantum-based explanation in 1905 of the photoelectric effect, which won him the Nobel Prize in Physics, was one of the key events in the early development of the quantum theory.

Just as the old Newtonian mechanics should have been, in theory, able to explain everything we observe in terms of the forces on moving particles, quantum mechanics should also be able to explain all physical phenomena from basic principles (and do a better job of it than Newtonian mechanics could have done). In practice, most complex phenomena are too complicated for Newtonian mechanics, and they are also too complicated for quantum mechanics to explain. One example is the evaporation of a liquid. In theory, a physicist should be able to use quantum mechanics to explain the evaporation of a particular liquid under specified circumstances, but in practice there are too many atoms involved.

Evaporation could be described more simply if it were made exactly analogous to the photoelectric effect. To do this, it is necessary to use pressure waves to cause the evaporation rather than general heating of the liquid. Heating causes evaporation because it increases the average speed of atoms or molecules in a liquid. If the average speed is increased, a few atoms or molecules gain enough energy to fly off the surface of the liquid. A pressure wave provides the same additional burst of energy, but more selectively than heating does.

A pressure wave through the liquid is a wave and light is a wave, so there is a basis for using pressure waves to produce effects similar to those of light waves. In quantum mechanics, a small pressure wave should also carry energy in packets, just the way a wave of light does. A pressure wave viewed this way is called a *phonon*, partly in analogy with *photon* and partly because such a wave or packet is a small-scale version of the waves that carry sound. To complete the analogy, a phonon striking the surface of a liquid (from within the liquid) should behave like a photon striking a metal in the photoelectric effect; one phonon should give up its energy to a single atom, causing the atom to evaporate from the surface of the liquid.

Until 1983 this effect had not been demonstrated, however. That year, M. J. Baird, F. R Hope, and A. F. G. Wyatt of the University of Exeter (United Kingdom) were able to show that this quantum effect existed at the atomic level. It was a classic piece of experimental physics.

The reason that the effect could not be seen previously was that the phonons dissipated before they reached the surface of the liquid. The higher the energy a phonon has, the shorter the distance it will travel in a liquid. When the frequency of the phonon was high enough to knock atoms out of the liquid, the phonons would tend to break apart before they reached the surface.

Here is where helium-4 enters the picture. Helium-4 permits quantum effects that other liquids do not. In particular, above a certain critical range of frequencies, phonons can travel for quite long distances without breaking apart.

The actual experiment used a heater in liquid helium-4 that had been cooled to nearly absolute zero. The heater, by starting a pulse traveling though the frigid helium-4, produced the phonons. A detector was suspended above the surface of the helium-4. By varying the distances of both the heat source and the detector from the surface, Baird, Hope, and Wyatt were able to tell, from measuring the time it took for the detector to register, that they were detecting single atoms ejected from the surface by individual phonons. Because this was a simplified setup, compared with ordinary evaporation, they were able to compute the times directly from first principles. The computed times agreed with the measured times. It was the first time evaporation of any liquid was understood from first principles.

Another Criterion for Controlled Fusion Met

It is clear that we will someday run out of such fossil fuels as oil, coal, and natural gas (although the date for this situation keeps being pushed farther away), but we will continue to need sources of energy. A common high-technology hope for the energy that will be needed in the future is controlled nuclear fusion, the process that powers the sun, stars, and hydrogen bombs. The idea is to create a miniature sun inside a power plant. Work has been going on for about 30 years, but the technology has proven to be very difficult, and about another 30 years are likely to be needed before the idea becomes practical.

The basic idea is to produce an extremely hot mass of hydrogen, so hot that the electrons and protons are separated; that is, the hydrogen is ionized. In this form, the hydrogen becomes a phase of matter called a *plasma*. If the plasma is hot enough and dense enough, the hydrogen nuclei will fuse together, forming helium and releasing large amounts of energy. In the core of a star, such as the sun, the density is enormous, which ignites the fusion. The fusion then creates heat, which, so to speak, adds more fuel to the fire.

Hydrogen comes in three different isotopes: ordinary hydrogen with one proton and one electron; deuterium with one proton, one neutron, and one electron; and radioactive tritium with one proton, two neutrons, and one electron. Ordinary hydrogen is more common than water (two atoms of hydrogen in every molecule of water). Deuterium, although much less common, is also in what amounts to a limitless supply, as 0.02 percent of all hydrogen is deuterium. You can get more energy out of a fusion reaction if you start with one of the heavier isotopes of hydrogen. The sun's enormous mass makes the "burning" of ordinary hydrogen sufficient for heating, but in smaller amounts the extra energy must be obtained by starting with deuterium or, even better, tritium. Tritium is rather hard to come by, but controlled fusion can theoretically produce sufficient energy by using a combination of a lot of deuterium and some tritium.

Because the temperature required for this deuterium-tritium mixture to achieve fusion is in excess of 100,000,000°C (180,000,000°F), the question becomes, What will you keep it in? All substances will melt or vaporize at that temperature, so the standard method used by workers trying to develop controlled fusion is magnetic confinement. The plasma can be kept inside a magnetic "flask" while it is being heated. In that way, it does not touch any material substances. Electrons and nuclei, the components of plasma, normally travel in straight lines. Since they are both charged particles, however, a magnetic field deflects their paths into circles, forming the "flask."

A key development in the effort to attain controlled fusion came during the 1950s and 1960s when Lev Artsimovich and other Soviet physicists developed the Tokamak, a form of magnetic "flask" that is shaped somewhat like a doughnut. Both Soviet and American efforts since that time have relied mainly on variations of the Tokamak design.

As noted, there are two criteria required for the plasma to begin to produce energy: heat and sufficient density of the plasma. The key point is, of course, the break-even point, the point at which the energy used to power the reaction is equal to the energy produced by the reaction. It takes a lot of energy to run the magnets and to heat the plasma. For the reaction to produce as much energy as is required to run the equipment, the heat must be there along with enough ions. In 1978 a group at Princeton achieved the required heat, but they did not have very many ions confined in their magnetic "flask." Even the best magnetic "flasks" are subject to instabilities, which can cause them to leak heavily after a certain density is reached. Typically, a successful confinement of plasma is measured in milliseconds.

The breakthrough in 1983 was in meeting the other requirement. Workers at the Massachusetts Institute of Technology (MIT) managed to get enough plasma into the flask for the first time ever. Previous efforts at MIT had resulted in the plasma breaking out of confinement when more plasma was added. The success was achieved by shooting frozen pellets of deuterium into the plasma. After there was as much plasma in the flask as the flask could normally hold, the pellets were used to add more fuel. Because the pellets entered the flask rapidly, then turned quickly to plasma themselves, for about 50 milliseconds the necessary density was achieved before the confinement broke.

Of course, the plasma was not hot enough, so true breakeven has not been reached. However, at 3:06 A.M. on December 24, 1983, the Princeton group got its Christmas present—a new $314,000,000 fusion reactor. There is good reason to believe that, when the bugs have been worked out, the new reactor at Princeton will become the first to reach both break-even criteria at once. This is expected in 1986.

After that, it is still another problem to produce more energy than is used to keep the reaction going, even on a laboratory scale. When that goal has been achieved, the technology will have to be developed far enough to make large commercial plants possible, quite another thing from laboratory successes. Most people do not think that the commercial production of fusion power will occur in this century.

Is It a Coincidence, or Are We Drifting Apart?

Astronomers and physicists work with a lot of different numbers. Some of the numbers are easy to explain because they can be calculated from a basic theory. Other numbers are just given as the result of an experiment. For example, the gravitational constant G and a number known as the "fine structure constant" are found by measuring them very precisely. There is not at present a theory that would have as, say, a solution to an equation, the gravitational constant or the fine structure constant. Consequently, scientists remain on the lookout for something that would explain these numbers from first principles.

Sometimes the explanations get a little silly. An astronomer who shall be nameless once noticed that the number of electrons in his estimate of the size of the universe involved the number 137. He knew that the fine structure constant, which is the measure of a property of electrons, was almost exactly $1/137$. So he concluded that the fine structure constant was what it was because of the size of the universe.

Other explanations have to be taken seriously. Paul A. M. Dirac has a proven track record in making predictions based on studying the mathematics of physical theory. For example, Dirac successfully predicted the characteristics of the positron in 1930, two years before it was found, and of other forms of antiparticles 25 years before the particles were discovered experimentally. So when Dirac noticed in the late 1930s that certain numbers seemed too close to be a coincidence, physicists started planning experiments to find out whether Dirac's hunch was right.

What Dirac noticed was that if you measured the age of the universe in natural time units, as an atomic clock would measure it, the age is about 10^{40} such units. (10^{40} is 1 followed by 40 zeros.) Similarly, the ratio between the electromagnetic force and the gravitational force is also about 10^{40}. Other fundamental constants also yielded numbers around 10^{40} when they were put into natural units. Could this be a coincidence? Dirac was not sure, but he did suggest that what seemed to be tying these numbers together is the age of the universe. If these ratios were the same now as they had been in the past, and if they were to continue to be the same in the future, then at least one of the values involved had to be changing with time. Dirac chose to suggest that the gravitational constant was changing slowly, getting weaker, in fact. This suggestion has come to be known as Dirac's Large Number Hypothesis.

If Dirac were right, then time as measured by atomic clocks would be different from time as measured by the revolutions of planets around the sun. In effect, the planets would be gradually getting farther from the sun, lengthening their periods.

The change predicted by Dirac was so small, however, that at first no one could measure it. To be on the safe side, however, physicists developed the habit of including the possibility that gravity was changing in estimates of possible error when they published their theories or experiments.

Since 1948 a number of studies have yielded different versions of whether or not Dirac could have been right. In 1948 Edward Teller calculated that Dirac was wrong, since greater gravity 4,000,000,000 years ago would have brought Earth too close to the sun for life to have evolved. Not everyone agrees with Teller's calculations, however. Thomas Van Flandern has been analyzing the period of the Moon (the length of time for the Moon to orbit Earth) since 1970. As of 1981 his conclusion was that Dirac was right; changes in the period of the Moon are about the size that would be expected if gravity were weakening at the Dirac rate.

Two different groups in 1983 came forth with new calculations based on data gathered with the help of the Viking lander on Mars and the Mariner 9 Mars orbiter. The accuracy of their method was good enough to tell the distance from here to Mars within about 10 meters (33 feet). This enabled the groups to make very accurate comparisons of the orbit of Mars over a period of several years. If it were changing at the rate predicted by Dirac's Large Number Hypothesis, they would detect it.

What they detected was that the change in gravity for the time period they used was much less than Dirac's prediction, although they did find some change. What this means depends on who you ask. Some would abandon the Large Number Hypothesis altogether, but others think that it still might prove true in some modified version. After all, the large numbers are still all about the same size, so it might not be a coincidence.

KNIT ONE, PURL TWO

P. A. M. Dirac has degrees in both mathematics and physics, making him a true mathematical physicist. As a result, his interest in patterns is intense. One day he became fascinated watching a woman knitting a sweater. He stared at the knitting with intense concentration for several minutes before announcing that he had discovered another pattern that could be used to take a ball of yarn and turn it into cloth. Borrowing the knitting needles, he awkwardly demonstrated his new discovery. The knitter was surprised to see that Dirac had discovered on his own the second basic stitch of knitting, known since time immemorial as purling.

1984

Update 1984: Physics

UNPREDICTED PARTICLES

The same set of experiments that won Carlo Rubbia of Harvard University and his team at CERN fame in 1983, when collisions of protons and antiprotons were used to prove the existence of W and Z particles, continued to reveal exciting new results.

Perhaps the most important events that were detected are those that had been predicted by no theory. In the case of the W and Z particles, theory led experiment, but in physics experiment leads theory more often than the other way around. That is certainly the case with five events that were recorded in the collisions. These events show jets of particles flying off in one direction, and nothing at all flying off in the other. By conservation of momentum, the "nothing at all" has to be an undetected particle.

Other events from the same set of experiments revealed evidence for a particle that was predicted but which had previously escaped detection. The particle is the top quark, the last of the six predicted quarks (see "Naked Beauty or Bare Bottom Found," page 508).

LIFE IN THE ELEVENTH DIMENSION

Most people who follow physics have gotten used to the idea that we live in four dimensions, not three (time added to the three dimensions of length, width, and height).

The idea that there might be more than four dimensions goes back to the German-Polish mathematician Theodor Kaluza and the Swedish physicist Oskar Klein, who proposed in the 1920s that a universe with five dimensions could, to some degree, unify relativistic gravity and electromagnetism.

In 1984, a number of physicists put forth their own Kaluza-Klein theories at a conference at Fermilab in Batavia, Illinois. One group represented people who are trying to find a quantum theory of gravity. The other group were particle physicists, trying to make more sense out of the fundamental properties of subatomic particles. While none of the speakers seems to have brought the others around to his or her particular universe (for one thing, quantum gravity physicists like even numbers of dimensions, while particle physicists like odd numbers, especially eleven), it appeared likely that one of the Kaluza-Klein theories might eventually become our standard notion of the universe.

Smashing Particles

Some day in the next five to ten years a great ring, 60 kilometers (37 miles) to 200 kilometers (125 miles) in diameter, will occupy a part of the desert that was previously home to cacti, jackrabbits, and pack rats. Building this ring will cost at least $1,500,000,000—and probably much more. The effort will consume the energies of a whole generation of specialized scientists to the extent that the building process alone may redirect the course of American science, ironically directing physics away from the kinds of problems the machine is being built to solve. When the project is complete, the purpose of this giant ring will be to toss two invisible streams of tiny objects toward each other with the intent of breaking the objects into small bits.

There is some chance that this giant "toy" for physicists will never be built. Even if it is not, other, smaller rings and structures with similar goals in mind are strewn around our planet. Expensive particle smashers are the principal tool of one branch of physics.

Physics, like the other major sciences, is divided into separate branches that often barely communicate with each other. A specialist in the properties of materials has about the same relation to a high-energy physicist as a student of animal behavior does to a molecular geneticist. And yet, the different branches of physics (and the different branches of biology) are increasingly coming together in fundamental ways. Today, the physics of high energy—smashing particles—is central to the development of cosmology, unification theories, relativity theory, astronomy, and perhaps much more.

The executive director of the New York Academy of Sciences, Heinz R. Pagels, has compared particle smashers to microscopes. Just as more and more powerful microscopes reveal objects of smaller and smaller size, so higher and higher energies in particle smashers reveal more and more of the basic realities of nature.

In 1983 one of the major events was the confirmation of the theory unifying two of the basic forces of physics (see page 515). The confirmation was achieved by the development of a controversial new particle smasher. The theory, which had already reached a level of acceptance that had given its creators the Nobel Prize, predicted that at high enough energies, certain new subatomic particles would be produced.

Another recent event of significance was the discovery of the bottom quark in the form of naked beauty or bare bottom (see page 508). Although the first clue—hidden beauty—was discovered in 1977, the particle smasher that produced hidden beauty was not quite powerful enough to produce naked beauty. When particle physicists got a new particle smasher at Cornell University in 1979, the Cornell Electron Storage Ring (CESR), they were excited to note that, although it had not been planned that way, the CESR energy levels were exactly right for finding naked beauty. And that is what they did.

Increasingly, physicists are going to Europe, most notably to the CERN facility in Geneva, Switzerland, because that's where the energy is. The CERN Super Proton Synchrotron, converted to shove antiprotons into protons, was the only place with the energy to make the particles needed to confirm the unifying electroweak theory. In fact, it is clear that the lead in particle physics will always go to the country with the most powerful particle smashers. Until recently, that was the United States, with occasional attempts by the Soviet Union to achieve dominance. Today, there is every reason to believe that the lead could switch to Europe or even to Japan. American scientists have banded together to ask their government for help to prevent this from happening.

From Balloons to Underground

In the beginning, there were no machines, only detectors. Around the turn of the century, radiation was the "new kid on the block." One of the many people studying radioactivity, Victor F. Hess, calculated correctly that most of the background radiation found at the surface of Earth is caused by sources within the rocks and soil. But his calculations also showed that the amount of radiation decreases too slowly with altitude; therefore, some radiation must be coming from space. Hess took seven balloon flights to check out his calculations. As a result, he was the discoverer of cosmic rays.

High-energy physicists, as particle physicists are also known, go where the energy is. Balloon flights with particle detectors became standard in the 1930s and 1940s. As a result, Carl Anderson, one of the cosmic-ray physicists, discovered both the positron and the muon by observing their paths in photographs of Wilson cloud-chamber detectors that were taken on balloon flights. C. F. Powell and G. P. Occhialini discovered the pi meson in cosmic rays the same way a few years later. In fact, people continued to discover strange new particles in cosmic rays well after the first Earth-bound atom smashers were built, because cosmic rays are more powerful than any machine-produced radiation.

During the early period, however, the only other method of obtaining energy at levels high enough to involve new particles was to use natural sources. For example, James Chadwick discovered the neutron in 1931 by bombarding samples of beryllium with alpha rays (nuclei of helium-4) from polonium. In fact, his earliest experiments failed because he did not have enough sufficiently pure polonium.

Around the same time, it occurred to a number of different researchers that better results might be obtained if particles were accelerated in some way before being used to bombard targets. Calculations showed that protons were the most suitable particles then known (particle physicists were familiar with only electrons, protons, alpha particles, and other ionized nuclei, so they did not have a lot of choice). The first physicists to get a machine that produced results were John Douglas Cockcroft and Ernest Thomas Sinton Walton, who simply used rectified alternating current to accelerate protons along a 3-inch brass tube. With this fairly simple apparatus, they obtained particles with energy as high as 380,000 electron-volts by 1929. Robert Jemison Van de Graaf was next with the "Van de Graaf generator," the apparent inspiration for the machines in Dr. Frankenstein's laboratories in the famous motion picture. The ions accelerated by his 1931 machine reached five times the energy of the particles accelerated in the Cockcroft-Walton machine.

The real trail, however, was laid out by Ernest Orlando Lawrence, who built his first accelerator, called a cyclotron, in 1929, as a test model. It was less than 28 centimeters (11 inches) in diameter, but it produced energies of 8,000 electron-volts, and it was the predecessor of the particle smashers of today.

The idea behind the cyclotron was not original with Lawrence, but he was the one who made it work. In a cyclotron, the ions are sent around in a circle. Each time they go around the circle, they are given a boost. When they reach the desired velocity, they are shot from the cyclotron toward the target. Magnets were used to keep the ions in a circular path.

Cyclotrons in their original form eventually reached energies of nearly 15 MeV (million-electron-volts) in the 1940s, but modifications such as the sector-focused cyclotron reached energies ten times that high by the 1970s. Other designs led to even higher energies. For example, one of the first of the new generation of particle accelerators, the Bevatron at Berkeley, California, was designed to reach energies above 1 GeV (a GeV is a giga-electron-volt, or 1,000,000,000 electron-volts; in the United States this unit was formerly known as the BeV, for billion-electron-volts, which accounts for the name *Bev*atron). These particle accelerators cleverly worked their way around the limitations of the cyclotron, and produced energies as high as 20 GeV.

One solution was to abandon the circular design, which then required very long accelerators to reach high energies. The 3-kilometer (2-mile) accelerator at Fermilab, near Chicago, Illinois, was a successful major investment in such a machine; for a time it kept the United States at the forefront of high-energy physics.

New design insights began to come from Europe, especially from the European Organization for Nuclear Research, or CERN, in Geneva, Switzerland. In 1972 CERN pioneered the concept of intersecting proton beams with their Intersecting Storage Rings. This system could generate energies that were equivalent to 1 TeV (1 tera-electron-volt, or 1,000,000,000,000 electron volts), although its pure energy production was less than that. In 1981, however, CERN became the absolute leader with the Super Proton Synchrotron, which produced energies as high as 270 GeV by causing a beam of protons to collide with a beam of anti-protons. This was the device that located the W and Z particles, confirming the electroweak theory that unified the weak force and electromagnetism.

The United States struck back, but too late, with the Tevatron, a superconducting-magnet-based accelerator at Fermilab. On August 15, 1983, the Tevatron set what was then the world record for high energy: 700 GeV or 0.7 TeV.

Another U.S. project to enter a new area was the creation of the Stanford Linear Collider (SLC), designed to reach energies of 100 GeV by colliding electrons with positrons. Construction on the SLC started on Halloween Day 1983; the SLC survived its first limited test as early as January 8, 1984, when its parent, the Stanford Linear Accelerator Center (SLAC) passed an electron beam through an existing linear accelerator. Incidentally, the pronunciation of SLC at SLAC will be "slick at slack."

Despite these advances, the rate of progress at CERN caused the American particle-physics community to wonder if their current projects could keep them ahead—or even in the game. The physicists decided to leapfrog CERN if they could.

THE DESERTRON

On July 11, 1983, the High Energy Physics Advisory Panel of the U.S. Department of Energy strongly recommended the construction of the new particle accelerator described in the first paragraph of this article. At the same time, they made the controversial recommendation that the United States shut down another project, so that funds could be diverted toward the new accelerator.

The project that was to be shut down started life as ISABELLE at Brookhaven National Laboratory. Somewhere along the way, ISABELLE got into so much trouble that it was decided to rename her Colliding Beam Accelerator, or CBA. At the point the project shutdown was recommended, in 1983, CBA was half finished and had cost $150,000,000. Nevertheless, it was clear that CBA would be abandoned.

The new machine was to be officially called the Super Superconducting Collider, or SSC, but it quickly acquired the nickname "Desertron." Various views were expressed on what the name Desertron meant. The most obvious was that the planned underground storage ring was so large that the SSC could only be built in the desert, where land is cheap and where people would not have to be relocated away from the construction site, since few people live in the desert. Others suggested that the name referred to the accelerator's entering an area where nothing much was known, just as a person wandering into the desert would not know what lies ahead. The most cynical view, however, was that the planned range of 10 TeV to 20 TeV would turn out to be a "particle desert," where nothing interesting would be found. (This view became less tenable in April 1984, when CERN experiments at 150 GeV to 200 GeV, the edge of the "desert," produced not one, but two, inexplicable phenomena.)

The Desertron, for all its size and power, actually would capitalize on the design breakthroughs of the CERN accelerators. That is, it would be designed to collide two beams, either two beams of protons or a beam of protons with a beam of antiprotons.

The recommendation was endorsed by the Department of Energy on August 10, 1983, shortly after it was made. Plans were to conduct studies for a few years, then begin construction with an eye to finishing around 1995. Various states want the money that the project will generate, with Texas getting the first bid in shortly after the project was announced. No very early decisions will be made on the site, however.

WHAT'S OUT THERE?

Physics is always an interplay between theory and experiment. Particle physicists need higher-energy accelerators now in particular because the theorists are finding that they have too many conflicting hypotheses and not enough facts to decide among them. There are the various Grand Unification Theories, or GUTs, for example. All of them predict unification at high energies (just as the electroweak theory required a higher energy than had been previously available to confirm it).

The electroweak theory itself has one of the predictions that theorists would most like to test. Somewhere beyond the 1 TeV range should lie the Higgs particle, which is the progenitor of the Ws and the Z. Not enough is known about it to predict its mass exactly, but if it is a few TeV, the accelerator will need several times as many TeV to create the particle. Because not all protons will collide with each other when the two beams collide, higher energies are needed to guarantee the energies of individual collisions. To be on the safe side, to get a 1 TeV collision, you ought to have 5 TeV of energy, and 10 would be safer still. Thus, the Desertron is aiming for the range between 10 TeV and 20 TeV.

Theories about the Big Bang that started the universe on its way suggest that particles at that time were not the same as the ones we see now. When the Bang occurred, these "symmetrical" particles broke into the particles that now inhabit the universe. While the Desertron has no hope of approaching the energy of the Big Bang, it may be able to create some of the less energetic of the primeval particles.

Other physicists are looking in another direction. Some calculations suggest that quarks are made of some constituent particles, which the Desertron might just be able to find.

No one has found the top quark as yet. Just as locating the bottom quark required more energy than older machines possessed, perhaps the top quark is hiding away in higher-energy mesons.

Not all of these projects will require the Desertron, of course, and some may require even more energy than the Desertron will be able to supply. Also, some of the desired reactions simply may not occur for colliding beams of protons or protons and antiprotons, the planned particles to be accelerated by the Desertron, but require beams of electrons, nuclei, or some other particles. Nevertheless, these projects suggest some of the ways that higher energies are used to find out about smaller particles.

The Nobel Prize for Physics

Date	Name (Nationality)	Achievement
1901	Wilhelm K. Roentgen (Germany)	Discovery of X-rays.
1902	Hendrik A. Lorentz (Netherlands) Pieter Zeeman (Netherlands)	Discovery of effect of magnetism on radiation.
1903	A. Henri Becquerel (France) Pierre Curie (France) Marie Curie (France)	Becquerel's discovery of spontaneous radioactivity and Curies' later study of radioactivity.
1904	John Strutt, Lord Rayleigh (England)	Discovery of argon.
1905	Philipp Lenard (Germany)	Work on cathode rays.
1906	Joseph J. Thomson (England)	Discovery of the electron.
1907	A.A. Michelson (U.S.)	Spectroscopic and metrologic study of light.
1908	Gabriel Lippmann (France)	Color photography.
1909	Guglielmo Marconi (Italy) Karl Ferdinand Braun (Germany)	Wireless telegraphy.
1910	J.D. van der Waals (Netherlands)	Equation of state for gases.
1911	Wilhelm Wien (Germany)	Laws of radiation of black bodies.
1912	Nils Gustaf Dalen (Sweden)	Automatic gas regulators for lighthouses and sea buoys.
1913	H. Kamerlingh-Onnes (Netherlands)	Low temperature physics and liquefaction of helium.
1914	Max von Laue (Germany)	Crystal diffraction of X-rays.
1915	Sir William Henry Bragg (England) Sir William Lawrence Bragg (England)	Study of crystals with X-rays.
1916	No award	————————
1917	Charles Barkla (England)	X-ray scattering by elements.
1918	Max Planck (Germany)	Quantum theory of light.
1919	Johannes Stark (Germany)	Spectral lines in electrical fields.
1920	Charles Guillaume (Switzerland)	Special nickel-steel alloy.
1921	Albert Einstein (Germany)	Law of photoelectric effect.

Date	Name (Nationality)	Achievement
1922	Niels Bohr (Denmark)	Atomic structure and radiation.
1923	Robert Millikan (U.S.)	Elementary electrical charge and the photoelectric effect.
1924	Karl Siegbahn (Sweden)	X-ray spectroscopy.
1925	James Franck (Germany) Gustav Hertz (Germany)	Discovery of the laws of electron impact on atoms.
1926	Jean-Baptiste Perrin (France)	Discontinuous structure of matter and equilibrium of sedimentation.
1927	Arthur Holly Compton (U.S.) Charles Wilson (England)	Compton's effect of wavelength change in X-rays and Wilson's cloud chamber.
1928	Sir Owen Richardson (England)	Effect of heat on electron emission.
1929	Prince Louis de Broglie (France)	Wave character of electrons.
1930	Chandrasekhara Raman (India)	Laws of light diffusion.
1931	No award	———
1932	Werner Heisenberg (Germany)	Discovery of quantum mechanics.
1933	Paul A.M. Dirac (England) Erwin Schrödinger (Germany)	Discoveries in wave mechanics.
1934	No award	———
1935	Sir James Chadwick (England)	Discovery of the neutron.
1936	Victor F. Hess (Austria) Carl D. Anderson (U.S.)	Hess's discovery of cosmic radiation and Anderson's of the positron.
1937	Clinton Davisson (U.S.) George P. Thomson (England)	Discovery of electron diffraction by crystals.
1938	Enrico Fermi (Italy)	Work with thermal neutrons.
1939	Ernest Lawrence (U.S.)	Invention of the cyclotron.
1940	No award	———
1941	No award	———
1942	No award	———
1943	Otto Stern (U.S.)	Magnetic momentum of molecular beams.
1944	Isidor Rabi (U.S.)	Magnetic properties of molecular beams.

Date	Name (Nationality)	Achievement
1945	Wolfgang Pauli (Austria)	Pauli exclusion principle for subatomic particles.
1946	Percy W. Bridgman (U.S.)	Laws of high-pressure physics.
1947	Sir Edward Appleton (England)	Discovery of ionic layer in atmosphere.
1948	Patrick Blackett (England)	Discoveries with and improvements upon the Wilson Cloud Chamber.
1949	Hideki Yukawa (Japan)	Discovery of pi meson.
1950	Cecil Powell (England)	Photographic study of atomic nuclei.
1951	Sir John Cockcroft (England) Ernest T.S. Walton (Ireland)	Transmutation of atomic nuclei in accelerated particles.
1952	Felix Bloch (U.S.) Edward Mills Purcell (U.S.)	Measurement of magnetic moment of neutron.
1953	Fritz Zernicke (Netherlands)	Phase-contrast microscope.
1954	Max Born (England) Walther Bothe (Germany)	Born's work in quantum mechanics and Bothe's in cosmic radiation.
1955	Willis Lamb, Jr. (U.S.) Polykarp Kusch (U.S.)	Lamb's measurement of hydrogen spectrum and Kusch's of magnetic momentum of electron.
1956	William Shockley (U.S.) Walter H. Brattain (U.S.) John Bardeen (U.S.)	Studies on semiconductors and invention of the electronic transistor.
1957	Tsung-Dao Lee (China) Chen Ning Yang (China)	Discovery of violations of law of conservation of parity.
1958	Pavel A. Cherenkov (U.S.S.R.) Ilya M. Frank (U.S.S.R.) Igor E. Tamm (U.S.S.R.)	Principle of light emission by electrically charged particles moving faster than speed of light in a medium.
1959	Emilio Segrè (U.S.) Owen Chamberlain (U.S.)	Demonstration of the existence of the antiproton.
1960	Donald Glaser (U.S.)	Bubble chamber for subatomic study.
1961	Robert Hofstadter (U.S.) Rudolf Mössbauer (Germany)	Hofstadter's measurement of nucleons and Mössbauer's work on gamma rays.

Date	Name (Nationality)	Achievement
1962	Lev D. Landau (U.S.S.R.)	Superfluidity in liquid helium.
1963	Eugene Paul Wigner (U.S.) Maria Goeppert Mayer (U.S.) J. Hans D. Jensen (Germany)	Wigner for mechanics of proton-neutron interaction and Mayer and Jensen for theory of nucleic structure.
1964	Charles H. Townes (U.S.) Nikolay G. Basov (U.S.S.R.) Aleksandr Prokhorov (U.S.S.R.)	Development of maser and laser principles in quantum mechanics.
1965	Julian S. Schwinger (U.S.) Richard P. Feynman (U.S.) Shinichero Tomonaga (Japan)	Research into the basic principles of quantum electrodynamics.
1966	Alfred Kastler (France)	Study of atomic structure by optical pumping.
1967	Hans A. Bethe (U.S.)	Study of energy production of stars.
1968	Luis W. Alvarez (U.S.)	Discovery of "resonance particles".
1969	Murray Gell-Mann (U.S.)	Classification of elementary particles.
1970	Hannes Alfvén (Sweden) Louis Néel (France)	Alfvén for plasma physics and Néel for antiferromagnetism.
1971	Dennis Gabor (England)	Invention of holography.
1972	John Bardeen (U.S.) Leon N. Cooper (U.S.) John R. Schrieffer (U.S.)	Theory of superconductivity without electrical resistance at temperature of absolute zero.
1973	Leo Esaki (Japan) Ivar Giaever (U.S.) Brian Josephson (England)	Theories on superconductors and semiconductors important to microelectronics.
1974	Anthony Hewish (England) Martin Ryle (England)	Hewish's discovery of pulsars and Ryle's improvements in radiotelescopy.
1975	James Rainwater (U.S.) Ben Mottelson (Denmark) Aage Bohr (Denmark)	Studies proving asymmetrical structure of the atomic nucleus.
1976	Burton Richter (U.S.) Samuel C.C. Ting (U.S.)	Parallel discovery of subatomic particle that established the existence of charm.

Date	Name (Nationality)	Achievement
1977	Philip W. Anderson (U.S.) John H. Van Vleck (U.S.) Nevill F. Mott (England)	Mott and Anderson for work on amorphous semiconductors; Van Vleck for the magnetic properties of atoms.
1978	Arno A. Penzias (U.S.) Robert W. Wilson (U.S.) Pyotr L. Kapitsa (U.S.S.R.)	The Americans for discovery of microwave radiation from the Big Bang and Kapitsa for work in low-temperature physics.
1979	Steven Weinberg (U.S.) Sheldon L. Glashow (U.S.) Abdus Salam (Pakistan)	Link between electromagnetism and the weak force of radioactive decay.
1980	James W. Cronin (U.S.) Val L. Fitch (U.S.)	Studies showing that "charge-parity" and time symmetry could be violated.
1981	Nicolaas Bloembergen (U.S.) Arthur L. Schawlow (U.S.) Kai M. Siegbahn (Sweden)	Advances in technological applications of lasers for the study of matter.
1982	Kenneth G. Wilson (U.S.)	Theory of phase transitions.
1983	William A. Fowler (U.S.) Subrahmanyan Chandrasekhar (U.S.)	Investigations into the aging and ultimate collapse of stars.

Fermi Awards

Year	Name	Year	Name
1954	Enrico Fermi	1969	Walter H. Zinn
1955	No Award	1970	Norris E. Bradbury
1956	John von Neumann	1971	Shields Warren
1957	Ernest O. Lawrence		Stafford L. Warren
1958	Eugene P. Wigner	1972	Manson Benedict
1959	Glenn T. Seaborg	1973	No award
1960	No award	1974	No award
1961	Hans Bethe	1975	No award
1962	Edward Teller	1976	William L. Russell
1963	J. Robert Oppenheimer	1977	No award
1964	Hyman G. Rickover	1978	Harold M. Agnew
1965	No award		Wolfgang Panofsky
1966	Otto Hahn	1979	No award
	Lise Meitner	1980	Alvin M. Weinberg
	Fritz Strassman		Rudolf E. Peierls
1967	No award	1981	No award
1968	John A. Wheeler	1982	W. Bennett Lewis

WEIGHTS AND MEASURES

Of the two common systems for weights and measures, the International System (known both as SI or metric, although the SI system is not exactly the same as the metric) and the customary system, the International System is by far the more significant in scientific pursuits. The following tables show first the International System, then the customary (in less detail), and finally some important conversion factors between the two.

Length or Distance

The basic unit for length is the *meter*, which is slightly longer than the customary yard. In 1983 the meter was redefined in terms of the speed of light. One meter is exactly the distance light in a vacuum travels in 1/299,792,458 second. Other units of length are decimal subdivisions or multiples of the meter.

For the measure of small objects:

1 decimeter (dm) =	10 centimeters = 0.1 meter	
1 centimeter (cm) =	0.01 meter	
1 millimeter (mm) =	0.1 centimeter = 0.001 meter	
1 micron (μ) =	0.001 millimeter = 0.0001 centimeter = 0.000001 meter	
1 angstrom (A) =	0.0001 micron = 0.0000001 millimeter	
=	0.0000000001 meter	

For the measure of large objects and distance:

1 dekameter (dam) =	10 meters
1 hectometer (hm) =	10 dekameters = 100 meters
1 kilometer (km) =	10 hectometers = 100 dekameters = 1,000 meters

To put this in reverse, a meter is:

10 decimeters	10,000,000,000 angstroms
100 centimeters	0.1 dekameter
1,000 millimeters	0.01 hectometer
1,000,000 microns	0.001 kilometer

One very large measure of distance useful in astronomy is the *light-year*. It is defined as the distance light travels in a year (approximately 365.25 days) at the rate of exactly 299,792,458 m per second, or about 299,792 km per second. It is approximately equivalent to 9,460,000,000,000 km (9.46 x 10^{12} km). A smaller unit, for measurements within the solar system, is the *astronomical unit*, which is the average distance between Earth and the sun, or about 150,000,000 km (1.5 × 10^8 km).

Area and Volume

Measures of area and volume are derived directly from units of length by defining a plane region 1 meter on a side as 1 square meter and a 3-dimensional region 1 meter on a side as 1 cubic meter. Consequently, the numerical relationship between the area units is the square of the relationship between the linear units; likewise the relationship is cubed for volume units.

Square Measure

For the measure of small areas:

1 square millimeter (mm^2) = 1,000,000 square microns
1 square centimeter (cm^2) = 100 square millimeters
1 square meter (m^2) = 10,000 square centimeters

For the measure of larger areas:

1 are (a) = 100 square meters
1 hectare (ha) = 100 ares = 10,000 square meters
1 square kilometer (km^2) = 100 hectares
= 1,000,000 square meters

Cubic Measure

Cubic measure is similarly derived from units of length:

1 cubic centimeter (cm^3) = 1,000 cubic millimeters (mm^3)
1 cubic decimeter (dm^3) = 1,000 cubic centimeters
1 cubic meter (m^3) = 1,000 cubic decimeters = 1,000,000 cubic centimeters

Cubic centimeter is sometimes abbreviated *cc* and is used in fluid measure interchangeably with milliliter (mL). See below.

Fluid Volume

Fluid volume measurements are directly tied to cubic measure. One *milliliter* of fluid occupies a volume of 1 cubic centimeter (cm^3 or cc). A *liter* of fluid (slightly more than the customary quart) occupies a volume of 1 cubic decimeter or 1,000 cubic centimeters.

The relationships for fluid measure are as follows:

           1 centiliter (cL) = 10 milliliters (mL)
           1 deciliter (dL) = 10 centiliters = 100 milliliters
               1 liter (L) = 10 deciliters = 1,000 milliliters
        1 dekaliter (daL) = 10 liters
        1 hectoliter (hL) = 10 dekaliters = 100 liters
          1 kiloliter (kL) = 10 hectoliters = 1,000 liters

Mass and Weight

Mass and weight are often confused. Mass is a measure of the quantity of matter in an object and does not vary with changes in altitude or in gravitational force (as on the Moon or another planet). Weight, on the other hand, is a measure of the force of gravity on an object and so it does change with altitude or gravitational force.

The International System's basic unit for measurement of mass is the gram, which was originally defined as the mass of 1 milliliter (= 1 cubic centimeter) of water at 4°C (about 39°F). Today the official standard of measure is the kilogram (1,000 grams). Since mass and weight are identical at standard conditions (sea level on Earth), grams and other International System (SI) units of mass are often used as measures of weight or converted into customary units of weight. Under normal conditions, the gram is equal to about $\frac{1}{28}$ of an ounce in customary weight units. The kilogram is equal to about 2.2 pounds.

The relationships between SI units of mass are as follows:

       1 centigram (cg) = 10 milligrams (mg)
        1 decigram (dg) = 10 centigrams = 100 milligrams
              1 gram (g) = 10 decigrams = 100 centigrams = 1,000 milligrams

       1 kilogram (kg) = 10 hectograms (hg) = 100 dekagrams (dag)
                         = 1,000 grams
          1 metric ton (t) = 1,000 kilograms

Time

The International System has adopted the customary measures of time by which the day is divided into 24 hours; the hour into 60 minutes; and the minute into 60 seconds. Decimal fractions of time are used to measure smaller time intervals:

       1 millisecond (ms) = 0.001 second (10^{-3})
     1 microsecond (μs) = 0.000001 second (10^{-6})
       1 nanosecond (ns) = 0.000000001 second (10^{-9})
        1 picosecond (ps) = 0.000000000001 second (10^{-12})

Temperature

For the common range of temperature, the International System measures in the Celsius (formerly called the centigrade) scale denoted by the abbreviation C. In this scale, water freezes at 0°C and boils at 100°C.

For very low temperatures, measurement is made on the Kelvin scale (abbreviated K). The temperature change denoted by 1 degree is the same as in the Celsius scale, but the zero point is set at *absolute zero*. Thus, the equivalencies between Kelvin and Celsius are:

	K	C
Absolute zero	0	−273.15
Freezing point, water	273.15	0
Boiling point, water	373.15	100

In measuring very high temperatures, the differences between the Kelvin and Celsius scales become insignificant.

Force, Work/Energy, Power

In physics, compound measurements of force, work or energy, and power are essential. There are two parallel systems using International System units: The *centimeter/gram/second* system is most often used for small measurements, and the *meter/kilogram/second* system is most often used for larger measurements. They are described below

Measurement of Force

c/g/s unit	dyne (dy)	The force required to accelerate a mass of 1 g 1 cm/s²
m/k/s unit	newton (new)	The force required to accelerate a mass of 1 kg 1 m/s²

Measurement of Work or Energy

c/g/s unit	erg	The dyne-centimeter; i.e., the work done when a force of 1 dy produces a movement of 1 c
m/k/s unit	joule (j)	The newton-meter; i.e., the work done when a force of 1 newton produces a movement of 1 m (= 10,000,000 ergs)

Measurement of Power

c/g/s unit	erg/second	A rate of 1 erg per second
m/k/s unit	watt (w)	The joule/second; i.e., a rate of 1 joule per second (= 10,000,000 erg-seconds)

Heat energy is also measured using the *calorie* (cal), which is defined as the energy required to increase the temperature of one cubic centimeter (1 mL) of water by 1°C. One calorie is equal to about 4.184 joules. The *kilocalorie* (Kcal or Cal), is equal to 1,000 calories and is the unit in which the energy values of food are measured. This more familiar unit, also commonly referred to as a Calorie, is equal to 4,184 joules.

Electrical Measure

The basic unit of quantity in electricity is the *coulomb*. A coulomb is equal to the passage of 6.25×10^{18} electrons past a given point in an electrical system.

The unit of electrical flow is the *ampere*, which is equal to a coulomb-second; i.e., the flow of 1 coulomb per second. The ampere is analogous in electrical measure to a unit of flow such as gallons-per-minute in physical measure.

The unit for measuring electrical potential energy is the *volt*, which is defined as 1 joule/coulomb; i.e., 1 joule of energy per coulomb of electricity. The volt is analogous to a measure of pressure in a water system.

The unit for measuring electrical power is the *watt*, as defined in the previous section. Power in watts is the product of the electrical flow in amperes and the potential electrical energy in volts.

P (in watts) = I (electrical current in amperes) × V (potential energy in volts)

Since the watt is such a small unit for practical applications, the *kilowatt* (= 1,000 watts) is often used. A *kilowatt-hour* is the power of 1,000 watts over an hour's time.

The unit for measuring electrical resistance is the *ohm*, which is defined as the resistance offered by a circuit to the flow of 1 ampere being driven by the force of 1 volt. It is derived from Ohm's law, which defines the relationship between flow or current (amperes), potential energy (volts), and resistance (ohms). It states that the current in amperes (I) is proportional to potential energy in volts (E), and inversely proportional to resistance in ohms (R). Thus, where voltage and resistance are known, amperage can be calculated by the simple formula

I (in amperes) = E (in volts) ÷ R (in ohms)

Customary measure

Customary measure in the United States is derived from English measure, but in some details, the American system has deviated from the English. In general, it is far more difficult to use for scientific pursuits since it is not based on decimal subdivisions or multiples of a standard unit. The following summary gives only the most common of customary measures. Many others are used in particular industries.

Length or Distance

Customary length measurements use units that divide easily into common fractions rather than the decimal system. Thus, for example, the 1,760 yards in a mile can be evenly divided down to 32nds (55 yards), 16ths (110 yards), etc. The 36 inches in a yard break evenly into thirds, fourths, sixths, and other common fractions.

```
      1 foot (ft)  =  12 inches (in.)
      1 yard (yd)  =  3 feet = 36 inches
       1 rod (rd)  =  5 ½ yards = 16 ½ feet
  I furlong (fur)  =  40 rods = 220 yards = 660 feet
      1 mile (mi)  =  8 furlongs = 1,760 yards = 5,280 feet
```

An International Nautical Mile has been defined as 6,076.1155 feet.

Area

Areas are derived from lengths as follows:

```
        1 square foot (sq. ft. or ft²)  =  144 square inches (sq. in.)
       1 square yard (sq. yd. or yd²)  =  9 square feet
              1 square rod (sq. rd.)  =  30¼ square yards = 272¼ sq. ft.
                          1 acre  =  160 square rods = 4,840 sq. yd.
                                  =  43,560 sq. ft.
               1 square mile  =  640 acres
                  1 section  =  1 mile square
                1 township  =  6 miles square = 36 square miles
```

Cubic Measure

Cubic measures are similarly derived.

```
   1 cubic foot (ft³)  =  1,728 cubic inches (in³)
   1 cubic yard (yd³)  =  27 cubic feet
```

Fluid Measure

Customary fluid measure has no even relationship to cubic measure. A gallon is equal to 231 cubic inches of liquid or capacity.

```
          1 cup  =  8 fluid ounces (fl. oz.)
      1 pint (pt)  =  2 cups = 16 fl. oz.
     1 quart (qt)  =  2 pt = 4 cups = 32 fl. oz.
   1 gallon (gal)  =  4 qt = 8 pt = 16 cups
```

Weight

The most common customary system of weight is the avoirdupois:

```
              1 pound (lb) = 16 ounces (oz)
 1 (short) hundredweight = 100 lb
          1 (short) ton = 20 hundredweights = 2,000 lb
   1 long hundredweight = 112 lb
             1 long ton = 20 long hundredweights = 2,240 lb
```

A different system, called troy weight, is used to weigh precious metals. In troy weight, the ounce is slightly larger than in avoirdupois, but there are only 12 ounces to the troy pound.

Time

The basic units of time are the same in both the International System and in customary measure. See above.

Temperature

Customary measure generally uses the Fahrenheit scale of temperature measurement (abbreviated F). In this system, water freezes at 32°F. and boils at 212°F. Absolute zero is −459.7°F.

Force, Work/Energy, Power

The foot/pound/second system of reckoning uses the pound as a fundamental unit of force. Sometimes, to avoid confusion with the use of *pound* for mass another name, the *poundal* is used when force is meant, reserving the pound for mass.

```
       1 slug = the mass to which a force of 1 pound will give an
                acceleration of 1 foot per second ( = approximately
                32.17 pounds)
    1 poundal = fundamental unit of force
 1 foot-pound = the work done when a force of 1 pound
                produces a movement of 1 foot
1 foot-pound-second = the unit of power equal to 1 foot-pound per second
```

Another common unit of power is the *horsepower*, which is equal to 550 foot-pounds per second.

Thermal work or energy is also measured in *British thermal units* (Btu), which are defined as the energy required to increase the temperature of a pound of water 1°F. The Btu is about 0.778 foot-pound.

Electrical measure

There are no separate measures for electricity in the customary system.

Conversions

Length

In 1959 the relationship between customary and International measures of length was officially defined as follows:

> 0.0254 meter (exactly) = 1 inch
> 0.0254 meter × 12 = 0.3048 meter = 1 international foot

This definition, which makes many conversions simple, defines a foot that is shorter (by about 6 parts in 10,000,000) than the *survey foot*, which had earlier been defined as exactly $1200/3937$, or 0.3048006 meters.

Following the *international foot* standard, the major equivalents are

> 1 inch = 2.54 centimeters = 0.0254 meter
> 1 foot = 30.48 centimeters = 0.3048 meter
> 1 yard = 91.44 centimeters = 0.9144 meter
> 1 mile = 1609.344 meters = 1.609344 kilometers

> 1 centimeter = 0.3937 inch
> 1 meter = 1.093613 yards = 3.28084 feet = 0.00062137 mile

Area

> 1 sq. inch = 6.4516 cm^2
> 1 sq. foot = 929.0304 cm^2 = 0.09290304 m^2
> 1 sq. yard = 8361.2736 cm^2 = 0.83612736 m^2
> 1 acre = 4046.8564 m^2 = 0.40468564 hectares
> 1 sq. mile = 2,589,988.1 m^2 = 258.99881 hectares
> = 2.5899881 km^2

> 1 cm^2 = 0.1550003 sq. in.
> 1 m^2 = 1550.003 sq. in. = 10.76391 sq. ft. = 1.195990 sq. yd.
> 1 hectare = 107,639.1 sq. ft. = 11,959.90 sq. yd. = 2.471044 acres
> = 0.003861006 sq. mi.
> 1 km^2 = 247.1044 acres = 0.3861006 sq. mi.

Volume

> 1 in^3 = 16.387064 cm^3
> 1 ft^3 = 28,316.846592 cm^3 = 0.028316847 m^3
> 1 yd^3 = 764,554.857984 cm^3 = 0.764554858 m^3

> 1 cm^3 = 0.06102374 in^3
> 1 m^3 = 61,023.74 in^3 = 35.31467 ft^3 = 1.307951 yd^3

Fluid Measure

$$1 \text{ fl. oz.} = 29.573528 \text{ mL} = 0.02957 \text{ L}$$
$$1 \text{ cup} = 236.588 \text{ mL} = 0.236588 \text{ L}$$
$$1 \text{ pint} = 473.176 \text{ mL} = 0.473176 \text{ L}$$
$$1 \text{ quart} = 946.3529 \text{ mL} = 0.9463529 \text{ L}$$
$$1 \text{ gallon} = 3,785.41 \text{ mL} = 3.78541 \text{ L}$$

$$1 \text{ milliliter} = 0.0338 \text{ fl. oz.}$$
$$1 \text{ liter} = 33.814 \text{ fl. oz.} = 4.2268 \text{ cup} = 2.113 \text{ pint}$$
$$= 1.0567 \text{ quart} = 0.264 \text{ gallon}$$

Mass/Weight

The following conversions between units of mass (grams, kilograms) and units of weight (ounces, pounds) are valid only under standard conditions (at sea level on Earth).

$$1 \text{ ounce} = 28.3495 \text{ grams}$$
$$1 \text{ pound} = 453.59 \text{ g} = .45359 \text{ kilogram}$$
$$1 \text{ short ton} = 907.18 \text{ kg} = 0.907 \text{ metric ton}$$

$$1 \text{ milligram} = 0.000035 \text{ oz}$$
$$1 \text{ gram} = 0.03527 \text{ oz}$$
$$1 \text{ kilogram} = 35.27 \text{ oz} = 2.2046 \text{ lb}$$
$$1 \text{ metric ton} = 2,204.6 \text{ lb} = 1.1023 \text{ short ton}$$

Temperature

Common benchmarks for comparison of temperature scales are:

	F	C	K
Absolute zero	−459.7	−273.15	0
Freezing point, water	32	0	273.15
Normal human body temperature	98.6	37	310.15
Boiling point, water	212	100	373.15

The simplest means of converting is by formula:

To convert a Fahrenheit temperature to Celsius, subtract 32 from the temperature and multiply it by ⅚ (or 0.5555…)

$$C = \frac{5}{9} \times (F - 32)$$

To convert a Celsius temperature to Fahrenheit, multiply the temperature by ⅘ (or 1.8), then add 32.

$$F = \frac{9}{5} C + 32$$

To convert to Kelvin, find the temperature in Celsius and add 273.15.

$$K = C + 273.15$$

Force, Work/Energy, Power

Measurement of Force

1 poundal	= 13,889 dynes	= 0.13889 newtons
1 dyne	= 0.000072 poundals	
1 newton	= 7.2 poundals	

Measurement of Work/Energy

1 foot-pound	= 1356 joules	
1 British thermal unit	= 1,055 joules	= 252 gram calories
1 joule	= 0.0007374 foot-pounds	
1 (gram) calorie	= 0.003968 Btu	
1 (kilo) Calorie	= 3.968 Btu	

Measurement of Power

1 foot-pound/second	= 1.3564 watts	
1 horsepower	= 746 watts	= 0.746 kilowatts
1 watt	= 0.73725 ft-lb/sec	= 0.00134 horsepower
1 kilowatt	= 737.25 ft-lb/sec	= 1.34 horsepower

INDEX